D1288749

ANIMAL TOOL BEHAVIOR

ANIMAL TOOL BEHAVIOR

THE USE AND MANUFACTURE OF TOOLS BY ANIMALS

REVISED AND UPDATED EDITION

ROBERT W. SHUMAKER

KRISTINA R. WALKUP

and BENJAMIN B. BECK

Foreword by **Gordon M. Burghardt**

THE JOHNS HOPKINS UNIVERSITY PRESS | BALTIMORE

Printed in the United States of America on acid-free paper
9 8 7 6 5 4 3 2 1

The Johns Hopkins University Press
2715 North Charles Street
Baltimore, Maryland 21218-4363
www.press.jhu.edu

Library of Congress Cataloging-in-Publication Data
Shumaker, Robert W.
Animal tool behavior : the use and manufacture of tools by animals / by Robert W. Shumaker, Kristina R. Walkup, and Benjamin B. Beck; foreword by Gordon M. Burghardt. — Rev. and updated ed.
p. cm.
Rev. ed. of: Animal tool behavior / Benjamin B. Beck. 1980.
Includes bibliographical references and index.
ISBN-13: 978-0-8018-9853-2 (hardcover : alk. paper)
ISBN-10: 0-8018-9853-6 (hardcover : alk. paper)
1. Tool use in animals. I. Walkup, Kristina R. II. Beck, Benjamin B. Animal tool behavior. III. Title.
QL785.B32 2011
591.5'13—dc22 2010019753

A catalog record for this book is available from the British Library.

Special discounts are available for bulk purchases of this book. For more information, please contact Special Sales at 410-516-6936 or specialsales@press.jhu.edu.

The Johns Hopkins University Press uses environmentally friendly book materials, including recycled text paper that is composed of at least 30 percent post-consumer waste, whenever possible. All of our book papers are acid-free, and our jackets and covers are printed on paper with recycled content.

To Anne, William, and Carly. You brighten my life every day.

RWS

To Dustin, and my parents, Dan and Merry. Thank you for your patience, love, and support.

KRW

To Beate Rettberg-Beck—and thanks to my dog, Heidi. Both felt ignored during writing bouts.

BBB

CONTENTS

FOREWORD

Ground squirrels kick sand into the faces of venomous snakes to deter attacks. Ant lions engage in a similar behavior in their sand pits to incapacitate prey. Degus (small rodents) use rakes to access food, an ability shared with many birds and non-human primates. Some mice set out markers to aid in finding their way home. Birds use small food items to bait fish, but crocodiles have turned the tables, using fish to attract birds, which they then attack. New Caledonian crows sometimes travel with a toolkit of proven implements for probing for food (including lizards in crevices). Crabs use all sorts of objects, animate and inanimate, to affix to themselves or to the shells they inhabit, for camouflage against predators. Apes are able to use tools of all kinds in both captivity and the wild. Through observation and practice they crack open nuts, apply herbal medications, open locks and doors, use sticks to stir liquids, saw wood, and even dig with a shovel. In fact, while tools are mostly used in foraging for food, they also are employed in many other contexts, such as to deter predators, facilitate courtship and copulation, mark territories, and intimidate competitors of their own species.

Yet what do these fascinating examples mean? How scientifically valid are the observations? Is tool use evolutionarily important? Is tool use, even tool making, an indication of highly intelligent and cognitively complex behavior? What, in fact, really counts as tool use? Do not merely eating, breathing, and walking on the ground involve using some "thing" as a tool? Thirty years ago, in 1980, Ben Beck wrote the first comprehensive book on tool use throughout the animal kingdom, helping to foster theory and research on this topic by collecting and systematizing observations from myriad sources, providing definitions and conceptual clarification, and exploring the evolutionary ramifications of using tools in ethology, comparative psychology, and human evolution.

Now we have the long-awaited update and revision of Ben Beck's seminal book, *Animal Tool Behavior*. As editor of the Garland Ethology book series in which it appeared, I was privileged to write a foreword to that volume. The last paragraph of my foreword stated that "*Animal Tool Behavior* will become a widely consulted reference and starting point for future work in the area of animal tool use and manufacture—an area fascinating in itself and for its relationship, real and imagined, to intelligence, language, cognitive processes, and human evolution." I do not consider myself highly prescient, but this claim was right on.

Ben was been most fortunate to have engaged Rob Shumaker and Kristina

Walkup as coauthors for this brilliantly realized revision. Together they have about tripled the number of examples and greatly expanded the diversity of species and phenomena involving tool use and manufacture. The definition of tool use is comprehensively evaluated in chapter 1. The durability of Beck's original formulation after thirty years, needing only minor revision, is remarkable. The authors carefully review many of the critiques of Ben's original definition, as well as many proposed alternatives. I have had some experience with definitional issues, namely with "communication" and "play," in efforts to develop definitions that also are truly useful. Much work needed to go beyond the "I know it when I see it" attitude of some scientists, who disparage the value of conceptual rigor in behavioral concepts. The history of Ben's book vindicated amassing and analyzing putative examples to test ideas about the provenance of a phenomenon.

The second accomplishment of this book is to comprehensively review examples of tool use throughout the animal kingdom, among both vertebrates and invertebrates. The expansion of taxonomic diversity is significant. Table 3-1 in the first edition depicts examples of phyletic differences in tool use among animals across eight taxonomic groupings, with only one involving invertebrates. The comparable table 7.1 in the new edition uses twenty-three groupings, and six of these are invertebrates! Furthermore, in this table all birds are still grouped together. It took another table in the new edition (3.1) to show differences in modes of tool use across the thirty-three avian families in which tool use has been described. While the complexity and ingenuity in non-human primate tool use and manufacture continues to astound, I am most impressed with the remarkable growth of knowledge regarding birds. Alas, for this herpetologist, the meager to nonexistent records of tool use in amphibians and reptiles continue to disappoint.

The authors often elaborate on evolutionary, cognitive, developmental, and ecological issues and evaluate experimental studies in a most insightful manner. But given the volume of studies on such topics, especially in non-human primates (let alone humans), they could not be explored in length. However, the final chapter on seven myths about tool behavior provides closure to a mind-bending ride, nicely delineates the provocative implications posed by the comparative evidence, and supplies essential tools for future research.

Tool use is a major cognitive and behavioral component of "mental evolution," to use an old term that, I believe, is as relevant today as when G. J. Romanes wrote the first books with this term in the title in the late nineteenth century. In fact, it is a tribute to the authors' scholarship that so many examples from this era, and earlier, are included. While these early works were, by themselves, scholarly suspect and gathered without photographic support, subsequent studies are attesting to the past perceptiveness of people in diverse cultures across the world. This should impress and humble professional scientists who follow in their wake.

Tool making and tool use have often been held up, along with language, as pinnacles of cognitive achievement and progressive creativity. It may be better to view tool use and manufacture as sometimes necessary, but not sufficient, ad hoc solutions for the survival and subsequent evolution of animals when they confront novel problems. Today, when oil spills threaten oceans and bombings are a near daily occurrence, understanding the origins of tool behavior may be highly relevant to the future of the earth. Is it possible to reap the benefits of tools while controlling the negative, indeed tragic, consequences of their misuse?

Gordon M. Burghardt

PREFACE

This book provides precise definitions of tool use and tool manufacture, a complete catalog of all reported cases of tool use and tool manufacture by extant non-human animals, and a discussion of seven prevailing myths about animal tool behavior. Space limitations have forced us to defer detailed discussions of ontogenetic, ecological, evolutionary, and cognitive aspects of tool behavior to a later publication(s). This is the second edition of a monograph by the same title originally published by Beck in 1980 and later translated into Italian (Beck 1986).

Alcock (1981, 231) wondered if tool use is worthy of monographic treatment, given that it "plays a minor to trivial role in the foraging economy and social life of all but a tiny number of species" except our own. Hansell and Ruxton (2008, 77) made the same argument (without acknowledging Alcock), commenting that "tools are more special to researchers than to the animals that use them." We disagree on the "minor to trivial role" and "tiny number of species," and obviously think a monograph was warranted in 1980 and is even more warranted now. The first edition referenced approximately 670 articles, books, and films. This edition cites about 1,750 sources from roughly 3,000 articles we reviewed on the general topic. Most of the additional sources appeared after 1980, attesting to the growth and vitality of the field. The increases are not a result of sampling effort; we made similar efforts to cite every known source in both editions, although, as stated above, this edition does not fully address ontogeny, cognition, ecology, and evolution. While we are now three authors, whereas the first edition was the work of one, and Internet search tools helped us locate sources more easily for the second edition, these alone do not account for the increase in publications and, by proxy, scientific interest. Despite our best efforts, we acknowledge the likelihood that we have missed some citations, particularly those in non-English books or journals. Bentley-Condit and Smith (2010) recently provided a catalog of animal tool behavior that they claimed to be comprehensive, but it contained only about 600 references, despite their using search methodologies that were similar to ours. We include about 700 references for the tool behavior of great apes alone, while they had only about 100 for apes in general.

The references are representative of the published observational record of tool behavior, but this record is biased toward more studied species and toward those taxa where an observer might be more likely to look for tool behavior. We did not seek out unpublished film or video archives to search for cases of tool behavior. We used personal communications when they were offered to

This antique print depicting an orangutan with a walking stick reveals early fascination with and misperceptions about apes and their tools. Artist unknown.

us, but we did not request unpublished observations except to clarify published ones.

In this edition, unlike the first, we include reports of tool behaviors that result from imitation or emulation of humans. Species appear to differ in their predisposition to imitate or emulate. Captive chimpanzees (*Pan troglodytes*) and orangutans (*Pongo* spp.) are notable imitators and emulators (e.g., Call and Tomasello 1994; Greenfield et al. 2000; K. Hayes and Hayes 1952; Russon 1999b; Russon and Galdikas 1993, 1995; Tomasello et al. 1987). Russon (2000, 76) described an orangutan in a rehabilitation camp who "hammered nails, sawed wood, sharpened axe blades, chopped wood, dug with shovels, siphoned fuel, swept porches, painted buildings, pumped water, blew blowguns, fixed blowgun darts, lit cigarettes, (almost) lit a fire, washed dishes and laundry, baled water from a dugout by rocking it side to side, put on boots, tried on glasses, combed her hair, wiped her face with Kleenex, carried parasols against the sun, and applied insect repellent to herself," in each case using the technique that "matched the one used [by humans] in the camp." Socially mediated learning, particularly imitation and emulation, is relevant to recent scientific discussions of cultures among wild chimpanzees and orangutans (e.g., Tomasello 1990; van Schaik et al. 2003, 2009; Whiten et al. 1999). In 1980, ape cultures were only dimly anticipated, and Beck underestimated the potential theoretical importance of human-based imitation and emulation.

Of course, a few species other than apes also imitate humans. For example, captive bottlenose dolphins (*Tursiops aduncus*) were reported by Taylor and Saayman (1973) to wipe the interior of aquarium windows with a feather. Species differ not only in their predisposition to imitate or emulate, but also in their opportunities to do so. Those that humans like to keep captive and those that tolerate close proximity to people will have more opportunity to closely observe and reproduce human behavior. Thus our inclusion of tool behavior acquired by observing humans does result in a non-representative sampling across taxa, but there is much to be learned about capacity and ontogeny from such cases. This is especially true with individual animals that are *enculturated*, that is, have been reared or kept for long periods in close association with humans. Such animals, most typically apes (Furlong, Boose, and Boysen 2008; Russon 1999a; Tomasello, Kruger, and Ratner 1993), are exposed to many of the manipulable items normally associated with human life (*affordances*). As detailed above, orphan orangutans living in rehabilitation stations show a remarkable array of tool behavior modeled after human activities (e.g., Russon 1999a, 2000). Some enculturated apes, such as Kanzi, Washoe, Koko, and Chantek, simultaneously learn to communicate linguistically, which might potentiate imitation because the ape can be asked to "do as I do" (as reviewed by Hillix and Rumbaugh 2004). These individuals may not be typical, but they may be particularly instructive.

Through the use of the operant techniques of successive approximation and differential reinforcement, all vertebrates, and probably many invertebrates, can be trained to emit any response of which they are anatomically capable. Indeed, many captive vertebrates have been trained to perform complicated and unlikely behaviors, for example, a pig-tailed macaque (*Macaca nemestrina*) using a toilet (Bertrand 1976). More tritely, bears are trained to ride bicycles, porpoises to play basketball, and chickens to play pianos. Such training sometimes produces behaviors that appear more cognitively complex than they are. Given the intricacy and precision with which animal behavior can be produced by operant conditioning, it is likely that a skilled animal trainer could greatly lengthen our catalog of tool behavior. Thus, as in the first edition, we have not tried to include all cases of tool behavior produced intentionally through operant training by humans, but we do include some cases of purposively trained tool behavior where they illuminate capacity, ontogeny, or evolution. We are not arguing that operant (trial-and-error) learning plays no role in the acquisition of tool behavior by wild or untamed captive animals. Beck (1980) and Visalberghi and Limongelli (1996) showed that quite the opposite is true. However, tool behavior produced by human training has little relevance to natural history or behavioral evolution, and the inclusion of such cases in our catalog would make it needlessly and impractically lengthy.

Tool behavior of both wild and captive animals is included. As Beck noted in 1980 (and as Boinski, Quatrone, and Swartz 2000 and Candland 1987 repeated), since captive animals need not search for food and water or guard against predators, they have more available time and energy to explore their environment and manipulate inanimate objects. The acquisition of tool patterns is, in part, a function of exploration and manipulation (Beck 1972, 1976; Birch 1945; Jackson 1942). Thus the likelihood of acquisition of tool behavior, even without imitation or training, is higher among some captives, at least those living in environments with many objects to explore and manipulate (Kummer and Goodall 1985). However, the foraging demands and specific foraging opportunities that exist for wild chimpanzees might select for a greater frequency and variety of tool use than is seen among captives, despite the captives' greater "free time." Wickler and Seibt (1997) cite Kurt and Hartl (1995) as showing that wild Asian elephants (*Elephas maximus*) use a greater variety of tools than captives. In any event, we subscribe to the belief that the types of behavior and the nature of social relationships of captive animals will be similar to those of wild conspecifics when they are maintained in an intraspecific social environment approximating that of wild populations and in a physical environment that contains functional simulations of essential physical elements (see Kummer and

A chimpanzee studied by Wolfgang Köhler stands on a tower of three boxes to reach a suspended banana. Reprinted from Köhler 1917.

Kurt 1965 for a specific example). The frequencies of some behavioral patterns are altered, but the patterns themselves are similar. However, the functional and evolutionary contexts of behaviors are often obscured in captivity. Beck (1977) compared the tool behavior of the captive chimpanzees in the colony of the Prussian Academy of Science on the island of Tenerife, as observed by Wolfgang Köhler from 1914 to 1916, with the tool behavior of wild chimpanzees as reported by many field workers. Köhler (1925) observed practically every mode of tool use and tool manufacture that was later reported for wild chimpanzees. Köhler appreciated the cognitive significance of the chimpanzees' tool behavior, but nobody could have anticipated the functional and evolutionary significance of his observations until they were replicated in nature. Nonetheless, observations of tool behavior of captive animals have proven to be very instructive and are included here.

Throughout this book we will, for brevity, drop the word "non-human" before "animal" and "primate." It should be fully recognized, and with considerable pride, that we humans are both. While Groves (2001) classified all species of the genera *Homo*, *Pan*, *Gorilla*, and *Pongo* as hominids, for convenience we will refer to members of the genera *Pan*, *Gorilla*, and *Pongo* as great apes, and members of the genus *Homo* as humans. We use the pronoun "who" in referring to great apes (e.g., "the gorilla who") and "that" for all other animals ("the wolf that").

ACKNOWLEDGMENTS

Very early in the process of researching and writing this second edition of *Animal Tool Behavior*, two important facts quickly became apparent. First, our initial estimate of the literature involved in a comprehensive review of tool use among all non-humans was overly conservative. We quickly learned that our colleagues have been remarkably productive, far more than anticipated, and the scope of this revision expanded exponentially. Second, although the original files for the first edition in 1980 had been meticulously organized and maintained, there was no manuscript from which to draw except the published book. Working from Beck's personal and only copy, the entire book was manually transformed into an electronic format. We are most grateful for the small army of patient and wonderful people who entered text manually, scanned page after original page, searched out obscure references, filed, and performed an endless number of invaluable tasks. Katie Klag, Erin Wessling, Stephanie Bogart, Gina Cozad, Nick Gould, Jackie Mobley, and Samantha Dunn, we could not have done this without you.

As the manuscript developed into a book, we called upon friends and colleagues for advice, favors, and assistance. Without exception, all of them were remarkably generous with their time and talent. Beate Beck, Carole Villeneuve, and Peter Clay effortlessly translated complex journal articles. Elizabeth Fox relayed critical details of her field work with orangutans. Our remarkable colleague and friend Hirata Satoshi was our guide through the Japanese literature and the world of rehabilitant bonobos. Anne Russon provided critical insights, unending support and enthusiasm for our efforts, generous helpings of collegiality, remarkable friendship, and a treasure trove of personal photographs. William McGrew offered welcome insights as well. We solicited constructive criticism and critical review from Serge Wich, Jill Pruetz, and Marina Vančatová, who greatly improved our work. Isobel Osius, Gloria Morris, Gwen Cox, and Dana Watson graciously helped us to fix our errors. We thank all of you for your invaluable contributions.

Vince Burke has been inspiring and remarkably patient and accommodating. Kathleen Capels is the most efficient and helpful copy editor we have ever encountered; her work was meticulous, scholarly, sensitive, and humorous.

This book was conceived and written while we were on the staff of Great Ape Trust; we are grateful to our colleagues and Trust founder Ted Townsend for their encouragement and support.

By coincidence, the first and second editions of this book were published 30 years apart. By design, this most recent effort involved three generations of scholars. Ben Beck served as graduate advisor and mentor for Rob Shumaker, who did the same for Kristina Walkup. As co-authors, each of us championed different sections of the book as the manuscript developed, but the process has been intensely collaborative and we now happily share responsibility for every word of this revision. However, Shumaker and Walkup would be remiss in not acknowledging the important role that Beck has played in the continuity of critical thinking that has shaped this volume and, through succession, our respective academic and professional careers. Ben Beck's generosity to both of us has been profound, far more than can be acknowledged with a simple "thank you." We hope that the zeal with which we have approached the writing of this book is an indication of the admiration and affection that we have for you.

We are three generations of scientists, each with our own academic interests and points of view, yet sharing far more than we differ. Our common interest in understanding the behavior and cognition of other species is passionate and profound, and so is our commitment to their welfare in captivity and their conservation in the wild.

ANIMAL TOOL BEHAVIOR

1 Introduction

VISITORS TO THE THINK TANK EXHIBIT ON ANIMAL COGNITION AT THE SMITHsonian National Zoo in Washington, D.C., are greeted with a life-size cutout of a cow bearing the question "Is a Cow a Tool?"—that is, a tool for producing and transporting meat and milk. The cow's purpose is to acquaint visitors with the difficulties of defining tool use, which will doubtlessly persist long after the exhibit is dismantled. Definition is essential to intelligent scientific discussion, and the process of definition frequently results in an improved understanding of the phenomenon itself. Definition also facilitates rendering human concepts into machine-readable form, which would greatly aid topical searches (C. Catton and D. Shottoen, unpub. report). In addition, definition facilitates the design of tool-using robots (St Amant and Wood 2005). Tool use would seem to be easy to define. Theoretically, a scientist or non-scientist could intuitively and unequivocally identify certain behaviors as tool use and exclude others. Further, it would seem easy to achieve a high degree of agreement among observers in such categorization. Many authors do not even bother to provide an explicit definition in their reports and discussions of tool use.

To illustrate some of the difficulties of definition, we offer a list of fifty-three behaviors. Make a note of which you think are, and are not, tool use and which you think are questionable, using whatever definition you favor:

1. A monkey discouraging human intruders by passively dislodging dead branches as it flees through the canopy.
2. An orangutan discouraging human intruders by breaking off, aiming, and dropping branches from the canopy.
3. A chimpanzee discouraging human intruders by defecating and urinating from the canopy.
4. A gorilla discouraging human intruders by tearing up and waving saplings.
5. A chimpanzee discouraging human intruders by throwing feces.
6. A chimpanzee driving away another chimpanzee by hitting him with the body of a colobus monkey.
7. A macaque discouraging aggression from another macaque by brandishing an infant.

8. An archerfish spitting water droplets to fell a flying insect.
9. A chimpanzee beating a leopard with a stick.
10. A human killing an antelope with a spear.
11. A human shooting an antelope with a rifle.
12. An Egyptian vulture opening an egg by dropping stones on the egg.
13. A herring gull opening a mussel by dropping the mussel on stones.
14. A sea otter opening an abalone by pounding it on a stone that is balanced on the otter's ventrum.
15. A wasp closing its subterranean nest by tamping the soil over the opening with a pebble.
16. A wasp obscuring its nest opening by arranging dead leaves and sticks as surrounding litter.
17. A finch using a cactus spine to extract insect larvae from a crevice.
18. A baboon using a stick to reach food.
19. A monkey using its tail to reach food.
20. A gibbon pulling in a vine to get fruit that is growing at the end.
21. A crow pulling in a string to get food that has been tied to the end.
22. An ant crossing a streamlet over a bridge of linked fellow ants.
23. A spider monkey infant crossing an arboreal gap by walking across its mother while she holds on to branches on each side.
24. An oryx scratching its flank with its horn.
25. A horse scratching its flank with a stick that it holds in its mouth.
26. An elephant scratching its flank by rubbing against a tree.
27. A chimpanzee stacking boxes to get suspended food.
28. A chimpanzee propping and climbing a stick to get suspended food.
29. A chimpanzee swinging a stick overhead to get suspended food.
30. A chimpanzee climbing a fence to get suspended food.
31. A chimpanzee climbing on to the shoulders of a conspecific to get suspended food.
32. A rat pressing a lever in a Skinner box to get food.
33. A chimpanzee stabbing at a chicken with a wire.
34. A shrike skewering an insect on a thorn attached to a tree.
35. A human digging roots with a stick.
36. A human plowing a field with a horse-drawn plow.
37. A human plowing a field with a tractor-drawn plow.
38. A bowerbird using a decorated bower to attract a mate.
39. A deer accumulating mud on its antlers to attract a mate.
40. A moth secreting a pheromone to attract a mate.
41. A lemur rubbing scent gland secretions on its body.
42. A chicken ingesting grit to facilitate eating.
43. A chimpanzee ingesting medicinal plants to combat endoparasites.
44. A heron attracting fish by scattering bits of biscuits on the water.
45. An elephant spraying water on its back to cool off.
46. A rhinoceros wallowing in water to cool itself.
47. A lion sleeping in the shade of a tree to cool off.
48. A chimpanzee covering itself with cloth and vegetation for warmth.
49. A bird using a nest to incubate eggs and raise its young.
50. A bonobo sleeping in a nest.
51. A beaver building a dam to provide a pool suitable for den construction.
52. A hermit crab inhabiting an abandoned shell for protection.
53. A human wearing a helmet for protection.

Many of these behaviors have been considered in the scientific literature as being true tool use or related in some way to tool use. The difficulty of defining tool use is demonstrated if you found that you categorized even one of these behaviors as questionable. From our experience, we would guess that you ended up with more than one question mark and are beginning to feel uncomfortable with the intuitive definition that you carried with confidence only a few minutes ago.

When Beck (1975, 1980) first tried to categorize this list, he had the benefit of influential definitions that had been provided by two pioneers in the field: Jane van Lawick–Goodall and John Alcock. Van Lawick–Goodall (1970, 195) defined tool use as "the use of an external object as a functional extension of mouth or beak, hand or claw, in the attainment of an immediate goal." She later stipulated that the object must be "held" (J. Goodall 1986), but this came after Beck's original definition. Alcock (1972, 464) defined tool use as "the manipulation of an inanimate object, not internally manufactured, with the effect of improving the animal's efficiency in altering the form or position

A spider monkey infant crosses a gap in the trees by using its mother's body as a bridge. Photographs courtesy of Anne Shumaker.

of some separate object." Beck concluded that both definitions were inadequate.

The first problem had to do with the portability of the external object or inanimate object. For example, consider examples #12, an Egyptian vulture opening an egg by dropping stones on the egg, and #13, a herring gull opening a mussel by dropping the mussel on stones. The vulture picks up the stone and drops it on the egg (van Lawick–Goodall and van Lawick 1966.) The herring gull picks up the mussel and drops it on the stones (Beck 1982). Chimpanzees provide a similar contrast. At many study sites they have been observed hammering tough-skinned fruits, seeds, and nuts with sticks or stones to gain access to the edible interior. Pounding of food with sticks and stones has not been observed at all chimpanzee study sites, but chimpanzees at some of those sites have been seen to open tough-skinned fruits similarly by pounding them against tree buttresses and stones (Gašperšiè and Pruetz 2005; Marchant and McGrew 2005). However, van Lawick–Goodall (1970) claimed that pounding fruit *on* trees is not tool use, while pounding fruit *with* a stick or stone is. Presumably she would also argue that example #13 (gull dropping mussel on stones) is not tool use. Most would agree with the distinction. If a tree against which a fruit is pounded, or stones on which a mussel is dropped, are to be considered tools, then tool use as a category of behavior becomes meaninglessly broad. A chimpanzee climbing a tree to reach fruit growing in the canopy would have to be considered to be using the tree as a tool. A gull walking on a stone seawall would have to be considered to be using the wall as a tool. Yet these distinctions were not derivable from van Lawick–Goodall's definition. One could infer the distinction from Alcock's use of the term "manipulation," since it seems valid to say that the chimpanzee manipulates the stick or stone hammer, but not the tree buttress. The vulture manipulates the stone, but the gull does not.

A second problem had to do with the connection of the tool to the "goal" or the "separate object." A rat pressing a lever in a Skinner box to obtain a food pellet (example #32) would fit both van Lawick–Goodall's and Alcock's definitions perfectly. Nonetheless, few would view this as tool use, because the connection or orientation of the lever to the food pellet is not established or even perceived by the rat.

A third problem was that van Lawick–Goodall restricted tools to "external objects," while Alcock required that they be "inanimate" and "not internally manufactured." We have observed captive apes aiming and throwing feces (example #5). Some spiders actually throw their webs (M. Robinson and Robinson 1971). Many think that these are forms of tool use. Are feces and spider silk external? They are when they are used. Are they inanimate? They are certainly internally manufactured. Beck et al. (in prep.) observed a chimpanzee hitting a conspecific with a dead colobus monkey; was the monkey inanimate?

A fourth problem resulted from Alcock's definition being concerned solely with tool use in feeding or drinking. It needed to be expanded to include tool use in the contexts of self-maintenance (example #25, horse scratching flank with a stick) and social interactions (example #39, deer accumulating mud on antlers to attract a mate [Beck, Leja, and Zemanek 1978]). Animals use tools not only to alter the form or position of objects, but also the condition of objects or other organisms. Further, sometimes those objects can be conspecifics, or even the user itself.

Thus Beck (1975; 1980, 10) proceeded to define tool use as "the external employment of an unattached environmental object to alter more efficiently the form, position, or condition of another object, another organism, or the user itself, when the user holds or carries the tool during or just prior to use and is responsible for the proper and effective orientation of the tool." For the most part, this definition has been widely adopted and has stood the test of time. St Amant and Horton (2008, 1199) described it as having "served the field well," being "widely used," "the most influential," "the current standard," "straightforward," "robust," and a "good match to consensus judgements."

Others have been more critical, calling the definition "circuitous," "ambiguous," "grammatically complex," "arbitrary," "a definition of convenience rather than of biological distinctness," "extremely narrow," "too typological," "overly rigorous," "fundamentally inadequate," not self-evident, logical rather than psychological or social, focused on objects rather than on behavior, "not very usable," and "unelaborated" (e.g., Hansell 1984; Lestel and Grundmann 1999; B. Preston 1998). Some favored simpler definitions, such as J. Goodall (1986, 536) and Boesch and Boesch (1990, 86); in both publications a tool was defined as an object "held in the hand, foot or mouth and used so as to enable the operator to attain an immediate goal." Chevalier-Skolnikoff (1989, 564) proposed that tool use was "the use of one unattached object to effect a change in another." Matsuzawa (1999, 650) defined a tool as "a detached object that is used in some way to arrive at an apparent goal." P. Becker (1993, 13) classified tools as "all non-bodily objects, with which an animal achieves an immediate goal" (translated from German by B. Rettberg-Beck). Without adequate explanation, Becker excluded nests and other constructions as tools but disagreed that tools must be held or carried. We have also rediscovered an early definition by Hall (1963, 479): "the use by an animal of an object or of another living organism as a means of achieving an advantage." Hall further stipulated that the mediating object may not be part of the user's bodily equipment, and the advantage is in extending the user's range of movements or increasing their efficiency. Beck (1980) did not cite Hall's definition, though in retrospect it may have influenced his own. All of these definitions have the benefit of being simple and reflecting an intuitive notion of what tools and tool use are.

S. Parker and Gibson (1977, 624–625) drew on the definition originally provided by Beck (1975) to define tool use as "manipulation of an object (the tool), not part of the user's anatomical equipment and not attached to a substrate, to change the position, action, or condition of another object or of the object on the tool, or through action at a distance as in aimed throwing." The meaning of the clause "or of the object on the tool" was not clear and seemed to be confounded with what the authors called "proto-tool use" in the same paper (see below).

Some authors have favored broader and more complicated definitions than Beck's over the years. Pierce (1986, 96) defined tool use "as the active external manipulation of a moveable or structurally modified inanimate environmental object, not internally manufactured for this use, which, when oriented effectively, alters more efficiently the form, position, or condition of another object, another organism, or the user itself." This, he claimed, allowed inclusion of nests and burrows, but not spider webs. It would also eliminate other animals as tools. We disagree about the inclusion of nests and burrows, since the completed nest or burrow is not held or manipulated. Nor would we eliminate other animals as tools when the animal is physically held or manipulated. Reed (1985, 95) defined tools as a form of "secondary energy trap," or "an object used for acquiring or conserving energy more efficiently than an individual or social group could without that object." Reed included nests, burrows, spider webs, any features used as places to hide or avoid environmental extremes, symbiotic hosts, mutualistic partners, and even a "well-functioning spouse." Reed would not require that a tool be held and manipulated. Asano (1994) provided an even broader definition, defining tools as objects that aid natural sense organs (a telescope), provide artificial stimuli (a map), help perform expanded or non-natural response topographies (a hammer, a hang glider), alter the impact of natural consequences (pain medication), or provide artificial consequences (money). As in Reed's definition, social partners can be tools. This is a functional, not a descriptive, definition and is very human biased. Asano would not require that tools be held and manipulated, used externally, or effectively oriented by the user.

Lestel and Grundmann (1999, 372) proposed discarding "tool use" altogether, preferring to substitute "mediated actions," defined as "those material or intellectual aids that enable an animal to alter its performances or competences, either by changing the nature of its operation or by increasing its field of action." Mediated actions incorporate much of what Beck called "borderline cases," such as nest building, as well as including social interactions where animals use or exploit each other, such as coalitions, cooperation, domestication, slavery, parasitism, symbiosis, and trading and sharing. Even culture (as a "container" for knowledge) and scavenging ("using predators" to get food) are mediated actions. Yet Lestel and Grundmann's concept of mediated actions also has definitional problems: What exactly are "material or intellectual aids"? What are "nature of operation" and "field of action"? We could possibly see tool behavior as a subset of mediated actions, without altering our definitions of tool use and manufacture, but our survey of the literature suggests that the concept of mediated actions has not been broadly accepted as a supercategory of behavior. The scientists we read preferred to use terms such as "theory of mind," "cooperation," "sharing," "culture," and, yes, "tool use" and "tool manufacture." Indeed, Lestel and Grundmann themselves continually referred to "tools" in their article.

St Amant and Horton (2008, 1203) defined tool use as "the exertion of control over a freely manipulable external object (the tool) with the goal of (1) altering the physical properties of another object, substance or medium (the target, which may be the tool user or another organism) via a dynamic interaction or (2) mediating the flow of information between the tool user and the environment or other organisms in the environment." There are three critical points here, which we acknowledge and incorporate into our definition and catalog. First, tools do not have to be "unattached" (as in Beck's original definition), just "freely manipulable." Second, "exertion of control" refers to both purposiveness and whether the user's responsibility for establishing the proper and effective orientation of the tool is "controlled" rather than "accidental" or "incidental." Third, they stressed the importance of tool use in mediating the flow of information between the user and its environment. We include mediation of information flow as a function of tool use, rather than adding it to the definition itself. St Amant and Horton's paper was thus very influential in our revision of Beck's original definition.

Bentley-Condit and Smith (2010) adopted a definition that they said was "very similar" to St Amant and Horton's, but they never did provide their definition. They excluded "social tool use," by which they appear to mean the actual use of another animal as a tool rather than the use of a tool for social purposes. Bentley-Condit and Smith also excluded nests and burrows but included many examples of proto-tool use (see below). They concluded that any definition of tool use will be subjective and will have "shortcomings [*sic*]."

B. Preston (1998, 542, fn 20) characterized Beck's search for a clear-cut definition of tool use as a "disease," a "mistake," and a "problem," all in a single paragraph. Her article is a scholarly and astute consideration of "trying to draw sharp definitional lines" around tool use. She focused on the requirements that tools must be external objects, that tools cannot be parts of the user's own body, that tools must be held or carried, that they must be unattached, and that the user must be responsible for the proper and effective orientation of the tool. Preston recounted Beck's concerns about these requirements, and she added some even more compelling examples of ambiguity and arbitrariness in the definition. She suggested that Heidegger's (1927/1962) concept of "equipment" might be a suitable cure for Beck's disease and a philosophically more satisfying alternative to the "Beckian approach." Equipment more closely approximates the term "material culture" favored by anthropologists and used by McGrew (1992), and it certainly would include tools as defined by Beck. But Preston's equipment encompassed much more: houses, the ground we walk on, and the air we breathe. Preston assured her readers that the "continua" were not limitless, but she never provided the definitional limits. Plainly put, her treatment is not sufficiently pragmatic and operational to be of much use to behavioral and evolutionary scientists. Nonetheless, her thoughts did affect our revision of Beck's original definition, although it would still not meet all of her objections.

OUR PRESENT DEFINITION OF TOOL USE IS: The external employment of an unattached or *manipulable attached* environmental object to alter more efficiently the form, position, or condition of another object, another organism, or the user itself, when the user holds *and directly manipulates* the tool during *or prior* to use and is responsible for the proper and effective orientation of the tool.

The altered portions are italicized. We have added

two words ("manipulable attached"), changed "holds or carries" to "holds and directly manipulates," and subtracted one word ("just"). These small changes in wording reflect volumes of discussion and opinion and, in themselves, deserve detailed consideration.

The definition now includes manipulable attached objects, as well as unattached objects, as possible tools. Attached objects, of course, cannot be held in their entirety or carried, so attached objects were excluded as tools in the first edition. As noted above, many authors strongly suggested that manipulable attached objects could be tools. For example, St Amant and Horton (2008) cited a hypothetical experiment in which a raven is presented with a deep vertical tube with a small bucket at the bottom. The bucket contains food. There is a string attached to the tube, with a hook at the end. To get the food, the raven has to drop the string into the tube, hook the bucket, and pull it up to the top of the tube. That the string is attached, they argue, is not critical to its being a tool. The raven must manipulate the string and the hook and would be responsible for their proper and effective orientation with the bucket. Likewise, an orangutan using an attached branch to pull in a piece of fruit growing out of reach on another branch is using the first branch as a tool, even though the first branch is attached. The orangutan had to manipulate the first branch and had to establish the proper and effective orientation between the branch and the fruit. Russon (1998) and S. Thorpe, Holder, and Crompton (2009) described the use of attached but flexible trees as "sway poles" by orangutans. The orangutan climbs the tree and uses its weight to set the top of the tree swaying in a gradually increasing arc, until the orangutan can reach and transfer into a neighboring tree. We would now consider the string, branch, and tree in these examples to be tools because, despite their being attached, they are directly manipulated and the other elements of the definition are met.

As St Amant and Horton correctly pointed out, most unattached objects are manipulable, and it is manipulability that is an essential attribute of tool use. Attachment per se is irrelevant. The quality of manipulability distinguishes between attached objects that can be used as tools and those that cannot. As noted above, we would not consider a large stone or tree buttress on which a nut is pounded to smash it open as a tool. Attached objects can be tools only if they are manipulated and used in a way that meets the other elements of the definition. Thus we have deleted "carries" (which entails supporting the entire weight of the carried object) and added "holds and directly manipulates" to the definition.

With the addition of manipulable attached objects as potential tools, we must reiterate Hall's (1963) and Beck's (1980) stipulation that a tool cannot be an attached part of the user's body. A tail or a horn (while still attached to the monkey or antelope) cannot be a tool.

Many (most notably Boinski, Quatrone, and Swartz 2000; Panger 1998, 1999; and S. Parker and Gibson 1977) have argued that the non-manipulated large stones, tree buttresses, and stout branches used as anvils for nut smashing are functionally analogous to the manipulable stones or detached branches used as hammers for nut smashing. The hammers are tools, not because they are unattached, but because they are manipulated. Anvils that are not moved or manipulated prior to use are sometimes called "proto-tools," to acknowledge their functional similarity to small anvils that are moved or manipulated (S. Parker and Gibson 1977). Parker and Gibson defined proto-tool use as altering a detached object by manipulating it relative to (on, against) a fixed substrate. Many (e.g., P. Becker 1993) questioned why we should distinguish proto-tools, which are not held or manipulated, from true tools, which are. Bentley-Condit and Smith (2010) equated proto-tools with borderline tools and differentiated borderline from true tool use, but they included borderline tool use in their taxonomic comparisons of types and frequencies of tool use in general. Beck's (1980) exclusion of proto-tools is probably the most controversial aspect of the original definition, but he defended the distinction because it enabled the exclusion of trees as tools for locomotion and caves and houses as tools for shelter. In the absence of such a distinction, the concept of "tool" becomes so encompassing as to be meaningless. As St Amant and Horton (2008) concluded, there appears to be a consensus that manipulation of some sort is an essential property of tool use. *Proto-tools*, then, are functionally analogous to tools but are not held and directly manipulated during or prior to use.

Proto-tool use has been called "object use" (Panger 1998, 1999) and "substrate use" (Boinski, Quatrone, and Swartz 2000). "Object use" is easily confused with "object manipulation," which we will discuss below. "Substrate use" is often employed in a different sense by those who study animal locomotion. To avoid confusion, we will refer to "proto-tool use." None of these

An anteater's tongue is an anatomical adaptation for extracting social insects from nests, but the tongue is not a tool, because it is part of the anteater's body. Many birds and primates use tools to extract social insects. Photograph courtesy of Lydia Möcklinghoff.

authors felt that proto-tool use should be included in the definition of tool use. Incidentally, there is no evidence that proto-tool use can be seen as a more primitive form of tool use, or as an evolutionary or ontogenetic precursor to tool use, as Panger (1999) implies. There is also no empirical evidence that true tool use is cognitively more demanding than proto-tool use, as claimed by S. Parker and Gibson (1997) and Marchant and McGrew (2005). The distinction between tool use and proto-tool use is purely logical.

Returning to the new definition, there is a qualifier to our term "direct manipulation." "Manipulation" refers to mechanical, not psychological, force. We do not consider threat, incitement, solicitation, or deception as forms of manipulation for purposes of the definition of tool use.

Panger (1999) and Urbani and Garber (2002) argued that in tool use, both the tool ("agent of change") and the object the tool alters ("object of change") must be manipulated. Thus these authors would say that an elephant throwing a branch at a human intruder would not be using a tool, because the human is not manipulated. In comparing tool use and proto-tool use, S. Parker and Gibson (1977, 625) said that tool use involves "detachment and manipulation of both the object of change and the agent of change," but they included aimed throwing as tool use, where the object of change is not manipulated. Our reading of this paper as a whole is that Parker and Gibson did not require manipulation of both the tool (agent of change) and the object of change in true tool use, but their words might suggest that they did. In any event, we reject this requirement; tool use requires only that the tool, the agent of change, be manipulated. In our opinion, manipulation of only the altered object (object of change) is proto-tool use or object use. Manipulation of only the tool, or the tool *and* the altered object, is tool use.

The word "hold" has many meanings. It can mean "carry," which requires support of the entire weight of the object being held. This was the sense in which the phrase "holds or carries" was used in the first edition. Yet the weight of an *attached* object cannot logically be borne in its entirety. Even the weight of some unattached objects cannot be borne by some animals. With the addition of "manipulable attached" objects as potential tools, we added "grasp" as a second meaning of the word "hold." Thus "hold and directly manipulate" in our new definition can mean "to carry" (unattached objects only) or "to grasp and move" (unattached or attached objects).

Intuitively, we are most comfortable viewing unattached objects that are carried and manipulated as potential tools. Yet the orangutan sway pole described above is an attached object that is grasped and moved, but not carried. Similarly, orangutans or gorillas that slide (unattached) boxes or other objects into position for use as climbing tools are grasping and moving them, but not carrying them.

We subtracted "just" before "prior" in the new definition because we consider the latency between holding or manipulating an object and then later using it as a tool to be irrelevant. Consider an example in which a chimpanzee props a stout branch against a fence and then climbs it. The chimpanzee is not moving or manipulating the branch during use, but it does move and manipulate it prior to use. And it is using the branch as a tool, whether climbing it seconds or days after propping it. Thus we have deleted the word "just." Of course this raises the question of whether a chimpanzee that climbs a tool propped by another is using the branch as a tool. The definition would imply that it does not, because the climber does not hold or manipulate the tool prior to use and is not responsible for the proper and effective orientation of the tool.

We considered deleting "external" before "employment," because self-medication and consumption of rare nutritional elements could be seen as forms of tool use and, when the plant compound or trace nutrient is swallowed, it is internally employed. Yet if we recognize internal employment as tool use, we would also have to include as tool use the ubiquitous ingestion of food and water to maintain metabolic balance. This would again render tool use to be so inclusive as to be meaningless. We chose to retain "*external* employment" in the definition and assign self-medication and consumption of trace nutritional elements to the borderline category (see chapter 7).

Candland (1987) and St Amant and Horton (2008) correctly pointed out that the simple phrase "to alter" implies a goal to be achieved or an incentive to be gained. In other words, the phrase implies intentionality or purposiveness. Every case in our exercise implies purposiveness. Beck (1975, 1980) intended that the original definition was to be operational, free of cognitive considerations and unobservable mentalistic inferences. But "to alter" implies purposiveness, and purposiveness is a richly cognitive term. We could finesse this issue by saying in the definition that tool use "is the external employment of an . . . environmental object *with the effect of altering* more efficiently the form, position, or condition . . . " That would seem to strip the definition of cognitive considerations, but doing so would over-objectify the definition. There appears to be consensus around goal-directedness or purposiveness as inherent in the notion of using an object as a tool. An object cannot be a tool unless it is used purposively to achieve a goal. This is St Amant and Horton's "exertion of control."

A gorilla stands on a ball that she has Repositioned so that she can see out of a window. Photograph courtesy of The Revealed Project and Marina Vancatova.

Purposiveness in itself does not imply consciousness or causal understanding, and we will list many examples of tool use where an inference of consciousness or understanding would seem unsupported, although few would disagree that the tool is used purposively, to achieve a goal. We decided to retain this part of the original definition, understanding that the phrase "to alter" does indeed imply goal-directedness. Candland (1987) was comfortable with this, as long as the increased "efficiency" of altering "the form, position, or condition" is strikingly obvious or, better yet, measured and compared with a similar behavior in the same species without tools (see Nishida and Hiraiwa [1982, 95] for an explicit example of this method). Purposiveness is currently an unobservable mental construct that can be studied by the Nishida and Hiraiwa type of comparison, by inferential prediction of what an animal or machine would do in a specific situation if it were purposive, or intuitively through "folk psychology" (Dennett 1988).

When reviewing the literature for our catalog, we took the observer's word about purposiveness. If an observer inferred purposiveness in tool use, we included the case as tool use. Panger (1998, 1999) and Visalberghi and Fragaszy (2006) required that a behavior have an unambiguous immediate goal to be considered as tool use. Wherever one sets the bar on purposiveness, it is the only criterion that separates tool use, on the one hand, from simple object manipulation or object play, where an object might incidentally be altered. We define *object manipulation* (and by extension *object play*) as "holding or directly manipulating unattached or manipulable attached environmental objects with no evident proximate purpose." Object manipulation and play undoubtedly serve an ultimate or adaptive function (e.g., Fagen 1981), but we use "purpose" here as immediate and tangible. We do consider objects used to stimulate or enhance social play as tools.

St Amant and Horton (2008) cited examples of unsuccessful tool use in which "the form, position, or condition" of something is not altered even though the user appeared to try to do so. When learning to crack nuts, juvenile chimpanzees (Matsuzawa 1994) and capuchin monkeys (Ottoni and Mannu 2003) frequently fail in their attempts. While unsuccessful, this is tool use, because there are enough purposeful successful cases to differentiate an unsuccessful attempt from simple object play or object manipulation. This may seem trivial, but it becomes important in the consideration of the ontogeny of tool use, where younger or unskilled individuals first begin to use a tool, but without the success of more skilled individuals.

A zoo-living male Grevy's zebra plays exuberantly with his food pan. Because there is no evident purpose, this is object manipulation, not tool use. Photograph courtesy of Benjamin Beck.

We can also envision lucky behavioral accidents, resulting perhaps from simple object manipulation, in which playing with an object produces an alteration of form, position, or condition that is beneficial to an animal, and the resulting reinforcement could lead to purposeful repetitions that would clearly be tool use. Bard and Vauclair (1989) concluded that the first, accidental case was not goal directed, and thus in itself would not be considered tool use. Again, this is important in the study of the ontogeny of tool use. Of course there are many reports of a single occurrence of tool use, and it is not possible to know if this was the first case for that individual or a repetition. Unless otherwise specified by the author, we treated single-occurrence reports as tool use.

Although we explicitly acknowledge that our definition of tool use incorporates the cognitive construct of purposiveness, we reject the requirement (Urbani and Garber 2002) that tool use must evidence the user's causal understanding of the relationships or mechanical dynamics of the objects involved. Indeed, Beck (1980) argued, and we still agree, that there is no requisite relationship between tool use and causal understanding or, for that matter, consciousness or intelligence. We will revisit this in chapter 7.

Visalberghi and Fragaszy (2006, table 27.2) made a related point about cognitive complexity: they criticized Beck's definition because it does not discriminate among cases of tool use that appear to differ in complexity because of the number of objects involved. Matsuzawa (1996, 1997, 2001) also argued for a relationship between the number of objects involved in a case of tool use and cognitive complexity, but he did not criticize the definition. Again, we acknowledge that some cases of tool use are cognitively more complex than others, but we feel that cognitive complexity is at best irrelevant and at worst misleading in the definition of tool use (see chapter 7).

As a final step in this reconstruction of how we arrived at our revised definition of tool use, we considered adding a second sentence to the definition: "The condition of another organism or the user itself may include its sensory input or its knowledge." However, we decided that while this is a qualifying phrase, and a function of some tool use, it need not be part of the definition itself. Using an object to alter sensory input or knowledge of another organism would include modification of a social signal, which we would consider to be a function of some tool use. This is St Amant and Horton's (2008) "mediation of the flow of information." Nishida (1981) was the first to introduce "augment the signal value of display" as a function of some tool use.

To "carve the definition at the joints," to use B. Preston's (1998, 513) evocative phrase, our current definition has four distinct elements:

1. a tool user externally employs an unattached or manipulable attached environmental object
2. to alter more efficiently the form, position, or condition of another object, another organism, or the user itself (with "to alter" implying purposiveness)
3. when the user holds or directly manipulates the tool during or prior to use and
4. when the user is responsible for the proper and effective orientation of the tool

Returning to the exercise and using this definition, we would confidently categorize as tool use examples #2 (orangutan aiming and dropping branches), #4 (gorilla waving saplings), #5 (chimpanzee throwing feces), #6 (chimpanzee hitting another with dead monkey), #7 (macaque brandishing infant), #8 (archerfish spitting water at insect), #9 (chimpanzee beating leopard), #10 (human killing antelope with spear), #12 (vulture dropping stones on egg), #14 (sea otter opening abalone on stone held on chest), #15 (wasp tamping nest opening with pebble), #16 (wasp obscuring nest opening with leaves), #17 (finch extracting larva with cactus spine), #18 (baboon reaching food with stick), #25 (horse scratching flank with stick), #28 (chimpanzee propping and climbing stick), #29 (chimpanzee swinging stick), #33 (chimpanzee stabbing at chicken), #35 (human digging roots with stick), #39 (deer accumulating mud on antlers), #41 (lemur rubbing gland secretions on body), #44 (heron scattering biscuits), #45 (elephant spraying water), #48 (chimpanzee covering itself with cloth), #52 (hermit crab inhabiting shell), and #53 (human wearing helmet). Each of these examples conforms to all four segments of the carved-at-the-joints definition.

Examples #42 (chicken swallowing grit) and #43 (chimpanzee eating medicinally active plants) are not considered tool use because the grit and plants are internally, not externally, employed. Examples #19 and #24 are excluded because the tail and horn are attached parts of the monkey's and oryx's bodies (respectively).

Examples #13 (gull dropping mussel on stones), #22 (ant crossing bridge of linked fellows), #23 (spider monkey crossing gap on mother), #26 (elephant scratching on tree), #27 (chimpanzee climbing a tower of stacked boxes), #30 (chimpanzee climbing fence), #34 (shrike skewering insect on thorn), #38 (bowerbird attracting mate with bower), #40 (moth secreting pheromone), #46 (rhinoceros wallowing), #47 (lion sleeping in shade), #49 (bird nest), #50 (bonobo nest), and #51 (beaver dam) are not tool use, because the putative tool in its entirety is not held or manipulated by the user. Many of these examples (#13, #26, #30, #34, #38) could be called proto-tool use (see above). Example #31 (chimpanzee climbing on the shoulders of a conspecific to reach food) would not be considered tool use if the climber simply got on another chimpanzee who happened to be sitting or standing beneath the food, but it would be tool use if the climber pushed or pulled, that is, manipulated, the other chimpanzee into the proper position.

Example #1 (monkey passively dislodging branches) is not categorized as tool use, due to lack of purposiveness. Example #3 (fleeing chimpanzee defecating and urinating) is not tool use, both because it may not be purposive and because the user does not hold or directly manipulate the feces or urine.

Examples #20 (gibbon pulling in vine), #21 (crow pulling in string with tied-on food), and #32 (rat pressing lever to get food) are excluded, because the user is not responsible for the connection between the putative tool (vine, string, and lever, respectively) and the object on which it acts (fruit, food, and pellet, respectively), and in these examples that connection would be considered proper and effective orientation. Establishing the proper and effective orientation is what Visalberghi and Fragaszy (2006, 536) described as producing a "needed relation" between one object and another. Many mammals and birds can apparently recognize causal relationships in string-pulling experiments (e.g., Beck 1967; Heinrich and Bugnyar 2005; Schuck-Paim, Borsari, and Ottoni 2009), but for the behavior to be included as tool use, the animal must *produce*, not simply recognize, the proper and effective orientation between the tool and the incentive.

Examples #11 (human shooting antelope with rifle), #36 (human plowing field with horse-drawn plow), and #37 (human plowing field with tractor-drawn plow) are also not categorized as tool use, in part because the putative user is not totally "responsible for the proper and effective orientation." We call these Constructions (see table 1.2). These examples test the limit of the "hold and manipulate" and the "responsible for" criteria. One could argue that the humans manipulated each of these Constructions in its entirety: the rifle and bullet; the reins, horse, harness, and plow; and the steering wheel, motor, drive shaft, and plow. The humans might have been responsible for the proper and effective orientation of each of the links, and they may even have established all of the proper linkages. They certainly aimed the bullet and steered the plows, but they are *not responsible for the mechanical linkages* between the rifle, the bullet, and the antelope; or the reins and harness, the horse, the plow, and the soil; or the tractor controls, the motor and wheels, the plow, and the soil, respectively. These *linkages* are the proper and effective orientations in these examples. Since the putative tool users are not responsible for the linkages, we do not categorize instances involving such linkages as tool use. Our interpretation of "responsible for" becomes evident when comparing these examples to a human killing the antelope with a spear (example #10), or pulling the plow personally, or hitting the antelope over the head with the rifle. An additional reason for eliminating examples #11 (shooting antelope with rifle), #36 (using horse-drawn plow), and #37 (using tractor-drawn plow) is that the putative tool user is not holding or directly manipulating the tool *as a whole*. The user holds or manipulates parts of the Construction, but not the bullet or the plow. It is the bullet or the plow, not the human, which act literally upon the "form, position, or condition" of the antelope and soil, respectively.

Perhaps our new definition of tool use is less meaningful biologically than logically. Borderline cases still exist. We have learned that the act of definition consists of striving for consensus among experts, not elucidating an unambiguous categorical reality; experts do not agree on the definitions of "language" or "culture," either. Our definition is the result of considering and trying to incorporate all of the alternatives and thoughts about defining tool use. Our revised definition serves as the basis for assembling the catalog that makes up chapters 2–6. And it helps to resolve the question that greets visitors to Think Tank: a cow is not a tool for manufacturing and transporting meat and milk, both because the user does not hold or directly manipulate the cow, and because the user is not responsible for the proper and effective orientation of the cow to the meat or milk (the cow's own metabolism is responsible).

Happily, given the definition of tool use, defining tool manufacture is relatively simple:

Tool manufacture is simply any *structural* modification of an object or an existing tool so that the object serves, or serves more effectively, as a tool.

We added the word "structural" to Beck's original definition to clarify that "modification" means alteration of the dimensional physical characteristics of the tool, not simply its spatial orientation. A tool user can use a tool that is manufactured by another, provided the use meets our definition. Boesch and Boesch (1990, 94) defined tool manufacture as "all alterations actively accomplished on an object to modify its shape." The word "object," out of the context of the article, could be taken too broadly, that is, nest building could be seen as tool manufacture. Gruber (1969) suggested that true tool manufacture must involve the use of another object as an aid in production. He proposed such a stipulation to preserve a qualitative uniqueness in human tool behavior (we now know that it doesn't even do that, because some apes and monkeys sometimes use tools to make tools). We dismiss such a restriction as needlessly anthropocentric, since it discourages a truly comparative analysis of tool behavior. The use of a tool to make another tool is called Secondary Tool Use (see table 1.2).

G. Hunt and Gray (2004a, S88) and Pruetz and Bertolani (2007, 412) borrowed the term "crafting" from archeology literature to describe tool manufacture that involves a number (minimum not specified) of successive steps, including what we would call Detaching and Subtracting, as well as "fine three-dimensional shaping," which we infer to mean some fundamental changes that we call Reshaping. These authors were describing the manufacture of insect-probing tools by New Caledonian crows and hunting tools by chimpanzees. The term crafting invites comparisons with human tool manufacture and may represent stages of complexity analogous to Visalberghi's and Fragaszy's (2006) "relational stages" and Matsuzawa's (1996, 2001) "tree node" analysis for tool use. Nonetheless, we see Crafting as fitting comfortably within our definition of tool manufacture.

B. Preston (1998) noted that some feel that manufacture must occur for something to be considered a tool; for example, an unmodified natural object cannot be a tool. Westergaard and Suomi (1995d, 1017) stated plainly that a "tool is an object which has been modified." We obviously disagree with such an illogical and anthropocentric position. Oswalt (1973) distinguished between a "naturefact" (unmodified tool) and "artefact" (modified tool). For our present purposes, the critical point is that an object does not have to be modified to be a tool.

A New Caledonian crow uses a Crafted hook tool to extract an insect from a dead log. Photograph courtesy of Gavin Hunt.

MODES OF TOOL USE AND TOOL MANUFACTURE

The diversity of tool behavior begs for some categorization. Modes of tool use are a second-order elaboration of an "alteration of form, position, or condition" of an object, organism, or the user. Modes of tool manufacture are a second-order elaboration of "modification." In the 1980 edition, Beck extracted twenty-one modes of tool use and four modes of tool manufacture *a posteriori* from his catalog of tool use and then used them to arrange the catalog in a uniform format for each treated animal taxon. Modes are intended to be categorical aids rather than reflections of some underlying biological order. The names of most of the modes are operationally descriptive and, in themselves, do not describe function. We will capitalize the modes to distinguish their formal organizational role from conversational use.

While the names of the modes are primarily descriptive of motor actions, some include disparate actions that have the same general characteristics. For example, the mode we call Drag, Roll, Kick, Slap, Push Over involves actions that all move an object in space without either throwing it or dropping it (we call the latter modes Throw and Drop, respectively). Throw and Drop also result in objects moving through space, but the objects moved by tool behaviors assigned to the Drag, Roll, Kick, Slap, Push Over mode stay close to the substrate, while those of Throw and Drop have an aerial trajectory. They could all be assigned to a mode named "Move Objects through Space," but that would feel too removed from the actual motor patterns. We invite readers to experiment with their own higher-order classification (experience suggests that the invitation is not required). Although we acknowledge some arbitrariness of the modes as a classification scheme, we do provide very precise definitions in table 1.1 to at least allow scientists to make a preliminary assignment of a tool behavior they might observe.

A mode can include similar but different forms of tool use, for example, chimpanzees Insert and Probe for honey, ants, and termites. A mode can even subsume behaviors that are quite distinct. For example, included in the Brandish, Wave, Shake mode are the use of anemones in feeding by marine crabs, the use of fly whisks by elephants, the use of infants in agonistic buffering by macaques, and the use of branches in agonistic display by chimpanzees. Additionally, a few cases of tool use do not seem to fit comfortably into

Table 1.1
Modes of tool use and tool manufacture

Name of use mode	*Function*	*Definition and comments*
Drop	Create or augment signal value of social display; amplify mechanical force; extend user's reach	Cause an object to fall. The object is propelled primarily by gravity.
Throw	Create or augment signal value of social display; amplify mechanical force; extend user's reach	Propel an object through open space. Can be aimed or unaimed. The object is propelled by the user's own energy.
Drag, Roll, Kick, Slap, Push Over	Create or augment signal value of social display; amplify mechanical force	Drag: Pull an object while walking or running; the object touches the substrate. Roll: Propel an object such that it turns over repeatedly on the substrate. Kick: Propel an object with foot through space or on the substrate. Slap: Hit an object or another organism with a flattened appendage. Push Over: Deflect an attached vertical object.
Brandish, Wave, Shake	Create or augment signal value of social display; camouflage; bodily comfort	Brandish: Conspicuously display an object away from the body. Wave: Move an unattached object repeatedly up and down or side to side. Shake: As in Wave, but the object is attached.
Bait, Entice	Extend user's reach; create or augment signal value of social display	Manipulate or place, and conspicuously display, an object such that it attracts prey or a potential interactant to approach.
Club, Beat	Amplify mechanical force	Hit another organism agonistically with an object. The object (the tool) can be another organism or a part thereof.
Pound, Hammer	Amplify mechanical force	Hit an object or prey item forcefully, often repeatedly, with a second, relatively hard object (the tool).
Pry, Apply Leverage	Amplify mechanical force	Push or pull on an object (the tool), using a fulcrum.
Dig	Amplify mechanical force	Excavate the substrate, usually earth.
Jab, Stab, Penetrate	Amplify mechanical force; extend user's reach	Jab: Use the distal end of an object to forcefully push another object or organism away, or to animate another organism; synonymous with prod. Stab: Use the distal end of an elongate object to forcefully puncture or perforate another organism so as to harm it. Penetrate: Use the distal end of an object, usually elongate, to forcefully puncture or perforate a surface or an inanimate object. Often followed by Insert and Probe. May be combined with Reach to allow retrieval of the object.
Reach	Extend user's reach	Use an elongate object to touch or retrieve another object when the user's prehensive structures are too short, or use objects of any shape to avoid touching the target object. Tool held throughout use. Can be followed by Scratch or Rub when affected body surfaces are inaccessible.
Insert and Probe	Extend user's reach	As in Reach, but when the target object is embedded in a hole or behind a restricted opening. Can include Insertion in bodily orifices.
Scratch, Rub	Amplify mechanical force; bodily comfort	Move an object across a bodily surface, often repeatedly, while applying pressure. Includes tickling. Can follow Reach, if a body surface is inaccessible without the tool. Can be combined with Insert and Probe within a bodily orifice.
Cut	Amplify mechanical force	Movement, usually repeated and back and forth, of an object with a sharp edge to incise another object or surface.
Block	Amplify mechanical force; extend user's reach	Place an object to prevent or impede movement or action of another object, fluid, or organism.

(*continued*)

Table 1.1 (continued)

Name of use mode	*Function*	*Definition and comments*
Prop and Climb, Balance and Climb, Bridge, Reposition	Extend user's reach by expanding accessible three-dimensional space; bodily comfort	Prop and Climb: Place and stabilize an elongate object vertically or diagonally against another object or surface, and then move up or climb up the object. Distal end of propped object touches the other object or surface. Stable. Balance and Climb: Place an elongate object vertically and then move up or climb up the object. The distal end of the balanced object does not touch another object or surface. Unstable. Bridge: Place an elongate object or organism over water or open space such that each end rests on a surface on opposite sides of the water or spatial gap. User locomotes on the object. Stable. Reposition: Relocate and climb on an object or organism. Includes rafting (placing a buoyant object on water to support user's weight).
Hang	Extend user's reach by expanding accessible three-dimensional space	Suspend the object(s) from an overhead substrate or surface. User locomotes or rests on the object.
Contain	Effective control of fluids and small solid objects	Place fluids or objects into or on top of another object (the tool) to control and/or transport them.
Absorb	Effective control of fluids; bodily comfort	Soak up and move fluid. May be combined with Wipe.
Wipe	Effective control of fluids and small solid objects; bodily comfort	Mechanically displace and remove fluid, soft solids, or small solids. May be combined with Absorb.
Affix, Apply, Drape	Create or augment signal value of social display; effective control of fluids; camouflage; bodily comfort	Affix: Attach an object to the body, a surface, or another organism with adhesive. Apply: Attach a fluid or an object to the body, a surface, or another organism without adhesive. Drape: Place an object on the body or on a surface temporarily.
Symbolize	Abstract or represent reality	Carry, keep, or trade an object that represents another object, another organism, or a psychological state.

Name of manufacture mode	*Function*	*Comments*
Detach	Structural modification of an object or an existing tool by the user or a conspecific so that the object/tool serves, or serves more effectively, as a tool	Remove the eventual tool from a fixed connection to the substrate or another object.
Subtract	As above	Remove and discard a portion of a tool or an eventual tool so the tool can be used, or used more efficiently.
Add, Combine	As above	Join or connect two or more objects to make one tool that is held or directly manipulated in its entirety during its eventual use.
Reshape	As above	Fundamentally restructure material to make a tool. May include Detach, Subtract, Add/Combine.

any mode, such as an orangutan pulling a branch in front of itself to hide; others, such as Stabbing a chicken to Reach and retrieve it, appear to fit into more than one.

Despite the controversies over Beck's original definition of tool use, there has been little discussion of his modes. Still, we have made some changes. We started with Beck's twenty-one modes of tool use and then added more as our far larger catalog took form. The contents of the catalog drove the names and definitions of the modes, not vice versa. Some of the new modes were suggested by others (see below) and some, such as Block, by observations reported since 1980. Some changes in the names of the modes and the behaviors included within them resulted from our changes in the definition of tool use (mainly the inclusion of manipulable attached objects as tools), and some resulted from a more critical reexamination of Beck's categorizations. We now identify twenty-two modes and name and define them in table 1.1, before presenting our catalog, to help the reader better understand the catalog's organization.

We combined "Aimed and Unaimed Throwing" into Throw, since authors sometimes found it impossible to determine if an animal was aiming the object it threw (e.g., McGrew et al. 2003).

Penetrate in our mode Jab, Stab, Penetrate does not differentiate between "puncture" and "perforate," as used by Sanz and Morgan (2007) and Boesch, Head, and Robbins (2009).

We have removed "Stack and Climb" from Beck's list of modes. When animals (usually primates) stack objects, they make a tower, which they then usually climb to expand accessible three-dimensional space. Beck (1980) found this awkward, since the entire tower is not held or carried, but he included the tower as a tool nonetheless. Even with our revised definition, the completed tower is not held or directly manipulated, and thus is not a tool. Like a bird's nest, it is a Construction in which every component can be viewed as a tool, but the final product is not a tool because it is not held or directly manipulated in its entirety (see discussion below in regard to Associative Tool/Object Use). A single item, such as a box or limb that is moved as a locomotor affordance, is a tool and is included in the mode we call Prop and Climb, Balance and Climb, Bridge, Reposition.

Lethmate (1982) adopted Beck's original modes and added "Cover," "Float," and "Manipulate" to categorize cases of orangutan tool behavior described by others after 1980. His "Cover" categorized orangutans' use of leaves and other objects to protect their hands from spiny fruits and surfaces. We use Apply for this. Lethmate's "Float" categorized the behavior of rehabilitant orangutans commandeering canoes to cross rivers. We use Reposition for this and include it in a single mode with other patterns of tool use (Prop and Climb, Balance and Climb, Bridge) that allow the tool user to expand the three-dimensional space to which he/she has access. "Float" might also be confused with birds placing food on water to attract fish, a form of tool use that we put into a mode called Bait, Entice. Lethmate used "Manipulate" to categorize what appears to be augmentation of social display. We have chosen not to use this term, because it is used in our definition of tool use, and the use of objects to create or augment the signal value of display is better characterized as a function of several other, more descriptive modes, for example, Throw, Brandish/Wave/Shake, and Drop.

One of Köhler's chimpanzees Balances and Climbs a pole to reach a suspended incentive. Reprinted from Köhler 1917.

We added several new modes, but they include relatively few cases of tool use. Scratch, Rub is the movement of a tool across a body surface, usually repetitively. This was subsumed under Reach in the original twenty-one modes, but sometimes the point of contact with the body surface is within reach. The tool serves to moderate, and often concentrate, the force with which the tool is applied. We include tickling under this mode. Another new mode is Cut, to describe the use of sharp objects to create openings in surfaces or to sever string or rope (to date, only reported for monkeys and apes). A third addition is Block, in which the user places an object to obstruct the movement of an object or another organism. Block differs from Reposition in that the former impedes movement while the latter facilitates movement.

A fourth new mode of tool use is likely to be controversial. We have named this mode Symbolize, based on Sousa and Matsuzawa's (2006, 413) "symbolic tool." Symbolize is admittedly not a motor pattern. Apes may carry log "dolls," wadded leaves, and human toy dolls, sometimes obsessively (e.g., Wrangham and Peterson 1996). These objects appear to abstractly represent reality, and they might be symbolic. Carrying them appears to provide psychological comfort or possibly restore physiological homeostasis. If so, they are tools. These behaviors, however, could easily be confused with simple object manipulation or object play. The tools might simply be conditioned reinforcers with no verifiable symbolic property.

Symbolic tool use might be involved in some cases of trading tokens or other objects for a primary reinforcer, but only where the user chooses from among different types of tokens or objects to secure different types of rewards. Still other cases of symbolic tool use involve the use of abstract, manipulable, three-dimensional objects that are said to represent things, other individuals, and actions. These tools again seem to abstractly represent reality, but in these cases they help the user attain primary reinforcers. Some cases subsumed by this Symbolize mode appear to fit into Asano's (1994) Class 1-b tools, which, he asserted, included only human tool use. L. White (1942, 371) also argued that symbolic tools were uniquely human: "Among apes, the use of tools is a sensory, neuromuscular, and conceptual process. Among men [humans], it is a sensory, neuro-muscular, conceptual, and symbolic process." We include symbolic tool use because the behaviors literally fit our definition of tool use, but the mode might also point to a link between object-based and non-object-based representational cognitive operations.

We omit as cases of symbolic tool use the manipulation of symbols on a keyboard by birds and primates, and the deposit of tokens into a machine to get a reward. These behaviors can be symbolic, but by our definition they are not tool use, because the user is not responsible for the proper and effective orientation of the putative tool (key, token) and the reward. The computer/dispenser is a black box in which, in the words of Goldenberg and Iriki (2007, 286), "the transformation of the manual action into action on external objects is no longer transparent" to the user. Tokens and abstract symbols are symbolic tools only when the objects are "held or directly manipulated," when the user is "responsible for the proper and effective orientation of the tool," and when there is evidence that the token or object represents something else. Thus tokens can be tools and not be symbolic (trading with a human), and they can be symbolic and not be tools (insertion into a vending machine). Objects must first be tools, *and* then be symbolic, to qualify as symbolic tools. We would consider the chimpanzee token use described by Sousa and Matsuzawa (2001, 2006) to be symbolic, but not symbolic tool use, because by our different and more restrictive definition of tool use, the chimpanzees were not using the tokens as tools.

Apes, and sometimes monkeys and birds, will trade objects with people and each other in exchange for primary reinforcers (e.g., Brosnan and Beran 2009; Brosnan and de Waal 2005; Hyatt and Hopkins 1998; present authors, pers. obs.). We consider this tool use because the user entices the human to approach and trade, though it is not necessarily symbolic tool use. Entice also now includes the presentation of food and objects in courtship display.

We found no reason to revise or redefine Beck's original four modes of tool manufacture (Detach, Subtract, Add/Combine, Reshape), except to stress that a tool made by Adding/Combining two or more objects or tools must itself be held and directly manipulated after it is made. Lethmate (1982) adopted Beck's modes of tool manufacture. Lestel and Grundmann (1999) did so as well, to describe nest building, but they did so without attribution (compare Beck 1980, 105, with Lestel and Grundmann 1999, 375). Oswalt (1976) provided a taxonomy of tool "production": "Reduction" which is equivalent to our Subtract, "Conjunction" (our Add, Combine), and "Replication" and "Linkage," which are roughly equivalent to Metatools and Tool Composites (considered below). Oswalt based his classifications almost entirely on human tool behavior, and thus he may have regarded Detach as so ubiquitous as to be not worth mentioning.

Like humans, some animals manufacture tools with a progressive series of qualitatively different steps. An animal may first Detach the future tool, Subtract some portions, and then Reshape it. We suggest the use of the terms *first-order manufacture, second-order manufacture*, and so on to describe the steps in such a progression. Sometimes a modification that occurs after the first modification is called "secondary modification" (e.g., Sanz, Morgan, and Gulick 2004, 579), but this could be confused with Secondary Tool Use (use of a tool to make a tool; see table 1.2).

One can reasonably infer from many cases of tool use, especially by great apes, that they had also manufactured the tools beforehand, but that the manufacture was not observed. In this volume, we report only those cases of manufacture where the authors of the original reports specifically mentioned or described manufacture. In some of those cases the authors specified that tool use and manufacture were not actually observed, but inferred by the recovery of discarded, obviously manufactured, tools (McGrew, Tutin, and Baldwin 1979).

Hansell (2000, 91) quite casually offered a provocative thought about tool manufacture: an animal transforms an object into a tool "simply by picking it up" before using it. That is, the most elemental form of tool manufacture may be psychological, the recognition that an object could be a tool, rather than behavioral or physical.

The modes of tool manufacture and their definitions also appear in table 1.1.

FUNCTIONS OF TOOL USE

Beck (1980) not only categorized all cases of tool use into modes, but he then placed the twenty-one modes into six overall functions. Functions are a third-order classification of tool use, also shown in table 1.1. Beck stressed that a vast majority of the myriad cases of animal tool use, among a wide variety of animals, in twenty-one modes, served only four general functions: (1) to extend the user's reach, (2) to amplify the mechanical force that the user could apply, (3) to augment agonistic display, and (4) to more effectively control fluids. Again, there was a bit of forced lumping with regard to these functions of tool use, and a few cases appeared to serve none of these. Here, we use these same four functions but elaborate on them, and also add three new functions. Nonetheless, the original four still apply to the great majority of animal tool use.

Following the suggestion of Nishida (1981), we changed Beck's function of "augment agonistic display" to "create or augment signal value of display." Observations since 1980 show that apes use objects in affiliative as well as agonistic display, and in a few cases the object *is* the display, rather than an enhancement.

We expanded the functions of Contain to include control of soft solids (e.g., vomit) and small solids (e.g., ants), as well as control of fluids. We also note that gaining access to otherwise inaccessible places in an animal's three-dimensional environment is a special form of the function of extending reach. We have added three more functions: to hide or camouflage the user, to provide or enhance bodily comfort for the user, and to abstract or represent reality (see above). "Camouflage, Hide" was mentioned in the first edition, but the cases were too few to warrant a special functional category. Now they are more numerous. Bodily comfort was forced into other functional categories, for example, "extend reach" in the first edition. Now we recognize it as a unique function.

Thus we now have twenty-two modes of animal tool use serving seven functional categories, and four modes of tool manufacture serving one function (to make an object serve more effectively as a tool).

Bentley-Condit and Smith (2010) separated tool use into ten functional categories that are organized by biological function (e.g., food preparation, mate attraction) rather than our motorically descriptive modes (e.g., Insert and Probe, Reach) and our physically descriptive functions (e.g., extend reach, amplify mechanical force). We feel that their categorization introduces more taxonomic bias (e.g., elephants would not be expected to construct nests) than ours.

MORE ON MODES AND FUNCTIONS

At a given moment in space and time, the extent of an animal's reach is limited by the length and shape of its prehensive structures and by locomotor affordances in its environment. Reach is further limited by environmental objects such as walls and trees, which may be close but are too dense for the animal to reach through. By moving, the animal can change its location, but locomotor abilities, physical restraint, or ambivalence may prevent an animal from moving close enough to an incentive to engulf it within the sphere of reach. Reaching is the mode of tool use most commonly employed to bring an incentive within the sphere of reach. Propping and Climbing/Balancing and Climbing/Bridging/Repositioning and Hanging may be used to reach an incentive that is above the user or separated from the user by a barrier, and Baiting, Enticing may be used if the incentive is animate and otherwise reluctant to approach the user.

Although the animal's prehensive structures may

be of sufficient length, they may be too thick to make contact with objects that are within the sphere of reach but are embedded in narrow crevices or holes. In such cases, Inserting and Probing allows extension of reach. Pound/Hammer, Pry/Apply Leverage, Dig, Penetrate, or Cut may be used to create and/or widen the mouth of the crevice or hole. Some objects of interest may create ambivalence; for example, they might be unpleasant or too dangerous to touch. Tools might be used in the Reach or Jab mode to contact such objects. Thus ten of the twenty-two modes of tool use can function to extend the user's reach.

Animals commonly use tools to amplify the mechanical force that they can apply to the environment. Dropping, Throwing, Clubbing/Beating, Pounding/Hammering, Prying/Applying Leverage, Digging, and Blocking all increase the mass and decrease the elasticity of prehensive structures, and thus increase force. Prying/Applying Leverage and, probably, Digging incorporate fulcrums to amplify force. Jabbing/Stabbing/Penetrating, Cutting, and Blocking concentrate force on a small point so as to deliver it more effectively. Clubbing/Beating, at least with an elongated object, increases the speed, and thus the force, of a swing. Thus ten of the twenty-two modes of tool use function to amplify mechanical force.

Drop, Throw, Drag/Roll/Kick/Slap/Push Over, Brandish/Wave/Shake, and Affix/Apply/Drape can (among other functions) serve to create or augment signal value in social display.

Scratch/Rub—as well as some cases categorized in Prop and Climb/Balance and Climb/Bridge/Reposition, Brandish/Wave/Shake, Affix/Apply/Drape, Absorb, and Wipe—serve to enhance bodily comfort. For example, an elephant might Wave a branch to repel biting flies, or an ape might Reposition a box to provide a dry seat on a wet substrate.

Contain and Wipe serve to more effectively control fluids, soft solids, or small solids. Absorb acts only to more effectively control fluids. Affix/Apply can also serve to more effectively control fluids, soft solids, and small solids.

Brandish/Wave/Shake and Affix/Apply/Drape can also operate to hide or camouflage the user.

Symbolize is a unique mode that serves a unique function, and it comprises a relatively small number of cases.

Thus twenty-one of our twenty-two modes of animal tool use, and therefore the vast majority of the cases of animal tool use, serve six broad functions (some modes can serve more than one function), and most cases of tool use and the modes to which they are assigned serve only four functions. This not only aids organization, but it also provides an additional perspective on the selective advantages of tool behavior.

ASSOCIATIVE TOOL BEHAVIOR

Since 1980, there has been a stimulating profusion of terms beyond animal "tool use" and "tool manufacture," among which are secondary tools, metatools, serial tools, tool sets, tool kits, tool assemblies, tool assemblages, constructions, associative tools, multi-function tools, tool composites, tool sequences, and tool crafting. All of these terms, except multi-function tool, refer to the use of more than one tool, and with the exception of tool kit, the tools are used in close temporal and spatial proximity. Table 1.2 provides our definitions of these terms. As with the modes of tool use and manufacture, we will capitalize these terms when referring to them as formal organizational categories. To our surprise, Bentley-Condit and Smith (2010, 186) completely ignored associative tool use despite claiming that their catalog was comprehensive.

These terms have sometimes been used without definition, or with differing definitions, and may overlap with or duplicate one another. For example, Sugiyama (1997) included Metatools, Tool Sets, Sequential Tools, and Tool Composites *sensu stricto* as all being Tool Composites *sensu lato*. Wimpenny et al. (2009) inversely included Tool Composites and Secondary Tools as subsets of Metatools, and then further divided Secondary Tools into Sequential Tools and constructive tools (constructive tools being equivalent to our Secondary Tools). They acknowledged that there are other classification possibilities but claimed theirs was "descriptively explicit" and separated "classes of acts of potentially different levels of cognitive demand" (e6471).

The major source of confusion has been the definition of Metatool. Matsuzawa appears to have been the first to use the term in conjunction with animal tool behavior. His initial definition of Metatool use was "use of a tool for another tool" (Matsuzawa 1991, 571), which he later amended to be use of "a tool that serves as a tool for another tool" (Matsuzawa 1994, 361). Sousa and Matsuzawa (2006, 412) amended the definition again as use of "a tool to produce or enhance the effectiveness of another tool." Wimpenny

Table 1.2
Associative Tool Use: tool use and tool manufacture incorporating more than one tool, and related terms

Term	*Definition*	*Pertinent references*
Associative Tools	Tools used in any combination to achieve an outcome. Includes all other categories in this table except Tool Kit, Tool Assemblage, Tool Crafting, and Construction.	Sugiyama 1997; present authors
Sequential Tool	A tool used to acquire another tool.	Wimpenny et al. 2009
Tool Set (Toolset)	Two or more tools used sequentially, usually each in a different mode, to achieve a single outcome. See Serial Tool, below.	Bermejo and Illera 1999; Brewer and McGrew 1990; Deblauwe et al. 2006; Mannu and Ottoni 2009; McGrew 2004; Sanz and Morgan 2007; Sanz, Schöning, and Morgan 2009; Suzuki, Kuroda, and Nishihara 1995; Wimpenny et al. 2009
Serial Tool	Synonymous with Tool Set.	Sanz and Morgan 2007
Tool Composite	Two or more tools used simultaneously, usually each in a different mode, to achieve a single outcome, where the first tool is not used to manufacture the second. May include Metatool(s).	Carvalho et al. 2008, 2009; McGrew 2004; Sugiyama 1997
Metatool	A tool used simultaneously with a second tool to increase the efficiency or effectiveness of the second tool, where the first tool (the Metatool) acts directly on the second. The second tool could function as a tool on its own; the Metatool makes it a better tool. Every Metatool is a part of a Tool Composite. The use of a Metatool is not tool manufacture.	Matsuzawa 1991, 1994; Sousa and Matsuzawa 2004; Sugiyama 1997
Tool Kit (Toolkit)	All of the various tools known to be used by a single individual, a single population, a species, or a genus, as specified by the author.	Mannu and Ottoni 2009; McGrew 1992, 2004; Russon 2006; van Schaik, Fox, and Sitompul 1996; Teleki 1974
Tool Assemblage (Tool Assembly)	All of the tools found in the same context at a particular site that are presumed to have been used by the same group or population. Assemblages may accumulate over time.	Carvalho et al. 2008; Odell 2004
Multi-Function Tool	A single, specific tool used in two different modes and/or for two different functions.	Bermejo and Illera 1999; Boesch, Head, and Robbins 2009; Fay and Carroll 1994; Sanz and Morgan 2009; Tutin, Ham, and Wrogemann 1995
Secondary Tool	A tool used to manufacture (structurally modify) another tool. The usual mode of manufacture is Subtract. Differs from Metatool, in which no tool manufacture is involved.	Kitahara-Frisch 1993; Kitahara-Frisch, Norikoshi, and Hara 1987; Mannu and Ottoni 2009; Schick et al. 1999; Sugiyama 1997; Toth et al. 1993; R. Wright 1972
Tool Crafting	Tool manufacture that involves a number (minimum not specified) of successive steps and three-dimensional Reshaping. Only one object is usually involved, but it could include Secondary Tool Use.	G. Hunt and Gray 2004a; Oakley 1967; Pruetz and Bertolani 2007
Construction	Two or more tools and/or objects physically linked to make a functional, semipermanent thing that, once completed, is not held or directly manipulated in its entirety. A Construction itself is therefore *not* a tool. Nor is it tool manufacture, because the product is not a tool.	Collias and Collias 1976; present authors

et al. (2009, e6471) defined Metatools as "tools [which] are directed at objects that are not the agent's ultimate goal . . . but which have a role in achieving the ultimate goal." These are all very general and can include many of the forms of Associative Tool Use shown in table 1.2. The question is whether Metatool is a general category that includes many other forms of Associative Tool Use or a specific category of Associative Tool Use.

Examining the cases that prompted Matsuzawa's original use of this term, and other cases described by Carvalho et al. (2008) and Sugiyama (1997), provides direction. Most of these cases involved wild chimpanzees using stone hammers and portable stone anvils as Tool Composites to Hammer open nuts. Because the anvil was not fixed in these cases, it was not sufficiently stable and level, and the nuts rolled off. The chimpanzee wedged a third, smaller stone (the Metatool) under the anvil to enhance its effectiveness as a tool. Carvalho et al. (2008) described the use of two wedges to level an anvil. We consider these uses of leveling wedges to be the "archetypal" cases of Metatool Use (see below). The Metatool(s) did not "produce" the anvil, or "produce the effectiveness" of the anvil, which are two ways the Sousa and Matsuzawa (2006) definition can be read. Rather, it "enhanced the effectiveness of the anvil." Another informative case involves a wild chimpanzee who used a twig to push a "leaf tool" deeper into a tree cavity to Absorb water and then used the twig to retrieve the leaf tool (Matsuzawa in Sugiyama 1995a). R. Shumaker (pers. obs.) saw a captive orangutan use a stick to push a paper towel into an out-of-reach puddle of sweet liquid, thereby getting more liquid than it could with the stick alone. These cases lead us to a more specific definition of a Metatool: a tool used simultaneously with a second tool to increase the efficiency or effectiveness of the second tool, where the first tool (the Metatool) acts directly on the second, without being used in the manufacture of the second tool. By this definition a Metatool is part of a Tool Composite ("two or more tools used simultaneously, usually each in a different mode, to achieve a single outcome, where the first tool is not used to manufacture the second").

Consider the orangutan using a stick and paper towel in combination to reach a puddle of sweet liquid. In the case observed by Shumaker, the stick was long enough to reach the puddle, but the ape could only get a small amount of liquid with each use. When the ape used the stick to push a paper towel into the liquid and then retrieved the saturated towel, she could get more with each use. Thus the towel is a Metatool, because it was used simultaneously with the stick, increased the effectiveness of the stick, and acted on the stick directly. However, the stick and towel together are also a Tool Composite, because they were used simultaneously to get the liquid. Now consider a hypothetical case where the stick is too short to reach the liquid, as is the towel. The ape can Reach and Absorb the liquid only by using the tools together. The towel in this case does not "increase the efficiency and effectiveness" of the stick, because the stick was not effective at all without the towel. This is a Tool Composite, but not a Metatool. Carvalho et al. (2008) and Wimpenny et al. (2009) use Tool Composite synonymously with our Metatool, but we feel this is not exact. In our scheme, all Metatools are part of a Tool Composite, but not all Tool Composites involve Metatools (see chapters 6 and 7 for examples).

Metatool Use is different from Sequential Tool Use in that a Metatool is not used to acquire another tool, but rather to increase the efficiency or effectiveness of the second tool. The distinction in the case of Secondary Tool Use is that a Metatool, by our definition—and contrary to Tomasello and Call (1997) and Jalles-Filho, Teixeira da Cunha, and Salm (2001)—is not used to manufacture the second tool. A tool manufactured by Adding or Combining two or more elements becomes one object, rather than remaining more than one, as is the case with Metatools.

We prefer the use of Metatool as a specific category of Associative Tool Use rather than as a general category that includes many other forms of Associative Tool Use. First, the definitions and distinctions among the forms of Associative Tool Use are clarified. But more important is that until the cognitive operations underlying these various forms are operationally characterized, we should resist any hierarchical categorization based on putative, weakly inferred cognitive distinctions. One of the assumptions of the classification by Wimpenny et al. (2009, e6471) was that it separated classes of "acts of potentially different levels of cognitive demand." We agree that Associative Tool Use is rich ground for cognitive investigation (indeed, the experiments described by Wimpenny et al. [2009] on Sequential Tool Use by New Caledonian crows, and analyses by Carvalho et al. [2008] of hammer and anvil use by chimpanzees are brilliant examples of such investigation). Many researchers, such as Matsuzawa (1996, 1997, 2001), Russon (1999a), and Visalberghi and Fragaszy (2006), felt that the number of objects involved in a case of tool use was an index to the cognitive complexity of the behavior and thus assigned special cognitive and evolutionary significance to the behaviors subsumed in table 1.2 (although they did not explicitly consider Constructions). We agree that Associative Tool Use is cognitively more complex than simple tool use, in part because of the number of tools and objects involved in a case. Yet the categories of Associative Tool Use can be equivalent with regard to the number of tools and objects involved. Thus there is no current evidence, based on numbers of tools or objects involved or on other characteristics, for a cognition-based hierarchical organization of the categories subsumed within Associative Tool Use. We will thus treat each form of Associative Tool Use as hierarchically equivalent (except for Metatools being a form of Tool Composite, without implying that Tool Composites are cognitively more complex). We anticipate that future investigations will reveal which forms of Associative Tool Use reflect greater cognitive complexity.

Nonetheless, some definitional damage control is needed now, and it shows why these deliberations are not biologically trivial. The definitions of Metatool by Matsuzawa (1991, 1994) and Wimpenny et al. (2009) have been adopted in many of the laboratories working with corvids (crows, rooks, ravens, and New Caledonian crows) and have been applied to what we would call Sequential Tool Use, use of a tool to get another tool. Bird and Emery (2009a), Taylor et al. (2007), and Wimpenny et al. (2009) described captive New Caledonian crows using a short stick to retrieve a long stick, which they then used to Reach a food incentive. The authors called this Metatool use but did not note the difference between Metatool use *sensu lato* and Metatool use *sensu stricto*. N. Clayton (2007, R894) used the general definition of Metatool and thus confusingly (we think mistakenly) made a general assertion that these studies of crows "open up the whole issue of Metatool Use beyond the domain of primates." We know of no cases where a non-primate used a Metatool in the specific sense, like a chimpanzee, orangutan, or capuchin monkey did.

We have identified a particular form of Associative Tool Use that we call a Construction, defined as "two or more tools and/or objects physically linked to make a functional, semipermanent thing that, once completed, is not held or directly manipulated in its entirety." To be clear, a Construction is *not* a tool. Others have used tool construction synonymously with tool manufacture and with Tool Composite, both of which differ from our use of Construction. Included among Constructions are nests, beaver dams and food caches, bowers, and the towers of stacked boxes built by apes, since in all of these the builders do not hold or directly manipulate the final product in its final constructed form. The nest, dam, bower, and tower builders do hold and manipulate most of the individual elements at some point in construction, and each element can be considered a tool that is Repositioned, Applied, and/or Affixed to alter the form and/or condition of the final Construction. Although some Constructions involve tools, others, for example, burrows, do not.

We have not added cooperative tool use to table 1.2, because the interesting aspect of cooperation is the relationship between and among two or more tool users, rather than among two or more tools. If one crow used a short stick to retrieve a longer stick that a second crow then used to Reach food, would this be Sequential Tool Use or simply two crows each using a different reaching tool? If one chimpanzee Penetrated a termite mound with a stout stick, and another then Inserted a twig and withdrew termites, would we classify this as a Tool Set or simply two animals each using a different tool to reach a single outcome? The permutations are endless and fascinating.

SITE-SPECIFIC CATEGORIES OF TOOL BEHAVIOR

Another set of terms arises from comparisons of patterns of tool use and tool manufacture between and among study sites or populations of a particular species. To our knowledge, these terms have been applied only to great apes, New Caledonian crows, and killer whales, but we will point out some other candidate cases in our catalog. All definitions used here are from Whiten et al. (1999, 682). A particular form or variant of tool behavior is said to be *customary* when it is shown at the site by "most able-bodied members of at least one age-sex class." Of course, all genetically based tool behaviors would be customary by this definition, but we infer that the discussants of culture in apes are referring only to learned behavior. The behavior is said to be *habitual* when the behavior is not customary at the site but has been shown there repeatedly by several individuals. The behavior is categorized as *present* when it is neither customary nor habitual but has been shown by at least one individual on at least one occasion. The behavior is said to be *absent* when it has never been recorded at the site, even though there have been extensive observations at the site, and the requisite tool materials (stones) and the incentives (hard-shelled fruits) are present, or it can be *absent for ecological reasons* if requisite tool materials and/or the incentives are not present. The occurrence of the tool variant is said to be *unknown* if sampling efforts or observational opportunities have been inadequate.

Customary and habitual tool behaviors at one site are sometimes said to be *cultural* when the same pattern is absent in the same species at another site, despite the requisite materials and incentives being present at the second site.

Our esteemed colleague and expert on animal tool behavior, Bill McGrew (1989, 458), once wrote about the subject that "to survey all primary sources would be exhausting," and he was just talking about apes! The following chapters (2 through 6), which survey the tool behavior of all animals, are intended to be exhaustive, but we would welcome reports of oversights or misinterpretations.

2

Invertebrates

TOOL USE: INSECTS

Drop

Pavement ants (*tetramorium caepitum*) Drop soil onto both alkali (*nomia melanderi*) and halictid (*Lasioglossum zephyrum*) bees (Lin 1964–1965; G. Schultz 1982). The ants, on finding a bee nest, enter the nest to raid pollen. If a bee is guarding the entrance, one or more ants will first pick up pieces of soil and Drop them down onto the bee. The ants do not Drop soil if a guard bee is not present. The soil dropping often ends in the death of the bee, as the bee eventually approaches and is attacked by the ants (G. Schultz 1982).

Dolichoderine ants (*Conomyrma bicolor*) Drop objects and stones into the entrance of *Myrmecocystus* ant nests (Moglich and Alpert 1979). As many as thirty ants may surround a nest entrance for hours, bombarding the entrance and any emerging ants with stones. Moglich and Alpert argued that the ants do this to prevent competitor ants from leaving the nest to forage.

Rau (1937) described an observation of a wasp (*Podium* sp.) that carried water in her mouth and Dropped it to soften the earth as she excavated her burrow.

Throw

Sand-throwing behavior has been documented among ant-lions (larvae of *Myrmeleon* spp.) and worm-lions (larvae of *Vermileo* spp.) (de Beer 1948; Pierce 1986; Skaife 1957; T. Tanaka and Ono 1975; Thompson in Romanes 1892; W. Wheeler 1930). The ant-lions excavate funnel-shaped pitfalls in friable soil and burrow into the bottom of the pit to await prey that enters or falls in. They then seize the prey, drag it beneath the soil, and consume its soft tissue and body fluids. If the prey attempts to avoid capture and escape from the pitfall, the ant-lion showers it with sand, which hinders its escape. The ant-lion Throws the sand by rapid dorsiflexion of its head and mandibles, similar to the behavior used by the ant-lion in excavating and removing debris from its pitfall. Tanaka and Ono (1975) described the sand throwing as an instinctive behavior that also seems to include a learned component, as sand-throwing efficiency and prey-recapture rates increase with age. Pierce (1986) reported that worm-lions are known to knock prey into their pit by Throwing sand at approaching prey.

Bait, Entice

McMahan (1982, 1983a, 1983b) described Baiting of prey by the neotropical assassin bug (*Salyavata variegata*), which uses previously captured termite carcasses to capture additional termites. By holding and shaking the carcass over the nest's entrance hole, the assassin bug lures a termite into attempting to retrieve the carcass for its own consumption. Once a termite grasps the lure, the assassin bug slowly pulls the termite out of the nest entrance toward itself. Once the termite is within reach, the assassin bug quickly kills and eats its new victim. McMahan (1983a, 1983b) reported that the assassin bug, if not disturbed, will repeat this process an average of seven or eight times. The author once saw an assassin bug capture thirty-one termites in this manner over the course of three hours.

Males in some insect species present and transfer "nuptial gifts" during courtship or copulation (see reviews by Boggs 1995; Vahed 1998). Such gifts may include edible food as well as stones, balloons of silk, or twigs (Kessel 1955; Osten-Sacken 1877). For example, male empid flies capture an insect, spin a cocoon around it, and present it to a potential mate. In some closely related species the cocoon contains an inedible object, and in others it is empty. However, all seem to distract the female sufficiently for the male to copulate without being eaten by her (Kessel 1955).

Pound, Hammer

Females of the wasp genera *Ammophila* and *Sphex* (commonly called digger wasps) excavate subterranean burrows to deposit eggs and a prey insect for their larvae to feed on as they develop. When the female places prey in the burrow, she fills the entrance with pebbles and soil. When the burrow is closed for the final time, the female sometimes holds an object in her mandibles and uses it to press (Pound or Hammer) the soil, compacting it and making it less conspicuous (Armbruster 1921; Berland in W. Thorpe 1963; Brockmann 1985; H. Evans 1959, 1965; H. Evans and Eberhard 1970; Frisch 1940; Haeseler 1985; Hartman 1905; C. Hicks 1932a, 1932b; Hungerford and Williams 1912; Iwata 1976; Minckiewicz in Thorpe 1963; Molitor 1931; Peckham and Peckham 1898; Rau and Rau 1918; Rayment in Frisch 1940; Tilden 1953; Tsuneki 1968; G. Wheeler and Wheeler 1924; Williston 1892). The wasps were observed using pebbles, clods of earth, bits of twig, bark, seeds, and the tarsus and tibia of what appeared to be a locust for Pounding.

Insert and Probe

Some female digger wasps Insert small twigs into nest burrows they have closed with soil and Probe with them (J. Brockmann, pers. comm. 1977; Hartman 1905). This behavior may settle and pack the soil and provide the female with sensory information regarding the adequacy of the closure.

Contain

Several species of ants from the genera *Aphaenogaster*, *Pogonomyrex*, *Novomessor*, and *Solenopsis* use bits of leaf, wood, soil, mud, grass, and sand to transport soft or liquid foods such as jelly, honey, fruit pulp, and the body fluids of prey (Agbogba 1985; Banschbach et al. 2006; Barber et al. 1989; Fellers and Fellers 1976; H. Fowler 1982; McDonald 1984; Morrill 1972; Tanaka and Ono 1978). The ants place or drop the objects into or over the soft or liquid foods with their mandibles. The fluid adheres to or is absorbed by the tool (if absorption is the primary dynamic, some cases may properly be placed in the Absorb mode). After an interval, the tool is withdrawn and carried back to the colony where both the user and other workers feed from it. A worker can retrieve approximately ten times as much food (by weight) with this type of tool than it can carry without one (Fellers and Fellers 1976). In a detailed examination of this behavior in the common forest ant (*Aphaenogaster rudis*), Banschbach et al. (2006) found that this tool use is a specialized task, performed by a small subset of workers within the colony. Kitabayashi, Kusunoki, and Gunji (1999) reported the use of a food "cart" to transport food by wood ants (*Formica japonica*). However, we do not consider this as tool use, because the experimenters placed the food on the cart.

Affix, Apply, Drape

The larvae of many different types of insects, including species of Ascalaphidae, Chrysomelidae, Chrysopidae, Derodontidae, Geometridae, Reduviidae, and Sphindidae, carry debris on their bodies. Debris may include dead insect or arachnid bodies or parts, wax produced by other insects, slime mold, spider webs, aphid "wool," vegetation, lichens, and eggs. The debris may provide a physical and chemical defense against predators and conspecifics and hide the user

A female digger wasp Hammers the closure of her burrow with a pebble. Drawing courtesy of H. Jane Brockmann.

A cassidine beetle larva Applies a fecal shield over itself to deter predators. Photograph courtesy of Thomas Eisner, Cornell University.

from prey (K. Anderson, Seymour, and Rowe 2003; N. Brown and Clark 1962; Butler 1923; Eisner and Eisner 2000, 2002; Eisner, Eisner, and Siegler 2005; Eisner and Silberglied 1988; Eisner, van Tassell, and Carrell 1967; Eisner et al. 1978, 2002; N. Gómez, Witte, and Hartmann 1999; Hingston 1932; Killington 1936–1937; Leschen and Carlton 1993; Mason et al. 1991; McHugh and Kiselyova 2003; Morton and Vencl 1998; Nakahira and Arakawa 2006; New 1969; Nogueira-de-Sá and Trigo 2002, 2005; Olmstead 1991, 1996; Olmstead and Denno 1992, 1993; Pergande 1912; Shelford 1917; Skorepa and Sharp 1971; R. Smith 1926; Tauber and de León 2001; Tauber et al. 2000; Tsukaguchi 1995; Vencl and Morton 1998). The shield does not seem to deter all predators; some predators and parasitoids may even be attracted to it (Müller and Hilker 1999, 2003, 2004; Schaffner and Müller 2001). However, in one study on the function of this behavior, Bacher and Luder (2005) found that the shield in one form of cassidine beetles was effective against parasitoids, and they hypothesized it also protected the beetles from wind and desiccation.

Larvae of the Chrysopidae (green lacewings) are referred to as "trash-carriers," as they habitually camouflage themselves by carrying "packets" of debris dorsally (R. Smith 1926). In several subfamilies of the leaf beetles (Chrysomelidae), the larvae carry their exuviae (shed "skin") and feces on two abdominal spines known as "fecal shields" (Eisner, van Tassel, and Carrel 1967; Gressitt 1952; Maulik 1919; Olmstead 1994, 1996; Rye 1866). Nymphal forms of Reduviidae (assassin bugs) Apply sand, soil, or dust to their bodies and may carry "backpacks" of plant and prey parts (Ambrose 1999; Brandt and Mahsberg 2002; Louis 1974; McMahan 1982; N. Miller 1956; Odhiambo 1958; Zeledon, Valerio, and Valerio 1973). This camouflage may help the larvae to capture prey and provide protection against predators and cannibalistic conspecifics (Ambrose 1986; Livingstone and Ambrose 1978).

When worm-lion larvae reach maturity and are about to pupate, a sticky liquid is secreted between the final larval cuticle and the pupal cuticle beneath. This liquid is not secreted during previous larval molts. As soon as the final larval cuticle is shed, the liquid causes grains of sand to adhere to the pupal cuticle. The sand anchors the cuticle when the pupating worm-lion struggles free of it. W. Wheeler (1930, 162) noted, "One might maintain that the [worm-lion] pupa in thus crudely selecting and fixing to its cuticle the sharp sand grains of its immediate environment and in employing them as an aid in locomotion, is really behaving as a tool-using animal."

Formicine weaver ants of the genera *Oecophylla*, *Camponotus*, *Dendromyrmex*, and *Polyrhachis* use their own larvae as tools in nest construction and repair (Hölldobler and Wilson 1977, 1990; Jacobson and Wasmann 1904–1905; W. Wheeler 1910). The leaves are rolled and/or pulled together by the cooperative efforts of the workers. When the leaves are appropriately positioned, other workers pick up the larvae, hold them in their mandibles, and move them back and forth like shuttles between the leaf edges. The larvae secrete silk that cements the leaves together to produce a sturdy nest. The workers themselves are incapable of producing silk; only larvae can do so, usually to spin pupal cocoons. The workers, however,

A weaver ant worker uses an ant larva as a shuttle to weave a nest. Reprinted from W. Wheeler 1910.

A digger wasp Applies a leaf to camouflage her burrow. Drawing courtesy of H. Jane Brockmann.

utilize the larvae before pupation in this unique form of tool use.

Among the order Trichoptera, larvae of the hydroptilid caddisflies spin a cocoonlike tube to which they Affix bits of leaf and stick, shells, pebbles, and other objects. The larvae carry this casing around as they mature, and the behavior thus meets our definition of tool use. Larvae of other caddisfly families construct casings that are fixed to the substrate and are not carried (Hanna 1960; Milne and Milne 1939; H. Ross 1956, 1964); this would not be tool use.

Females of the water strider (*Gerris remigis*) carry a copulating male for hours. Wilcox (1984) found that during the mating season, females could only forage effectively if carrying a male. The male's presence allowed the female to move and forage without harassment from other males.

Applying or placing objects to cover, protect, and/or conceal eggs or a nest has been reported for numerous types of insects (Matthews and Matthews 1978). For example, the female of the digger wasp, after Pounding the soil of her burrow entrance with a stone (see above), sometimes repositions the stone on the soil over the nest, blocking the entrance to potential invaders. These wasps also camouflage their nests by Applying debris, such as leaves or sticks, over the nest entrance (Brockmann 1985). Choe and Rust (2007) reported that females of the bee assassin bug (*Apiomerus flaviventris*) collect the resin of brittlebush (*Encelia farinosa*) by scraping the plant surface with their forelegs. A heavy coating of this resin is then Applied to the periphery of their egg masses. In captive experiments, Choe and Rust found that this behavior was essential for depositing eggs; otherwise, eggs were scattered, leaving them vulnerable to both parasites and desiccation.

TOOL USE: ECHINOIDS

Affix, Apply, Drape

Many types of echinoids, or sea urchins, Drape themselves with objects. These objects, which include algae, shell fragments, or pebbles, may serve as camouflage or protection from light (Millott 1956). The urchin moves the material onto itself, using its tube feet and spines, and the material is then held with the tube feet (Millott 1956). This "covering response" may also assist in capturing food, particularly algal debris, with the urchin grasping whatever is available and eventually consuming the algae on the materials with which it covered itself (Dix 1970; Douglas 1976). However, Dix (1970, 193) noted that "the nature of the covering response differs between urchin species." For example, Millott (1956) concluded that light avoidance was the primary reason for debris placement in the variegated or green sea urchin (*Lytechinus variegatus*), but Dix (1970) found no evidence for this in a different urchin (*Evechinus chloroticus*). Protection from predation through camouflage may be an important function of covering. For example, Dayton, Robillard, and Paine (1970) reported that covered Antarctic urchins (*Sterechinus neumayeri*) were often able to escape predation from the anemone *Urticinopsis* sp., while uncovered urchins were always captured. Levin, Gooday, and James (2001) described a specimen of spatangoid urchin (*Cystechinus loveni*) Draped in rhizopod protists. They argued that in this species the behavior is likely associated with camouflage, or to increase the relative density of the urchin to reduce transport by currents. Dumont et al. (2007) concluded that the covering response in green sea urchins (*Strongylocentrotus droebachiensis*) is likely due to multiple factors associated with protection from environmental stresses, mainly wave surges and UV radiation.

TOOL USE: CRUSTACEANS

Brandish, Wave, Shake

Boxer crabs, also known as "pom-pom crabs" (*Lybia tesselata* and *L. edmondsoni*), Detach small anemones from the substrate and Brandish or Wave one in each cheliped (Bürger 1903; Duerden 1905). The crab moves with its chelipeds extended and waving. If the crab is mechanically disturbed, the chelipeds with the anemones are directed at the source of irritation. This behavior would presumably facilitate the discharge of stinging nematocysts by the anemones toward the threat. However, the crab's use of anemones is not limited to protection or defense. If food is placed near the oral disc of the anemone, the crab immediately seizes the food with its anterior ambulatory appendages. Thus any food ensnared by the anemone in its own tentacles is apt to be appropriated by the crab. The crab also removes debris adhering to the body of the anemone and ingests the edible bits.

Block

D. Ross (1983) noted that the hermit crab (*Diogenes edwardsi*) carries an anemone in its cheliped for protection. When the hermit crab withdraws into its shell, the anemone it holds fills the shell aperture, Blocking access to predators.

Affix, Apply, Drape

Hermit crabs are decapod crustaceans of the superfamily Paguridea, which lack the hard abdominal exoskeleton that provides protection for other types of crabs. The coenobitid hermit crabs are semiterrestrial, while the diogenid and pagurid hermit crabs occupy a variety of marine microhabitats. Most hermit crabs live in the shells of dead gastropods for substantial portions of their lives (Hazlett 1981). Laboratory studies demonstrate that hermit crabs exhibit considerable sensory discrimination and selection in choosing shells for occupancy (Reese 1962). Once a suitable shell is located, it is occupied by the crab until the crab's growth necessitates a search for a larger shell (see photo below). Hermit crabs of the genus *Pagurus* sometimes

A hermit crab occupies a marine snail shell for protection against predators and environmental extremes. Photograph by Jessie Cohen, courtesy of the Smithsonian National Zoo.

aggregate in the vicinity of predation on a marine gastropod, engage in intense agonistic behavior until a single dominant individual emerges, and then wait for the predator to drop the empty shell. The dominant crab occupies the newly available shell, if it is suitable, and the other crabs begin a frantic sequence of shell exchange. While the occupation of a vacant shell is a deliberate affair, usually lasting several minutes, occupation of new shells dropped at a predation site takes only a few seconds (R. McLean 1974). Living in empty gastropod shells is a behavioral adaptation that apparently provides increased protection from predation and environmental extremes. The crab clings tenaciously to the inside of the shell with a specialized hooklike pleopod (appendage). Although the gastropod shell is not carried or held with an appendage, it is literally carried by the crab and thus fits our definition of tool use. Hermit crabs may also live in and carry sponges, termed "portable" or "mobile" sponges, which grow over empty gastropod shells (F. Stanford 1995).

A moss crab that has decorated itself with branched bryozoans that it collects from the sea floor. Photograph courtesy of J. Stachowicz.

Sea anemones sometimes attach themselves to the shells of hermit crabs, thereby hitching a ride with no obvious benefit to the crab. However, many types of hermit crabs actively transfer anemones to their shells (D. Ross 1983). Laboratory simulations demonstrate that the crabs gain significantly enhanced protection against octopus predators by carrying the anemones (D. Ross 1971). Even hermit crabs without shells may form associations with anemones. For example, the shell-less hermit crab *Paguropsis typica* uses modified claws to hold anemones to its abdomen (D. Ross 1983).

Many different species of brachyuran crabs collect shells, rock chips, algae, coral, and sessile organisms such as sponges, tunicates, and anemones and either carry them or Affix or Apply them to their bodies (Duerden 1905; Guinot, Doumenc, and Chintiroglou 1995; Meglitsch 1972; D. Ross 1983; Wicksten 1986). Wicksten (1986) described species that carry other organisms as having specialized legs that are shorter and positioned subdorsally. She reviewed carrying behavior in crabs, noting that this behavior was present in at least five crab families, including Dromiidae, Homolidae, Tymolidae, Dorippidae, and Latreilliidae (Wicksten 1982, 1986).

Sponge crabs (Dromiidae) cut sponges and hold them over their carapaces (Portmann 1959). The sponge may grow over the crab, providing an ever-present portable shelter as well as physical and chemical camouflage. The crab becomes difficult to see and the sponge is distasteful to predators (Wicksten 1986). While Ross (1983) described this as a symbiotic relationship, sponge crabs will also perform this behavior with non-living objects, such as cardboard, in an experimental context. However, biotic material is preferred. Some dromiids not only cover their carapace, but their entire dorsal surface by carrying pelecypod shells (Wicksten 1986).

Carrier crabs (Homolidae) carry fragments of sponge, driftwood, or coral, or entire organisms such as anemones, jellyfish, or sea urchins, for defense against predators (Chintiroglou, Doumenc, and Guinot 1996; Wicksten 1982). Wicksten (1982, 478) speculated that these objects "might be thrust at an oncoming predator, discouraging an attack." Confirmation of this is lacking. However, holding large objects over the carapace (such as live jellyfish), has been confirmed to play a role in camouflage and defense. The crab may even bury its body into the sand, leaving only its carried object exposed at the surface, effectively hiding itself.

Decorator or masking crabs of the family Majidae (spider crabs) Drape themselves in algae, bryozoa, sponges, or anemones (Acuña, Excoffon, and Scelzo 2003; Dixon and Dixon 1891; Garstang 1890; Getty and Hazlett 1978; Holmes 1911; Martinelli, Calcinai, and Bavestrello 2006; Stachowicz and Hay 2000; Thanh et al. 2003; Wicksten 1978, 1979, 1980, 1993). In addition to providing camouflage from predators and prey, the camouflage materials may be used as a food store by the crab (C. Woods and McLay 1994). The crabs are described to "plant [the material] on limbs, rostrum, and carapace, which can grow to cover the crab completely" (D. Ross 1983, 169). Wicksten (1986) describes the mechanics of masking as distinct from the behavior of the crabs that carry objects, presumably because masking, in our terms, involves actually Affixing or Applying the objects to the body. Spider crabs are confirmed to Apply the objects on the body with specialized hooked appendages, rather than carry or hold them. Fernández et al. (1998, 728) noted that decorating by these crabs "reveals a sophisticated adaptation to the environment in which they live . . . [because they] adapt the appearance of their carapace to the substrate that they inhabit in order to make the individual less evident against predators."

TOOL USE: ARACHNIDS

Throw

Spiders of the family Deinopidae hold a small part of their web, a "net," between their front legs and Throw it over passing prey (Akerman 1926; Coddington and Sobrevila 1987; Robinson and Robinson 1971). These spiders are commonly known as ogre-faced or net-casting spiders. When prey approaches, the spider lunges, expanding the size of the net in the process, thus covering and capturing its prey in the net (Robinson and Robinson 1971). Akerman (1926, 418) described the attack, stating, "like a flash the spider stretches the elastic snare to its full expansion . . . and hurls herself forwards, throwing the net over the moth."

Bolas spiders (*Mastophora* spp.) do not build webs but instead carry a sticky globule, the bolas, made of the same secretions used by orb weavers in web construction (Gertsch 1955; Hutchinson 1903; Longman 1922). To position the bolas, the spider first constructs a horizontal trapeze line between vegetation and then creates the bolas, detaches it from its body, and hangs it on a silk thread from the trapeze line. The spider then positions itself so that it can use its mouthparts and front legs to manipulate the swinging bolas (Eberhard 1980; Yeargan 1988, 1994). When prey approaches (the spider may actually bait the prey by secreting a substance that mimics the prey's sex-attractant pheromone [Eberhard 1977]), the spider swings the ball. If struck, the prey sticks to the ball.

Bait, Entice

Male spiders of the genus *Pisaura* capture and present flies to females during courtship. If the female is receptive, she begins to eat the fly and the male has about an hour to copulate without danger of being devoured by the larger female (Bristowe 1929, 1971).

A net-casting spider awaits its prey. Photograph courtesy of Jonathan Coddington.

Affix, Apply, Drape

Corolla spiders (*Ariadna* spp.) place small stones in a single-layered circle around the entrance to their burrows (Costa et al. 1993; Henschel 1995). The preferred stones are small quartz pebbles, similar in size and shape to one another. The spiders typically place seven or eight stones in the ring. Costa et al. (1993) suggested that the presence of the stone ring might facilitate prey detection when the spiders are in their burrows. They noted that because the stones are Affixed to the burrow by silk threads, the stones might serve to transmit vibrations of prey movements to the spider. Henschel's (1995) research supported this hypothesis, and he argued that the stones do serve as foraging tools, allowing the spiders to extend their sensory range by monitoring the vibrations passed through the stones. Henschel further argued for selectivity in tool choice: while the spiders could physically carry larger stones,

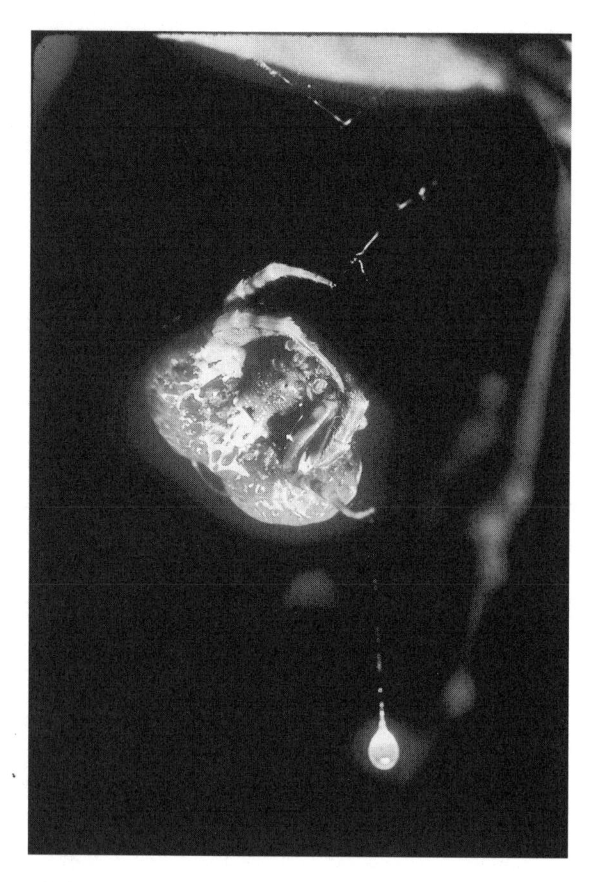

A bolas spider awaits its prey. Photograph courtesy of Jonathan Coddington.

they select smaller stones because they provide more accurate information about the prey's location.

The camouflaging and self-burying behavior of some spiders may be considered a form of tool use when the spider actively moves the sand or camouflaging material onto its body with its legs. For example, females and immature wandering spiders (Homalonychidae) may cover their bodies in fine soil. Female wandering spiders may also incorporate sand or solids into their egg sacs, possibly to prevent desiccation and for camouflage (Domínguez and Jiménez 2005; Vetter and Cokendolpher 2000). Other spiders that manipulate soil or sand to camouflage or bury themselves include the genera *Sicarius*, *Cryptothele*, *Paratropis*, *Microstigmata*, *Bradystichus*, and *Trogulus* (Domínguez and Jiménez 2005).

The Texas brown tarantula (*Rhechostica hentzi*) uses its silk to bind soil together during burrow construction (Formanowicz and Ducey 1991). The spider deposits its silk onto loosened soil and then rakes the soil and the silk together to form a mass. It may then roll the mass away, or pick it up and carry it out and away from the burrow it is constructing. Similar behavior has also been described in wolf spiders (*Lycosa* spp.) (Emerton 1912; Gertsch 1949), and in several different families of sand-burrowing spiders (Henschel 1998).

Black widow spiders (*Latrodectus hesperus*) produce a very sticky and viscous web for defense when threatened (Vetter 1980). The web is either held toward the predator or is placed or dropped on the predator. Vetter found that when deployed against predators, such as mice, the defensive web did increase the likelihood of the spider's survival. The web seems to irritate predators. Vetter noted that after contacting the web, a mouse retreated to clean the material off itself, allowing the spider to escape.

TOOL USE: CEPHALOPODS

Throw

Lane (1957, 43) reported that octopuses (genus unspecified) sometimes deliberately squirt water, stating that they are "able to 'fire' like archer-fishes." He described experiments on the "marksmanship" of captive octopuses. In one study, an octopus shot a jet of water at the experimenter, who was five feet away, drenching his shirt. He also described a report of a pet octopus expelling water toward its owner when it was due for feeding. Lane (1957, 44) added that Charles Darwin is noted to have experienced this "water barrage," as he was struck in the face while observing wild octopuses in shallow water. Octopuses may also shoot jets of water at other marine animals, including sea anemones (Boycott 1954) and scavenging fish (Mather 1992), to repel the animals. Mather (1995) mentioned that the water jet might also be used for capturing prey, altering the landscape, or moving away unwanted items, such as feces or debris.

Dig

Octopuses use a water jet to excavate shelters by blowing out debris and sand (Mather 1995).

Block

Thorpe (1963) cited reports by Pliny and Power of tool use by octopuses (*Octopus* spp.). The cephalopods are said to have used stones to prop open the shells of large bivalves while they consumed the flesh. Thorpe also relayed a personal communication from Berry that *O. digueti* uses a clamshell as an artificial operculum to close gastropod shells inhabited by the

Seven quartz pebbles placed by a corolla spider around the entrance to its burrow. Photograph courtesy of Gobabeb Training and Research Center, Namibia. Reprinted by permission from Henschel 1995.

octopus. Lane (1957) reported an observation of an octopus that dropped coral into open oysters to Block open their shells. Wells (1962; 1978, 243), however, considered these observations "nonsense," noting that captive experiments (Bierens de Haan 1926; Boycott 1954) have failed to replicate the phenomenon.

Octopuses place objects such as large stones, shells, glass fragments, crab claws, and crab shells in front of the entrance to their shelters to Block or conceal the entrance from other animals (Mather 1994). Mather concluded, and we agree, that these stones qualify as tool use according to our definition.

Affix, Apply, Drape

Octopuses (Octopoda) and squid (Sepiolida) habitually bury themselves in sand for camouflage. If they actively place the sand on their bodies with their appendages, this would meet our definition of tool use. R. Anderson, Mather, and Steele (2004, 15) described how the squid *Rossia pacifica* Applies sand to its body by throwing, "'cupping' a bit of sand in each arm tip and then throwing the sand straight back onto its head and body by curling the arm upward while retracting it."

Finn, Tregenza, and Norman (2009) reported defensive tool use by the veined octopus (*Amphioctopus marginatus*). The octopuses frequently carried coconut shell halves and, when threatened, assembled them into a shelter by aligning the two halves of the coconut and hiding inside. These authors argued that the behavior is significant, as the octopuses carry the shells for future use as a shelter, despite the immediate energetic and locomotor costs. During travel, the octopus carries the shells under its body, in a form of locomotion termed "stilt walking," which the researchers described as "ungainly and clearly less efficient than unencumbered locomotion" (R1070).

TOOL USE: GASTROPODS

Pry, Apply Leverage

Weldon and Hoffman (1975) studied the righting behavior of the Pacific intertidal prosobranch snails *Tegula brunnea* and *T. funebralis* in captivity. When a snail is inverted, it typically extends the propodium ("foot") and probes for a solid substrate such as a large rock. When one is encountered, the propodium adheres to the substrate and the snail pulls itself over. This technique is not effective, however, on substrates of sand or gravel. On gravel, the snail grasps one pebble with the

An octopus without shelter is vulnerable to predation. Photograph by Mark Norman, courtesy of M. Norman / Museum Victoria.

An octopus creates a shelter by repositioning one half of a coconut shell above and one half below itself. Photograph courtesy of Roger Steene.

prehensile anterior tip of the propodium and conveys it to the medial and posterior aspect of the foot with the same undulatory movements used in locomotion. Other pebbles are transferred in the same manner until enough pebbles are Applied or accumulated to shift the center of gravity and tip the snail toward an upright position. If an elongate pebble is grasped, it is not transferred to the posterior portion of the foot. Instead, it is inserted into the gravel or sand and used to Pry or Apply Leverage that also effects righting. Since an inverted snail is especially vulnerable to predators, the use of tools in righting is undoubtedly adaptive.

Affix, Apply, Drape

The Caribbean gastropod *Xenophora conchyliophora* Affixes empty shells, primarily of bivalves, and other debris to the exterior of its own shell (Berg 1975). According to Berg, the resultant doubling of shell diam-

An octopus carries two shell halves to the point of use by holding them beneath its body and "stilt-walking" along the ocean floor. Photograph by Mark Norman, courtesy of M. Norman / Museum Victoria.

eter adds stability and provides a protective canopy under which the animal grazes. This behavior may also provide camouflage.

Hobbie (1993) reported several examples of his aquarium snails using rocks and coral fragments as tools. He reported that the shell of one snail was accidently damaged, resulting in part of the shell dangling from the snail. The snail was seen to manipulate two rocks with its foot, placing them on each side of the attached shell fragment. It then rotated the rocks on the shell fragment until it was able to break the shell fragment free. Hobbie experimented with additional snails, and observed one drag a coral fragment across its broken piece of shell until it came off.

TOOL MANUFACTURE: INVERTEBRATES

Of the invertebrate tool-using species reported above, only ants and crabs, and perhaps octopuses, were reported to manufacture tools. We found no cases of tool manufacture by echinoids, gastropods, or arachnids.

Detach

Drop

Pavement ants may Detach a piece of soil from a crusted surface before Dropping it onto an alkali bee guarding a nest entrance (G. Schultz 1982).

Brandish, Wave, Shake

Boxer crabs Detach small anemones from the substrate before Brandishing or Waving them as aids in defense and feeding (Duerden 1905). Duerden emphasized that the anemones are useless as tools if they are injured when they are Detached from the substrate.

Contain

Aphaenogaster rudis ants that use objects to transport and Contain liquids may Detach debris from nearby leaves or sticks before dropping it into a liquid food source (Banschbach et al. 2006).

Affix, Apply, Drape

W. Wheeler (1930) noted that ascalaphids Detach fragments from a brick before Affixing or Applying them to their bodies as camouflage. Normally, however, they simply pick up and Apply loose sand grains. Green lacewing larvae pluck (Detach) "wool" from aphids before Applying it to themselves as camouflage (Eisner et al. 1978). Other species of chrysopid larvae were observed Detaching objects such as sycamore leaf scales to use in forming their dorsally carried trash packets (Eisner et al. 2002).

Hermit crabs, sponge crabs, carrier crabs, and decorator crabs must Detach sessile organisms such as sponges or sea anemones from the substrate before carrying them or Applying or Affixing them to their bodies. Duerden (1905) specifically noted that decorator crabs Detach sponges before application. Wicksten (1979) reported an observation of a moss crab (*Loxorhynchus crispatus*) Detaching and reusing these "decorations" from its shed exoskeleton after molting. Brooks (1990) described hermit crabs Detaching sea anemones from the surrounding environment with a combination of tapping, pulling, and rubbing. In addition, Brooks (1990, 55) stated that a hermit crab that actively places sea anemones on its shell "usually transferred them again when it moved to a larger home, just as people move old furniture." Sponge crabs and spider crabs use their chelipeds to Detach parts of sponges, algae, or hydroids, by cutting out or wrenching off a section of the attached organism (McLay 1983; Wicksten 1980; C. Woods and Page 1999).

Add, Combine

Absorb

Florida harvester ants (*Pogonomyrmex badius*) Combine grains of sand to form small pellets that they place in honey. The pellets Absorb the honey and are transported to the nest (Morrill 1972).

Reshape

Affix, Apply, Drape

Decorator crabs may Reshape their decorations before Applying them to their bodies (Wicksten 1980). Mastro (1981) described the decorator crab known as the kelp crab (*Pugettia producta*) softening algae with its mouthparts before attaching it to its body for camouflage. C. Woods and Page (1999) described a spider crab (*Thacanophrys filholi*) Detaching a section of sponge from the substrate and then transferring it to its mouthparts, possibly to roughen the surface to assist in its attachment to the setae on its carapace.

ASSOCIATIVE TOOLS: INVERTEBRATES

Tool Composite

Hermit crabs, such as *Pagurus prideauxi*, that carry gastropod shells and, further, attach sea anemones to their shells may be considered to be using a Tool Composite, as both the shell and the anemone serve to protect the crab. As many hermit crabs carry gastropod shells for protection, but not all carry anemones, the anemone can be seen as increasing the effectiveness of the gastropod shell in protecting the hermit crab, and may also be considered a Metatool.

We consider the veined octopuses' use of two coconut-shell halves to form a shelter as a type of Tool Composite. Finn, Tregenza, and Norman (2009) classified the assembled shelter as a single functioning tool, in which case the octopus would have Combined the halves to manufacture a tool. This is the most parsimonious categorization if the two halves are held or manipulated as a single tool, which does not appear to be the case. If the assembled shell halves are not held or manipulated as a single tool, but rather each is used as a tool separately and simultaneously, they would constitute a Tool Composite. A third alternative is that each shell half is a tool, and the two are linked by the octopus to make a Construction. In this case, neither the tools (shell halves) nor the assembled shelter would be held or manipulated by the octopus, but it seems that the octopus does hold both halves during use. The completed shelter therefore is most likely a Tool Composite. We elaborate on these distinctions in chapter 7.

Multi-Function Tool

Boxer (or "pom-pom") crabs may be considered to use sea anemones as Multi-Function Tools. The crab holds an anemone in each cheliped and Brandishes or Waves it when disturbed (Bürger 1903; Duerden 1905). The anemone serves a second function as a foraging tool, as the crab takes food that the anemone ensnares or accumulates on its body.

3 Fish, Amphibians, Reptiles, Birds

WE ORGANIZE THIS CHAPTER BY TAXONOMIC CLASS, SINCE THERE ARE TOO FEW cases to warrant organization by order or family. This is especially true of fish, amphibians, and reptiles. Reports of tool behavior by birds are more numerous, perhaps due to observation effort or ease of observation. Hawks, eagles, and buzzards; cockatoos; parrots; bowerbirds; and crows, ravens, jays, and their allies are the most common bird tool users. Galápagos woodpecker finches (*Camarhynchus pallidus*), scientifically famous for their tool use, use tools in only three modes, all to extract embedded insects. There are no or only isolated cases for other avian families. Gulls, skillful proto-tool users, have only rarely been reported to use a tool. No overall taxonomic trends or ecological correlates are apparent.

TOOL USE: FISH

Throw

Toxotid and anabantid fish from Southeast Asia project spouts of water toward prey above the surface of the water (Bekoff and Dorr 1976; Dill 1977; T. Gill 1909; Herald 1956; Lüling 1958, 1963; Milburn and Alexander 1976; Schuster et al. 2006; H. Smith 1936; Vierke 1973; Vierke and Lüling 1972). Of the toxotids, *Toxotes jaculatrix* and *T. chatareus* are known shooters. The anabantids *Colisa chuna*, *C. fasciata*, *C. lalia*, and *Trichogaster trichopterus*, as well as a *C. fasciata* × *lalia* hybrid, have been observed to shoot. The most extensive published investigations have concerned *T. jaculatrix*, the true archerfish. Wild and captive archerfish have been observed to shoot water spouts at insects, spiders, an infant lizard, bits of raw meat, models of appropriate prey, observers' eyes, and lit cigarettes. The target objects are either perched on nearly vertical surfaces above the water, or suspended or flying above the water. The spout is produced by forceful compression of the gill covers, which projects water through a tube composed of the tongue compressed against a groove in the roof of the mouth. The spout leaves the mouth as a jet that soon breaks up into a fine spray and several large droplets. It is the droplets that strike the prey, and they may travel linear distances as great as 3 meters. Accuracy varies from an observed 25 percent at 30 centimeters to an estimate of 100 percent at 1 meter. Some

An archerfish spitting a water drop at an insect. Reprinted from T. Gill 1909.

of this variation may be due to age differences and individual proficiency. Experienced archers may actually aim just below prey on a vertical substrate so as to knock it back into the water toward the fish, rather than soak it or knock it away from the fish. In addition to the specialization of the mouth for forceful water projection, vision is acute and the eyes are sufficiently large and mobile to provide binocular vision, which presumably allows precise visual localization of prey. Immediately before shooting, the archerfish pivots vertically so that the long axis of its body forms an angle of 140 to 170 degrees with regard to an imaginary straight line connecting the mouth to the target. This appears to minimize light refraction and parallax that would affect accuracy adversely. Despite these elaborate morphological and behavioral specializations for shooting, *Toxotes* is not totally dependent on tool use for capturing prey. In fact, shooting may be employed with less frequency and preference than simply leaping or searching for food on or below the surface of the water.

Hueter et al. (2004) and Kuba, Bryne, and Burghardt (2010) reported that various types of sharks and rays use jets of water to reveal prey hidden in sand. Kuba, Byrne, and Burghardt (2010) conducted an experimental study, finding stingrays (*Potamotrygon castexi*) able to use water as a tool to extract a food reward from a pipe. The stingrays used undulating fin movements to create a water current, which forced the food out of the pipe. Non-tool-using strategies were also used, such as "employing their disc-like body like a suction cap" (Kuba, Byrne, and Burghardt 2010, 4). One subject also used a water jet to blow the food reward out of the pipe.

Contain

Bshary, Wickler, and Fricke (2002) provided a review of cognition in fish. They noted the use of leaves to transport eggs as a form of tool use. Some fish, including a South American cichlid (*Aequidens paraguayensis*) and the brown hoplo catfish (*Hoplosternum thoracatum*), lay their eggs on leaf litter and use the leaves as "tablets" to move their eggs into more protected areas when threatened (Armbrust 1958; Keenleyside and Prince 1976).

TOOL USE: AMPHIBIANS

We have found only a few reports of amphibian or reptilian behavior that conforms to our definition of tool use.

Gadow (1901) reported an example that we include under the mode Affix, Apply, Drape in horned frogs (*Ceratophrys* spp.). Gadow (1901, 217) described how these frogs bury themselves into the ground, and added, "If there is not enough green vegetation, they throw, with their feet, little lumps of earth upon their backs."

TOOL USE: REPTILES

In their report on birds fishing with bait, Davis and Zickefoose (1998, 141), citing unpublished data from N. Mooney, stated that "baiting for birds with fish by crocodiles has been previously observed." W. Davis, Jr. (pers. comm.) confirmed that he observed an Australian saltwater crocodile (*Crocodylus porosus*) eating a dead fish, slapping it on the surface of the water and consuming the large pieces. He noted that after the fish was consumed, a few fragments remained on the water's surface. The crocodile then submerged itself under the water, with only its eyes above, and watched a black kite (*Milvus migrans*) that was interested in the fish scraps. Davis also relayed Mooney's observation of similar behavior and concluded that the crocodiles were ambushing the birds that came to eat the scraps. Unfortunately, from these observations, it is not possible to confirm that the crocodiles were responsible for the placement of the fish remnants or if they were just a byproduct of consumption. To be tool use, the crocodile would have to purposefully place the food on the water.

TOOL USE: BIRDS

Bird taxonomy, as reported in cited sources, has been updated to follow F. Gill, Wright, and Donsker (2009).

Drop

A bird referred to as a "black eagle" (species not reported) in Rhodesia Dropped eleven sticks toward researchers investigating its nests (Dick and Fenton 1979). The sticks fell within an average of 3 meters from the researchers. A ferruginous hawk (*Buteo regalis*) Dropped a stone on a human who was intruding near its nest (Blair 1981). Wood (in Boswall 1983b) described a zoo-living wedge-tailed eagle (*Aquila audax*) that often Dropped bones and other objects on its caretakers.

Berney (1905) and A. Campbell and Barnard (1917) provided circumstantial evidence that black-breasted buzzards (*Hamirostra melanosternon*) of Australia Drop stones in flight in order to break emu eggs (*Dromaius novaehollandiae*). They found buzzards on or near emu nests with freshly broken and consumed eggs. In each case, a round stone about the size of a chicken egg was found in the nest. The authors noted the rarity of such stones in undisturbed emu nests. J. Gould (1865) and Legge (1917) relayed reports of this behavior from indigenous Australians. Boswall (1977b) and Chisholm (1954) provided similar reports by several other observers, as well as from aboriginal nature lore. Leitch (1953) actually observed these birds Dropping stones on an emu nest.

Alexander (1838), Andersson (in van Lawick-Goodall 1970), Sciater and Stevenson-Hamilton (both in Schaller 1973), and J. Wood 1877 (also see Baxter, Urban, and Brown 1969) reported that Egyptian vultures (*Neophron percnopterus*) carry stones aloft in their talons and Drop them onto ostrich eggs (*Struthio camelus*). Having cracked the thick shells, the vultures feed on the contents. Boswall (1978) and Skead (1971) relayed reports, each from anonymous sportsmen, of similar behavior by Egyptian vultures in South Africa. Boswall (1978) also described a third-hand report stating that the same behavior is documented in East African tribal natural history. Brooke (1979b) reported an account of what he presumed to be an Egyptian vulture Dropping a stone from the air on a nest. He concluded that using stones to open eggs should be accepted as a widespread behavior of Egyptian vultures; "customary" might be used in the current vernacular (see chapter 1).

It is not always clear from the early accounts if the birds open eggs by aerial bombardments while they are flying or by Throwing down or Dropping stones while standing (Brooke 1979b). Brooke reported that early accounts of the birds Dropping stones were actually incomplete accounts of stones being Thrown down while standing on the ground. An Egyptian vulture was observed to take a stone and Drop it on the carapace of a turtle (Auffenberg 1981).

Griffen (in Montagu 1970) and Lockwood (1963) noted accounts of tool use by black kites, termed "fire hawks" by indigenous Australians. They were told that the birds picked up smoldering sticks from areas burned by brush fires, Dropped the sticks on unburned grass, and then fed on small animals fleeing the newly started fire. Although the birds were frequently seen feeding on insects, and possibly other small prey, at fire fronts (Andersen, Cook, and Williams 2003), there appear to be no scientific reports of deliberate fire-starting and Dropping behaviors. Accounts of birds inadvertently starting fires from dropping cigarettes and embers have been speculated (Burton 1959; J. Goodman 1960).

A male osprey (*Pandion haliaetus*), flying in the vicinity of his nesting mate, Dropped stones on an intruding osprey (Roche 1996).

Finch (1982) reported that a sulfur-crested cockatoo (*Cacatua galerita*) Dropped leafy twigs onto a pair of bat hawks (*Macheiramphus alcinus*) sitting on a lower branch in a tree.

A bird termed a "white-necked crow" was reported to Drop stones on ostrich eggs to open them (A. Martin 1890). Brooke (1979a) believed this was probably a pied crow (*Corvus albus*). Rolando and Zunino (1992) reported that a hooded crow (*C. cornix*) Dropped twigs on humans as they approached her young, which had fallen out of the nest. An American crow (*C. brachyrhynchos*) Dropped pine cones three times onto a human's head as the intruders climbed up to her nest (Caffrey 2001). Another American crow dropped flower petals twice onto her brood mate in what Caffrey interpreted as play. The brood mate jumped back from the petals, looked up at her, and then moved out of her dropping range. One of a pair of northern ravens (*C. corax*) defended its nestlings against intruding scientists by Dropping or Throwing down rocks (Janes 1976). The bird stood on a cliff above its nest site, picked up a rock in its beak, and Dropped it down toward the scientist with "a slight flip of its head"(Janes 1976, 409). Eight golf-ball-sized rocks were Dropped, the largest of which was 8 centimeters in diameter and 2.5 centimeters thick, striking the author once. Rock Dropping was not observed on subsequent visits to the nest, but the supply of rocks may have been exhausted. Montevecchi (1978) saw a fish crow (*C. ossifragus*) Drop a marshgrass stalk on an incubating laughing gull (*Leucophaeus atricilla*), and a northern raven Dropped a tuft of dry grass on a nesting black-legged kittiwake (*Rissa tridactyla*). Montevecchi inferred that the corvids were trying to harass and displace the nesting birds to get at their eggs. Heinrich (1988) expressed uncertainty as to whether these behaviors were tool use. He argued that an alternative explanation might be that it was a displacement of anger or frustration. Whether the birds Drop the twigs to discourage intruders or to express frustration, we consider both examples as tool use.

Bird and Emery (2009a) described an experimental investigation of tool use by captive rooks (*Corvus frugilegus*). In the first experiment, a worm was suspended on a platform out of reach of the birds. The birds learned through trial and error to Drop a stone down a tube to collapse the platform and retrieve the worm. When provided with a selection of stones, all of which fit into the tube, the birds preferred the largest stones, even though all were heavy enough to collapse the platform. The researchers also found that the birds preferred long, thin stones to round stones. The birds retrieved stones outside the testing area when stones were not provided in their immediate vicinity. When sticks were provided instead of stones, Bird and Emery noted that the birds used them immediately for this task, demonstrating behavioral flexibility. When provided with functional (proper size and weight) and non-functional sticks and stones, the birds all choose the functional tools.

Bird and Emery (2009b) also conducted an experiment in which four captive rooks Dropped stones into a container in order to raise the water level enough to move a floating worm within reach. They noted that two of the birds solved the problem on the first trial, and that the birds Dropped the exact number of stones necessary to raise the water level to reaching distance. The birds also showed a preference for Dropping larger stones over smaller stones. When presented with an additional tube with a worm inside on a bed of sawdust rather than water, the birds quickly learned that the sawdust (a novel material to them) could not be manipulated by Dropping rocks and showed a preference for Dropping the stone in the tube with the water.

Judd (1975) reported that a captive pet blue jay (*Cyanocitta cristata*) Dropped bits of food or trash into its water bowl when the water level dropped. This served to raise the water to drinking level.

Throw

Bindner (1968) reported aimed Throwing of stones by captive bald eagles (*Haliaeetus leucocephalus*). Bindner's eagles aimed and Threw stones at crickets and a turtle, and a leash at a human to gain his attention. The Throwing had a more horizontal than downward trajectory, and hits were scored at distances up to 60 centimeters.

Aumann (1990) reported that a captive black-breasted buzzard Threw stones at a chicken egg in an experimental setting. In the final two trials of the study, the chicken eggs were replaced with a dead chicken. In the first of these trials, the buzzard also Threw stones at the dead bird before consuming the chicken. Aumann reported that the buzzard was selective, choosing 40-gram stones from a selection of stones that ranged from 15 to 45 grams. Debus (1991) and Pepper-Edwards and Notley (1991) each found that a (different) captive black-breasted buzzard could open eggs by Throwing stones down from a standing position. Pepper-Edwards and Notley (1991) specifically noted that the bird raised the stone vertically in its bill before hurling it downward.

Egyptian vultures in Tanzania were observed to break open ostrich eggs by aimed Throwing (van Lawick–Goodall 1970; van Lawick–Goodall and van Lawick 1966, 1968; and discussed by Alcock 1970). During this behavior, the bird stands on the ground within 1 meter of the egg. A stone is held in the bill, the neck is extended and dorsiflexed so that the stone

A black-breasted buzzard Throwing a stone on an egg. Photograph courtesy of Stephen Debus.

is maximally elevated, and the stone is Thrown toward the egg. Eggs were broken with four to twelve hits in periods ranging from two to eight minutes. Accuracy was estimated to be 50 percent. One or more stones were used by a bird to break an individual egg. The stones, which weighed from 45 grams to more than 500 grams, were carried to eggs that were up to 10 meters away. The birds attempted to break model eggs, even when they were brightly painted or much larger than real ostrich eggs. However, a white egg-sized cube did not elicit stone throwing. Boswall (1977b) conducted an experiment with an Egyptian vulture in which an ostrich egg was placed near the bird, which quickly began to hurl stones at it. Iankov (1983) also found that captive Egyptian vultures Threw stones to open eggs.

Mayaud (1983) concluded that stone Throwing was not a universal behavior in the species, finding that some individuals did not Throw when given the opportunity. Cramp and Simmons (in Thouless, Fanshawe, and Bertram 1989) and Veselovsky (1970) presumed that social learning was necessary for the acquisition of stone throwing. However, Thouless, Fanshawe, and Bertram (1989) found no evidence of the need for social learning or cultural transmission in experiments with captive Egyptian vultures, since a naive bird Threw stones to open an egg. They also found that when the birds were provided with a variety of stones, varying from 27 grams to 232 grams, the birds preferred to Throw 46-gram stones. They argued that "the origins of aimed stone throwing are probably related to the unaimed throwing of small eggs, since the actions are similar and all tested birds strongly preferred to throw rounded or egg-like stones than jagged [ones]" (Thouless, Fanshawe, and Bertram 1989, 9). Note that this speculation implies that this form of tool use evolved from or originated with proto-tool use.

A bristle-thighed curlew (*Numenius tahitiensis*) picked up a piece of coral with its beak and repeatedly Threw it at an albatross egg (Marks and Hall 1992).

Hobbs (1971) observed tool use by white-winged choughs (*Corcorax melanorhamphos*) while feeding on freshwater mussels (*Velesunio ambiguus*). The choughs held pieces of previously consumed mussel shells or other intact mussels in their bills, extended their necks, and Dropped or Threw the tools down on other mussels to break them.

Drag, Roll, Kick, Slap, Push Over

Several wild Australian brush turkeys (*Alectura lathami*) were reported to Kick sand, stones, and debris backward with accuracy at a lace monitor lizard (*Varanus varius*) (Dow 1980). Fleay (in Dow 1980) described a captive Australian brush turkey "showering" him with leaves and dirt as he approached the nest, we assume by Kicking.

Hays and Donaldson (1970) documented sand kicking in black skimmers (*Rynchops niger*) to dig a hole in which they hide, and to distribute sand on their backs for camouflage.

Andersson (1989) reported an observation of a fan-tailed raven (*Corvus rhipidurus*) attempting to open a ping-pong ball with a stone as if the ball were an egg. The bird selected a stone it could not pick up due to its large size, Dragged it over to the ball, and flipped the stone over the ball. After the ball rolled away from the stone, the bird selected a smaller stone and attempted to open the ball by Hammering (see description under Pound).

Bait, Entice

"Bait-fishing," using a lure to attract fish, has been documented in many types of birds (Davis and Zickefoose 1998; Kushlan 1978). Bait-fishing may be passive or active. In passive bait-fishing, the bird does not manipulate the bait, but waits near bait that has been placed on the water's surface by another agent, such as a human throwing bread into the water. In active bait-fishing, the bird actively manipulates and places the bait (Zickefoose and Davis 1998). Only active bait-fishing is included here as tool use, because only then is the bird responsible for the proper and effective orientation of the tool (the bait).

The most well-documented bait-fishers are the green-backed herons, including the green heron (*Butorides virescens*) and the striated heron (*Butorides striata*). Lovell (1958) reported that a green heron placed bits of bread on the surface of the water and seized fish attracted by them. The heron sometimes placed the bait where fish had surfaced, and it retrieved and replaced a piece of bait that had drifted away. The bird gathered bread placed some distance from the water and carried it to the point of use, as well as displacing other birds that attempted to eat the bait. Sisson (1974) reports identical behavior by several biologically related captive green herons. Pelleted fish food was used as bait by these herons. Norris (1975) observed another green heron Baiting fish with a feather. The subjects of both Lovell's and Sisson's papers used bait sufficiently often to suggest that the behavior was well established in their repertoires, which could be inferred to be evidence of habitual tool use.

Boswall (1977b, 1983a), English (1987), R. Harvey (1999), Keenan (1981), and Oake (1992) each reported accounts of green-backed herons bait-fishing with either insects or spiders. Higuchi (1988a) saw a green-backed heron bait-fishing with bread and popcorn. He suggested that bait-fishing might be culturally acquired in these birds, since they do not all exhibit the behavior. Foxall and Drury (1987), C. Preston, Moseley, and Moseley 1986, S. Robinson (1994), Walsh, Grunewald, and Grunewald (1985), and P. Wood (1986) reported green-backed herons Baiting fish by placing objects including winged seeds, flowers, live insects, leaf fragments, or twigs on the water surface.

Higuchi (1986, 1988b) observed striated herons bait-fishing, arguing that the Japanese bird differs from the American green heron in its use of many different types of lures, including leaves, insects, moss, roots, feathers, berries, and foam. Higuchi's (1986) work also suggested that social learning may be involved. He noted that juveniles are rarely successful, failing both because they do not crouch down after they have tossed the bait (allowing the fish to see them), and because their lures are often too large.

Prytherch (1980) saw a squacco heron (*Ardeola ralloides*) placing insects on the surface of a pond and then watching the lure. Although the bird was unsuccessful in getting a fish, the author presumed the bird was Baiting.

A green heron prepares to Bait fish with bread. Photograph courtesy of Mark Swanson / Swanson Media.

D. D. McCullough and Beasley (1996) and Riehl (2001) reported that black-crowned night herons (*Nycticorax nycticorax*) used bread to bait fish. The birds they observed removed and repositioned the bread back onto the water's surface after it had become soaked and heavy. A goliath heron (*Ardea goliath*) used a small stick in an attempt to attract fish. The bird repeatedly picked up the twig, dropped it, and then watched and waited (Hunter, Calhoun, and Wilcove 2004).

Roberts (1982) documented a black kite picking up a piece of bread with its talons, dropping it into a river, and attempting to catch fish and crayfish attracted to the bread.

Boswall (1977b) relayed Alders's unpublished report of sunbitterns (*Eurypyga helias*) Baiting fish with maggots.

A captive lesser black-backed gull (*Larus fuscus*) used bread to successfully Bait and capture a fish from a pond (Sinclair 1984). Henry and Aznar (2006) observed several instances of a herring gull (*L. argentatus*) using bread to lure and capture goldfish.

Boswall (1983a) summarized observations of a pied kingfisher (*Ceryle rudis*) bait-fishing. On multiple occasions, the bird dropped bread onto the surface of a water, capturing the fish that came up to eat it.

In a different form of Baiting, burrowing owls (*Athene cunicularia*) collect and scatter dung near their nest chambers, in what has been interpreted as a way to attract dung beetles as prey (Levey, Duncan, and Levins 2004; M. Smith and Conway 2007). Levey,

A burrowing owl waits motionlessly after placing dung as Bait for dung beetles. Photograph courtesy of Ron Wolff and Doug Levey.

Duncan, and Levins (2004) argued that this behavior is akin to bait-fishing in herons, noting that the owls stand motionless near the burrow, waiting for beetles to approach the dung. M. Smith and Conway (2007) suggest that the dung may also signal to other owls that the chamber is occupied.

Males and, in a few cases, females of a number of avian species present food, nest material, stones or trinkets to mates during courtship, nest construction, and incubation (Armstrong 1965; Eibl-Eibesfeldt 1970; Lack 1940; Moreno et al. 1994). The presented objects presumably Entice the other bird to mate or cooperate. We thus consider courtship feeding to be tool use.

Broad-billed rollers (*Eurystomus glaucurus*) are known to bring objects, including shells, stones, fragments of plastic or glass, and aluminum foil, to their nests (H. Nakamura and Tabata 1988). These objects are apparently swallowed by the nestlings as "gizzard stones," which would not be tool use (Nakamura and Tabata do not claim that it is). The authors placed a selection of objects in a test area near a nest, and the birds moved the objects even before laying eggs, suggesting that the objects were used in courtship display. The peak of object removal corresponded to the peak in copulations. The researchers suspected that the objects were used during the birds' courtship displays.

Baiting by holding or tossing a tool is typical in bowerbirds of the genera *Sericulus*, *Ptilonorhynchus*, and *Chlamydera*. During the courtship display, males may use "enhancement props," holding bower "decorations" in their bills and tossing them toward the females (C. Frith and Frith 2004).

Club, Beat

Bindner (1968) reported that a captive bald eagle Clubbed a turtle with a stick held in its bill. Boswall (1983b) relayed an anonymous report of an Egyptian vulture holding a stone in its beak and Beating a monitor lizard (*Varanus* sp.) to death with it. Ilany (1982) reported on an Egyptian vulture that used a stone to beat an Egyptian spiny-tailed lizard (*Uromastyx aegyptius*) dead. Although Ilany did not report how the stone was held, we assume it was held in the bird's bill.

Pound, Hammer

Bindner (1968) also reported that the same captive bald eagle Hammered crickets and a scorpion with a rock held in its talons, presumably to subdue them as prey. Myers (in Boswall 1977a) reported observing two Egyptian vultures Hammering an egg with a stone held in the beak.

Hobbs (1971) reported that white-winged choughs used pieces of mussel shell or other intact mussels to Pound or Hammer on mussels in order to open them.

Boswall (1977b) cited a personal communication from F. Duvall who saw two American crows using stones to Hammer acorns. Andersson (1989) observed a fan-tailed raven using a stone (unsuccessfully) to open a ping-pong ball (the author presumed the bird thought the ball was an egg). The stone was held in the beak and used to Hammer the ball.

Pounding or Hammering may also be used for display. G. Wood (1984) documented three cases of a palm cockatoos (*Probosciger aterrimus*) holding either a branch or an unidentified object with its foot and using it to drum on a tree trunk in what the author presumed was a display. Boswall (1983a) provided a description of a captive Tanimbar corella (*Cacatua goffiniana*) performing an acoustic display by grasping a small metal container with his foot and using it to strike the bell hanging in his cage.

Pry, Apply Leverage

The woodpecker finch uses twigs or cactus spines to Pry out insects it finds while probing into holes or crevices (Bowman 1961; Eibl-Eibesfeldt 1961; Hundley 1963; Lack 1947, 1953). This behavior is further discussed below under the mode Insert and Probe. Greenhood and Norton (1999) observed a wild woodpecker finch that held bark or a wood chip in its beak and used it to scrape and flip over epiphytes on the surface of a branch. After inspecting the area, presumably for insects, the bird moved on, continuing to flip and search.

Powell and Kelly (1975) found that four American crows in an experimental setting with operant conditioning apparatus wedged matchsticks or rods between the edge of the response panel and the key, so that it acted as a lever. They noted that "each crow worked to wedge the tool (matchstick or metal rod) between the edge of the intelligence panel and the plastic key itself, with the tool being supported by the metal screen. When it was wedged in this position, the bird had only to peck the end of the rod in order to receive food" (Powell and Kelly 1975, 252). They inferred that this innovative behavior allowed the birds to operate the panel more efficiently.

A green jay (*Cyanocorax luxuosus*) was reported to Pry off pieces of bark with a short twig held in its beak (Gayou 1982). After removing a section of bark, the bird dropped the tool to consume the exposed insects. Jewett (1924) reported King's observations on a captive Northwestern crow (*Corvus caurinus*) that held a stick it its beak to Pry a peanut from a crack in its bamboo perch.

Morse (1968) observed brown-headed nuthatches (*Sitta pusilla*) using bark scales held in their bills to Pry firmly attached bark from tree trunks or limbs to expose invertebrate prey. Tool use was most common when pine seeds, a common food of these nuthatches, were not abundant. The birds appeared to use the Prying tools only on pine species with flaky bark, and carried the bark tools from tree to tree. The behavior was widespread (possibly customary) in Morse's Louisiana study population. Pranty (1995) and Withgott and Smith (in Kingery and Ghalambor 2001) also observed brown-headed nuthatches engaging in the same behavior. T. Mitchell (1993) observed this behavior in a white-breasted nuthatch (*Sitta carolinensis*) as well.

Salmons and Gough (2007) reported that a chestnut-backed chickadee (*Poecile rufescens*) used either a stick or a thorn to Pry seeds from a hanging suet feeder.

Merlen and Davis-Merlen (2000) observed a captive woodpecker finch using twigs to Pry off bark and to Pry buttons off a video camera. This same bird also used what the authors termed "long tools" to Pry apart human toes, apparently in exploratory play.

Dig

D. DeCourcey (pers. comm.) saw a captive female eclectus parrot (*Eclectus roratus*) use a segment of palm frond to excavate a burrow in the sand-and-gravel substrate of her cage when no nest box was available; the parrot then laid eggs in the burrow. Kawata (in Boswall 1983a) reported that a captive sulphur-crested cockatoo held a small stone in its beak and used it to scrape the soil away from an area on the floor of its enclosure. Boswall speculated that the bird might have used the tool to loosen the dirt to be consumed as "gizzard grit."

The rainbow bee-eater (*Merops ornatus*) excavates nest burrows that it decorates with natural and man-made objects. Chisholm (1954) cautiously relayed a second-hand observation of one bee-eater using a stick to Dig in the sand while excavating its burrow.

Potter (1970) observed an American robin (*Turdus migratorius*) Digging for ants in leaf litter with a twig held in its bill. The bird rubbed its plumage with the ants it uncovered, putting down and reusing the twig several times.

Wild New Caledonian crows (*Corvus moneduloides*) used dried, grasslike stems to probe through loose substrate (Rutz et al. 2007). One of these crows used at least three different tools in this manner. The crows also traveled with their tools, suggesting that they recognized good tools and kept them to reuse. Priddey (1977) saw a common blackbird (*Turdus merula*) using a twig to clear an area of snow in an apparent search for food. A starling (*Sturnus vulgaris*) used a twig held in its beak to rake through grass and dirt while foraging (Niemeyer and Kingery 2003). It is not absolutely clear from the descriptions if the crow's, blackbird's, and starling's tool use is properly categorized in the Dig mode or belongs under Reach or Insert and Probe.

Jab, Stab, Penetrate

Woodpecker finches use twigs or cactus spines to impale and extract insects they find while probing into holes or crevices (Bowman 1961; Eibl-Eibesfeldt 1961; Hundley 1963; Lack 1947, 1953). This behavior is further discussed below in the mode Insert and Probe. Millikan and Bowman (1967) reported that one bird used a probing tool to Prod or Jab aggressively at a conspecific.

Balda (2007) described the use of a stick as a weapon in an interaction involving a Steller's jay (*Cyanocitta stelleri*) and an American crow on a feeding platform. After trying repeatedly and unsuccessfully to chase the crow off the platform, the jay Detached a stick from a nearby bush and approached the crow while holding the stick in its bill, with the pointed end facing outward toward the crow. The jay thrust the stick at the crow, which jumped back at the jay, causing it to drop the stick. The crow then picked up the stick, also holding it so that the pointed end faced outward, and lunged at the jay. The jay flew off, with the crow following, the twig still in its bill. According to Balda, this is the first and only documented case of a bird using an object as a weapon against another bird.

Reach

Epstein and Medalie (1983) trained a captive common pigeon (*Columba livia*) to push a box toward a target and, separately, to peck at a plate. The plate was positioned at the base of a partial wall and gradually moved farther and farther behind the wall as training progressed. The wall did not extend completely to the floor or the ceiling, allowing the box to fit under it. The bird could peck over and behind the wall. During training, the box and the plate were never presented together. In the experimental session, the experimenters positioned the plate behind the wall, out of the bird's reach for the first time, and placed the box in the enclosure as well. In ninety-two seconds, after the bird had failed to reach the plate with its beak, it slid the box under the wall and used it to Reach and touch the plate.

Funk (2002) found in an experiment that yellow-crowned parakeets (*Cyanoramphus auriceps*) used twigs to rake in seeds they couldn't otherwise reach from their cage. T. Jones and Kamil (1973) reported that captive blue jays used pieces of paper, grass, and other objects to Reach food pellets outside their cage. MacDonell (2000) found that American crows were able to use sticks and hooks to Reach food rewards. Powell and Kelly (1975) trained American crows to use keys and matchsticks to operate a keypad for a reward. Once the birds acquired use of these tools, they also generalized to using nails and paper clips as tools without additional training.

Insert and Probe

The geospizine finches of the Galápagos Islands command prominence in scientific history because they inspired Charles Darwin's formulation of the theory of natural selection. Gifford (1919) provided the first published description of tool use among the finches. Seeing a bird break off a twig to Insert and Probe in holes in a dead tree, he assumed that the bird was searching for insects. Gifford's report was confirmed and extended by Bowman (1961), Eibl-Eibesfeldt (1961), Hundley (1963), and Lack (1947, 1953). Woodpecker finches pick up or detach twigs, cactus spines, or leaf petioles. They hold the tools lengthwise in their bills and use them to Probe in holes and crevices in trees or under bark that they are unable to reach unaided. When a finch encounters an insect, it is Stabbed with the tool and withdrawn, Pried out with the tool, or simply Jabbed with the tool to agitate it sufficiently to cause its exodus. The bird then drops the tool or holds it under its foot to eat the insect. Both Bowman (1961) and Hundley (1963) saw finches carry twigs from branch to branch while searching for prey. Eibl-Eibesfeldt and Sielmann (1962) succeeded in closely observing the behavior of a captive specimen. It not only picked up or Detached tool objects, but it also shortened them or Subtracted bits that impeded insertion. It used similar tools to Dig in loose earth and Pry bark from dead branches.

Millikan and Bowman (1967) observed tool use in a small captive colony of woodpecker finches and performed some simple experiments to explore the complexity of the behavior and the controlling stimuli. They found that the finches used a variety of elongate objects, adjusting their posture and their manipulation for optimal use according to the dimensions of the tools. The birds also broke long objects to make shorter, more usable tools, and transported tools up to 2 meters between Detaching and using them. Hunger increased the frequency of tool retrieval and, probably, of use. Millikan and Bowman also observed a Galápagos large cactus finch (*Geospiza conirostris*) Inserting and Probing with tools. Perhaps significantly, this individual had been caged next to woodpecker finches and had many opportunities to observe their tool use.

Bowman (1961) did an extensive field study of dietary and habitat preferences of the Galápagos finches and an analysis of their head morphology. He related the tool use of the woodpecker finch to its mostly insectivorous diet and to the functional morphology of its bill and tongue. This species lacks extreme morphological specialization in these features for the extraction or rapid excavation of insects from dead wood. The use of tools appears to be a behavioral adaptation for feeding on deeply embedded insects. Bowman noted that the feeding behavior of the mangrove finch (*Camarhynchus heliobates*) was not known, but, because of similarities in head morphology, it was probably similar to that of the woodpecker finch (*C. pallidus*). However, he argued that *C. heliobates* occupies a lower, wetter microhabitat than *C. pallidus*, and niche differences might preclude tool use in *C. heliobates* despite its close relationship to *C. pallidus*. Curio and Kramer (1964) investigated this question and found that *C. heliobates* did indeed use tools in a manner similar to *C. pallidus*. Hundley (1963) saw another Galápagos finch, *Certhidea olivacea*, briefly Insert a leaf petiole or flower stem into a crevice and Probe with it without securing an insect. *C. olivacea* is entirely insectivorous but usually catches its prey on the wing or picks it from foliage. Hundley also relayed a second-hand report of tool use by another Galápagos finch, but its identification as *Geospiza fulginosa* or *G. fortis* is too imprecise to allow evaluation in terms of feeding behavior or bill morphology.

Merlen and Davis-Merlen (2000) observed a captive woodpecker finch using a variety of tools, including twigs, a feather, a grasshopper leg, fragments of worn glass, slivers of wood, and pieces of shells. The twigs and grasshopper leg were used to Probe into cracks. The bird also Inserted a twig between a sock and boot being worn by a human.

Tebbich et al. (2001) studied the ontogeny of tool use in wild woodpecker finches, finding that juveniles developed the ability to use tools whether or not a skilled conspecific model was present. They concluded that social learning was not involved in the acquisition of tool use in these birds. Tebbich et al. (2002) found that tool use is important for the subsistence of these birds; during the dry season, half of all prey was acquired using tools. Tool use in these birds allows for a variety of feeding techniques, as summarized by Tebbich et al. (2004, 95), who argued that "the use of tools extends the morphological properties of the beak temporarily without limiting behavioral versatility and flexibility." Tebbich and Bshary (2004) also studied tool use in these birds experimentally. The six birds they tested were able to use stick tools to Reach and retrieve food rewards from a tube, and one of the birds was able to avoid the trap placed in the tube more often than chance (the trap-tube experimental paradigm is discussed more fully in chapters 5 and 6). Tebbich and Bshary also found that several of their finches were able to choose a tool of a sufficient length to rake the reward from a tube on their first try. This experiment was patterned after Chappell and Kacelnik's experiment (2002) in which New Caledonian crows chose Reaching tools (twigs) of appropriate length (see below).

B. Marshall (1982) observed a marabou stork (*Leptoptilos crumeniferus*) use a stick to Probe into a hole in a fallen tree. Although the bird did not extract anything, the author presumed the bird was using the stick to dislodge prey inside the hole.

Richards (1971) saw a crested shriketit (*Falcunculus frontatus*) Probe for food by Inserting a twig into a crevice. Noske (1985) observed one possible case of twig tool using by a crested shriketit, although he did not elaborate on the details. In a personal communication from M. Coombes, cited by Boswall (1977b), a blue tit (*Cyanistes caeruleus*) was reported Inserting a twig into a nut hopper to dislodge a nut. A great tit (*Parus major*) was observed using a pine needle to Probe into crevices in bark to remove insect larvae (Duyck and Duyck 1984). H. Mitchell (1972) reported that a grey shrikethrush (*Colluricincla harmonica*) used a twig held in its bill to Probe in a hole in a brick. After a few probes, an insect crawled out of the brick and was caught. Olney (in Boswall 1978) reported that a zoo-housed Eurasian oystercatcher (*Haematopus ostralegus*) consistently Inserted a stick into crevices, dislodging and eating an insect in one instance.

Berg (in Skutch 1996) observed a pale-mandibled aracari (*Pteroglossus erythropygius*) using a twig in an attempt to extract an insect from a crevice.

Another bird of the south Pacific, the New Caledonian crow, has been reported to Insert and Probe with twigs; bamboo stems; leaf stems, petioles, or midribs; dry grass stems; the branching stems of ferns; and lengths of vine. They Probe with these tools to extract insect or reptilian prey from crevices, under bark, under leaf litter, in the ends of hollow branches, in upright stumps or logs, and around the bases of palm and *Freycinetia longispica* plants (Cunisset 1909 in G. Hunt 2000b; Hannécart and Létocart 1980 in G. Hunt 2000b; G. Hunt 1996, 2000b; G. Hunt and Gray 2002, 2006, 2007; Le Goupils 1928 in G. Hunt 2000b; Orenstein 1972; Rutledge and Hunt 2004; and Troscianko, Bluff, and Rutz 2008). G. Hunt (1996) reported that the crows carried their tools with them between foraging stops.

The crows use multiple forms of tools to Probe for and extract insects. In addition to leaf stems, bamboo stems, or twigs, they also use tools having hooks or barbs, and what are known as *Pandanus* tools (G. Hunt 1996; G. Hunt and Gray 2002). Hook tools and *Pandanus* tools are manufactured by the crows; each is highly standardized, but each is distinctly different from the other. These birds are claimed to be highly selective in the types of materials they use to make their tools. G. Hunt (1996) reported that the crows of the Pic Ningua study site prefer to make their hooked tools with freshly Detached twigs from *Elaeocarpus dognyensis*, while birds more than 50 kilometers away at the Parc Rivière Bleue use fresh twigs from *Cunonia vieillardii* to construct the same types of tools (G. Hunt and Gray 2006). This would be an interesting case of customary tool behavior if all adult birds at both sites use tools in the site-specific fashion, and cultural tool behavior if both plant species are present at both sites. G. Hunt (2008) reported that the birds also demonstrate behavioral flexibility in their ability to use barbed twigs from the introduced *Lantana camara* plant as tools to extract insects.

G. Hunt and Gray (2004b, 119) described the hooked-twig tools as having a "crochet-hook-like" projection at the end. The birds hold the tool by the non-hooked end, Insert it under debris or dead wood or into crevices, and move the tool rapidly back and forth, so that the hook catches a prey item (G. Hunt 1996; G. Hunt and Gray 2004a, 2004b). G. Hunt (2000a) argued that the manufacture and use of hook-type tools is unknown in other types of animals. He noted that the crows' hooked tools are not comparable with the branch-hook tool, also called a branch-hauling tool, used by chimpanzees (Sugiyama and Koman 1979). G. Hunt added that branch-hook tools have not spread throughout an observed chimpanzee population (have not become customary) as hooked-twig tools have among New Caledonian crows. He also argued that there is no evidence that chimpanzees know how to manufacture hooks, but the crows do. Elsewhere, G. Hunt (1996, 250–251) stated that the manufacture

A New Caledonian crow with a hook tool that it has manufactured to extract embedded prey. Photograph courtesy of Gavin Hunt.

and use of tools with hooks suggested, "an appreciation of tool functionality."

The *Pandanus* tools used by crows are short sections of leaf detached from the barbed leaf edges of the screwpine (*Pandanus* spp.). Three different but related tool designs for *Pandanus* tools have been documented and classified as wide, narrow, or stepped (G. Hunt and Gray 2003). Both the wide and narrow tools are not tapered, while the stepped tools contain multiple ridges, or steps, tapering to a point. Each step requires an additional cut-and-rip action by the bird with its beak to Detach the tool from the leaf. G. Hunt and Gray could not find ecological correlates for each of these forms; they argued that the tools have a common historical origin and that the different tool forms are a result of cumulative changes.

Most *Pandanus* tools are of the multi-stepped form (G. Hunt and Gray 2003). When used, the tool is held by the non-barbed side, with the sharp, rigid barbs facing up and away from the bird's head. As the tool is cut out and removed from a leaf, a matching counterpart remains on the leaf in the tree. G. Hunt (2000a) examined these and found that 99 percent of stepped tools were tapered to a point, which he argued provides a sturdy tool. Most *Pandanus* tools were removed from the left edge of the leaf (G. Hunt 2000a, 2004b;

G. Hunt, Corballis, and Gray 2001, 2006). G. Hunt (2000a) argued that this indicated laterality, or "handedness," at the population level in making and using the tools. These results contrast with the results of studies of captive New Caledonian crows in which Weir et al. (2004) found individual laterality preferences in ten crows but no population preference: five preferred the right and five left.

Wild New Caledonian crows also forage for lizards using tools (Troscianko, Bluff, and Rutz 2008). The bird they observed manufactured two tools from dry grass stems and then used each tool to Probe into crevices in a fence post for hidden lizards (species unnamed).

The tool-using techniques of wild New Caledonian crows have also been studied experimentally. G. Hunt (2000b) placed long-horned beetle (Cerambycidae) larvae in artificially created holes in dead wood. Several birds were observed using leaf stems to extract the larvae, generally selecting their tools from surrounding ground litter. G. Hunt noted that these tools were less flexible and drier than fresh stems. He also reported that one bird used a Detached section of tree root as a probe. The tool was held in the beak, and moved quickly up and down into the hole until a larva clamped onto the probe with its mandibles. The crows were sometimes seen to reuse tools that had been left in the holes. Although G. Hunt did not observe the birds foraging outside of the experimental apparatus for larvae, he believed there was ample evidence that they do so, based on tree wood damage and on tools that were recovered in the field.

In another experimental study with wild crows, G. Hunt, Rutledge, and Gray (2006) observed that the birds manufacture tools from leaves, using the resulting leaf stems to Probe for food placed in vertical holes at various depths.

Taylor et al. (2009) documented that six wild-born, captive crows were able to Insert a stick tool into a tube to pull out a reward. Three of the birds learned to avoid a trap in the tube through which the food could be lost. These three birds were also able to use tools to solve two of three variations of tests involving the trap tube, and they were alert to a trap when the problem was presented on a table. Taylor et al. cited this as evidence supporting the presence of causal understanding in these birds.

G. Hunt, Lambert, and Gray (2007) examined tool use in a hand-fed rescued New Caledonian crow, naive to tool use. Without benefit of training or social learning, the bird learned to use a stick tool to acquire food from a tube. The authors found that two of four captive New Caledonian crows, including the abovementioned naive crow, learned to use provided *Pandanus* tools to extract meat from a tube. G. Hunt, Lambert, and Gray (2007, 6) suggested "that social learning and a disposition to develop basic tool use without social input are both essential cognitive requirements for cumulative technological evolution."

A New Caledonian crow named Betty spontaneously bent and used a wire to lift a food-laden bucket from a container (Weir, Chappell, and Kacelnik 2002). Betty also modified and used aluminum strips, feathers, and twigs to pull in a food bucket from a tube and to obtain a food reward from an apparatus.

Chappell and Kacelnik (2002) studied tool selectivity in two captive New Caledonian crows, first by placing a food reward in a tube at different distances from the openings and providing sticks of various lengths to rake out the reward. Both birds selected tools whose length matched the distance to the food more often than expected by chance. When the tools were placed out of sight of the apparatus, only one of the birds chose a tool of sufficient length significantly more often than would be expected by chance. Chappell and Kacelnik (2004) then studied tool selection on the basis of diameter with these two crows. The apparatus in the second experiment was fitted with a hole of one of three different diameters, with a food reward behind it that the crow could get by pushing it out of the apparatus with the tool. When the birds were given three sticks with different diameters as potential tools, they preferred to use the thinnest stick. The authors suggested that the thinnest stick could always be used successfully on holes of any of the three diameters, and that thin sticks were lighter and perhaps easier to handle. However, the choice of the thinnest stick may not have been the optimal choice, since thinner sticks tend to bend more easily during the task. They authors did not systematically examine the role of tool diameter, rigidity, weight, and ergonomics. In a subsequent experiment, the crows used tools they had manufactured to Insert into the holes to get food. They tended to make tools of a diameter that closely matched the diameter of the hole, rather than only make a thin tool that would fit into holes of all diameters. The authors claimed that in these cases the crows were making tools that were optimal in terms of diameter and rigidity.

Holzhaider et al. (2008) presented captive crows

An array of recovered *Pandanus* tools manufactured and used by New Caledonian crows. Photograph courtesy of Gavin Hunt.

with premanufactured *Pandanus* tools. They found that the birds did not consistently orient the tools in a way that indicated an understanding of the function of the barbs in the probing tasks. The authors argued that the successful use of such tools by wild crows likely relies on associative trial-and-error learning. Among captive juvenile crows, Kenward et al. (2005) observed the spontaneous manufacture and use of tools from leaves to obtain meat from an apparatus. Kenward et al. (2006) also studied the ontogeny of use of a twig tool for probing by captive crows. They concluded that "individual learning, cultural transmission and creative problem-solving all contribute to the acquisition of the tool-oriented behaviours in the wild, but inherited species-typical action patterns have a greater role than has been recognized" (Kenward et al. 2006, 1329).

Many other corvids use tools to Probe for insects in crevices. An Indian house crow (*Corvus splendens*) Inserted a leaf into a cavity six times to extract ants (*Sima* spp.) (Rajan and Balasubramanian 1989). Caffrey (2000) saw a wild American crow use a small piece of triangular-shaped wood it had Detached and Reshaped from a fence rail to Probe into a hole in the same fence. The bird held the wood by its wider end, Inserting the narrow end into the hole. Although the bird was not successful, closer examination revealed a large spider in the hole. Gayou (1982) documented two green jays holding a small twig by the beak and Inserting the twig under bark to extract insects that adhered to the twig. In a captive study, Bird and Emery (2009a) found that rooks could modify stick tools for Probing into a tube and could modify a straight wire into a bent form to hook a reward-filled bucket placed in a tube to pull it within reach.

An African grey flycatcher (*Bradornis microrhynchus*) retrieved grass stems and Inserted them into holes in a concrete foundation to fish out winged termites (Boswall 1983b). The bird withdrew its tool when a termite gripped the stem with its mandibles.

A pygmy nuthatch (*Sitta pygmaea*) used a twig to Probe into crevices in a pine tree (S. Russell et al., pers. comm., in Pranty 1995). Green (1972) saw varied sittellas (*Daphoenositta chrysoptera*) Insert strips of wood into cavities to dislodge insect larvae. The tool was held in the bill while Probing and beneath the foot while eating a dislodged larva.

A mountain chickadee (*Poecile gambeli*) was seen using a wooden splinter to Probe into a tree crack, although it did not appear to acquire an insect (Gaddis 1981).

Scratch, Rub

G. Smith (1970, 1971) observed a little corella (*Cacatua sanguinea*), two yellow-crested cockatoos (*Cacatua sulphurea*), and two African grey parrots (*Psittacus erithacus*) Scratching their heads, necks, throats, backs, and sides with sticks, twigs, and other elongate ob-

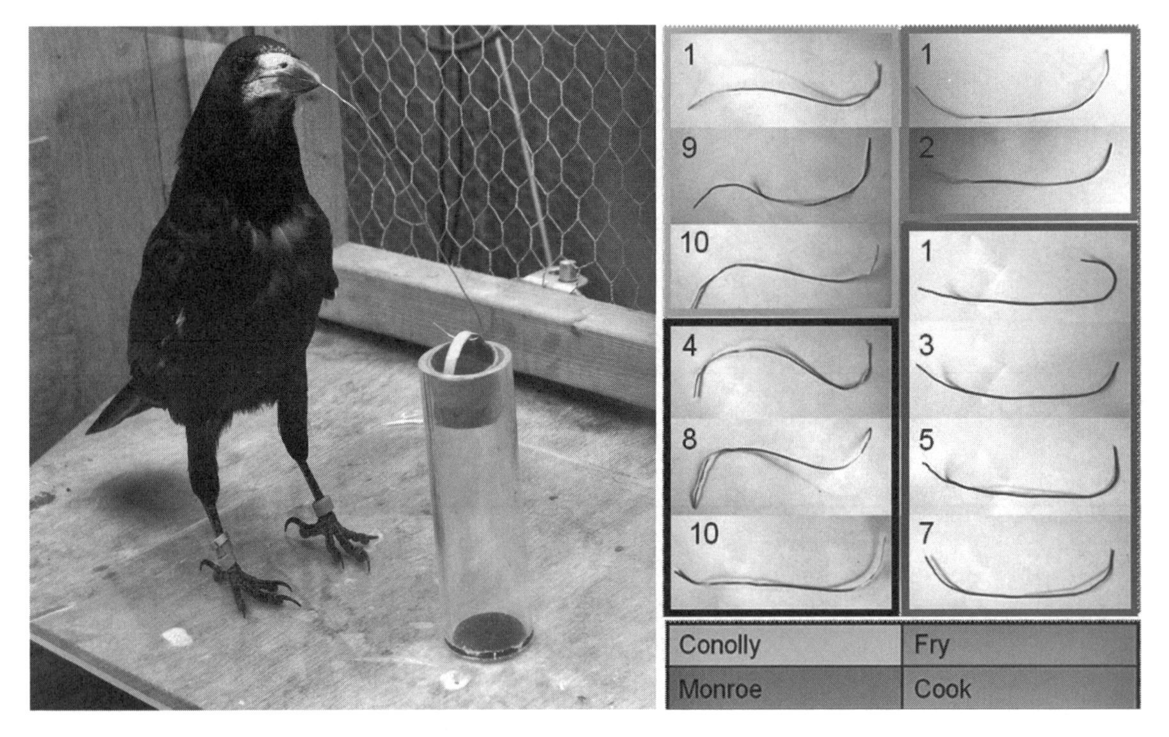

Left: A rook lifts a bucket of food from the bottom of a tube with a wire hook. The bird had manufactured the hook by bending a straight piece of wire. *Right*: A sample of wire hooks manufactured by rooks is shown. Photograph courtesy of Christopher Bird.

jects. Boswall (1977b, 1978, 1983a, 1983b) and Janzen, Janzen, and Pond (1976) relayed several observations of captive African grey parrots scratching with twigs, a feather, a bone, spoons, wire, wood splinters, keys, matchsticks, and pieces of biscuit and bread. Boswall (1977b) also cited Blanden's observation of a blue-fronted Amazon parrot (*Amazona aestiva*) preening its head feathers with a stick, as well as Plath's observations of unspecified parrots Scratching with sticks. Boswall (1983b) reported that a salmon-crested cockatoo (*Cacatua moluccensis*) Scratched its body with twigs, a detached feather, or bent wire, and he described a yellow-crested cockatoo and an Amazon parrot (*Amazona* sp.) that used a spoon to Scratch themselves. He also described a blue-and-yellow macaw (*Ara ararauna*) that used a bone for Scratching.

A captive American crow wedged a plastic slinky toy onto his perch and used it to Scratch his head (P. Cole 2004).

Block

The male satin bowerbird (*Ptilonorhynchus violaceus*) uses a tool held in the beak during bower "painting" (Frith and Frith 2004; Gannon 1930). A. Marshall (1954) relayed a number of unattributed observations of this painting behavior. According to these, the birds employ a wad, consisting of collected fragments of bark that are manufactured into a small ovoid pellet approximately 10 by 6 millimeters and 4 millimeters thick. It is not used as a brush, but seems to wedge the bill open so that a proper amount of the solution flows from the sides during painting. It also absorbs some of the solution, although this may not be its function. The wads may be reused, since they are left in the bower, whereas other objects are promptly removed. Not all satin bowerbird males that paint use the tool, and there are no available observations of tool use in painting by either spotted or regent bowerbirds. Chisholm (1971, 183) wrote of these tools, also termed "wads," that it is "now generally accepted that they are 'corks' or 'stoppers' grasped in the beak to prevent the mixture from escaping while the head is held sideways." Van Lawick–Goodall (1970, 198) referred to this bark as a "plug."

Reid (1982) reported that on several occasions, a captive rook inserted a plug into the drain hole of the aviary floor to Block water from draining, creating a drinking and bathing pool. When the plug was present in the aviary, the bird would carry it to a drain, drop it in, and then tap it into place with its bill. Faced with six drain holes in the aviary floor; the bird was selective in its behavior, choosing to plug holes nearest to the water source.

Prop and Climb, Balance and Climb, Bridge, Reposition

Epstein et al. (1984) also used pigeons in replicating Köhler's (1925) famous experiment in which chimpanzees were presented with a banana suspended overhead, out of reach. The chimpanzees Repositioned a box or boxes below the banana to reach it. Epstein and colleagues' pigeons were first trained to push a box and then, separately, to climb on a box and peck a banana. Although the birds were never trained to push the box under the overhead banana, they later Repositioned it so they could peck at the banana, in what the authors called an insightful solution.

P. Cole (2004) reported that a captive American crow Repositioned a small lightweight perch, which allowed it to reach a previously inaccessible object.

Contain

Rekasi (in Lefebvre, Nicolakakis, and Boire 2002) described a white stork (*Ciconia ciconia*) that provided water to its chicks by wringing out damp moss with its beak.

G. Smith (1971) reported that an African grey parrot baled water from its dish with a tobacco-smoking pipe held in its bill. Boswall (1983a) and G. Smith (1971) each reported that a yellow-crested cockatoo scooped up water with either nutshells or bottle tops held in its foot, so that the bird could drink from the container. Boswall (1983a) provided an account of a yellow-crowned Amazon parrot (*Amazona ochrocephala*) using a toy bell to scoop up and hold seeds, and possibly water as well. Boswall noted that the bird ate all of the food it was given in this manner, and that items were never placed in the cup for him. Pepperberg and Shive (2001) observed an African grey parrot using a bottle cap to scoop water from a dish and then drink from the cap. Porter (1936) saw keas (*Nestor notabilis*) bale water from a large container with cans and cups in what seemed to be play. Boswall (1977b) and van Lawick–Goodall (1970) cited a report by Fyleman of a captive cockatoo (species unspecified) using half a walnut shell to scoop water from a nearly empty dish.

Two researchers reported on an American crow named Clarence, kept at the Allee Laboratory of Animal Behavior at the University of Chicago, that was regularly supplied with various toys and trinkets for novelty (B. Beck, pers. obs.; E. Hess, pers. comm.). Among these was a small plastic cup. Part of Clarence's diet consisted of dried mash, which was wetted slightly to make it more palatable. If the laboratory assistants forgot to wet the mash, the crow compensated for the oversight by dipping the plastic cup in a watering trough, carrying the filled cup as far as 5 meters to the food pan, and emptying the water on the mash. Clarence repeated the behavior until the mash was saturated and then began to eat. If the water spilled in transit, the crow returned to the trough for a refill rather than proceeding to the food pan with an empty cup. The behavior appeared spontaneously, with no training or shaping.

P. Cole (2004) reported that a captive American crow transported water, food, and other items using containers that included a small Frisbee, a plastic cup, and a water bottle nozzle.

A captive marsh tit (*Poecile palustris*) used paper stickers as tools to acquire and store food (N. Clayton and Jolliffe 1996). After detaching a sticker from its feeder, the bird dipped it into the food bowl, where powdered food adhered to it, and carried the sticker to its perch. The bird returned with the sticker and dipped again, then folded the sticker in half and stored it. Clayton and Jolliffe reported that the bird repeated the behavior of storing six more times over the next twenty minutes. The next morning, the bird in the neighboring cage engaged in this behavior.

Absorb

A male gila woodpecker (*Melanerpes uropygialis*) employed a different mode of tool use to transport fluids (Antevs 1948). Nesting woodpeckers were provisioned with honey thinned to a watery consistency. The male dipped bits of bark into the honey and, when they had Absorbed the fluid, carried them to the fledglings. The honey could not otherwise have been transported efficiently to the nest.

Over thirty different species of birds have been observed submerging or soaking their food in water before ingestion, possibly to lubricate and ease the swallowing of prey (Morand-Ferron, Veillette, and Lefebvre 2006; Morand-Ferron et al. 2004; see also chapter 7). However, since the water is not manipulated, this is not considered tool use. Morand-Ferron et al. (2004) observed this behavior in Carib grackles (*Quiscalus lugubris*) and concluded that the behavior made consumption easier. Morand-Ferron, Veillette, and Lefebvre (2006, 342) called this behavior a "proto-tool food-processing technique that speeds up ingestion." However, Morand-Ferron et al. (2004, 1272) were unable to exclude the alternative possibility that "dry food acts as a better 'sponge' to aid water ingestion for the individual itself, or for its nestlings." If the food items are being submerged to Absorb water, they would be classified as tools. Koenig (1985) reported such a case in Brewer's blackbirds (*Euphagus cyanocephalus*). The birds submerged grasshoppers in water before carrying the insects to their nestlings. Koenig found experimental evidence that the grasshoppers had Absorbed a substantial amount of water, thus providing a water source for the young birds.

Wipe

Bartlett and Bartlett (1973) reported that a captive sandhill crane (*Grus canadensis*) habitually Wiped itself with a towel after swimming.

A captive female striped owl (*Pseudoscops clamator*) was observed to Wipe bloody food residue from her chicks with dry leaves (A. Goodman and Fisk 1973).

T. Jones and Kamil (1973) noted that a blue jay placed a piece of paper into water and then used it to Wipe its empty food bowl. The powdered remains of food in the bowl, otherwise unavailable to the bird, adhered to the wet paper so that they could be eaten.

Boswall (1978) cited a personal communication from Gibson describing the daily use of green leaves by a captive red-billed leiothrix (*Leiothrix lutea*) to Wipe itself after bathing.

Affix, Apply, Drape

"Anting" by birds has received an extraordinary degree of attention by naturalists and scientists; even a cursory survey of the literature yields numerous references that include over 200 species (see reviews in Bentley-Condit and Smith 2010; Burton 1959; Chisholm 1959; Craig 1999; Lozano 1998; McAtee 1938; Potter 1970; Rajchard 2008; Weldon 2004). Nonetheless, the definition, occurrence, and function of anting in birds are still disputed (Osborn 1998). Simmons (1966) differentiates between "true anting," which involves ants, and "anting with substitutes." Such substitutes include beetles, millipedes, mealworms, caterpillars, snails, onions, limes, cigarette butts, sawdust, mothballs, beer, and even fire and smoke from a lit match. A further distinction is made between "active anting" and "passive anting." In active anting (also known as "ant-application behavior," or "self-anointing"), the bird picks up ants in its bill and Applies them to its plumage; in passive anting (also called "ant-exposure behavior") the bird simply postures near ants and allows them to move into its plumage (Potter 1970; Simmons 1966; Weldon 2004; Whitaker 1957). Birds engaging in both active and passive anting exhibit characteristic stereotyped behavior patterns during anting. Active anting with either ants or substitutes appears to fit our definition of tool use, while passive anting does not.

Potter (1970) and Whitaker (1957) believed that anting is a behavior that is characteristic of several avian orders. Most active anting has been documented in passerine birds (Clark and Clark 1990; D. Clayton and Vernon 1993; Clunie 1976; Colahan 1981; Dater 1953; Dennis 1985; C. Dubois 1969; Eisner and Aneshansley 2008; Hailman 1960; Johnston 1985; Judson and Bennett 1992; Kelso and Nice 1963; W. King and Kepler 1970; Lunt, Hulley, and Craig 2004; Osborn 1998; Post and Browe 1982; Sick 1957; Simmons 1957; Skutch 1948; Someren 1996; Terres 1962; VanderWerf 2005; Wenny 1998; Whyte 1981). It has also been documented in other avian orders (as reviewed in Whitaker 1957), including Galliformes (Thomas 1957), Psittaciformes (Chisholm in Whitaker 1957; Lewington in Whitaker 1957), and Piciformes (Allsop 1949; F. G. Stanford 1949; Stone 1954).

Several researchers believe that anting plays a role in feather maintenance (Potter 1970; Simmons 1957; Southern 1963). The formic acid and other acidic compounds found in the defensive secretions and body fluids of ants are decidedly thermogenic. It seems safe to conclude that the heat-producing properties are important in proximal stimulation and that anting serves some role in feather or skin maintenance, perhaps to ease irritation resulting from molting, facilitate structural maintenance of the feathers, control ectoparasites through insecticidal action of the applied compounds, inhibit fungal and bacterial growth, and/or dissolve lipids accumulated in the plumage (Ehrlich, Dobkin, and White 1986; Osborn 1998; Simmons 1966). If the last role were a primary function, anting might be considered functionally analogous to dustbathing (Borchelt 1975). On the other hand, Judson and Bennett (1992) suggested that anting might serve as a way to remove toxic formic acid before ingestion of the ants. However, birds would not eat anting substitutes, such as cigarette butts, sawdust, or mothballs.

Burton (1959) recounted fire and smoke wielding by his captive rook, which he argued was a form of anting (see the review by J. Goodman 1960). The bird's frequent behavior of standing in and near open flames reminded Burton of the myth of the phoenix rising from the ashes, leading him to speculate that the myth may have arisen from this behavior. While standing near the flames, the rook also actively Applied burning embers to its feathers. Burton also observed two captive rooks and a Eurasian jay (*Garrulus glandarius*) that would hold lit matches in their talons and Apply the burning or smoking end to their feathers. J. Goodman (1960) cited additional observations of corvids using smoking cigarettes or matchsticks for anting. Burton (1969) also reviewed observations of wild birds carrying burning embers or cigarettes to their nests and implicated that these anting birds were most likely responsible for accidentally starting building and pasture fires.

Birds sometimes carry ants to perches before Applying them to their bodies. When Applying the ants, they adjust their behavior for maximal efficiency. For example, Simmons (1966) noted that the red-billed leiothrix usually seizes ants by the thorax, leaving the gaster protruding from the bill, and Applies the ants to the side of the body that the gaster faces. He also stated that many birds form and apply wads of ants. Formicine and dolichoderine ants, which do not sting but exude or spray toxic compounds when agitated, are frequently used for anting, while myrmicine ants, which sting, are rarely used.

Many birds oil-preen: they rub their bill, head, and neck first on the uropygial gland and then on the plumage. In so doing, a bird literally transports and Applies the oily glandular secretion to its body and thus could be said to be using the secretion as a tool. Preen oil may contribute to waterproofing and feather maintenance, and may also serve as an antibacterial agent (Stevens 1996; Thompson 1964).

Meyerriecks (1972) described a case of tool use by a double-crested cormorant (*Phalacrocorax auritus*). The bird held one of its own molted secondary wing feathers in its bill, Applied the tip of the feather to its uropygial gland, and Wiped the glandular secretion on its wings with the feather. The sequence was repeated several times, although no further repetition by this or other cormorants was seen. It is unclear why this bird preened with a feather rather than with its head and bill as usual. Meyerriecks had seen this individual bird preen in typical fashion shortly before it used the feather, but he noted that the bird stretched its head and bill toward the gland in what "seemed like an awkward manner." Perhaps the mobility of its neck was reduced due to injury or illness, and the cormorant used the feather to Reach the gland more easily.

A. R. Wallace (1869) described a unique mode of tool use by the palm cockatoo of the Aru Islands south of New Guinea. The cockatoo usually feeds on the hard nuts of the canary tree (*Canarium commune*), first making a notch in the nutshell with the lower mandible, then taking a piece of leaf and placing it between the nut and the upper mandible. Finally, the bird inserts its shorter lower mandible into the notch and breaks off a piece of the shell. The leaf apparently keeps the nut from slipping during the final step. Bertagnolio (1994) and Hebel (2002) observed hyacinth macaws (*Anodorhynchus hyacinthinus*) placing wood, nutshells, leaves, or grass in the beak to prevent slippage of foods from the beak during extraction. Schneider, Serbena, and Guedes (2002) described a hyacinth macaw that rolled a leaf around a fruit before placing the fruit under its upper mandible. The bird then scraped off the mesocarp from the fruit by moving its lower mandible. The authors concluded that the leaf placed around the fruit prevented it from slipping from the bird's mouth. Captive hyacinth macaws were reported to use small wood pieces as wedges, placed in the mouth against nuts (Borsari and Ottoni 2005). Borsari and Ottoni hypothesized that they do so to prevent movement or slippage of the nut in the beak, to assist with the application of force, or/and to reduce the impact of opening the nut. The birds position the tool under the upper mandible, placing the nut below it, and move the lower mandible to make a groove in the nutshell. Several pieces of wood were used in the process, but only one piece of wood was placed in the mouth at a time.

Male bowerbirds of the genera *Sericulus*, *Ptilonorhynchus*, and *Chlamydera* are reported to paint their bowers (Frith and Frith 2004). The paint consists of masticated charcoal, fruit pulp, grass, mud, or wood that is Applied to the bower with the bird's bill (A. Marshall 1954). Frith and Frith (2004) noted that the presence of the fresh paint on the bower might serve to indicate both the active maintenance of the bower and the quality of the male and the bower. We consider the paint to be a tool. If the paint truly attracts females, this might better be classified in the Bait, Entice mode.

The male satin bowerbird uses a tool held in the beak for bower painting (Frith and Frith 2004; Gannon 1930). Gannon (1930) found small pieces of bark saturated with a solution of charcoal and saliva in a bower

A hyacinth macaw stabilizes a nut in its beak with a wood wedge. Photograph courtesy of Andressa Borsari.

that had been freshly painted with the same solution. He tentatively concluded that the bark wad was held in the bill to facilitate the mixing or application of the solution. Chaffer (1931) provided observations of the preparation and use of the tool. Bark is nibbled into a wad that is held in the distal half of the bill during painting. Chaffer suggested that the wad absorbs the paint solution, thus allowing retention of more of the solution in the bill than would otherwise be possible. He added that the wad is probably also used as a brush to apply the paint. As will be noted below, these brushes are manufactured tools. The brush and the paint, both tools, are used simultaneously and thus are a Tool Composite. If the brush increases the efficiency of the paint, the brush can be said to be a Metatool. The bower itself, however, is not a tool but a Construction (see table 1.2 and chapter 7).

White-breasted nuthatches engage in a nest defense behavior termed "bill sweeping" (Kilham 1968, 1971; Pravosudov and Grubb 1993). The birds hold crushed insects in their bills and sweep the inside or the outside of the nest with the insects. Pravosudov and Grubb (1993) examined a beetle (*Meloe angusticollis*) used for this purpose and found that the bugs secrete an oily fluid when handled. Kilham (1968, 1971) argued the scent might serve to deter tree squirrels from the bird's nest.

P. Cole (2004) reported that a captive American crow placed wads of paper towel over containers holding foods or non-food objects, obscuring the containers' contents. C. Savage (1995) noted that several species of corvids, including crows, magpies, and ravens, hide food by burying it in a hole in the ground, sometimes carrying and placing stones or leaves on these caches to further obscure the food. Heinrich (1999) described the same behavior in ravens, which either placed food in crevices or holes, or dug holes with their beaks and then picked up nearby debris such as grass, leaves, or sticks to place over the hole. Several different species of nuthatches use bark to cover their food stores (Kilham 1974; D. Petit, Petit, and Petit 1989; Pranty 1995). We do not regard food caches as tools, but objects placed or Draped over caches to obscure them are considered tools.

Bekoff (2009) observed several black-billed magpies (*Pica hudsonia*) retrieving grass and placing it on the body of a dead magpie. In what he interpreted as grieving, Bekoff described how the birds "stood vigil" by the corpse before flying off, leaving the body and the grass behind (Bekoff 2009, 82).

Skutch (1996) reported that some gulls and terns (Laridae) placed a stone in their nest when there were only two eggs in the clutch instead of the typical three. Conover (in Skutch 1996) reported that the stone "eggs" were similar in size and appearance to true eggs. He found that gulls with three eggs, even if one was fake, stayed seated on the eggs longer than if only two were present. Skutch conjectured that the presence of three objects makes the incubation more conformable for the mother, and results in an increase in incubation efficiency.

TOOL MANUFACTURE: FISH, AMPHIBIANS, REPTILES

We found no cases of tool manufacture by fish, amphibians, or reptiles.

TOOL MANUFACTURE: BIRDS

Tool manufacture has been reported in 13 families of birds: Ardeidae, Cacatuidae, Corvidae, Pachycephalidae, Paridae, Picidae, Psittacidae, Ptilonorhynchidae, Ramphastidae, Sittidae, Sturnidae, Thraupidae, and Timaliidae.

Detach

Drop

Finch (1982) reported that a sulphur-crested cockatoo Detached small, leafy twigs from a tree before Dropping them on a pair of bat hawks. A hooded crow Detached twigs to Drop on humans approaching her young. Caffrey (2001) observed two cases of American crows Detaching objects, including pine cones later Dropped on a human, and flower petals Dropped on a brood mate. The ravens described by Janes (1976) apparently dug up stones to Drop on intruders. Another raven Detached a tuft of grass and Dropped it on an incubating black-legged kittiwake (Montevecchi 1978). Heinrich (1988) reported that ravens Detached loose twigs and bark and let them fall to the ground near humans in what may have been an expression of frustration.

Bait, Entice

S. Robinson (1994) reported that an Amazonian striated heron Detached a twig from an overhanging branch before using it as Bait to capture a fish.

Pry, Apply Leverage

Brown-headed nuthatches Detach bark scales from tree trunks or limbs, which they then use to Pry up other bark scales while foraging for insects (Morse 1968).

Jab, Stab, Penetrate

A Steller's jay Detached a stick from a bush and used it to Jab at an American crow (Balda 2007).

Reach

Captive blue jays observed by T. Jones and Kamil (1973) Detached strips from newspapers and used them to Reach food.

Insert and Probe

Galápagos woodpecker finches and mangrove finches Detach twigs from branches before using the twigs to Probe and Pry in holes and crevices (Beck in Bowman 1961; Curio and Kramer 1964; Eibl-Eibesfeldt 1961; Eibl-Eibesfeldt and Sielmann 1962; Gifford 1919; Hundley 1963; Lack 1947; Millikan and Bowman 1967). Curiously, although cactus spines are used as tools in the same manner as twigs, there are no reports of the spines being Detached.

A pale-mandibled aracari was observed Detaching a twig to Probe for insects (Berg in Skutch 1996). A crested shriketit Detached a twig before using it as a Probe (B. Richards 1971). A mountain chickadee Detached a wooden splinter from a tree to Probe into a crack (Gaddis 1981). An American crow pecked to Detach a triangular piece of wood from a fence rail, further modified it through Reshaping, and used it to Probe into a hole for a spider (Caffrey 2000).

New Caledonian crows Detach portions of the leaves of screw pines to make their *Pandanus* Probing tools. G. Hunt and Gray (2004b) provided the first observations of the manufacture of *Pandanus* tools. The birds make a series of cuts with their beaks across the fibers of the leaf, along with a ripping action moving parallel to the leaf fibers and stem. The cutting and ripping begins at what will be the tapered end of the tool, and then the bird shapes the wide end of the tool. G. Hunt (2000a) argued that these crows shape their stepped tools by two-dimensional sculpting, following a "rule system" allowing them to create tools with a high level of standardization in terms of size and shape. The creation of the tapered, stepped tool from the leaf requires very complex Detachment. Careful positioning of the bill combined with control of the length and width while creating the tool are essential. G. Hunt argued that the final shape of the stepped tool is determined by the bird before manufacture begins, and that this shape is functional for birds in Probing. The leaf from which a *Pandanus* tool is Detached is a matched counterpart of the tool and, as noted above, can provide information about manufacturing techniques in much the same way that stone cores can be examined for information regarding the manufacture of flakes.

New Caledonian crows also Detached dead leaf stems from trees and used the stems to extract long-horned beetle larvae from dead wood (G. Hunt 2000b). Troscianko, Bluff, and Rutz (2008) observed a New Caledonian crow Detaching grass stems before using the tools to drive lizards from crevices in a fence post. Kenward et al. (2005) saw a captive New Caledonian crow use a cut-tear-cut action to Detach part of a leaf to make a tool for retrieving a food reward, but they noted that the behavior did not resemble the stepped method used by wild New Caledonian crows to make *Pandanus* tools.

Scratch, Rub

A captive African grey parrot was reported to "prise" splinters off the pine posts in his cage in order to use them for Scratching (Janzen, Janzen, and Pond 1976).

Table 3.1

Modes of tool use across bird families. Families with no reported cases of tool use are not shown. The mode Symbolize is not shown; no bird is reported to use a tool symbolically.

Family	*Common names*	*Drop*	*Throw*	*Drag, etc.*	*Brandish, etc.*	*Bait, Entice*	*Club, Beat*	*Pound, Hammer*	*Pry, Apply Leverage*	*Dig*	*Jab, etc.*	*Reach*	*Insert and Probe*	*Scratch, Rub*	*Cut*	*Block*	*Prop and Climb, etc.*	*Hang*	*Contain*	*Absorb*	*Wipe*	*Apply, etc.*
Megapodiidae	megapodes (includes Australian brush turkey)			X																		
Ciconiidae	storks												X						X			
Ardeidae	herons, bitterns					X																
Phalacrocoracidae	cormorants, shags																					X
Accipitridae	kites, hawks, eagles, Old World vultures, buzzards	X	X			X	X	X														
Pandionidae	ospreys	X																				
Eurypygidae	sunbitterns					X																
Gruidae	cranes																				X	
Scolopacidae	sandpipers, snipes, bristle-thighed curlews		X																			
Haematopodidae	oystercatchers												X									
Laridae	gulls, terns, skimmers			X		X																X
Columbidae	pigeons, doves											X					X					
Psittacidae	parrots									X		X		X					X			X
Cacatuidae	cockatoos	X						X		X				X					X			X
Strigidae	owls					X															X	
Alcedinidae	kingfishers					X																
Coraciidae	rollers					X																
Meropidae	bee-eaters									X												
Ramphastidae	barbets, toucans, aracaris												X									
Picidae	woodpeckers																			X		
Ptilonorhynchidae	bowerbirds					X										X						X
Neosittidae	sittellas												X									
Pachycephalidae	whistlers (includes crested shriketits)												X									
Corvidae	crows, jays	X		X				X	X		X	X	X	X		X	X		X		X	X
	New Caledonian crows									X			X									
Corcoracidae	Australian mudnesters (white-winged choughs)		X					X														
Paridae	tits, chickadees								X				X						X			
Timaliidae	babblers, parrotbills, leiothrix																				X	

Table 3.1 (continued)

Family	*Common names*	*Drop*	*Throw*	*Drag, etc.*	*Brandish, etc.*	*Bait, Entice*	*Club, Beat*	*Pound, Hammer*	*Pry, Apply Leverage*	*Dig*	*Jab, etc.*	*Reach*	*Insert and Probe*	*Scratch, Rub*	*Cut*	*Block*	*Prop and Climb, etc.*	*Hang*	*Contain*	*Absorb*	*Wipe*	*Apply, etc.*
Sittidae	nuthatches								X				X									X
Sturnidae	starlings									X												
Turdidae	thrushes (includes robins, common blackbirds)									X												
Muscicapidae	chats, Old World flycatchers												X									
Icteridae	orioles, blackbirds																			X		
Thraupidae	tanagers (includes woodpecker finches)								X		X		X									

Contain

N. Clayton and Jolliffe (1996) reported that a captive marsh tit Detached paper stickers from its feeder, which it used to transport and store food that adhered to the stickers.

Absorb

A wild male gila woodpecker Detached bits of bark before soaking them in a water and honey solution (Antevs 1948). The woodpecker then carried the soaked bark to his nestlings.

Wipe

A red-billed leiothrix Detached leaves before Wiping itself with them (Gibson in Boswall 1978).

Affix, Apply, Drape

Palm cockatoos and hyacinth macaws Detached leaves that they used to open nuts or fruits (Schneider, Serbena, and Guedes 2002; A. R. Wallace 1869). Captive hyacinth macaws sometimes Detached pieces of wood from their perches to place in their mouths for use as wedges to open nuts (Borsari and Ottoni 2005).

Subtract

Bait, Entice

Higuchi (1986) documented two cases of "lure-making" by striated herons that involved Subtraction. On both occasions, the bird held a twig under its feet and used its bill to break it in half. The bird then used one of the halves for bait-fishing.

Pound, Hammer

The branches used as "drum sticks" by palm cockatoos observed by G. Wood (1984) had freshly colored wood exposed on each end. Wood presumed that the tools had recently been part of a living branch, with the tool first being Detached from the tree and further modified by the Subtraction of the distal end.

Insert and Probe

According to several observers, woodpecker finches Subtract portions of twigs and cactus spines to produce Probing tools of suitable length (Eibl-Eibesfeldt 1961; Eibl-Eibesfeldt and Sielmann 1962; Millikan and Bowman 1967). Further, they Subtract protruding twiglets that would interfere with insertion of the twig (Kastdalen in Bowman 1961), and they remove splinters from pieces of wood used as tools (Eibl-Eibesfeldt and Sielmann 1962). Tebbich and Bshary (2004) conducted an experimental study with captive woodpecker finches, requiring the birds to Subtract obstructions from toothpicks or twigs so that the tools fit into a tube containing a reward. They found that three birds were able to make appropriate modifications by removing the transverse obstructions. However, the birds did not demonstrate the ability to plan the required actions prior to use.

A pale-mandibled aracari Subtracted twiglets from a twig prior to using it to Probe for insects (Berg in Skutch 1996).

Chappell and Kacelnik (2004) found that captive New Caledonian crows disassembled bundles of sticks if their preferred tools were part of the bundle. In their study, the birds also Detached probing tools from a leafy bough, tending to choose tool material whose diameter matched the diameter of the hole through which it had to be Inserted. When first confronted with the task, the crows examined the holes and then flew to the bough and made their tools, suggesting that the crows were matching the diameter of the tool to that of the hole. The crows also Subtracted twigs and other side projections from branches before using them. In a different tool-manufacturing experiment, wild New Caledonian crows Subtracted leaf blades from the stems to use the stems to Probe in holes of varying depth for food (G. Hunt, Rutledge, and Gray 2006). While the birds were able to retrieve the food, there was no evidence that they were matching the length of their tools to the depth of the holes. Instead, the researchers argued that the birds were using a "two-stage heuristic strategy." In the first stage, the (usually unsuccessful) tool was not closely matched to the depth of the hole. In the second stage, the tool used was consistently longer than the first tool. Wild New Caledonian crows snip leaves off stems with their beaks and then use the stems to extract larvae from holes in dead wood (G. Hunt 2000b). G. Hunt (2008) found barbed twigs of *Lantana camara*, used for insect extraction, from which the crows had Subtracted leaves and side branches.

G. Hunt and Gray (2004a) described the highly standardized manufacture of hooked tools made from twigs by New Caledonian crows. Recall that these are distinct from the *Pandanus* tools whose use and manufacture we have described above. To make a hook tool, the birds begin by selecting a branch with a fork formed by two twigs. Using a "snapping technique" with their beaks, the birds Subtract one of the side twigs at a point above the junction (first-order manufacture) and then Detach the final twig tools just below the junction (second-order manufacture). The birds then use their beaks to sculpt and finally sharpen the hook on the Detached piece by precisely Subtracting bits of the wood (third-order manufacture). In some cases, leaves were also Subtracted from the tools during the sculpting process.

Troscianko, Bluff, and Rutz (2008) observed a New Caledonian crow Subtracting material from grass stems to shorten them before using the tools to expel lizards from crevices in a fence post.

In an experimental study, captive rooks Subtracted side branches from sticks before using them to Probe for food rewards from a tube (Bird and Emery 2009a).

Scratch, Rub

An African grey parrot Subtracted bark from a twig before Scratching itself with the twig (G. Smith 1970, 1971).

Add, Combine

Affix, Apply, Drape

From the descriptions provided by Chaffer (1931) and A. Marshall (1954), it appears that satin bowerbirds first Detach fibers from bits of bark and then Combine the fibers to form small ovoid pellets that they use as wads to paint their bowers. Both modifications are done with the beak.

Crows and starlings commonly collect large numbers of ants and Combine them into balls or wads to maximize their usefulness in anting (Simmons 1966).

Reshape

Bait, Entice

A black-crowned night heron observed by D. D. McCullough and Beasley (1996) broke up the piece of bread that it used for bait-fishing. We consider this a form of Reshaping a tool. Similarly, a bait-fishing herring gull shook its bread lure to break it up into smaller pieces (Henry and Aznar 2006).

Insert and Probe

G. Hunt et al. (2007) documented an innovative type of folding of *Pandanus* tools by two distantly related wild New Caledonian crows. The birds were each observed to fold the tool into a shape described as a "boomerang" before using it for extractive foraging. The authors argued that the folding technique emerged separately in each bird and hypothesized that it placed the working tip of the tool directly in the bird's line of sight, making the tool easier to handle.

Betty, a captive New Caledonian crow, spontaneously bent a straight piece of wire to use as a hook to pull in an out-of-reach bucket in nine of ten trials (Weir, Chappell, and Kacelnik 2002). They noted that

A Probing tool made by a wild New Caledonian crow by Subtracting leafy side branches from a stem. Photograph courtesy of Gavin Hunt. Reprinted by permission of the Nature Publishing Group from G. Hunt 1996.

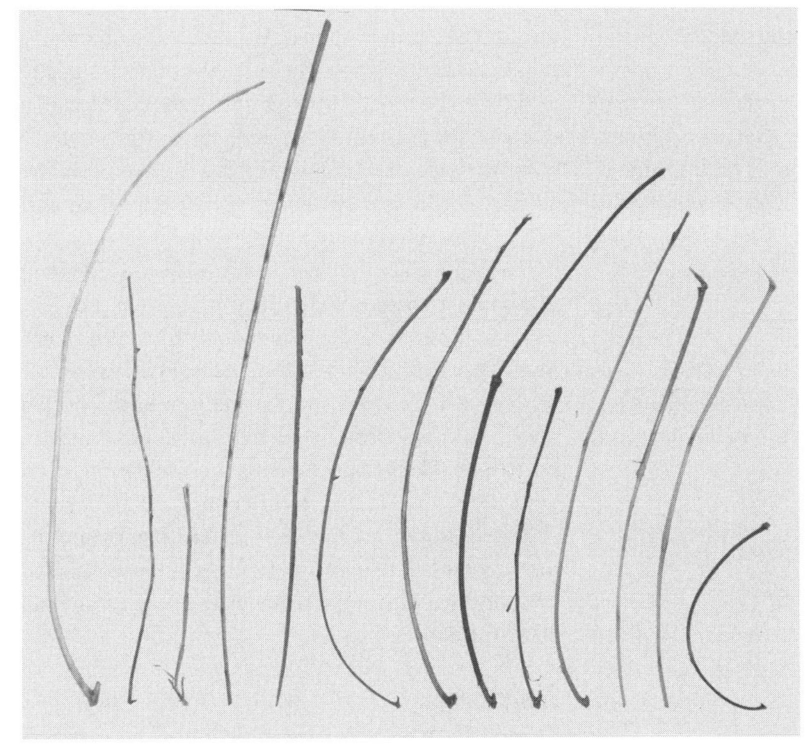

A sample of hooked Probing tools manufactured by New Caledonian crows. Photograph courtesy of Gavin Hunt. Reprinted by permission of the Nature Publishing Group from G. Hunt 1996.

in all but one trial, she attempted to use the tool in its straight form first, before bending it. To bend the straight wire, Betty either wedged one end of the wire into the sticky tape on the apparatus, or held the wire in her foot and pulled the other end with her beak. Weir and Kacelnik (2006) conducted a follow-up study with Betty involving the Reshaping of aluminum, a novel material, into a functional form. In the first of their experiments, she modified the aluminum strip by bending the end she held in her beak and then rotated it to use the modified end to retrieve a bucket in twenty-five of thirty-four trials. In the second experiment, she was presented with an aluminum strip that had been loosely bent on both ends to prevent it from being inserted into a narrow tube holding a food reward. Betty modified the tool on the first trial by squeezing the ends of the strip together, so that the tool was able to fit in the tube, although this may have been done to make the tool easier to pick up, rather than reflecting an understanding of functionality. Weir and Kacelnik (2006, 328) pointed out that after she modified the end, Betty first attempted to Probe with the opposite end, although she "almost instantly" remedied this by rotating it. In the final experiment, Betty was required to unbend a tool in order to increase its length to reach a reward inside a horizontal tube. She succeeded on the second trial. The authors stated that she "raised her head and beak (still holding the tool) in a distinctive and unusual manner, causing the end of the tool to bend backward against the lip of the tube" (Weir and Kacelnik 2006, 329). Despite these modifications, the authors noted that they could not argue that she planned or even understood her actions, as she often attempted to Probe before modifying the tool, or she used the modified tool incorrectly.

Caffrey (2000) observed an American crow that first Detached a wooden splinter with its beak and then used its beak to create a tapered end. Bird and Emery (2009a) documented four captive rooks that spontaneously bent straight pieces of wire into hooks to retrieve buckets containing a food rewards.

Scratch, Rub

Boswall (1983b) described a salmon-crested cockatoo that Reshaped its own detached primary feathers by bending them before using them to Scratch under its bill and on its head.

ASSOCIATIVE TOOL USE: FISH, AMPHIBIANS, REPTILES

The use of Associative Tools is undocumented in fish, amphibians, or reptiles.

ASSOCIATIVE TOOL USE: BIRDS

Sequential Tool Use and Tool Crafting are the only forms of Associative Tool Use unambiguously documented in birds. We do not believe that Tool Sets, Metatools, and Multi-Function Tools have been documented in birds, although several researchers reported what they term Metatool Use (Bird and Emery 2009a; Taylor et al. 2007). Bowerbirds that use bark to apply paint to their bower Construction are using a Tool Composite.

Taylor et al. (2007) described what they called Metatool Use, but we consider as Sequential Tool Use, by seven captive New Caledonian crows. Six of the seven birds first used a short stick to rake in a longer stick and then Inserted the longer stick into a container to acquire a food reward. The authors argued that analogical reasoning might explain these sophisticated tool-using abilities.

Wimpenny et al. (2009) studied Sequential Tool Use in New Caledonian crows. The birds were required to use a stick tool to retrieve a longer tool from inside a tube. The retrieved tool could be used to obtain either a longer tool or a food reward from other tubes. If the retrieved tool was long enough, it could be used to get food. If the retrieved tool was shorter, it could be used to get a longer tool that could then be used to get the food. The most complex condition required the use of three tools sequentially. Each subsequent step required obtaining a longer tool, eventually leading to acquiring the food. Although the authors found that many of the birds could use three tools sequentially to solve the task, they argued that associative processes such as chaining might be at work, rather than analogical reasoning, as suggested by Taylor et al. (2007).

Bird and Emery (2009a) conducted an experiment with captive rooks on what they called Metatool Use, but we again consider Sequential Tool Use. They found that all of the birds, from the first trial, were able to Drop a large stone into a tube to release another stone, and then Drop the second stone into a tube to release a food reward. The birds had the option of using a large stone to access either another larger stone or a small stone, but only the smaller stone fit into the tube containing the reward. Bird and Emery observed that the smaller stone was preferred more often than predicted by chance.

Tool Crafting

As already described, New Caledonian crows Craft their hook tools in a series of three, sometimes four, progressive and different steps that involve Detaching, Subtracting, and Reshaping. Their *Pandanus* tools do not qualify as being Crafted, since the same mode, Detach, is used repetitively, and there appears to be no Reshaping.

4 Non-Primate Mammals

WE HAVE ORGANIZED THIS CHAPTER BY ORDERS, ALTHOUGH NO TOOL BEHAVIOR is reported for some mammalian orders, such as the Chiroptera (bats). Even within the orders for which tool use is observed (rodents, carnivores, insectivores, cetaceans, proboscids, perissodactyls, and artiodactyls), many families are unrepresented among tool users, or many of the reports of tool use for an entire family are one or two isolated cases, such as the dexterous procyanids, where only one example is noted. Sampling effort is dramatically uneven, and we must remember that absence of evidence is not evidence of absence. There are no definitive taxonomic trends with regard to tool behavior among the non-primate mammals: such diverse forms as elephants, bears, sea otters, and beavers exhibit the most frequent and diverse tool use and manufacture. As with birds, there are also no apparent ecological correlates. The taxonomic organization of the chapter is thus simply one of convenience.

TOOL USE: RODENTS

Throw

In experimental settings, California ground squirrels (*Spermophilus beecheyi*) and rock squirrels (*Otospermophilus beecheyi*) Threw substrate at predators such as a domestic cat (*Felis catus*), rattlesnakes (*Crotalus* spp.) and a Pacific gopher snake (*Pituophis melanoleucus catenifer*) (Coss and Biardi 1997; Owings et al. 2001).

Drag, Roll, Kick, Slap, Push Over

Captive California ground squirrels Kicked sand into the faces of garter snakes (*Thamnophis* spp.), predatory gopher snakes, and rattlesnakes with "a forward-thrusting movement of their forepaws" (Owings, Borchert, and Virginia 1977, 229; Owings and Coss 1977). Squirrels also Kicked sand into burrows in which interactions between squirrels and snakes had recently occurred (Coss and Owings 1978; Hennessy and Owings 1978), and they Kicked sand in response to the recorded rattles of rattlesnakes (Rowe and Owings 1978). When directed toward snakes, the behavior seemed to cause the snakes to retreat. It was also recorded in the only observed interaction between a wild snake and a wild squirrel in this study.

Brandish, Wave, Shake

Eurasian beavers (*Castor fiber*) use objects during agonistic displays with rivals, most often at territorial borders. Thomsen, Campbell, and Rosell (2007) described a behavior termed "stick display" in which a beaver picks up an object, most often a stick, stands on its back legs, and then rapidly moves its body up and down as it holds the object with its mouth and front paws. This frequently occurs while the beaver is in shallow water, resulting in some splashing sounds. However, the authors concluded that this is primarily a visual, rather than an acoustic, display. During a study period of 175 nights involving twenty-eight beavers, stick displays were documented 131 times. The displays were performed by eleven different individuals, both male and female.

Dig

Holding a tool in her forelimbs, a female pocket gopher (*Thomomys bottae*) used stones and hard chunks of food, about 15 centimeters in diameter, to repeatedly Dig in soil during burrow excavation. The behavior served to both loosen and move the soil (S. Katz 1975, 1980).

Reach

In an experimental setting that involved extensive training, degus (*Octodon degus*) learned to use a rake to acquire food that was beyond their reach (Okanoya et al. 2008). The authors reported that the degus demonstrated functional understanding of the task. We include this behavior despite its being trained, because of the rarity of rodent tool use.

Block

Several species of ground squirrels (*Urocitellus columbianus, Ictidomys tridecemlineatus, I. mexicanus*) are known to plug their own burrows (MacClintock 1970 in I. McLean 1978). However, McLean reported that female Columbian ground squirrels (*U. columbianus*) selectively plugged some holes but not others, and also plugged holes that would be used in the future. Pregnant females were observed to carry nesting material into an underground chamber and then subsequently plug the entrance hole to that chamber with dirt. Plugs were left in burrow entrances for varying intervals, but they might be removed during the day and replaced at night. McLean (1978, 439) proposed that this Blocking behavior was not a strategy to protect infants from predators or weather, but rather was a means of deterring attacks by "non-resident conspecifics."

Captive beavers (*Castor canadensis*) successfully Blocked an overflow drain in their exhibit, causing their enclosure to be flooded with 15 centimeters of water (Pilleri 1983). The overflow pipe was fitted with a cap that contained three small holes, each measuring 0.8 centimeters in diameter. The beavers inserted one stick to fit exactly into each hole, and Pilleri (1983, 99) noted that the calibration of these tools "was so accurate that the sticks could be replaced in order to dam up the pool again." Even though the sticks were removed by the caretakers, the beavers made new plugs and repeated this performance several nights in a row. Ultimately, they abandoned the sticks and covered the drain with grass, leaves, and mud. Their persistence required the caretakers to replace the drain with a large filter instead.

Naked mole-rats (*Heterocephalus glaber*) use tools as they excavate their tunnels and chambers. These rodents remove substrate with their procumbent incisors, and captive individuals have been observed to place a barrier (such as a wood chip or piece of husk from a tuber) "behind their incisors and in front of their lips and molars" as they work (Shuster and Sherman 1998, 72). The authors concluded that this Blocking behavior likely prevented choking or aspiration of the debris. If the tool moved while being used, it was repositioned or replaced. This use of a Blocking tool was only associated with substrates that resulted in "fine particulate debris" (Shuster and Sherman 1998, 71).

Prop and Climb, Balance and Climb, Bridge, Reposition

Zimmerman (1952) described a recently captured female harvest mouse (*Micromys minutus*) repeatedly Propping and Climbing an oat stalk to escape from its aquarium. No training was involved.

D. Barnes (2005) reported finding some willow (*Salix* sp.) branches that had been cut 1 meter above the ground by a wild North American beaver. He characterized this as an "extraordinary height," given that most beavers cut at a height of around 30 centimeters from the ground (Barnes 2005, 441). Closer inspection revealed a willow stem, approximately 12 centimeters in diameter and cut by a beaver at both ends, Propped at a 45 degree angle against the willow clump. Barnes surmised that a beaver cut, transported, and Propped the stem in place and Climbed on it to facilitate reaching desirable branches. In a case that shows that tool use is not always the best solution, P. Richard (in Chauvin and Muckensturm-Chauvin 1980) hung a piece of bread in a tree in a captive beaver's enclosure and made a pile of sticks beneath the treat so that the beaver could reach it. The beaver readily climbed the stack and got the bread. Then Richard disassembled the pile, leaving the unstacked branches beneath the bread in the tree. He expected the beaver to restack the branches to get the bread. Instead, the beaver simply cut down the tree!

Affix, Apply, Drape

While foraging, wild wood mice (*Apodemus sylvaticus*) placed small, conspicuous objects around their home range. Stopka and Macdonald (2003) concluded that these objects serve as visual reference points for the mice, allowing them to accurately orient themselves in a large area. Additionally, the use of objects rather than scent may be more durable and less likely to attract predators.

Rice-field rats (*Rattus rattoides*) self-anoint, but they only Apply the anal gland secretions of the weasel (*Mustela sibirica*), one of their predators (Xu et al. 1995). The behavior is not correlated with either age or sex and is presumed to function as a means of concealment from the weasels. California ground squirrels, rock squirrels, and Siberian chipmunks (*Eutamias sibiricus*) anoint themselves with the scent of rattlesnakes (Clucas, Owings, and Rowe 2008; Clucas et al. 2008; Kobayashi and Watanabe 1986; Owings et al. 2001). The California ground squirrels acquired the scent from chewing the shed skin of the snakes (Clucas, Owings, and Rowe 2008; Clucas et al. 2008), rock squirrels Applied substrate that had been in contact with the snakes (Owings et al. 2001), and Siberian chipmunks anointed themselves with nibbled snake skin from a carcass or with snake urine. In all cases, this anointing behavior likely produced a form of "olfactory camouflage" (Clucas, Owings, and Rowe 2008, 855), functioning to prevent predation from rattlesnakes (see also Clucas et al. 2008; Kobayashi and Watanabe 1986).

TOOL USE: CARNIVORES

Throw

Harington (1962) and Perry (1966) reviewed accounts in Eskimo nature lore and explorers' logs of polar bears (*Thalarctos maritimus*) Throwing down chunks of ice or rocks on resting walruses and seals. The missiles either kill the pinnipeds or injure them sufficiently to prevent their escape. Kiliaan (1974) provided persuasive circumstantial evidence of the same behavior and relayed Jonkel's inference that polar bears use rocks to spring foot snares. Perry (1966) noted that young polar bears at the London Zoo were seen to Throw ice chunks in a context that was unclear, and a polar bear at the Edinburgh Zoo was observed to Throw a horse femur repeatedly at its mate. Captive-born polar bears at the Brookfield Zoo Threw a variety of objects, including logs and large aluminum beer kegs weighing about 7 kilograms, from a bipedal stance. A 20-month-old female was once observed to carry a keg up a steep stair from a deep moat, transport it several meters across the grotto surface, and then lift and Throw it down forcefully. The keg never touched the substrate during the sequence and was carried deftly between the medial aspect of her left forelimb and the left lateral aspect of her head and neck. The zoo observations demonstrated that polar bears are strongly disposed and physically able to manipulate, pick up, and Throw inanimate objects. The frequency of Throwing and the size and weight of thrown objects increase with age (Pluta and Beck 1979). Further, the bipedal Throwing posture of the Brookfield bears is virtually identical to the Throwing postures of bears portrayed in an ivory carving and an engraving included in Harington's paper (1962).

Van Lawick–Goodall (1970) reported that two wild-caught banded mongooses (*Mungos mungo*) Threw stones between their hind legs at a model of an ostrich egg.

Drag, Roll, Kick, Slap, Push Over

Eskimo hunters claim that polar bears camouflage their black noses by pushing along a piece of ice in front of themselves as they stalk. However, observation of a number of polar bear hunts provided no confirmation of this behavior (Stirling 1974; Stirling and Latour 1978). Kiliaan (1974) convincingly described circumstantial evidence of a polar bear that Rolled a 20-kilogram chunk of ice 6.5 meters to a seal breathing hole, where it appeared to be used as a tool for hunting (see Throw for a description).

Bait, Entice

Perry (1966) suggested an unconfirmed form of tool use by polar bears. The bears are said to capture and

An antique illustration of a wild polar bear Throwing down a rock or chunk of ice on a walrus. Reprinted from C. F. Hall, *Arctic researches, and life among the Esquimaux.* New York: Harper and Brothers, 1865.

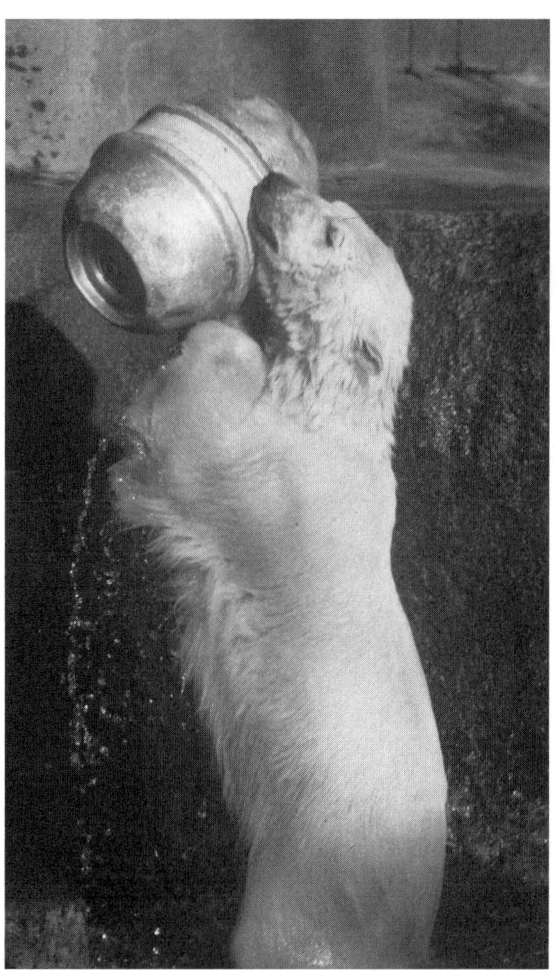

A polar bear at the Brookfield Zoo picks up a beer keg from a bipedal position. Photograph courtesy of Benjamin Beck.

injure sleeping seal pups and then let them struggle in the water. The bear captures the pup's mother as she is Baited toward her wounded young.

In the course of observations of a pack of wolves (*Canis lupus*) at the St. Louis Zoo, an adult female was seen to approach one of her pups that had a freshly killed rat. The pup defended the prey and bit the mother on the nose. The mother moved to another area of the enclosure where she dug up a piece of meat, which she carried to the pup and dangled over his head. As relayed by the author, "the pup attempted to seize the meat, but the bitch did not give it up" (M. Fox 1971, 641). With the pup following, the female moved to a far corner of the enclosure and dropped the meat, which the pup took. The mother then returned to the rat and ate it.

Pound, Hammer

Sea otters (*Enhydra lutris*) provide one of the best documented forms of animal tool use. Sea otters, particularly from California coastal populations, Pound hard-shelled mollusks on rocks that they balance on their chests or abdomens. Hall and Schaller (1964) and van Lawick–Goodall (1970) stated that sea otters are the only non-primates that use tools habitually, but we now know that this is not true in general, and for sea otters this claim may have to be qualified by study site or population.

Sea otters are the only mustelid carnivores to have retractile claws on their front feet. This is considered an adaptation for manipulation (Kenyon 1969). Indeed, sea otters utilize their forelimbs minimally in locomotion (E. Fisher 1939; Kenyon 1969). There is also a loose flap of skin between each forelimb and the chest, forming axillary pouches in which food and inanimate objects can be transported (Kenyon 1969).

E. Fisher (1939) first described tool use by sea otters, noting that California sea otters Hammered small mollusks on rocks resting on their chests. Several hard blows were required to break a mollusk. The otters got the rocks from the ocean floor. Murie (1940) confirmed Fisher's observations of tool use by California sea otters but did not observe the behavior among Aleutian archipelago populations. However, he felt in retrospect that he had heard Aleutian otters breaking mussels on stones. Limbaugh (1961) also described the Pounding of food on rocks and flat stones balanced on

A polar bear at the Brookfield Zoo Throws a beer keg from a bipedal position. Her posture and technique are similar to that of the wild bear (see illustration on previous page). Photograph courtesy of Benjamin Beck.

the abdomen by California sea otters: the rocks were 6.4 to 8.9 centimeters in diameter, and flat stones were up to 15.2 centimeters in diameter. Limbaugh noted that stone tools were retained and carried on multiple dives.

Hall and Schaller (1964) made detailed quantitative observations of the behavior near Monterey, California. The primary food exploited by this tool use was the California mussel (*Mytilus californianus*). The otter surfaced with a single mussel and a stone, rolled so that it was floating on its dorsum with the stone balanced on its ventrum, and then immediately began to Pound the flat side of the mussel against the stone. The mussel was held with both forefeet. The stones were smooth, but varied considerably in shape, and were estimated to weigh between 468 and 666 grams.

Hall and Schaller also noted several variations on the typical pattern. Sea urchins (*Strongylocentrotus* spp.), unidentified crabs, and spiny lobsters (*Panulirus interruptus*) occasionally were Pounded on stones. An unidentified food item was Pounded on an abalone shell. Food was infrequently pounded directly on the ventrum, and coral algae (*Corallina* spp.) were balanced on the ventrum and pounded with the forefeet in seven cases. On eleven occasions an otter pounded on its ventrum with its forefeet without having either a stone or a food item.

Wade (1975) reported that California sea otters opened pismo clams (*Tivela stultorum*) by Pounding them against another clam shell balanced on the ventrum.

Murie's (1940) speculation that Aleutian sea otters also open mollusks by Pounding them on stones was substantiated by Krear and by Lensink (in Hall and Schaller 1964). These correspondents saw only immatures use tools. Hall and Schaller hypothesized that since the mussels available to Aleutian otters are relatively small, adults can break the shells with their teeth. Younger animals, however, must resort to Pounding them on stones. R. Jones (1951) noted that aged Aleutian otters begin to Pound urchins on their chests when their teeth are worn and eroded. Thus Pounding in this population appears to be limited primarily to very young and very old individuals, and therefore is observed infrequently. Krear (in Hall and Schaller 1964) saw one adult carry a stone during a prolonged feeding period without using it as a tool. Kenyon (1969) and Lensink (in Hall and Schaller 1964) saw subadult Aleutian otters Pound bits of coralline algae or rocks together in what appeared to be play. Otters captured in the North Pacific readily opened large clams in captivity by Pounding them on rocks, against the side of their cement pool (both proto-tool use), or against shells balanced on the ventrum (tool use) (Kenyon 1969).

Additionally, sea otters may employ two rocks simultaneously as hammer and anvil, or use a piece of broken shell or crab carapace as a hammer on the same prey from which it was removed. While rocks are the tool used most often, "empty shells, driftwood, empty glass or plastic bottles, aluminum beverage cans, and other manmade objects discarded by humans" are also employed (Riedman and Estes 1990, 33). Both Alaskan and Californian populations have been observed to Pound two live clams together to break the shells (Riedman and Estes 1990). From a developmental perspective, successful tool use was first observed to occur at nineteen-and-one-half weeks of age when an otter cracked open a turban snail (*Tegula* sp.) with a rock hammer (Davis 1979 in Payne and Jameson 1984). Pups ranging from twenty-two to twenty-four weeks of age were also documented to successfully use a tool with prey that they captured on their own (Payne and Jameson 1984).

It would seem that shellfish Pounding is (or at least was at the time of these observations) customary in the California sea otter population and habitual in the Aleutian population. More information on prey and stone availability, and on the ontogeny of the behavior, is required before the behavior could confidently be designated as cultural.

Abalones (*Haliotis* spp.) are common in the diet of California sea otters. The otters need not use stones to open abalones because they are not bivalves; the edible portion is directly accessible with teeth and forefeet. However, abalones cling tenaciously to subsurface rocks and the manner in which otters remove them has been debated. E. Fisher (1939) presciently suggested that the otters Hammer the abalones with rocks to loosen their grasp. Cox (1962) and Ebert (1968) noted that the shells of abalones eaten by otters have characteristic breaks and scars that might have resulted from Hammering. Cushing (1939) was skeptical and suspected that the otters approached the abalones stealthily and simply pulled them off the substrate when they were not firmly attached. Houk and Geibel (1974) finally supplied first-hand scientific observations of a sea otter Hammering an abalone from the underwater substrate with a stone. Thus sea otters use stones not only as portable anvils, but also as true Hammers (see observations in the Pry, Apply Leverage mode). Kenyon (1969) noted that a captive sea otter loosened the cover of an underwater drain by Pounding it with a rock and had similarly damaged the side of its cement pool.

These accounts are striking for the otters' frequency of retaining and transporting rocks between separate instances of use on the surface (Houk and Geibel 1974). If suitable rocks were scarce, retention would eliminate the need to locate a new tool for each food object. The retention and reuse of hammer stones and anvils by chimpanzees is frequently cited as an indicant of cognitive complexity (see chapters 6 and 7).

Pry, Apply Leverage

Riedman and Estes (1990) noted that sea otters used rocks, slabs of flat concrete, and a glass bottle to Pry urchins, abalone, and oysters off of rocks or out of crevices.

Reach

Spectacled bears (*Tremarctos ornatus*) at the Basel Zoo used long sticks to Reach leaves and fruit growing in trees outside of their enclosure, as well as to Reach bread and a heron in the surrounding water moat (Lang 1974). Dittrich (1983) observed a bear (genus unspecified) at the Basel Zoo use a stick to retrieve a piece of bread from a water moat.

Scratch, Rub

An adult female lion (*Panthera leo*) living in the Waza National Park in Cameroon was immobilized and fitted with a radio collar. During her recovery from anesthesia, an *Acacia* thorn became imbedded in a front paw. Two days later, the thorn was causing her obvious discomfort, and she attempted to remove it with her teeth. Failing to do so, she was observed to move her mouth along the ground until she picked up a large thorn, which she held between a pair of her canines. For at least thirty minutes, she Rubbed this large thorn against the hurt paw, presumably attempting to remove the painful thorn. When subsequently observed, her paw appeared healthy and she showed no signs of discomfort (H. Bauer 2001).

Hart's observations, relayed by van Lawick-Goodall (1970), described an unusual form of tool use by a female cocker spaniel (*Canis familiaris*). The dog placed small hard objects, such as marbles, against her maxillary incisors and pulled the matted hair of her paws between the object and her tongue or teeth, presumably as a form of personal grooming. P. Larson (pers. comm.) reported a pet dog that routinely used a paperclip to Scratch between her toes.

Eisenberg and Kleiman (1977) noted that giant pandas (*Ailuropoda melanoleuca*) at the Smithsonian National Zoo cleaned their undersides by Rubbing with clods of sod or soil held in their forepaws.

Block

North American badgers (*Taxidea taxus*) hunt Richardson's ground squirrels (*Urocitellus richardsonii*) and have been observed to actively Block entrances to the squirrels' burrows to assist in capturing them. These badgers used three specific methods to plug the entrances. In the first (72% of 391 blockages), they used material such as soil, vegetation, or snow that was immediately adjacent to the entrance. In the second (22% of 391 blockages), the badgers moved soil from a nearby mound. In the third (6% of 391 blockages), nearby objects such as wood blocks, clay bricks, or a rock were used (Michener 2004).

Prop and Climb, Balance and Climb, Bridge, Reposition

Dathe (1961) saw a captive brown bear (*Ursus arctos*) push a small keg against a wall of its enclosure and stand bipedally on top of it to get closer to visitors from whom it was begging food.

Wood blocks and soil moved by a badger to plug the entrance of a ground squirrel burrow, presumably to facilitate predation. The line shows the path over which the badger dragged the soil, and the circled numerals indicate the original locations of the blocks. Photograph courtesy of Gail R. Michener.

Contain

Sea otters wrapped (and therefore immobilized) live crabs in kelp that was draped over their abdomen as they consumed another food item (Riedman et al. 1988 in Riedman and Estes 1990).

A female springer spaniel used her Frisbee as a container to carry a hockey puck. When playing with these items, she regularly placed the puck on the Frisbee, allowing her to carry and monopolize both toys (S. Carter, pers. comm.).

Affix, Apply, Drape

Self-anointing with bodily substances might properly be called tool use when the substance is held and actually applied by the animal, and a purpose is claimed or clearly documented. European hedgehogs (*Erinaceus europaeus*) self-anoint with their own strong-smelling saliva (Brockie 1976; D'Have et al. 2005). Upon encountering an odorous object or substance such as a toad, rotten meat, or urine, they mouth or lick it vigorously. The substance is mixed in the mouth with copious amounts of saliva. The resultant frothy mass is then licked or flung onto the spines with the tongue (Burton 1957; Eisenberg and Gould 1966; Eisentraut 1953; Herter 1972). This behavior is known as *Selbstbespucken*, which has been politely but imprecisely translated as self-anointing. Eisenberg and Gould (1966) noted similar behavior in the arboreal tenrec (*Echinops telfairi*). Hedgehog self-anointing is sex dependent; males self-anoint twice as often as females (D'Have et al. 2005). Although the precise function of this behavior is unclear, it is generally assumed to allow for social signaling in various contexts, predator deterrence, or both (Brockie 1976; Brodie 1977; D'Have et al. 2005). Coatis (*Nasua nasua*) enthusiastically self-anoint with the resin of the *Trattinnickia aspera* plant, vigorously Applying the secretion into their fur. The function however, remains unclear (Gompper and Hoylman 1993).

E. Fisher (1939), Kenyon (1969), and R. Jones (1951) noted that sea otters Apply strands of kelp to their bodies to add buoyancy and stability during prolonged periods of floating rest or sleep.

Rensch and Dücker (1959 in Rensch 1973) reported

A pet spaniel uses its Frisbee toy to carry a hockey puck. Photograph courtesy of Sherry Carter.

that during play, a mongoose (*Herpestes edwardsi*) ran around after putting its head into a paper bag.

TOOL USE: PROBOSCIDS, ARTIODACTYLS, AND PERISSODACTYLS

Drop

Wild African elephants (*Loxodonta africana*) Dropped rocks and logs on electric fences. This either breaks the fence or short-circuits the wires by pushing them into contact with the earth (Poole 1996 in Holdrege 2001).

Throw

I. Douglas-Hamilton (Douglas-Hamilton and Douglas-Hamilton 1975) described how Boadicea, a large matriarch African elephant that was a familiar figure in his Tanzanian field study, Threw a log toward him. Although the log whizzed past his head and struck his vehicle, he did not conclude that the log was aimed. Throwing dust, branches, and mud mixed with pebbles at intruders has been reported for wild African elephants by Hendrichs and Hendrichs (1971). Poole (1996, 139, in Holdrege 2001) stated that African elephants aimed and Threw "large stones, sticks, a Kodak film box, my own sandal, and wildebeest bone." She added that "elephants have been known to intentionally throw things at each other . . . during escalated fights and during play." Large bull elephants have picked up young bulls and Thrown them at fences. The resulting damage to the fence allowed the elephants to cross (Chevalier-Skolnikoff and Liska 1993).

Wild Asian bull elephants (*Elephas maximus*) have been documented to Throw branches toward jackals and a leopard on several occasions (Kurt and Hartl 1995). A wild African elephant female agonistically aimed and Threw watery mud, clumps of grass, and a bundle of forbs at a white rhinoceros (*Ceratotherium simum*) that attempted to approach a small and temporary puddle being used by the elephant (Wickler and Seibt 1997). Hart, Hart, and Pinter-Wollman (2008, 88) noted that elephants have Thrown objects "at rodents competing for fruit beneath a tree."

Captive African and Asian elephants aimed and Threw stones, soil, branches, grass, and feces toward humans and large novel objects (Kühme 1962, 1963; B. Beck, pers. obs.). The missiles were thrown with the trunk. Kurt and Hartl (1995, 317) stated that when Throwing, elephants "can aim objects accurately up to distances of 20 meters." Kühme (1962, 1963) noted increased accuracy with practice. Kurt (1991 in Kurt and Hartl 1995, 317) stated that Asian bulls in musth Throw stones and "prepared sticks" at mahouts, sometimes causing injuries such as fractured bones. R. McDonald (pers. comm.) observed a captive African elephant female Throw mud at a rhinoceros in an adjacent enclosure in contexts that were unclear. A female Asian elephant was observed to Throw dung and straw into an adjoining stall (Proske in van Lawick-Goodall 1970). An Asian bull, apparently out of annoyance, Threw fecal boluses at people, machines, and an automobile in the public space beyond his dry-moated enclosure (B. Beck, pers. obs.). African elephants blow air (which we consider to be Throwing) or Throw dirt and objects at birds attempting to eat their food (Chevalier-Skolnikoff and Liska 1993) and Asian elephants blow sand in unspecified but aggressive contexts (Kurt and Hartl 1995).

Drag, Roll, Kick, Slap, Push Over

Grzimek (1956) and Kühme (1962, 1963) noted that captive African elephants propelled soil toward rats, dogs, and humans with their forefeet. Asian and African elephants aggressively Rolled objects at small animals (Chevalier-Skolnikoff and Liska 1993; Kurt and Hartl 1995).

Brandish, Wave, Shake

Hart, Hart, and Pinter-Wollman (2008) cited W. Harris (1838) as likely providing the first example of tool use in elephants when he reported "fly switching" (fly swatting) with branches. C. Darwin (1871) relayed observations of identical behavior by a wild Asian elephant and by Asian working elephants. Peal (1879) was riding an Asian elephant when he noticed its helpless torment due to the presence of large flies. He allowed the elephant to move to the side of the road where it broke off a long leafy branch and fanned itself to repel the flies for the remainder of the journey. Rensch and Altevogt (1954) saw Asian working elephants Wave leafy branches to chase flies from parts of their bodies that were unreachable with their trunks. Reid (1985) reported that a working Asian elephant grasped a small bundle of straw from the load it was carrying. The elephant flicked it against its ear, apparently driving away or swatting an annoying insect. Hart and Hart (1994) and Hart et al. (2001) described fly switching as a very common behavior among both wild and captive Asian elephants, as a means of repelling biting flies. Chevalier-Skolnikoff and Liska (1993) and Poole (1996 in Holdrege 2001) also characterized fly swatting by African elephants as a common behavior. Hart and Hart (1994) documented that elephants reduced the number of flies by 43 percent with fly switches. The authors also specifically noted that elephants appeared to save preferred branches for switching, and laid them down nearby if they performed other activities such as eating.

Both Asian and African elephants aggressively Brandish objects at small animals (Chevalier-Skolnikoff and Liska 1993; Kurt and Hartl 1995). Van Lawick–Goodall (1970) reported two cases in which African elephants tore off and Waved branches in their trunks in response to the approach of an automobile. When foraging, they Shake branches in order to loosen ripe fruit, which then falls to the ground and is eaten (Kurt and Hartl 1995).

A female wild horse (*Equus ferus*) was observed to grasp and lift sagebrush with her mouth and Brandish it at more dominant individuals, resulting in displacement. Among the 135 individuals in the population, two other horses within her band subsequently exhibited the same behavior (Berger 2008).

Club, Beat

Reid (1985) relayed an observation from 1852 (also recorded by Tuker 1953) of Asian working elephants refusing to cross a rapidly flowing river. Despite urgings and threats by humans, they refused to move. Eventually, the "master elephant" was brought to the scene. He was described as using the "tremendous thick chain" that was attached to his trunk to Beat the recalcitrant elephants in order to make them move (Reid 1985, 402). This account suggests that he did this of his own volition.

Chevalier-Skolnikoff and Liska (1993) noted that an African elephant held a stick in its trunk and used it to hit a human in an unspecified context.

Dig

Campbell (in Boswall 1977b) saw a horse, presumably *Equus caballus*, rake snow with a stick.

Reach

Peal (1879) reported that a newly captured elephant ripped up a fence stake with its trunk and broke off a piece beneath its foot. The elephant then Reached into its left axilla with the stick and used it to dislodge a leech. Peal felt that the animal could not have reached the area directly with its trunk to remove the leech. He later saw similar behavior by other elephants. Shoshani and Eisenberg (2000, 136) stated that "there are numerous observations of elephants using a tree branch to scratch themselves in places where their trunk and tail would not reach." This is substantiated by Chevalier-Skolnikoff and Liska (1993), Kurt and Hartl (1995), and Williams (1950 in Holdrege 2001).

Bierens de Haan (1931) and Rensch and Altevogt (1954) cited an observation by Zedtwitz of a captive African elephant Reaching toward distant food with a twig that it held in its trunk. Bierens de Haan (1931) skeptically noted an additional observation by an unnamed zoo visitor of similar behavior by a different elephant. Using a methodology that has been most often applied to monkeys and great apes, Nissani (2004) provided two elephants with out-of-reach food within a horizontal tube. Both subjects were able to either blow or suck with their trunks to obtain the reward.

While some may argue the logic of air being used as a tool, in this case it functioned in the same manner as a stick.

Insert and Probe

Laursen (1975) provided a third-hand observation that a captive elephant Inserted and Probed with a stick or piece of straw in its own temporal gland. Chevalier-Skolnikoff and Liska (1993) provided verification of this behavior, with a case of a captive elephant that used a stick to probe in its temporal gland.

Scratch, Rub

A wild African elephant Scratched its leg with a twig held in the trunk (Douglas-Hamilton and Douglas-Hamilton 1975). Moss (1988) and Poole (1996 in Holdrege 2001) both observed African elephants Scratch between their front legs with a stick to dislodge ticks. Hall (1963) and van Lawick–Goodall (1970) cited Williams' report of an Asian elephant Scratching itself with a twig held in the trunk. McKay (1973) observed captive elephants using objects such as sticks and twigs to "Rub" their bodies.

W. Thorpe (1963) cited Huxley's observation of a domestic goat, presumably *Capra hircus*, Scratching itself with a piece of straw that it held in its mouth.

Van Lawick–Goodall (1970) provided an account and photograph by Chapman of a horse Scratching its flank with a stick held in its mouth. It is of interest that this horse's tail had been docked, thus depriving it of a structure horses often use to respond to irritation on the hindquarters. Analogous behavior in zoo-housed ungulates has been documented three times. Grummt (1963) reported that a zoo-housed bull water buffalo (*Bubalus bubalis*) ripped a top rail from the wooden fence that surrounded its enclosure and balanced the rail on its head and back in the curve of one horn. It then began to move its head and neck up and down so that the tip of the rail was Rubbing and Scratching its back. The discovery explained the frequent detachment of fence rails that had previously puzzled zoo officials. Lau (1965) saw nearly identical behavior by another bull of the same species at the Berlin Zoo. This animal, however, could also Rub the back of its head and neck by hooking and holding one end of the rail with a horn and allowing the other end to rest on the ground. When a pole was leaned against the fence, the animal knocked the pole to the ground and then lifted it with a horn for use. When a pole was inserted between other fence rails, the buffalo first extracted it. A bull eland (*Taurotragus oryx*) in the Overton Park Zoo in Memphis, Tennessee, knocked over a telephone-pole-sized scratching post. Several days later the eland was noted to pick up the large, heavy pole by wedging it between its horns, and to Scratch its back with the distal end (D. Anderson, pers. comm.). In all three of these cases, the behavior was repeated often. It is of interest that the horns in these two species are not long enough to reach most areas on the back. Longer-horned bovids such as Siberian ibex (*Capra ibex siberica*), beisa oryx (*Oryx beisa*), and addax (*Addax nasomaculatus*) frequently use their horns directly to scratch their backs (B. Beck, pers. obs.), but this is not tool use, since the horns are attached parts of the animals' bodies.

Block

Gordon (1966) reported that during times of drought, African elephants dug narrow holes in search of water. Once water was found, the elephants drank and then Blocked the flow of water with leaves, grass, and dung. They then covered the hole with sand before leaving, preventing other animals from drinking from the hole. Asian working elephants filled the bells they wore around their necks with mud or clay to prevent the clapper from moving (Williams in Shoshani and Eisenberg 2000). The authors concluded that they do this as a way of avoiding detection when they raided garden plots. Work elephants have also been documented to spontaneously use a wedge to Block the movement of a log. In one case, an elephant placed a vertical wedge between his trunk and his tusk to Block a large log from rolling backwards toward his head and potentially harming his rider (Williams 1950). Kurt and Hartl (1995, 316) stated that "it is well known that working elephants use wedges to hinder logs from rolling down steep slopes."

Prop and Climb, Balance and Climb, Bridge, Reposition

A rogue elephant was described by Romanes (1892) as chasing several people up a tree. Unable to reach them, the elephant noticed timber that was stacked nearby. The elephant moved each of the thirty-six pieces in the stack, piled them at the base of the tree trunk, placed his rear feet on the pile and his front feet on the trunk of the tree, and attempted again (still unsuccessfully) to reach the people.

Wipe

A young wild African bull elephant was observed Wiping cuts on its back with a clump of grass held in its trunk (Douglas-Hamilton and Douglas-Hamilton 1975). Captive African elephants used vegetation or hay held in their trunk to sweep the ground or scrub walls or objects. Scrubbing was usually preceded by immersing the tool in water. Sweeping was accomplished by blowing air through the trunk. Elephants also Wiped dirt and debris from their bodies by blowing air (Chevalier-Skolnikoff and Liska 1993).

Affix, Apply, Drape

A dramatic aspect of African elephant behavior, reported as early as the third century A.D. by Aelian, is their exploration and manipulation of the remains of dead conspecifics (Douglas-Hamilton and Douglas-Hamilton 1975; Nissani 2004). These authors summarized unpublished observations of burying behavior, which is one component of this response: the elephants Draped branches, leaves, grass, and soil over the carcasses. Elephants also covered other dead animals, as well as dead or immobile humans (Douglas-Hamilton and Douglas-Hamilton 1975; Grzimek 1956; Kühme 1962, 1963; Nicholson 1955; Sikes 1971). The materials were thrown or placed with the trunk or kicked up with the forefeet. In one unusual example, an elephant Applied mud to a wound on a dead individual (M. Turner, cited in Douglas-Hamilton and Douglas-Hamilton 1975 and in Chevalier-Skolnikoff and Liska 1993).

Wild elephants frequently encounter and circumvent barriers designed to impede their movement. Grzimek (1970) described one tactic that may have been tool use, employed to neutralize a wire fence that the elephants could not destroy by force. They broke down many large trees and deposited them on the fence. The weight of the tree trunks caused the cable strands to sag sufficiently to allow the elephants to step over them. Chevalier-Skolnikoff and Liska (1993) provided a similar report. African elephants have also piled branches on a newly constructed road, essentially closing it off to vehicles. The elephants repeated this behavior four times when the branches were removed (Chevalier-Skolnikoff and Liska 1993). Asian elephants covered an open waterhole with branches (Kurt and Hartl 1995).

African (Douglas-Hamilton and Douglas-Hamilton 1975 in Chevalier-Skolnikoff and Liska 1993; Moss 1988; Sikes 1971) and Asian (Kurt and Hartl 1995; McKay 1973) elephants Applied water, mud, powdery earth, regurgitation, and vegetation to their bodies. The behavior is common in both wild and captive individuals. The material is Applied to the sides, head, back, and belly by squirting, throwing, or blowing it with the trunk. Frequently, the animals rub against a vertical feature such as a tree, boulder, or wall after Applying the material. The behavior may function in hygiene, thermoregulation, and/or control of ec-

An elephant Applies sand to its back. Photograph by Mehgan Murphy, courtesy of the Smithsonian National Zoo.

toparasites. Furniss (1879a, 1879b) described his own observations and those by Conklin of zoo-housed elephants carefully laying hay and grass over their backs. The behavior was seen only when biting flies were numerous. While many ungulates bathe and wallow to produce a functionally similar effect, elephants have the manipulative ability to hold and Apply the materials and may thus be said to be using them as tools. Asian bull elephants in musth have been observed to Apply mud and water to their temporal glands, and then Apply that mixture to trees in the environment. Kurt (1974, 1992 in Kurt and Hartl 1995) interpreted this as a form of scent marking intended to repel nonmusth bulls.

The horns of antelope and the antlers of deer are, in most species, larger in males than in females, as a result of sexual selection. They function both in ritualized intraspecific aggression between males and in male sexual displays toward females. Horn or antler size within a given species is correlated with success in agonistic encounters and with reproductive success (Geist 1966, 1971; Walther 1974). These structures often are effective merely as visual stimuli; for example, a male may submit simply upon seeing another male with larger horns or antlers. Beck (1980) observed captive male greater kudu (*Tragelaphus strepsiceros*) and male Père David's deer (*Elaphurus davidianus*) Applying mud and hay to their horns or antlers, respectively. Similar behavior has been observed among Père David's stags at the Hellabrun Zoo in Munich (Heck 1970), at the Bronx Zoo (Schaller and Hamer 1978), and at the Smithsonian National Zoo's Conservation and Research Center in Front Royal, Virginia (Wemmer et al. 1983). Mud and vegetation may make the horns or antlers look larger. The vegetation may be Applied passively, during feeding from elevated hay racks or while thrashing shrubs. Adornment with mud, however, appeared to be purposive, because mud was Applied to the antlers and back by distinctive behavior patterns. Mud adornment appeared to be motivated by an interaction of sexual receptivity of females, intraspecific male aggression, and thermoregulation (Beck, Leja, and Zemanek 1978).

The literature describing display by male artiodactyls suggests that homologous or analogous behavior might occur in other taxa, but we have been unable to locate precise descriptions of clearly comparable behavior. Pruitt (1954) noted that rutting male whitetail deer (*Odocoileus virginianus*) rake their antlers through leaves and loose soil and throw the leaves and soil on their backs. Schaller (1967) reported that barasingha (*Cervus duvauceli*) stags stick their antlers into grass and later walk around with tufts hanging from the rack. R. Harris and Duff (1970) provided photographs of red deer (*Cervus elaphus*) and sika deer (*Cervus nippon*) stags with mud and vegetation on their antlers. D. R. McCullough (1971) noted that rutting tule elk (*Cervus nannodes*) stags often have debris hanging from their antlers.

TOOL USE: CETACEANS

Throw

Indo-Pacific humpback dolphins (*Sousa chinensis*) Threw sea shells during social play (Saayman and Tayler 1979 in Parra 2007).

A rutting male Père David's deer that has Affixed mud to his antlers to augment sexual display. Photograph courtesy of Benjamin Beck.

Bait, Entice

In order to attract the attention of human observers, a captive Indo-Pacific bottlenose dolphin (*Tursiops aduncus*) presented objects such as feathers or stones at the underwater viewing area of her enclosure. She showed a preference for particular objects, and would present different items in sequence if she was ignored (Tayler and Saayman 1973). Wild Amazon river dolphins (*Inia geoffrensis*) carried and displayed objects such as sticks, clumps of grass, or hard clay in their mouths. During the behavior, the items were sometimes thrown about. This behavior is performed almost exclusively by adult males and, although not yet well understood, is likely a sociosexual display aimed primarily at females. This behavior could also be listed under the Throw mode as well as the Brandish/Wave/Shake mode, but the primary function appears to be one of attraction (A. R. Martin, da Silva, and Rothery 2008).

Dig

Some members of a longitudinally studied population of bottlenose dolphins (*Tursiops* sp.) in Shark Bay, Australia, use sponges (*Echinodictyum mesenterinum*) while foraging (Smolker et al. 1997). The sponge is worn on the rostrum and used to probe into the substrate for fish (Krützen et al. 2005). The behavior has been documented in fifteen of 141 mothers in the population, and in at least seven of their offspring. Except for one male, this form of tool use is a female behavior that the authors describe as having "an almost exclusive vertical social transmission within a single matriline from mother to female offspring" (Krützen et al. 2005, 8939). Thus this behavior in this population appears to be at least habitual, perhaps customary, and may be cultural. Indo-Pacific humpback dolphins in South Africa also carry sponges on the rostra, and preliminary evidence suggests that dolphins in a western Australian population do so as well (Parra 2007). However, the function of this behavior has not been confirmed in these populations.

Prop and Climb, Balance and Climb, Bridge, Reposition

A 26-year-old adult male bottlenose dolphin (*Tursiops truncatus*) at the Indianapolis Zoo exhibited a form of rafting. Near the end of a terminal and debilitating illness, he frequently held a basketball in his rostrum for prolonged periods. His caretakers interpreted this as a means of floating his head above water with minimal effort, allowing ease of breathing. Prior to his illness, this behavior had never been observed being performed by him or any other dolphin at the facility (J. Baker, pers. comm.).

Contain

A number of different types of cetaceans employ bubble nets as a means of catching prey (see Sharpe and Dill 1997 for a review). Humpback whales (*Megaptera novaeangliae*) demonstrate perhaps the most elaborate form of this behavior, employing bubbles in configurations such as curtains, nets, clouds, or cylinders.

A bottlenose dolphin carries a sponge on its rostrum that will be used for Digging in the ocean floor in search of fish. Photograph courtesy of Michael Krützen.

These are used most often to surround and capture schooling fish (Clapham 2000; Leighton, White, and Finfer 2008; Sharpe and Dill 1997). However, capture techniques vary among geographically separate populations (Clapham 2000). Several explanations for the way in which the bubbles trap prey have been offered, including a negative acoustical impact on fish swim bladders, a mechanical effect, or a fear response by the prey (Leighton, White, and Finfer 2008; Sharpe and Dill 1997). These hypotheses are not mutually exclusive.

Wipe

Tayler and Saayman (1973) reported that a captive bottlenose dolphin Wiped the interior of his aquarium window with a seagull feather and a variety of other items in imitation of divers who cleaned algae from the glass. As he Wiped the glass, the dolphin also emitted "sounds almost identical to that of the diver's air-demand valve" and released bubbles similar to the ones released by the diver's exhaust air (Tayler and Saayman, 290). We do not consider the sounds as tool use, but it is a good example of dolphins' propensity to imitate. The authors noted that the two dolphins, using 6 by 8 centimeter pieces of broken tile, repeatedly scraped the bottom of their pool to detach seaweed, which they ate.

TOOL MANUFACTURE: RODENTS

Detach

Block

Captive North American beavers used sticks to Block an overflow drain (Pilleri 1983). Each stick was "gnawed off obliquely at both ends," and the end had been "whittled" so that it perfectly filled the hole (Pilleri 1983, 99). Some manufacture in the Subtract mode may also have been involved in this behavior.

TOOL MANUFACTURE: CARNIVORES

Detach

Drag, Roll, Kick, Slap, Push Over

A wild polar bear Detached a 20-kilogram piece of ice from a much larger mass and then rolled it to a seal breathing hole, likely for use in hunting (Kiliaan 1974).

Reshape

Scratch, Rub

P. Larson (pers. comm.) reported that a pet dog, prior to using a paperclip to groom its paws, searched for a suitable clip, held it between its paws and Reshaped it so that the rounded ends were separated from each other and protruded in different directions. This process took up to an hour. The dog also carried the tool with her as she moved through the house.

TOOL MANUFACTURE: PROBOSCIDS

Detach

Reach

An elephant ripped up a fence stake with its trunk and Detached a piece by stepping on it and breaking it off in its trunk. The elephant then Reached into its left axilla with the stick and used it to dislodge a leech (Peal 1879).

Block

A wild African elephant dug a deep hole in search of water. After drinking, the elephant stripped a piece of bark from a nearby tree, rolled it into a ball, and plugged the hole.

Subtract

Brandish, Wave, Shake

Elephants often manufacture their fly switches. After they Detached or picked up a branch, elephants were observed to further modify it (Hart and Hart 1994; Hart et al. 2001). They Subtracted side branches or shortened the branch before use. Elephants as young as 18 months of age were observed to engage in this sort of Subtraction, but this appeared to be beyond the abilities of a youngster that was 9 months old. After Detaching the leafy bough (first-order manufacture), the elephant Subtracted a large piece before using the smaller piece as the tool (second-order manufacture).

ASSOCIATIVE TOOLS: CARNIVORES

Tool Composite

Sea otters used two rocks as a hammer and anvil to open prey items prior to consumption (Riedman and Estes 1990).

5 Prosimians and Monkeys

THIS CHAPTER IS ORGANIZED INTO THREE MAJOR TAXONOMIC GROUPINGS: PROsimians, New World monkeys, and Old World monkeys. Urbani and Garber (2002) asserted that under natural conditions, monkeys do not use tools. However tool use is defined, this is an erroneous statement that ignores decades of published research to the contrary for both New and Old World monkeys. One genus, *Cebus*, the New World capuchin monkey, accounts for the great majority of the cases of tool behavior among these groups. Capuchin monkeys have the greatest degree of thumb opposability and overall dexterity of all of the New World monkeys, and show considerable predisposition and skill for manual manipulation of objects.

TOOL USE: PROSIMIANS

The prosimians, literally translated as "pre-monkeys," are the most primitive of the primates and include the lemurs, lorises, pottos, and galagos. Some would also describe tarsiers as prosimians, but this is debated by taxonomists (Shumaker and Beck 2003). There are very few reports of prosimian tool use.

Brandish, Wave, Shake

Male ring-tailed lemurs (*Lemur catta*) actively compete with each other using scent. Males rub their tail between glands on their forearms, raise it over their head, and Shake or Wave it with a slight, rapid motion at a rival male, thus transmitting a scent in their direction (Jolly 1966). These "stink fights" relay specific information about each male during competition for females (Charpentier, Boulet, and Drea 2008). The males are manipulating the scent (the tool) with their tails.

Reach

J. Barnes et al. (2006) and Santos, Mahajan, and Barnes (2005) presented lemurs (*Lemur catta*, *Eulemur fulvus*, and *Varecia rubra*) with a task requiring them to pull in a cloth or a cane to obtain a food reward. The lemurs succeeded, and in both studies were described as using tools. But since the "tools" had been pre-positioned behind the food, the lemurs were not responsible for the proper and effective orientation of the tool with regard to the food, so their behavior does not conform to our definition of tool use.

Prop and Climb, Balance and Climb, Bridge, Reposition

Wild aye-ayes (*Daubentonia madagascariensis*) were observed as they foraged. On three separate occasions, individuals moved a hanging liana so that it hung nearer a feeding site and then used the relocated vine to access the food (Sterling and Povinelli 1999). One of the individuals moved the liana three successive times, covering 2.5 meters. The authors noted that experimental studies with captive aye-ayes failed to replicate this finding.

Affix, Apply, Drape

On two separate occasions, a wild adult female black lemur (*Eulemur macaco*) captured and bit a millipede (*Charactopygus* sp.) and then rubbed it on her ventrum and tail (Birkinshaw 1999). The purpose of this behavior could not be determined. According to Birkinshaw, a captive lemur (species not identified) performed this anointing behavior both when presented with the same type of millipede or with a cigarette.

Urine washing (Applying urine to the palms of the hands and soles of the feet) occurs in galagos (Galagonidae), lorises (Loridae), mouse lemurs (*Microcebus murinus*), and mustached tamarins (*Saguinus mystax*), a New World monkey (Heymann 1995; Oppenheimer 1968). The function of the behavior is debated, but it may be related to thermoregulation, olfactory communication, cleaning, or improving the animals' grip when climbing (Harcourt 1981; Heymann 1995).

TOOL USE: NEW WORLD MONKEYS

New World monkeys include all species of nonhuman primates native to the southernmost portion of North America (Mexico), as well as Central and South America.

Drop

Capuchin monkeys (*Cebus* spp.), saki monkeys (*Pithecia* spp.), and squirrel monkeys (*Saimiri* spp.) Drop branches on pursuing humans (anonymous in Kortlandt and Kooij 1963). Howler monkeys (*Alouatta* spp.) Drop branches, twigs, and foliage at pursuing humans (anonymous in Kortlandt and Kooij 1963; C. Carpenter 1934; Dampier in C. Carpenter 1934; A. Schultz 1961). C. Carpenter and Dampier noted that the howlers broke off the branches before Dropping them. Dropping branches and fruits has been observed for spider monkeys (*Ateles* spp.) by anonymous correspondents of Kortlandt and Kooij (1963) and by C. Carpenter (1935), Hernandez-Camacho and Cooper (1976), A. Schultz (1961), and Wagner (1956). C. Carpenter (1935) noted that the branches Dropped by spider monkeys weighed over 4 kilograms and were sometimes held until the intruder approached more closely.

Defler (1979a, 156) observed "intense branch breaking" by *Cebus* when tayras (*Eira barbara*) were present. Defler (1979b, 492) also reported that one adult male broke off and Dropped "a previously selected dead limb" during a display. J. Baldwin and Baldwin (1977) reported that *Cebus* monkeys passively dislodged branches during branch-shaking displays and that one such limb, 4 or 5 centimeters in diameter, struck and wounded the senior author. Oppenheimer (1973, 1977) confirmed this behavior for *Cebus* and noted that dead branches were snapped off with the hands, feet, or tail. Snagged pieces of dead wood were merely shaken loose, while green branches were chewed off and allowed to fall. Klein (1974, 112) observed capuchin monkeys Dropping branches toward humans and emphasized that these monkeys showed more "apparent intent and instrumentality" in this behavior than did howler and woolly monkeys (*Lagothrix* spp.). On seven separate occasions, Chevalier-Skolnikoff (1990) was the target of branches, bark, and fruit Dropped by capuchin monkeys. The monkeys moved directly overhead and Dropped the items within a half-meter of her. Chapman (1986) reported that adult male and female capuchin monkeys Dropped sticks on a boa constrictor (*Boa constrictor*) for approximately fifteen minutes after it had preyed on a juvenile member of the group. One male repeatedly retrieved and Dropped the same stick again.

Kaufmann (1962) saw capuchin monkeys Drop palm nuts and other debris on a coatis (*Nasua narica*). The monkeys chased the coatis from a palm tree to the ground before the barrage. Rather than leave the area, however, the coatis simply ate the palm nuts Dropped by the monkeys.

Throw

Hernandez-Camacho and Cooper (1976) noted that woolly monkeys (*Lagothrix lagotricha*) voided excreta into their hands and Threw the material on pursuing humans, and that tufted capuchin monkeys (*Cebus apella*) also Threw feces and urine down toward humans.

Bandres, Campos, and Llavona (1989) and Urbani (1998) relayed reports by early Spanish explorers in America from the sixteenth and seventeenth centuries. These included the earliest known report of tool use by capuchin monkeys, from Fernández de Oviedo, who wrote of "cat-monkeys" (capuchin monkeys) that could "throw a small stone of the size and weight that their strength is capable of, as a man would do." Kortlandt and Kooij (1963) relayed anonymous reports of Throwing by capuchin monkeys but did not specify objects, species, or contexts, or whether this Throwing was aimed or unaimed.

Capuchin monkeys are the only New World monkeys reported to engage in aimed Throwing, and they do so in both agonistic and non-agonistic contexts. Boinski (1988, 177) provided observations of capuchin monkeys Throwing "branches, epiphytic bromeliads, and large fruits." These objects were directed toward "coatis, tayras, opossums (*Didelphis marsupialis*), and spider monkeys (*Ateles geoffroyi*)." In one notable event, Boinski stated that a capuchin monkey Threw a squirrel monkey (*Saimiri oerstedti*) at her when detachable branches were unavailable. Chevalier-Skolnikoff (1990) described a capuchin monkey that aimed and Threw a stick at two peccaries (*Tayassu tajacu*).

One of the earliest accounts of tool behavior by captive capuchin monkeys is provided by Romanes (1892), who acquired the monkey and housed it at his sister's home. She kept a detailed diary of its behavior, and Romanes stressed that she was a careful observer who did not try to train the monkey. Romanes quite frequently saw his capuchin monkey Throw a variety of objects at the dog and the people who shared the house. Romanes noted that the monkey learned that Throwing objects at the feet of humans was relatively ineffective. Since it could not reach their heads from the floor, it climbed to the top of furniture from which it could Throw more effectively and for greater distances. Chevalier-Skolnikoff (1989) noted that one individual Threw an unspecified object at a conspecific in play. Gibson (1990) detailed aimed Throwing by a pet male capuchin monkey in a wide variety of circumstances. This individual Threw anything that was within reach, and did so in different social contexts. As part of an agonistic display when angry or frightened, the monkey Threw objects at humans, chimpanzees, cats, dogs, ducks, flies, roaches, and images on a television screen. To attract attention, he playfully Threw objects at humans, cats, and dogs. He also played catch with humans using objects or food, and, when playing alone, he Threw objects in the air and caught them.

A captive female Threw a branch at a model of a snake being manipulated by an experimenter (Vitale, Visalberghi, and DeLillo 1991). Another captive was observed Throwing sticks to dislodge suspended food (Bierens de Haan 1931). Others Threw sticks, belts, tethered rodents, and other objects behind or beside food and then used the thrown objects to retrieve the food (Cope in Bierens de Haan 1931; Klüver 1933, 1937). Moynihan (1976) saw members of a captive group Throw a handkerchief among themselves for more than an hour. The handkerchief was Thrown primarily by hand, but the tail was used occasionally; hands and tails were used with equal frequency for catching.

Experimental studies conducted in captivity confirm the ability of capuchin monkeys to utilize aimed Throwing. Westergaard et al. (1998a) studied two groups of capuchin monkeys that were housed about 1 meter from each other. When food was presented to one group but not the other, individuals in the provisioned group routinely Threw food to the unprovisioned group. The monkeys sometimes Threw food toward human observers and in the direction of an empty enclosure, as well. The ability of capuchin monkeys to learn very accurate aimed Throwing has been well documented. Westergaard and Suomi (1994a, 1995c, 1997a) and Westergaard et al. (2000) presented their capuchin monkey subjects with rocks and an out-of-reach container that held peanut butter or sweet syrup. A rock successfully Thrown into the container became coated with syrup or peanut butter and was then returned by an experimenter to the monkey that had thrown it. The capuchin monkeys readily acquired and demonstrated aimed Throwing. They were accurate when the container was as much as 60 centimeters away, readily distinguished between an empty or a baited container, and could accurately Throw a rock into a slow-moving target (Westergaard and Suomi 1994a). The authors also found that the monkeys were individually consistent in demonstrating a right- or left-hand bias for throwing, as well as exhibiting a preference for an over- or underhanded motion. When stones of varying weights were provided (either 10 to 20 grams or 50 to 60 grams each), individuals showed a clear and consistent preference for either a lighter or heavier tool. Additionally, an individual's accuracy in hitting the target was greater when he or she used stones of the preferred weight (Westergaard and Suomi 1995c). Perhaps not surprisingly, the monkeys' accuracy decreased as the distance to the target increased. When trying to Throw longer distances, the capuchin monkeys more frequently used a bipedal rather than a tripedal stance (Westergaard et al. 2000). Westergaard et al. (2003) and Cleveland et al. (2003, 159) investigated the "relationships among body mass, sex, stone weight preference, and accuracy in a throwing task." Results affirmed strong individual preferences for rocks of certain weights; females consistently chose heavier rocks, and all subjects showed greater accuracy when rocks of the preferred weight were used. Cleveland et al. (2003) also concluded that, overall, male capuchin monkeys were more accurate in their aimed Throwing than females.

T. Evans and Westergaard (2004) explored the degree to which capuchin monkeys discriminated between functional and non-functional tools. In their methodology, the monkeys learned to Throw a tethered projectile into an out-of-reach bowl containing peanut butter. The peanut-butter-covered tool could then be pulled back into their enclosure. Once competency was demonstrated, the experimenters provided the monkeys with a non-functional-tool task along with the functional task. Examples of non-functional properties included a short tether that prevented the tool from reaching the bowl, or a missing tether so that the projectile was not retrievable. In all experiments, the monkeys chose the functional-tool task significantly more often than the non-functional tasks.

Drag, Roll, Kick, Slap, Push Over

Chevalier-Skolnikoff (1990) described aimed snag crashing in her direction on two occasions when a capuchin monkey Pushed Over 2- to 3-meter-tall dead trees, one landing only centimeters away from her.

Brandish, Wave, Shake

A group of wild capuchin monkeys displayed at field researchers by Shaking branches at them (Langguth and Alonso 1997). During an aggressive chase, a juvenile male capuchin monkey picked up fallen branches and flailed them at a coati (Chevalier-Skolnikoff 1990). A captive observed by Chevalier-Skolnikoff (1989) was observed to rush at zoo visitors aggressively while Waving a long leaf as a threat.

Bait, Entice

Belt (1874) described a pet capuchin monkey that held out a piece of bread in one hand to Entice young ducks. When a duck came within reach, the monkey grabbed it with the free hand and killed it with a bite to the breast. Romanes (1892) recounted Belt's description of this Baiting behavior. Boulenger (1936) related a nearly identical description of this mode of tool behavior by another captive capuchin monkey, which also Baited young ducks with pieces of bread and then seized, killed, and ate them. Boulenger believed that the monkey learned the behavior by watching its master feed the poultry. Gibson (1990) described a pet capuchin monkey that used cloth towels to playfully attract a kitten that also lived in the house. The monkey tossed the towel toward the kitten, which would grab it with her claws. These play sessions typically lasted ten to fifteen minutes, and no aggression toward the kitten was noted.

F. Mendes et al. (2000) observed a group of six zoo capuchin monkeys that used small bits of food such as potato, tomato, or banana to Bait fish. The monkeys dropped the food on the surface of the water moat surrounding their enclosure, then tapped it side to side or, more often, held the food slightly below the surface to attract fish. All individuals were observed to fish, and fishing sessions lasted from less than ten seconds to several minutes. The authors documented 123 sessions, and numerous captures of fish.

Addessi, Crescimbene, and Visalberghi (2008), Brosnan and deWaal (2004), and Fontenot et al. (2007) demonstrated that capuchin monkeys can learn to trade tokens for a food reward. While the tokens were tools, there was no suggestion that the tokens in these examples were symbolic.

Club, Beat

A wild red howler monkey (*Alouatta seniculus*) used a stick to repeatedly hit the branch on which a two-toed sloth (*Choloepus didactylus*) was hanging, as well as to strike the claws and toes of the sloth (Richard-Hansen, Bello, and Vie 1998). The authors stated that the monkey's hitting action was gentle, and the sloth appeared uninjured after the encounter.

Chippendale (in Kortlandt and Kooij 1963) reported that wild capuchin monkeys Clubbed or hit a rattlesnake with sticks or other objects. Boinski (1988) provided a detailed account of an adult male capuchin monkey wielding a 70 by 7 centimeter branch as a club to strike a venomous snake. The *terciopelo*, or fer-de-lance (*Bothrops asper*), was struck by the monkey at least fifty-five times. A subadult male joined the attack and Clubbed the snake at least ten times with a thin 50-centimeter-long branch that he broke off of a nearby plant. The author determined that the snake was killed by the monkeys, and that the most serious injuries were delivered to the head. A nearly identical captive example, in which a snake was killed with a metal container, was provided by Giudice and Pavé (2007). Mannu and Ottoni (2009) witnessed a capuchin monkey attempting to Club a white-eared opossum (*Didelphis albiventris*) with a stick. In a separate incident, they observed a female trying to Club a scorpion (genus not reported) with a stick and a piece of wood.

Chevalier-Skolnikoff (1990) observed a 1-year-old capuchin monkey attempting to climb into its mother's lap to nurse. Rejected by the mother and harassed by a presumed older sibling, the 1-year-old broke off a half-meter-long branch, used it to hit the mother, and then dropped the tool. The mother and sibling threatened the youngster, which responded by breaking off a 20-centimeter-long branch and attempting to strike the mother again.

Romanes's (1892) pet monkey used a stick to Club humans. Cooper and Harlow (1961) described a captive male capuchin monkey that Clubbed a cage mate with a stick when the latter tried to prevent him from feeding. The same animal later struck attacking rhesus macaques (*Macaca mulatta*) with a stick but failed to hit them. Chevalier-Skolnikoff (1989) observed a capuchin monkey that aggressively struck a conspecific with a stick.

Pound, Hammer

Reports of Pounding or Hammering by wild capuchin monkeys are numerous and occur primarily in the context of extractive foraging. The most frequent of these involve the use of a stone hammer to open nuts. The first known report of this behavior is from the early 1500s by Gonzalo Fernández de Oviedo (Bandres, Campos, and Llavona 1989; Urbani 1998) who assumed that the monkeys learned this behavior by observing humans. Field observations continue to reveal that capuchin monkeys commonly Hammer nuts with stones, usually first placing them on nonportable anvils, such as logs, trunks, or stone outcroppings (Canale et al. 2009; de Resende and Ottoni 2002; Falótico 2006; Fragaszy et al. 2004; Langguth and Alonso 1997; Mannu and Ottoni 1999; Ottoni, de Resende, and Izar 2005; Ottoni and Mannu 2001, 2003; Ottoni, Mannu, and de Resende 2002; Souza, Mendes, and da Silva 2002; Visalberghi 2004, 2006; Visalberghi et al. 2004, 2006).

Stones are the most common material for a hammer, but other nuts are used on occasion (Visalberghi et al. 2009a). Struhsaker and Leland (1977) observed a wild capuchin monkey in a tree Pounding one hard-shelled nut against another. Each hand held a nut and the hands were "brought forcefully together in a scissor-like manner, the left hand moving rapidly downward and the right hand upward meeting at chest level" (Struhsaker and Leland 1977, 124). Although the nuts were Pounded together a number of times, the monkey moved away before either nut was opened.

Ottoni and Mannu (2001) reported that all adults, subadults, juveniles, and at least one infant at their study site at Tietê used stones and sticks to crack nuts. Adults were the most efficient, but juveniles engaged in the behavior most frequently. Expertise increased with age. Social learning played an important role in the acquisition of this behavior, and capuchin monkey observers preferentially watched users that exhibited greater proficiency (Ottoni, de Resende, and Izar 2005; Ottoni and Mannu 2001; Ottoni, Mannu, and de Resende 2002; Ottoni et al. 2001). Pounding nuts with hammers is customary and cultural in this population.

Visalberghi et al. (2008) examined the properties of the palm nuts (*Astrocaryum campestre*, *Attalea barreirensis*, *Orbignya* spp., and *Attalea* spp.) that are consumed by the capuchin monkeys of Boa Vista, Brazil. The authors concluded that the durability of these nuts ("high resistance foods") necessitated the use of tools for the capuchin monkeys, which could not otherwise open them (Visalberghi et al. 2008, 884).

Capuchin monkeys also open larger tough-skinned fruits with hammers. The hammers might be stones or branches, and fixed anvils are sometimes involved (Boinski, Quatrone, and Swartz 2000; Mannu and Ottoni 2005, 2009; Moura 2002; Waga et al. 2006). Capuchin monkeys also used hammers to crack open cacti, seeds, dry tubers, and dry hollow branches (Moura 2003; Moura and Lee 2004a, 2004b, 2004c). Moura and Lee (2004c) specifically noted that Hammering was used to Pound tubers into smaller pieces to facilitate consumption, to process cactus, and to break into potential hiding places of vertebrate prey such as lizards. Rocha, dos Reis, and Sekiama (1998) described capuchin monkeys Hammering seeds they placed on

A capuchin monkey places a nut on a stone anvil and breaks the nut open with a stone hammer. Another group member watches intently. Photograph courtesy of Tiago Falótico.

an anvil. Once the seeds were cracked, the monkeys consumed the endosperm as well as *Coleoptera* larvae found inside.

Capuchin monkeys demonstrate selectivity in the stones they use for Hammering. Schrauf, Huber, and Visalberghi (2008) reported that capuchin monkeys learn through trial and error to use weight as a measure of effectiveness when choosing a tool. Visalberghi et al. (2009b, 214) concluded that the monkeys focus on both material and weight when choosing a tool. These authors stated that "the capuchin monkeys always touched the functional stone first, suggesting that they discriminated the stones by sight." When a visual assessment was difficult, the monkeys manipulated or tapped the stones before making a selection. Although some variation was reported, all capuchin monkeys appeared to prefer tools that seemed extremely heavy. Fragaszy et al. (2004) reported that chosen stones could weigh more than 1 kilogram, while most of the capuchin monkeys themselves weighed less than 4 kilograms. Falótico (2006) and Falótico and Ottoni (2005) observed "semi-wild" capuchin monkeys and noted an overall preference for stones weighing about 1.3 kilograms, but the monkeys used some weighing as much as 1.7 kilograms. Regardless of weight, juveniles used stones closest to anvil sites, while adults transported preferred stones to the anvil sites. Visalberghi et al. (2007) reported similar data, noting that the stones might weigh 25 to 40 percent of the user's body weight. Liu et al. (2009) provided even more extreme examples, documenting monkeys that used stones weighing 33 to 77 percent of their body weight. In all reported cases the monkeys lifted, aimed, and lowered the tools while standing bipedally (Liu, Fragaszy, and Simpson 2006; Liu et al. 2009; Visalberghi et al. 2007, 2009b). As noted above, capuchin monkeys transport nuts and hammer stones to anvil sites (Falótico 2006; Falótico and Ottoni 2005; Visalberghi et al. 2007, 2009a). Visalberghi et al. (2007, 443) summarized this set of behaviors by stating that "nutcracking is for these small monkeys an enormously strenuous activity."

Dampier (in Hill 1960) reported that capuchin monkeys inhabiting Gorgona Island off the Pacific coast of Colombia came to the rocky shore to feed on oysters that they opened by Hammering with stones. Jennison (1927) relayed an unattributed observation of wild capuchin monkeys using stones to Hammer open oysters, but he may have been referring to Dampier's report. Romanes (1892) noted similar observations by Wafer. While working in a Brazilian mangrove swamp, Fernandes (1991) observed an adult male capuchin monkey opening oysters by Pounding on them with a tool. Fernandes noted that since stones were absent from the swamp, the monkey used a piece of the oyster colony as a hammer.

In addition to Hammering to assist with extractive foraging, capuchin monkeys also Pound on rocky outcroppings to generate acoustic displays. Moura (2007) recorded thirty-nine events of stone banging, of which 69.2 percent were aggressive, accompanied by other clearly threatening behaviors directed at the author. The meaning of the remaining 30.8 percent was unclear, but these events were not perceived as aggressive by the author.

During a forty-three-month study, J. D. Anderson, Owren, and Boinski (2007) recorded 498 events in which capuchin monkeys produced percussive strikes against tree branches. The hard fruit of *Phenakospermum guyannense* was used 36.7 percent of the time; other materials used by the monkeys were not specified. The authors suggested that among several functions for this banging behavior, the percussive strikes might allow "males to broadcast honest signals of their location, alliance size, individual strength, and skill level" to other capuchin monkeys (J. D. Anderson, Owren, and Boinski 2007, 64).

The earliest known report of Hammering by a capuchin monkey in captivity was provided by Eras-

Left: Having placed a nut on a branch anvil, a capuchin monkey raises a stone hammer. Photograph courtesy of Tiago Falótico. *Right*: Another capuchin monkey performs the same action. Photograph by Elisabetta Visalberghi, courtesy of Project EthoCebus. The tool users in both photographs stand bipedally, bracing themselves with their tails.

mus Darwin (1794), who observed an old individual housed at the Exeter 'Change in London. The monkey had lost many of his teeth, but cracked nuts "one by one" with a stone. Hammering is frequent in captivity, where hard substrates are commonly available and unattached hard objects are provided by researchers and caretakers (Bierens de Haan 1931; Eisentraut 1933; Klüver 1933; Kooij and van Zon 1964; Matschie in Armbruster 1921; Nolte 1958; Rengger 1830; Romanes 1892; Tobias 1965; Vevers and Weiner 1963). The tools have included stones, metal or plastic objects, copper nuggets, blocks of wood, bones, and a dish. In one case, capuchin monkeys Pounded with a stick pestle to extract fiber and sap from sugarcane (Westergaard et al. 1995). Nuts were the most frequently opened items, but insects, eggs, and ice cubes were also noted. Giudice and Pavé (2007) observed captives opening hard fruits by Hammering on them with sticks. Chevalier-Skolnikoff (1989) reported two forms of Pounding to open nuts by her captive subjects. In one case, two nuts were Hammered together to crack the shells. In the other, a nut was placed on an indentation in a rock and then hit with a stick. We consider both to be tool use.

Generally, capuchin monkeys demonstrated spontaneous proficiency in the use of hammer tools (J. R. Anderson 1990; Antinucci and Visalberghi 1986; Bortolini and Bicca-Marques 2007; Cleveland et al. 2004; Fragaszy and Adams-Curtis 1991; Fragaszy and Visalberghi 1989; Fragaszy, Vitale, and Ritchie 1994; R. Mittermeier, pers. comm.; Sutton 1982; Visalberghi 1986, 1987; Visalberghi and Antinucci 1986; Visalberghi and Vitale 1990; Westergaard 1993b; Westergaard and Suomi 1993a, 1993b, 1994e, 1997a; Westergaard et al. 1996). However, Vancatova (1984) reported that none of her subjects spontaneously used tools to crack open nuts.

In a study that explored the creative abilities of capuchin monkeys, ten individuals were given modeling clay, sticks, and stones (Westergaard and Suomi 1997b). On nine occasions, four of the subjects combined objects by placing either a stone or stick on the clay and then Hammered it with another stone. Since the purpose of this tool use is ambiguous, we cannot definitively describe it as a Tool Set.

Westergaard and Suomi (1994d, 1994e) and Westergaard et al. (1996) described a capuchin monkey placing a stone against a bone and striking it with another stone to remove a casing that covered the bone, using bone fragments as hammer and chisel to penetrate an acetate covering that prevented access to a food reward, and, in another instance, using bone fragments and a copper nugget as hammer and chisel to penetrate the acetate. Jalles-Filho (1995) and Jalles-Filho, Teixeira de Cunha, and Salm (2001) presented their subjects with a box covered by thin Plexiglas that could be broken with a Pounding tool to access the food reward inside. The monkeys successfully used stone or branch hammers to access the contents. Two individuals placed sticks into the seam of the lid and then Hammered these levers in an operation Jalles-Filho (1995) considered analogous to the hammer and chisel reports by Westergaard and Suomi (1994d, 1994e). The hammer and chisel combinations in all of these cases constitute Tool Composites.

Garner (1892) provided an unusual case of Hammering by a captive capuchin monkey that may be revealing because of its intermediacy between proto-tool use and tool use (see chapter 1). Garner gave the monkey a small metal box that could be opened by depressing a spring-loaded clasp. The monkey was not strong enough to open the clasp with its finger and so learned to pound the box against a wall to open it. When confined to its cage, the monkey could not open the box because the wire was too fine to provide an adequate Pounding surface, so it learned to pick up and hold a wooden block on which it Pounded the box. Although the block was not used as a hammer, it was held and thus can be considered a tool of the same type as the sea otter's portable anvil.

Gibson (1990) described a captive male capuchin monkey that opened hard foods, such as coconuts, by Hammering on them with objects. He also Pounded things, such as locks or chains, against walls to call caretakers for food, drink, or attention. Gibson noted that this behavior ended once the monkey's desires were met. Chevalier-Skolnikoff (1989) relayed the observation of a capuchin monkey hitting a bell with a bucket to make it ring. This event occurred after the monkey accidently rang the bell by running into it while wearing the bucket on its head.

Pry, Apply Leverage

Captive capuchin monkeys have been reported to Pry the lids from wooden boxes with sticks and with a spoon (Rengger 1830; Romanes 1892). There is some indication that the monkeys had previously observed humans using levers in similar contexts. Rengger's (1830) subject also used a stick as a lever to roll over a piece of wood. Rensch (1973) reported the use of the handle from a broken tennis racket as a lever by a male capuchin monkey, which carried the handle with his tail and placed the thinner part between a platform and a stationary food container. He climbed above the lever and pushed on the free end with both hands, presumably attempting to dislodge or break the container. Vitale, Visalberghi, and DeLillo (1991) presented their capuchin monkey subjects with a model of a snake that protruded from a hole in a box. One individual used a wooden stick as a lever by inserting it under the box and attempting to Pry the attached apparatus from the floor of the enclosure. Jalles-Filho (1995) reported that four of seven monkeys presented with a closed box used sticks to Pry open the lid and obtain the corn that was inside. Prying with a stick was a more common solution than either Pounding or Cutting. A zoo-living capuchin monkey used a stick as a lever in an attempt to Pry open the door of its enclosure (Sutton 1982), and another Pried the mesh floor away from the wall with a stick (Giudice and Pavé 2007).

In an experimental study of social learning (Fredman and Whiten 2008), ten human-reared and fourteen mother-reared capuchin monkeys were tested on their ability to learn a tool-using task from a demonstrator. Successful completion required the monkeys to Pry open the lid of a tin with a screwdriver to access the food rewards contained within. The human-reared monkeys observed a human demonstrator, while the mother-reared monkeys observed a high-ranking conspecific as their demonstrator. A portion of the human-reared subjects closely reproduced what they observed, while none of the mother-reared subjects did.

A captive female capuchin monkey received a wound from a conspecific that was closed with five metal sutures. When fully awake after anesthesia, this individual explored the wound with a stick she had coated in syrup. Subsequently, she placed the tool between the wound and each suture and used it as a lever, successfully removing all of the sutures (Westergaard and Fragaszy 1987b).

Dig

Wild capuchin monkeys living in dry Caatinga forest habitat in southeast Brazil used stones to dig up roots, tubers, or insects (Mannu and Ottoni 2009; Moura 2002, 2003; Moura and Lee 2004a, 2004b, 2004c). Digging also involved some Pounding of the ground to disrupt the soil so that it could be removed. Mannu and Ottoni (2009, 245) characterized this as use of a stone as a "hoe," since the tool was moved horizontal-

A capuchin monkey uses a stone to Dig in soil for tubers, roots, and insects. Photograph courtesy of Tiago Falótico.

ly across the loose soil. Moura and Lee (2004a) noted that Digging was the most frequently observed form of tool use in this population, accounting for 65 percent of all tool observations.

Captives Dug in the ground with sticks in an unspecified context (Giudice and Pavé 2007). Westergaard and Suomi (1995b) presented ten capuchin monkeys with peanut-baited containers, buried under 5 centimeters of soil, and several sticks measuring 10 to 40 centimeters long. Four of the subjects spontaneously used sticks as Digging tools to extract the nuts in forty of the forty-eight trials presented. The monkeys employed two different Digging techniques. They either Dug with the tool and scooped away the displaced soil afterwards, or Dug with the tool with one hand while simultaneously scooping away soil with the other. Chevalier-Skolnikoff (1989) reported that a captive capuchin monkey Dug in the dirt with a stick, occasionally finding insects that it ate. She also observed capuchin monkeys Digging with leaf fronds.

In the study described above that explored the creative abilities of capuchin monkeys, ten individuals were presented with modeling clay, sticks, and stones (Westergaard and Suomi 1997b). Pertinent to Digging, eight of the subjects used the sticks and stones to scrape the surfaces of twenty-two of the twenty-four clay forms that were presented.

Jab, Stab, Penetrate

Mannu and Ottoni (2009, 248) witnessed a capuchin monkey attempting to Stab a white-eared opossum by making "spear-like movements" with a stick. During play, one young capuchin monkey broke a dead branch from a tree and Jabbed the other with it (Chevalier-Skolnikoff 1990). Cooper and Harlow (1961) noted that their capuchin monkey used a stick to prod or Jab a cage mate that was trying to monopolize its food. Romanes's (1892) monkey used a stick to Jab the family dog.

Westergaard and Suomi (1994d) described the behavior of captive capuchin monkeys provided with an acetate-covered container that held a sweet liquid. Two of their three subjects used a chisel-and-hammer technique to access the reward. Each monkey placed a stone chisel on the acetate and then struck it with the hammer, creating an opening into the container. The hammer was used in the Pound, Hammer mode, and the chisel was used to Penetrate the cover. This was a Tool Composite.

In Fredman and Whiten's (2008) experimental study of social learning (see Pry, Apply Leverage, above), human-reared and mother-reared capuchin monkeys were also presented with a task that required the monkeys to Penetrate a paper covering with a screwdriver to get to food. Human-reared monkeys observed a human demonstrator, while mother-reared monkeys observed a high-ranking conspecific. As in the previous part of this study, a portion of the human-reared subjects closely reproduced what they observed, while none of the mother-reared subjects did.

Reach

Many studies have explored tool use in marmosets, tamarins, and capuchin monkeys, primarily through laboratory problem-solving tasks that involve pulling an object to obtain a food reward (Cummins-Sebree and Fragaszy 2001, 2005; Fujita, Kuroshima, and Asai 2003; Hauser, Kralik, and Botto-Mahan 1999; Hauser, Pearson, and Seelig 2002; Hauser et al. 2002; Santos et al. 2005, 2006; Spaulding and Hauser 2005; Vancatova 2008; Visalberghi 1997; Voelkl, Rainer, and Huber 2002). Although the authors described these tasks as involving tools, they do not fit our definition of tool use, because the subjects were not responsible for the proper and effective orientation of the implements.

The only report of true tool use among the marmosets and tamarins involves golden lion tamarins (*Leontopithecus rosalia rosalia*). Eight free-ranging captive subjects were observed, including some who wore radio-transmitter collars. All of the individuals were seen to use twigs and/or the antennae of their radio collars during grooming or when foraging for food (Stoinski and Beck 2001). During grooming, individuals used tools to part their own hair or that of a partner, or used the tool directly to groom the skin or hair. When foraging, they Reached under pieces of tree bark with twigs to extract insects to consume. Three of the subjects used the end of the antenna on their radio collars for the same purpose.

Chippendale reported to Kortlandt and Kooij (1963) that a squirrel monkey (*Saimiri sciureus*) Reached with a stick to sweep a piece of fruit across the ground to dislodge ants that were on the food. Garber and Brown (2004) presented wild capuchin monkeys with an apparatus that required use of a dowel to Reach and dislodge bananas from inside a clear box. Despite 702 visits to the apparatus by numerous individuals over fifty-five consecutive days, none was successful in solving the task.

In a non-experimental setting, Gibson (1990, 209) described a captive male capuchin monkey that used cloth towels to Reach for things outside of his cage, such as "adhesive tape, scissors, newspapers, books, shoes, tape recorders, and cameras." Other captives used sticks to assess the depth of a pond in their enclosure (Giudice and Pavé 2007).

An adult female capuchin monkey that was experienced in using tools to Reach and extract sweet syrup from an enrichment apparatus also explored, treated, and groomed specific areas of her body with tools (Westergaard and Fragaszy 1987b). This individual regularly used stick tools, some of which she had coated with syrup, to groom her vaginal area. On several occasions, she was also documented using syrup-coated tools to treat wounds received from conspecifics. She regularly modified one end of the tool into a brushlike instrument, although both the modified and unmodified ends were used to explore the wounds. Another captive female used a small piece of straw to remove splinters from her hand (Giudice and Pavé 2007).

In a separate but similar example, a capuchin monkey mother used modified sticks and straw to treat a severe head wound sustained by her offspring after an infanticidal attack. The brushlike tips of the tools were moved across the wound in a stroking motion. Ritchie and Fragaszy (1988, 347) ultimately concluded that the mother was probably attempting to remove broken bits of skull that were "vertically embedded in the exposed brain" and could not be extracted with fingers. The mother also used the non-modified end of the tool to pick or scrape at the wound. Despite the efforts of the mother and subsequent medical treatment provided by humans, the infant did not survive.

A golden lion tamarin uses the antenna of its radio-transmitter collar to groom a conspecific. Photograph by Tara Stoinski, courtesy of Zoo Atlanta.

The use of an object to Reach and rake in an otherwise unreachable incentive is a classic paradigm in laboratory studies of primate tool use. Capuchin monkeys of various species have been reported to use a variety of Reaching tools—such as sticks, cloth, wire, belts, cards, palm fronds, a swing, a spoon, and tethered rodents—to solve these tasks (J. R. Anderson and Henneman 1994; Bates in Hobhouse 1926; Belt 1874; Belt in Romanes 1892; Bierens de Haan 1931; Chevalier-Skolnikoff 1989; Cooper and Harlow 1961; Cope in Bierens de Haan 1931; Eisentraut 1933; Garner 1892; Giudice and Pavé 2007; Harlow 1951; Johnson 1996; Klüver 1933, 1937; Krieg 1930; Romanes 1892; Warden, Koch, and Fjeld 1940). In some of the experiments, a suitable stick had to be retrieved from areas that were not in the same visual field as the food incentive. Some of the monkeys used short sticks to Reach others of sufficient length to get the food. Two of Warden, Koch, and Fjeld's (1940) subjects used a series of eight sticks before getting the food. These data are difficult to evaluate, however, because of imprecise specifications of possible pretest training and inadequate data presentation. Watson (1908) reported that an unidentified capuchin monkey did not learn to Reach food with a stick or cloth despite repeated trials and some tuition. Success in using a Reaching tool began to emerge in an infant capuchin monkey studied by Potì et al. (1990, 1991) between 13 and 17 months of age, and consistent success became apparent between 17 and 19 months of age.

G. Byrne and Suomi (2004) reported that of fifty-eight capuchin monkey subjects, thirteen were successful in using tools to reach food rewards. On average, successful individuals required 21.8 trials before they first demonstrated full competence. The youngest individual that raked in a food reward was approximately 21 months old. More males were successful than females, and no significant difference was found in the rates of success between juveniles and adults.

A variant of the Reaching task involves using a stick to Reach toward and knock down food suspended out of the monkey's reach. Subjects observed by Bierens de Haan (1931), Harlow (1951), and Klüver (1933) performed this task successfully. In some cases, a capuchin monkey used a stick to Reach the food while standing on a box, another Tool Composite (Klüver 1933). A captive individual confronted with a model of a snake coming out of a box used sticks to Reach under the apparatus and into the cylinder through which the snake's head protruded (Vitale, Visalberghi, and DeLillo 1991).

A capuchin monkey Inserts and Probes in a tree hole with a tool. Photograph courtesy of Tiago Falótico.

Insert and Probe

Jay (1968) cites a personal communication from R. Thorington of wild *Cebus* monkeys breaking off twigs, peeling the bark from them, and then Inserting and Probing with them under tree bark to secure small insects. Jay's account has been widely cited. This early report has been substantiated by several recent field studies. Capuchin monkeys have been observed to Probe in tree holes, stumps, rock crevices, and bark to locate and extract arthropods, wax, honey, or water. Pieces of vines, branches, sticks, or other plant parts may be used (Chevalier-Skolnikoff 1990; Mannu and Ottoni 2005, 2009; Moura 2002, 2003; Moura and Lee 2004a, 2004b, 2004c). Mannu and Ottoni (2009) recorded 157 different Probing events during 701.5 hours of visual contact. Moura and Lee (2004a) described Probing as accounting for 14 percent of 677 episodes of tool use in their study. Overall, all forms of tool use accounted for 2.1 percent of their total visual contact time with the monkeys.

There are two general types of experiments testing abilities to use Inserting and Probing tools. *Dipping* refers to a type of task in which the monkeys Insert a stick into a narrow opening, sometimes to explore, but most often to acquire a desired incentive such as sweet syrup, peanut butter, yogurt, or ants. The food incentive adheres to the tool, and the tool is withdrawn and the incentive eaten. Inserting and Probing in this context has been documented in many individual capuchin monkeys in a variety of captive settings (J. R. Anderson and Henneman 1994; J. R. Anderson, Lamarque, and Fagot 1996; Bortolini and Bicca-Marques 2007; Cleveland et al. 2004; Costello 1987; M. Dubois 2001; M. Dubois et al. 2001; T. Evans and Westergaard 2006; Flemming, Rattermann, and Thompson 2006; Fragaszy and Adams-Curtis 1991; Fragaszy, Vitale, and Ritchie 1994; Lavallee 1999; Westergaard 1991; Westergaard and Fragaszy 1987a, 1987b; Westergaard, Kuhn, and Suomi 1998a, 1998b; Westergaard and Suomi 1994b, 1994c, 1994g, 1995a; Westergaard et al. 1997, 1998c; Wojciechowski 2007). J. R. Anderson and Henneman (1994) noted that when their capuchin monkey subjects were presented with several tools of varying thickness that were out of the visual field of the apparatus, both chose tools of the correct diameter. The authors suggested that this could be interpreted as evidence for the presence of the monkeys' mental representation of the correct tool and the holes.

Pushing, the other type of Inserting and Probing task, has also been richly documented. In these cases, the capuchin monkeys Insert sticks into one end of narrow pipes, tubes, and tunnels and push the food reward out the other end (Harlow 1951; Klüver 1933, 1937). Although they were not the first to employ this methodology, Visalberghi and her colleagues have produced the most creative and definitive studies, revealing some of the underlying cognitive processes (Visalberghi and Limongelli 1991, 1992, 1994; Visalberghi and Trinca 1989, 1990). Visalberghi and Trinca (1989) reported that three of their four subjects spontaneously solved the pushing task. Their remaining individual, like Watson's (1908), was unsuccessful. Despite

A capuchin monkey Inserts a stick into a tube to dip for honey. Photograph courtesy of Elisabetta Visalberghi.

this initial success, subsequent phases of the study revealed that the capuchin monkeys did not appear to utilize a mental representation of the proper tool for the task. Visalberghi and Trinca (1989, 520) stated that the monkeys "did not show beforehand modification of the tool, but mainly modified it when their action was unsuccessful." Further, they noted that "at this level, a major difference with chimpanzees emerges" (Visalberghi and Trinca 1989, 511). This was revealed in a modification of the tube task, known as the *trap tube*. The trap is a hole in the bottom of the tube into which the food reward can fall and thus be lost to the monkey. If the monkey Inserts and pushes with the stick from one end of the tube, it can push the food out successfully. But if the monkey pushes from the other end, the food falls into the trap. When tested with the trap tube, capuchin monkeys did not consistently push from the correct end. They seemed not to anticipate "the effects of their actions on the reward" (Visalberghi and Limongelli 1994, 15). Visalberghi and Limongelli (1994, 21) stated that "our study highlights the difference between success in using a tool and understanding of the cause-effect relations involved in its use."

Westergaard and Suomi (1993b) provided nine monkeys with walnuts, stones, and sticks. After cracking the walnuts with stones, three of the subjects used sticks as Probing tools to extract the meat from the shell. Because two tools were used sequentially to complete this task, Westergaard and Suomi referred to the monkeys' use of a Tool Set.

Scratch, Rub

On several different occasions, wild spider monkeys were observed using sticks to Scratch various parts of their bodies (Lindshield and Rodrigues 2009).

Cut

Wild capuchin monkeys have been observed using stones as Cutting tools. Mannu and Ottoni (2009) recorded six instances in which flat stones, held with the flat sides vertical, were used like hatchets by the monkeys to Cut through branches, tubers, or cacti (*Opuntia* spp.).

Westergaard et al. (1997) provided fourteen captive capuchin monkeys with an apparatus containing live ants (*Tetramorium caespitum*). Access to the ants was prevented by a piece of clear acetate. Three of the monkeys used rocks as Cutting tools to open the acetate, allowing Insertion of a probe to extract the insects.

Westergaard and Suomi (1994d) presented fifteen capuchin monkeys with quartzite stones and a container that held sweet syrup. The container was covered with clear acetate, preventing access to the reward. Three of the capuchin monkeys held a stone against the acetate and moved it back and forth while pushing, thus Cutting through the covering. The authors reported that the monkeys most frequently performed this behavior by holding the tool in one hand. Cutting behavior was also performed by capuchin monkey subjects in a later study using a similar apparatus (Westergaard and Suomi 1997c).

A capuchin monkey working on the trap-tube task. The trap is the cup suspended below the tube. If the monkey pushes the food reward over the cup, it falls into the cup through a hole in the bottom of the tube and is lost to the monkey. The trap tube has been widely used in experimental studies of primate tool use. Photograph courtesy of Elisabetta Visalberghi.

In still another similar study conducted with seventeen capuchin monkey subjects, three used stone tools to Cut through a plastic covering to get peanut butter (Westergaard and Suomi 1995d). In an experiment designed to assess aspects of food sharing, tool sharing, and cooperation, eleven capuchin monkeys were given the opportunity to use sharp-edged stones to cut through a piece of acetate that prevented access to hazelnuts. Three of the subjects demonstrated proficiency in this task, in the manner described in Westergaard and Suomi (1994d). Westergaard and Suomi (1996) examined capuchin monkey hand preferences in a Cutting task. The subjects as a group demonstrated a clear right-hand bias (95%). In another experiment, Westergaard and Suomi (1997a) identified four different grips that the monkeys used to hold stones while Cutting. Capuchin monkeys also used stones to remove sausage casings around bones that had been coated with honey. The bones were mounted inside an apparatus that prevented the monkeys from using their teeth. Subsequent microscopic examination of the bones revealed cut marks made by the stone tools (Westergaard and Suomi 1994e).

Stones were not the only Cutting tools used by these capuchin monkeys. In three studies, six of eighteen subjects modified and used bamboo tools, three of nine monkeys used bone fragments, and three of fourteen used copper nuggets to Cut through acetate barriers to get food rewards (Westergaard and Suomi 1994f, 1995a).

Jalles-Filho (1995) reported that four zoo-living capuchin monkeys each placed a stick against the seam along the top of a box and moved it in a "back-and-forth" motion described as "cutting" in an attempt to open the box.

Block

A pet capuchin monkey held a cloth towel vertically between himself and a toy dragon presented by a human, thus forming a "shield" (Gibson 1990).

Prop and Climb, Balance and Climb, Bridge, Reposition

The captive capuchin monkeys studied by Bierens de Haan (1931) and Harlow (1951) Balanced and Climbed sticks to reach suspended food. Each monkey stood a long stick on end beneath the food, then scrambled up the stick and snatched the food before both tool user and tool fell to the substrate. Bierens de Haan's (1931) subject also Propped and Climbed a long stick against a vertical surface, forming a simple ladder that could be Climbed to reach the incentive. A capuchin monkey studied by Klüver (1933, 1937) moved a box beneath suspended food on several occasions and also successfully Repositioned boards to obtain a reward. In some instances, the monkey also used a stick to Reach the food while standing on the box. Urbani (1999) reported that a juvenile male capuchin monkey stood a detached, straight branch vertically and used it repeatedly to support and balance his upper body as he observed birds at a distance.

Bierens de Haan's (1931) capuchin monkey subject stacked boxes and tins to procure suspended food. This capuchin monkey made towers containing up to three elements and carried the elements several meters to the point of use. Since the experimenter presented stacking problems in a series of gradually increasing difficulty, the possibility of shaping or successive approximation of the solutions to the most complex problems cannot be eliminated. As detailed in chapters 1 and 7, each of the boxes in a tower is a tool, but the finished tower is not, because it is not held or manipulated in its entirety. The tower is a Construction.

Hang

A female woolly monkey picked up a piece of rope, tossed one end over the top bars of her enclosure, pulled the end back down, and then held the two ends while swinging back and forth (Darrow 1979).

Contain

K. Phillips (1998) observed wild capuchin monkeys using leaves as "cups" to remove and drink water from a tree cavity. This behavior was observed on several occasions and was performed by multiple individuals. Containers may also be utilized during food processing. Nolte (1958) provided a description of brown capuchin monkeys (*Cebus* sp.) placing overripe, sticky bananas on bread, apparently to keep their hands clean while eating. D. Katz and Katz (1936) noted that captive monkeys wrapped sticky bananas in leaves before picking them up. Since these authors worked with a colony that included *Cebus* and *Cercopithecus* species, and they do not specify which species wrapped the bananas, we can't be sure. However, the similarity to observations by Nolte (1958) makes it probable that it was the capuchin monkeys. Fragaszy and Adams-Curtis (1991) observed capuchin monkeys that placed small objects into opened nutshells. Gibson (1990) described a pet male capuchin monkey that placed objects on paper or towels to transport them or to prevent them from falling through the floor of his cage. He also wrapped food or favorite toys in towels, which Gibson interpreted as purposeful concealment. When presented with hot foods, he covered them with paper or cloth before picking them up.

Westergaard and Fragaszy (1985) studied the behavior of a group of eleven capuchin monkeys that were provided with manipulable objects for environmental enrichment. Seven of the monkeys were observed using scoops, cups, and containers to hold drinking water or small objects. When using tools for drinking, the monkeys were observed carrying the objects from as far as 2 to 3 meters away to fill them from a water spigot. Individuals as young as 10 months old used tools to Contain drinking water. Urbani (1999) observed a juvenile male that repeatedly dipped half an orange rind into water and used it as a cup. Chevalier-Skolnikoff (1989) observed individuals that used cups to hold dry food and to scoop water from a moat to drink. Squirrel monkeys (*Saimiri sciureus*) used small PVC caps to Contain water or pieces of food (C. Buckmaster, pers. comm.).

Absorb

K. Phillips (1998) relayed an observation of drinking by a wild capuchin monkey, which used a crumpled leaf to Absorb water from a tree cavity. It is unclear from the report if the monkey found the leaf in the crumpled condition or modified it. Mannu and Ottoni (2005) noted that wild capuchin monkeys used vines to collect water from within inaccessible tree cavities. This observation might also be classified in the Reach mode.

On two different days, Urbani (1999) observed a captive juvenile female capuchin monkey licking water that had soaked into a leaf found floating on a pond. Gibson (1990) described sponging by a pet capuchin monkey that used paper towels to Absorb and consume liquids and runny eggs. Westergaard and Fragaszy (1985) documented spontaneous use of paper towels as sponges for drinking water by members of a group of capuchin monkeys in an experimental setting, and in a different study (1987a) they saw eight of eleven individuals using paper towels, sticks, leaves, and straw as sponging materials to soak up juice. Paper towels were the most frequently used tool material in another study (Westergaard et al. 1995); five of eighteen capuchin monkey subjects used paper towels to Absorb sap that had been squeezed from sugarcane.

Wipe

C. Carpenter (1934) quoted Félix de Azara's early nineteenth-century statement that howler monkeys Wipe wounds with leaves that they have masticated first. Azara, however, did not observe the behavior but merely relayed the reports of others. A pet capuchin monkey used cloth towels to sweep debris out of his cage (Gibson 1990).

Affix, Apply, Drape

Wagner (1956) reported that wild spider monkeys attempted to remove human-inflicted blow darts or arrows and then placed leaves into the wounds. Boulenger (1936) and C. Carpenter (1934) relayed Buffon's account of behavior that differed, in that the leaves were reportedly chewed and Applied not by the wounded howler monkey, but by its fellow group members. Buffon also acquired this information secondhand. The use by howler monkeys of small sticks and fruit as "toilet aids" is noted without further description by Chippendale (in Kortlandt and Kooij 1963). Modern scientific confirmation of these behaviors is lacking.

Panger et al. (2002) documented a behavior, termed "leaf wrap," in which capuchin monkeys placed foods such as caterpillars (*Automeris* spp.) or *Sloanea terniflora* fruits in leaves before rubbing them on a substrate. Both of these foods have chemical or mechanical defenses that make them painful to handle directly. Panger et al. concluded that this process allowed the monkeys to handle the items comfortably while removing the irritating substances.

Spider monkeys "fur rub," or self-anoint, with a wide variety of substances. A. Richard (1970) observed monkeys picking limes from trees, biting and licking but not eating them, and then rubbing them all over their chests. The function was presumed to be repelling insects or benefiting the skin or hair. C. Campbell (2000) reported that spider monkeys picked leaves from three Rutaceae species, chewed them, and then Applied and Rubbed the resulting salivation and chewed leaves on their chests and axillary areas. Rubbing the sternal area against a tree branch or trunk as a form of scent marking usually followed. Laska, Bauer, and Hernandez Salazar (2007) provided a nearly identical description of this anointing behavior, except that leaves from the genera *Brongniartia*, *Cecropia*, and *Apium* were used. The authors concluded that this behavior functioned as a form of social communication. Captive spider monkeys observed by Dare (1974) self-anointed when presented with foods such as celery, green peppers, or "spicy substances." Additionally, *Cebus capucinus* and uakaris (*Cacajao rubicundus*) were reported to Apply and scrub with lemons and other "aromatic stimuli." Dare suggested that these behaviors were correlates of agitation or social status. Captive owl monkeys (*Aotus* spp.) were reported to anoint various parts of their bodies with "onions, garlic, chives, live millipedes, millipede-produced benzoquinones, and an extract of *P*[*iper*] *marginatum* leaves" (S. Evans et al. 2003; Zito, Evans, and Weldon 2003, 160). Zito, Evans, and Weldon suggested that the most likely explanation for this behavior was to repel insects. S. Evans et al. (2003, 135) tested thirty-four owl monkeys by presenting them with "onions, millipedes, and millipede-derived benzoquinones," all of which elicited self-anointing. The strongest response occurred with millipedes, producing the behavior in more than 80 percent of the monkeys.

Both wild and captive capuchin monkeys also anoint their bodies vigorously with a wide assortment of odorous objects and substances (Baker 1996; B. Beck, pers. obs.; J. Buckley 1983; Dare 1974; DeJoseph et al. 2002; Fiedler 1957; Field 2007; Gibson 1990; Hill 1960, 1967; Ludes and Anderson 1995; Nolte 1958; Oppenheimer 1968; Quinn 2004; Simmons 1966; Valderrama et al. 2000; Weldon 2004; also see Huffman et al. 2007 for a review). Ants, millipedes, onions, alcohol, perfumes, ammonia, peat, and orange peels are typical of the materials that elicit anointing. Hill (1960) relayed several earlier reports, including one by Linnaeus, whose brown capuchin monkey distributed snuff or tobacco on its back. The hands and, less often, the tail are used to Apply the substances.

Various explanations have been offered for these self-anointing behaviors. Capuchin monkeys, like owl monkeys, frequently apply millipedes to their bodies. The benzoquinones produced by the millipedes have been demonstrated to be "potently repellent to insects" (Valderrama et al. 2000, 2781), specifically mosquitoes (Weldon et al. 2003). Essentially, these millipedes function as benzoquinone dispensers for the monkeys, and the frequency of their use is correlated with rainfall, which is an indicator of mosquito density (Valderrama et al. 2000). Therefore, repelling insects was reported as one function clearly associated with capuchin monkeys self-anointing with millipedes (Valderrama et al. 2000; Weldon et al. 2003). This behavior is similar to anting in birds, both in the nature of the eliciting stimuli and in the intense, exaggerated character of the movements. Capuchin monkeys have also been documented Applying ants to themselves and, as with benzoquinones from millipedes, the presumed function is the release of formic acid (Falótico et al. 2004; Verderane at al. 2007). Although anting by capuchin monkeys was originally presumed to soothe or stimulate the skin (Longino 1984), more recent research has concluded that repelling ectoparasites is the more likely function (Falótico et al. 2004; Verderane at al. 2007).

Self-anointing with plants rather than insects may serve a different function. Baker (1996) reported that some plants used by capuchin monkeys for self-anointing are considered medicinal by indigenous peoples and are used specifically to treat skin ailments. While acknowledging that other possibilities exist, Baker posited that the use of these plants by capuchin monkeys might confer resistance to bacterial and fungal infections, particularly during the rainy season. DeJoseph et al. (2002, 925) provided additional examples of self-anointing with plants that provide "antibacterial, antifungal, and antiarthropodal properties."

Social context may also encourage self-anointing by capuchin monkeys (Gilbert, Brown, and Boysen 1998; Leca, Gunst, and Petit 2007; Meunier, Petit, and Deneubourg 2008). Meunier, Petit, and Deneubourg (2007, 1) concluded that self-anointing is socially facilitated with a "mimetic underlying mechanism," and they hypothesized that synchronization of behavior may provide a "group barrier to ectoparasite propagation."

The selection of items used for self-anointing has been demonstrated to be influenced by individual experience. Juveniles were more likely to use novel substances, suggesting that learning is a factor in this behavior (Baker 1997, 2000).

Urine washing is common for some New World monkeys, such as wild and captive capuchin monkeys, howler monkeys, and squirrel monkeys. As with self-anointing, several functions have been proposed, including communication (Oppenheimer 1968), sexual behavior (C. B. Jones 2003; K. Miller, Laszlo, and Suomi 2008), thermoregulation (Ruiz 2005), and social cohesion (Miller 1975 in C. B. Jones 2003). Perhaps related to capuchin monkey self-anointing or urine washing is Chippendale's note (in Kortlandt and Kooij 1963) of the use of sticks or fruit as "toilet aids."

We place all of this self-anointing and urine washing in the Apply mode, although it might also be classified in the Rub mode.

In another form of the Apply mode, in which the object is Applied to another inanimate object, Westergaard and Suomi (1994d) noted that five of their six capuchin monkey subjects activated a watering device by placing stones and stone flakes against a drinking tube. Squirrel monkeys performed the same behavior with a PVC cap (C. Buckmaster, pers. comm.).

A captive male capuchin monkey Draped cloth around his neck in a behavior compared with "playful self-decoration" (Rensch 1973). Gibson (1990) described a pet capuchin monkey that Draped himself with towels when sleeping, cold, or frightened. This individual also covered fear-eliciting objects, such as a toy dragon. Once they were covered, he hit or Pounded them. Chevalier-Skolnikoff (1989) described a capuchin monkey that placed a bucket on its head, presumably in playful exploration.

A capuchin monkey mother Applied chewed bits of bark to her infant's head wound. Ritchie and Fragaszy (1988, 347) suggest that this could be interpreted as a "primitive form of wound dressing."

Symbolize

In one of the earliest references to symbolic representation in captive primates, C. Carpenter and Locke (1937) presented a male capuchin monkey with a set of tasks to assess his ability to form associations between colored poker chips and specific foods, for example a red chip with a piece of orange, a blue chip with a peanut, and a white chip with a piece of banana. Instead of being asked to present a "correct" chip when a food was presented, the monkey was rewarded with the corresponding food when he handed a chip to an experimenter. The experimenters noted different preferences by the monkey on different days, and overall preference was not consistent across experimental sessions. To more fully assess whether or not the monkey used the chips as symbols for what they wanted, the experimenters added a green chip paired with bread, a non-preferred food item. The number of trades with the green chip was low. In the final phase of the study, the green chip was eliminated and a yellow chip was presented. The yellow chip was not rewarded, and responses to it became extinguished.

Westergaard et al. (1998b) also presented a captive male capuchin monkey with tasks designed to assess his ability to trade color-coded chips as tokens. In one phase, the monkey was presented with red and white chips. Trading a white chip was reinforced with a grape, while no reward was provided for a red chip. Consistent with Carpenter and Locke's results (1937), the subject readily demonstrated competence with this discrimination (Westergaard et al. 1998b). In subsequent tests, the monkey was provided with two specific tool tasks. One required the use of a dipping tool to extract sweet syrup, and the other involved using a rock as a Hammer to open a nut. The experimenters provided a yellow chip that was paired with a dipping tool, a black chip paired with a rock, and a red chip that was paired with nothing. When presented with an apparatus requiring the use of one of these two tools, the capuchin monkey demonstrated the ability to trade a token for the corresponding tool. The authors concluded that the monkey "selected chips in response to the demands of specific tasks" (Westergaard et al. 1998b, 105). In a replication of the study with two capuchin monkeys, Westergaard, Evans, and Howell (2007) investigated the degree to which the monkeys were able to generalize the use of tokens with a human experimenter and apply this to a conspecific. Ultimately, both monkeys consistently gave the appropriate tokens to their partners, based on the tool task that was provided. However, neither monkey demonstrated the ability (or willingness) to trade the corresponding tools in return.

Capuchin monkeys traded low-value foods for more preferred foods that were offered by experimenters. The subjects also traded food for a dipping tool that could be used to obtain sweet syrup. The monkeys were less likely to make this exchange if syrup was not available, suggesting that the tool itself was not inherently valuable (Westergaard et al. 2004).

TOOL USE: OLD WORLD MONKEYS

The Old World Monkeys currently live in Africa and Asia (and one small population lives on Gibraltar in southern Europe). They are all members of the Cercopithecidae, which is divided into two subfamilies, the Colobinae and the Cercopithecinae. The colobines, or leaf monkeys, have greatly reduced thumbs, chambered stomachs, and no cheek pouches; they eat mainly leaves. Tool use by colobines is uncommon. The cercopithecines have partly opposable thumbs, simple stomachs, and cheek pouches; they are omnivorous. Cercopithecines account for the majority of Old World monkey tool use.

Drop

Wild colobus monkeys (*Colobus* spp.) Dropped down branches or twigs on humans (anonymous in Kortlandt and Kooij 1963). Struhsaker (1975) confirmed this behavior for the red colobus (*Piliocolobus* sp.) of the Kibale Forest of Uganda. However, he saw only four cases of this behavior, all by adult males, in nearly 2,000 hours of observation.

Boulenger (1936) cited the observations by anonymous travelers in western Africa of patas monkeys (*Erythrocebus patas*, cercopithecines) Dropping down sticks, stones, and other debris on the occupants of river boats. Wild macaques (*Macaca* spp., cercopithecines) Dropped objects on humans (anonymous in Kortlandt and Kooij 1963; Kinnaman in Hall 1963). Kortlandt and Kooij (1963) noted that guenons (*Cercopithecus* spp.) Dropped branches on humans who pursued them. Formosan rock macaques (*M. cyclopis*) were reported to Drop objects from trees on predators (Poirier and Davidson 1979).

Pickford (1975) recorded deliberate Dropping by unidentified baboons (undoubtedly *Papio anubis*) in Kenya. The author wisely sacrificed data collection for escape, but he noted that he was probably a stranger to these baboons and, since the baboons approached him prior to Dropping, the behavior seemed more offensive than defensive. Baboons are also cercopithecines. Hamilton, Buskirk, and Buskirk (1975) studied chacma baboons (*P. ursinus*) that slept on rocky canyon walls and retreated to these ledges in the presence of humans and potential predators. In a one-year study, they recorded twenty-three incidents involving "voluntary" Dropping of a total of 124 stones from the ledges toward humans. The stones were "aimed in the sense that the stone is released in such a way that it falls toward the observer, the stoning individual having moved to a position on the cliff directly above or opposite the observer" (Hamilton, Buskirk, and Buskirk 1975, 488). The stones were picked up or pried from the substrate and either Thrown underhand or simply Dropped over the ledge. They averaged 583 grams, more than six times the average weight of stones randomly collected from the ledges. Since the Dropped stones were also about twice as large, it is not known whether the baboons were selecting for size, weight, or both. (Readers consulting this source should note a later correction concerning the size of the dropped stones: the dimensions of 165 and 104 centimeters should actually be millimeters.) Subadult and adult males were disproportionately represented among the Droppers. Dropping did not occur during encounters with other baboon troops or in response to either totally strange or very familiar humans. The behavior seemed to be elicited only by humans to whom the baboons were partially habituated.

Captive Guinea baboons (*Papio papio*) were observed Dropping stones on the edge of a hole in a piece of concrete. Their behavior enlarged the hole, providing them with access to soil in which they dug and otherwise explored (O. Petit and Thierry 1993). The authors specifically noted that the Dropping was forceful, resulting in large pieces of concrete being broken off. Pounding and Hammering with a stone sometimes occurred, but Dropping was the preferred technique, recorded in 68.4 percent of the cases observed over three months for one individual, and 93.6 percent of the cases during two months for another.

The acquisition of Dropping by a young male captive pig-tailed macaque (*Macaca nemestrina*) was observed during an experiment to investigate the use of a Reaching tool (B. Beck, pers. obs.). The tool was a threaded rod attached to the cage front with a chain. The chain had been provided to prevent the monkeys from wielding the tool in ways that might prove dangerous to conspecifics. Although an adult male learned to use the rod to secure food, the young male usually was denied access to the food by more dominant group members. The tool, however, was left unattended while they ate. The young male once snatched the tool while they were eating and climbed rapidly up the cage front. As he approached the top, the chain reached its full length, causing the tool to be snapped from his hands. It dropped on the feeding dominants, which fled, and the young male had momentary access to the food. He quickly learned to snatch and purposely Drop the tool, thereby gaining an increased share of the food. The dominant male finally resorted to intense threats and retention of the tool to prevent further disruption of his eating. N. Geberer (pers. comm.) relayed that the brother of this tool user learned several years later to intimidate his group mates by Dropping down a length of stick.

Throw

Lydekker (1910) reported that silvery leaf monkeys (*Trachypithecus cristatus*) broke off and Threw down branches on human observers, MacKinnon (1971) alluded to similar behavior by a wild proboscis monkey (*Nasalis larvatus*, colobines) but provided no details.

Jennison (1927) reported an unattributed observation that Northern Plains gray langurs (*Semnopithecus entellus*) drove a human from a path by attacking him with stones. Since this species is a more terrestrial colobine, such behavior might have been aimed or unaimed Throwing. An anonymous correspondent of Kortlandt and Kooij (1963) also reported aimed Throwing by a wild langur, but further details are lacking.

Captive mangabeys (*Cercocebus* spp., cercopithecines) reportedly Threw objects (anonymous in Kortlandt and Kooij 1963), but details were not provided. Armbruster (1921) and A. Schultz (1961) saw captive drills (*Mandrillus leucophaeus*, cercopithecines) Throw stones and other objects at humans, and Kortlandt and Kooij (1963) cited anonymous reports of aimed Throwing by captive drills and mandrills (*Mandrillus sphinx*, cercopithecines). Beck (1975) reported that captive mandrills agonistically Threw sand in the presence of humans. Wild guenons were said to aim and Throw sand and gravel on humans, and a captive was reported to have Thrown an object, but details are lacking (Kortlandt and Kooij 1963). Jennison (1927) and Kortlandt and Kooij (1963) relayed reports of de la Brue and other anonymous travelers who stated that wild patas monkeys Threw stones at humans. Kortlandt and Kooij's correspondents also mentioned Throwing by captive patas monkeys.

Derr and Quiatt (1992) described Throwing of stones by Japanese macaques (*Macaca fuscata*) in several different contexts, such as locomotion or play, but did not suggest any apparent function for these activities. A detailed multi-group analysis of unaimed Throwing by Japanese macaques was conducted by Leca et al. (2008) at ten different sites. The authors described the throwing as "mainly underarm, performed on the ground from a tripedal posture, and often accompanied by repeated vertical leaps" (Leca et al. 2008, 995). They reported no evidence of aiming in these groups, and concluded that the Throwing behavior augmented a display sequence. Kinnaman (in Hall 1963) described Japanese macaques Throwing down pine cones. G. Eaton (1972) noted that a captive male Japanese macaque incorporated unaimed Throwing of stones into his courtship displays. I. Tanaka (1995, 230) considered Throwing by Japanese macaques to be socially learned, with transmission based on "maternal kin and the social system of dominance."

Kortlandt and Kooij (1963) referred to observations by anonymous sources of Throwing by captive macaques, but they provided no further information. Boulenger (1936) referred to a wild Barbary macaque (*Macaca sylvanus*) Throwing down roof tiles at pursuing humans and relayed a similar, unrelated observation of escaped rhesus macaques (*M. mulatta*) effectively keeping pursuing London policemen at bay on several occasions by Throwing roof tiles at them. A zoo-living Sulawesi macaque (*M. nigra*) learned to Throw rocks over a cliff on its island habitat during agonistic display (Nickelson and Lockard 1978). A captive male and a captive female pig-tailed macaque (*M. nemestrina*) also Threw stones without aiming, sometimes as part of agonistic display and other times in contexts that were ambiguous (I. Bernstein, pers. comm.). Thierry et al. (1994) noted that Tonkean macaques (*M. tonkeana*) employed unaimed Throwing, but they provided no further details. Armbruster (1921) relayed the observations by Heck of a stump-tailed macaque (*M. speciosa*) Throwing sand and stones at zoo visitors. Heck felt that the monkey had learned the behavior by observing unruly children. Poirier and Davidson (1979) described a Formosan rock macaque that Threw rocks at people when they threw rocks at him. A young, captive, male long-tailed macaque (*M. fasicularis*) threw sand and rocks at humans (Artaud and Bertrand 1984). Additionally, he spontaneously and playfully Threw a rubber ring back to an observer who had thrown it to him.

In two experimental studies (Beck 1976; Hobhouse 1926), pig-tailed macaques and a rhesus macaque aimed and Threw a rod and a stick toward a food incentive before pulling in the food. Adult, free-ranging Japanese macaques that learned to use a stick to extract food from a transparent tube subsequently began to substitute Thrown stones to successfully acquire the reward (I. Tanaka et al. 2001; Tokida et al. 1994). These monkeys were reported to select the stones based on weight, and they could vary the force with which the stones were Thrown to thwart thieves within the group that waited at the other end of the tube. Short sticks were also Thrown in attempts to dislodge the food, but they were not an effective tool for this task.

Wild baboons (*Papio* spp.) Threw by aiming dust, sand, gravel, and vegetation at humans (anonymous in Kortlandt and Kooij 1963; Hamilton, Buskirk, and Buskirk 1978), and sand and gravel at a crocodile (Owen in Kortlandt and Kooij 1963). Hamilton, Buskirk, and Buskirk's (1975) reports of aimed Throwing by chacma baboons drew corroborative responses from Pettet (1975) and Pickford (1975). Pettet observed underhand

Throwing down of stones by olive baboons (*P. anubis*) in the Sudan. He felt that most were Thrown by adults and noted "uncomfortably good" aim. He also felt that partial habituation to humans was operative.

Lydekker (1910) cited Ludolph's assertion that wild hamadryas baboons (*P. hamadryas*) Threw sand and dust into the eyes of non-human predators.

Rawlinson (in Hornaday 1922) stated that wild baboons (genus unspecified) Threw and Rolled stones at soldiers. Unfortunately, the soldiers retaliated with machine guns.

Captive baboons aimed and Threw rocks, sand, gravel, bananas, and other objects in agonistic contexts (anonymous in Kortlandt and Kooij 1963; Armbruster 1921; Bolwig 1961, 1963; Hornaday 1922). Romanes (1892) quoted an amusing eyewitness account by Smith of a chacma baboon retributively Throwing mud on the parade uniform of a taunting military officer. The baboon was presumably a captive.

Aimed Throwing by baboons also occurs in non-agonistic contexts. A captive chacma baboon Threw sticks at a suspended banana to set it moving in an arc that brought it into reach (Bolwig 1961, 1963). A captive hamadryas baboon (Beck 1972, 1973a), a captive Guinea baboon (Beck 1973b), and a captive chacma baboon (Bolwig 1963) Threw rods behind or beside food incentives prior to using the rods to pull in the food. Kats (1972a, 1972b) trained hamadryas baboons to Throw objects at a target that they succeeded in hitting about a third of the time from distances as long as 5 meters.

The frequency of Throwing by baboons dictates a comment on the topography of Throwing behavior. Baboons usually throw while standing quadrupedally or while sitting. Washburn and Jay (1967) asserted that baboons are anatomically incapable of overhand Throwing in human fashion. Beck (1980) reported never having seen baboons Throw overhand, noting that there is, in fact, relatively little movement at the shoulder when they do Throw. The propulsive force is generated mainly by the hand, wrist, and forearm. Frequently the object is held between the downward-facing palm and the second through fifth digits, which are flexed. In this case the object is Thrown with a rapid extension of fingers, wrist, and forearm, in a pattern of "underhand backhand." Alternatively, the object is held in the same way but is raised to chest level by flexion of the forelimbs at the elbow and Thrown by rapid extension of the arm(s) and simultaneous extension of the fingers. The former pattern tends to be used when the animal is standing and the latter when sitting. The object is rarely brought above the animal's head in either posture. The same description applies to Throwing by macaques.

Drag, Roll, Kick, Slap, Push Over

Ball (in Lydekker 1910) and Hingston and Joleaud (in Hall 1963) reported that macaques deliberately dislodged stones with their hands and set them Rolling down slopes toward humans.

The earliest report of agonistic tool use by baboons is by a Lieutenant Shipp (in Romanes 1892), who in

A hamadryas baboon Throws a rod behind a pan of food prior to raking it in. Photograph by Leland LaFrance, © Chicago Zoological Society.

1810 was one of a party of twenty soldiers dispatched to recover clothing stolen by a chacma baboon from their barracks in South Africa. The baboon rejoined its troop and, as the soldiers approached the cliffs on which the animals had taken refuge, the baboons began to Roll "enormous stones" down on them. Unlike the soldiers in Rawlinson's account (see above), these retreated. Lydekker (1910) relayed similar cases but suspected that the rocks were accidently dislodged as the animals fled. Zuckerman and Zuckerman (1932) cited reports by Forbes and others that wild gelada baboons (*Theropithecus gelada*) and hamadryas baboons also Rolled stones, sometimes during conflict between social groups of the two species. Zuckerman and Zuckerman also doubted that the behavior was purposeful. Hall (1963) considered the same data and concluded that while such behavior might originate accidently during flight or display, it would not be surprising if the animals learned subsequently to impart a level of directedness to the behavior.

Maple (1975) reported that two captive olive baboons Slapped or hit a metal dish in the direction of the author while threatening him. Maple felt that the behavior was intentional.

Thierry et al. (1994) noted that Tonkean macaques Dragged branches during play.

Brandish, Wave, Shake

Mehlman (1996) detailed the stereotypic branch-shaking behavior performed by many species of monkeys, both wild and captive. This behavior is exhibited most commonly, but not exclusively, by adult males. These displays may convey a threat or excitement; they are directed within and between groups, and also to potential mates. Branch Shaking has been specifically described in Barbary macaques (Mehlman 1996), Japanese macaques (Modahl and Eaton 1977; Wolfe 1981), and Formosan rock macaques (Poirier and Davidson 1979). Modahl and Eaton (1977) provided a summary of all species of primates in which this behavior has been documented.

Western red colobus (*Piliocolobus badius*) have been observed Waving a long leaf as part of social play. They also incorporated sticks and long leaves into agonistic leaping displays (Starin 1990).

Deag and Crook (1971) described a type of Barbary macaque behavior called *agonistic buffering*. In extreme instances, a male picked up and presented an infant to another male that he sought to approach. Since some characteristics of cercopithecine infants seem to inhibit intraspecific aggression, the behavior effectively permitted the male to achieve peaceful proximity. Captive Barbary macaques carried dead infants and used them in male–male interactions in much the same way that they used live infants (Merz 1978). Itani (1959) reported that male Japanese macaques carried infants in potentially agonistic interactions with conspecifics.

Like some male macaques, male olive baboons solicit infants to cling to their chest or, if this fails, they roughly pick up and carry the infants. The behavior occurs when a male seeks to approach another male without aggression (Ransom and Ransom 1971). Kummer (1967) reported that subadult male hamadryas baboons carried infants to mitigate potentially agonistic interactions with adult males. G. Boese (pers. comm.) observed captive Guinea baboon males picking up and holding out infants either to inhibit attack by other males or to gain proximity to females. Hamilton, Buskirk, and Buskirk (1978) reported similar agonistic buffering by wild male chacma baboons, but they did not note whether the infants were actually held or carried. In all of these cases, infants were clearly used to inhibit aggression or force affiliation. Agonistic buffering, in turn, is part of a larger set of social behaviors in which primates and probably other animals enlist the aid of others to further their interests in social interactions with third-party conspecifics. Only those in which a second party is actually carried or held conform to our definition of tool use.

Bait, Entice

Modahl and Eaton (1977) postulated that the stereotyped branch-shaking displays of male Japanese macaques influenced mate choice by females, that is, the displays Enticed females.

Western red colobus monkeys used pieces of vegetation to invite social interactions. Juveniles manipulated Detached objects while directing an obvious play face at a conspecific. In the observed cases, this quickly evolved into a play session between the juveniles. The object was then either abandoned or incorporated into the play bout. On at least one occasion, a juvenile manipulated a piece of Detached termite nest while directing a play face at a human observer, which was interpreted as a cross-species invitation to play (Starin 1990). Starin also noted that adult male red colobus monkeys Detached branches and used them to challenge other males. If the invitation was accepted, the males engaged in a ritualized tug-of-war with the tool. These contests appeared to function as definitive tests of dominance, replacing physical aggression.

Van Lawick–Goodall (1970) saw a captive guenon Bait a dog with bread.

A male olive baboon held and carried an infant when he sought to be groomed by the infant's mother (Ransom and Ransom 1971).

Club, Beat

Juvenile western red colobus monkeys Detached long leaves and hit each other in play. This behavior was part of a larger "king of the mountain" game in which players attempted to topple each other from the leaf cluster where the tool was originally Detached. Hitting with the leaf also prevented others from acquiring leaves for use as play weapons (D. Starin, pers. comm.).

A long-tailed macaque Pounds shellfish with a stone. Photograph courtesy of Michael D. Gumert.

Pound, Hammer

A. Carpenter (1887) frequently observed wild crab-eating, or long-tailed, macaques (*Macaca fascicularis*) Pounding or Hammering oysters with stones. While Carpenter felt that the stones were relatively small (to a human, if not to a monkey), the behavior did result in shell breakage and access to the edible interior. The monkeys carried suitable rocks over distances as great as 75 meters to the low tide line where the oysters were captured. Malaivijitnond et al. (2007) observed the same behavior but noted that in addition to oysters (*Saccostrea cucullata*), the monkeys cracked and consumed detached gastropods (*Thais tissoti*), bivalves (*Gafrarium divaricatum*), and crabs (*Thalamita danae*). Despite being ubiquitous, barnacles (genus not identified) were avoided. The monkeys were observed transporting their tools as they hunted, utilizing hammers weighing as much as 1.78 kilograms. Adult long-tailed macaques generally weigh between 4 to 6 kilograms (Fooden 1995).

These macaques employed different forms of Hammering. Foods were opened by Pounding on them with hammer stones on horizontal surfaces, but they also "axed" shells to remove them from more vertical substrates. Hammer stones with different properties were chosen for each behavior. The monkeys also Hammered with the pointed end of auger shells to remove oysters from rocks (Gumert, Kluck, and Malaivijitnond 2009). Wild chacma baboons Pounded or Hammered tough-skinned fruits with stones (Marais 1969).

Hladik (1973) relayed observations by de Turckheim of Barbary macaques Pounding scorpions with stones. Prior to eating scorpions, unidentified wild baboons Pounded them with stones to immobilize them (Davison in Kortlandt and Kooij 1963; Vachon in Hladik 1973; Watson in Oakley 1961).

A Guinea baboon Pounds concrete with a stone to get access to underlying soil. Photograph courtesy of Odile Petit.

Yerkes (1916) reported that a captive crab-eating macaque Pounded nails and a lock with a stick and a hammer. The monkey had not been trained, but it did have the opportunity to observe carpenters. A captive Guinea baboon was observed using a stone hammer held with both hands to repeatedly Pound the edge of a hole in a piece of concrete (O. Petit and Thierry 1993).

Pry, Apply Leverage

Oyen (1978) observed wild olive baboons ingesting pebbles, which we do not consider to be tool use, because whatever the purpose of the behavior, the pebbles are internally employed. But one adult male used a stick to scatter or sort small stones lying on soil. Next, he Inserted the stick into the clay matrix, Probed for pebbles, and used it to Pry the pebbles from the clay. The sorting behavior was then repeated, and the pebbles were ingested. The whole sequence with the Prying stick tool was observed several times on one occasion but was never seen again, even though ingestion of pebbles was subsequently observed.

A captive chacma baboon used sticks as levers while Digging (Bolwig 1963). Captive hamadryas baboons have been observed to Pry or Apply Leverage with steel rods in the crevice between the wall and door of their enclosure (B. Beck, pers. obs.).

Dig

Jobaert reported to Kortlandt and Kooij (1963) that wild colobus monkeys, wild mangabeys, and wild baboons (genera not specified) Dug with sticks to widen the entrances of subterranean insect nests. Messeri (1978) reported that yellow baboons (*Papio cynocephalus*) on the Somali coast ate crabs that they dug out of the sand by hand. A cuttlefish bone partially covered in sand was found near two holes dug by a baboon. Messeri proposed that the baboon used the bone as a Digging tool.

Oyen (1979) observed an olive baboon using a twig to separate small stones from clay prior to swallowing them. The author noted that the baboon first held the twig "pencil-like, then shifted to a scalpel-like modified precision grip" (Oyen 1979, 595). He concluded that these manipulations allowed the monkey to Dig a larger number of stone fragments from the compacted soil.

A captive chacma baboon Dug with sticks in a context that was unclear (Bolwig 1963). A male guinea baboon living in a zoo used a 40-centimeter-long branch to scrape the soil inside an existing hole. A juvenile male long-tailed macaque used a variety of items, such as a tumbler, stick, bowl, stone, or lid, to Dig in the sand (Artaud and Bertrand 1984).

Jab, Stab, Penetrate

Bhat (1990) relayed an observation by Sadananda, who witnessed a lion-tailed macaque (*Macaca silenus*) puncturing a red ant (*Oecophylla smaragdina*) nest with a stick. The monkey ate the ants that swarmed onto the stick. Once the nest was nearly empty of adult ants, the macaque opened it and ate the immature ants that were inside.

Reach

Roth (in Kortlandt and Kooij 1963) reported that wild baboons (genus unspecified) Reached with sticks to flush insects from beneath stones. Long-tailed macaques were observed to use leaves during bouts of social grooming (Fuentes et al. 2001).

Guillaume and Meyerson (1934) reported that a captive agile mangabey (*Cercocebus agilis*) learned to use a stick to Reach food, but it did so awkwardly. Pollack (1998) observed several grivet monkeys (*Chlorocebus aethiops*, formerly referred to in older publications as vervet monkeys, *Cercopithecus aethiops*) using sticks and other provided tools to obtain pieces of grass growing outside of their enclosure. R. Horwich (pers. comm.) observed a zoo-living male grivet using a leaf vein approximately 20 centimeters long to Reach a potato chip outside of his enclosure. Subsequently, the same male persisted in using the stem to Reach a banana peel, which he ultimately obtained after much effort. Then he quickly discarded the peel, and it became apparent that his goal had been another potato chip that was hidden beneath it.

Gatinot (1974) described the spontaneous use of a stick by a zoo-housed patas monkey to Reach bits of food thrown by visitors. Simple experiments revealed that the monkey could alter its tool use to adapt to subtle situational variations.

In captivity, hamadryas baboons (Beck 1972, 1973a; Kats 1972b), chacma baboons (Bolwig 1961, 1963; Hall 1961), Guinea baboons (Beck 1973b; Guillaume and Meyerson 1934), a yellow baboon (Nellman and Trendelenburg 1926), an olive baboon (D. Choudhury, pers. comm.), and unspecified baboons (Protopopov in Klüver 1937; Rushkevich in Ladygina-Kohts and Dembrovskii 1969) used sticks and rods to Reach toward and rake in food. Bolwig's (1961, 1963) subject used short sticks to procure other sticks that were long enough to reach the food (Sequential Tool Use), and retrieved and used sticks that were not in the same visual field as the incentive. Beck's (1972, 1973a, 1973b, 1974) baboon tool users learned to use the tools by trial and error, and there was no evidence for socially mediated learning by other group members, despite their watching hundreds of successful trials. Beck (1974) hypothesized that the limited ability of baboons for socially mediated learning of tool use accounted for

A hamadryas baboon uses a rod to Reach a pan of food. Photograph by Leland LaFrance, © Chicago Zoological Society.

the lack of customary tool use in wild baboon populations.

Two captive olive baboons kept in laboratory cages with mesh floors learned independently to use a rod to secure food placed out of reach below their cages (Benhar and Samuel 1978). These animals had previously learned to use their tails to assist in obtaining objects through the mesh and generalized this skill to use of the rod when it was provided. The observers did provide some tuition, but it did not seem to have been critical in acquisition of the tool behavior.

Another variant on the use of a Reaching tool was produced spontaneously by a zoo-housed female Guinea baboon (D. Garvey, pers. comm.). This baboon was one of a group that occupied a cage that was elevated about 70 centimeters above the floor of the building. Food thrown by the visitors or scattered by the baboons themselves often ended up on the floor, where it was out of reach. The female had an infant and, at the age of about 6 months, it began to squeeze through the cage bars, drop to the floor, and eat the food found there. The mother learned to hold the infant by the tail during these forays, and when the infant had stuffed its cheek pouches with food, she pulled it back into the cage and stole the food for her own consumption. Watson (1908) saw similar behavior by a rhesus macaque.

R. Pietsch (pers. comm.) observed captive lion-tailed macaques using sticks as Reaching tools to capture frogs and toads that were floating or swimming in the deep areas of a moat that surrounded their enclosure. J. R. Anderson (1985) described Tonkean macaques that learned to acquire honey from a pan by manipulating a metal rod attached to their enclosure. Acquisition of this behavior involved considerable assistance from the experimenter, such as placing the rod into the honey.

A pet Tonkean macaque spontaneously used a leaf stalk as a Reaching tool. He consistently chose the longer of two tools when the reward was placed farther away. The authors concluded that the monkey used planning and foresight to solve the task (Ueno and Fujita 1998). Veino and Novak (2003, 47) also cited "planning strategies" in response to the successful performance of their rhesus macaque subject, which transported his Reaching tool between different locations.

Artaud and Bertrand (1984) reported that a captive long-tailed macaque mastered a traditional raking task in six sessions of five trials each. The monkey spontaneously threw objects beyond his reach and then used a rake to get them back, leading the authors to conclude that the task itself was inherently rewarding. Zuberbühler et al. (1996) described a dominant male long-tailed macaque using a Reaching tool to obtain fruit that had fallen from a tree outside of his enclosure. Subsequently, other group members performed the same behavior. When this alpha male was involved in using a tool, conspecifics were more likely to handle objects from the same object class, providing evidence for socially mediated increased interest in the task (stimulus enhancement).

The results of experiments designed to test the capacity of macaques to learn to use tools to Reach food incentives are discrepant. In one experiment, captive pig-tailed macaques learned without training to use a rod to get food, but the first solution occurred only after more than eight hours had elapsed (Beck 1976). Hobhouse (1926), Klüver (1937), Shepherd (1910), and Verlaine and Gallis (in Hooton 1942 and in Spence 1937) reported successful solutions by crab-eating macaques and rhesus macaques of such problems using sticks, cloth, rope, and wire. Hobhouse's (1926) rhesus macaque learned to use a short stick to secure another of sufficient length to Reach the food (Sequential Tool Use). Warden (1940) reported that rhesus macaques learned to use one rake to obtain an out-of-reach food reward. Subsequently, a sequential task was introduced in which several rakes of gradually increasing lengths were required to secure the food. Ultimately, one rhesus macaque completed a task that involved four rakes, while another individual solved a task with seven rakes. Shurcliff, Brown, and Stollnitz (1971), as well as Metevier, Stonemetz, and Novak (2006), reported that fifteen rhesus macaques were able to use Reaching tools to obtain rewards. Two of these mastered a Sequential Tool task in which a rake was used to obtain a rod that was next employed to obtain a food reward from a transparent tube. Warden, Koch, and Fjeld (1940) concluded that shaping was necessary for rhesus macaques to learn to use a Reaching tool. Negative results, even with some shaping in some cases, were reported by Buytendijk (in Nellman and Trendelenburg 1926), Drescher and Trendelenburg (1927), Nellman and Trendelenburg (1926), Watson (1908), and Yerkes (1916). A group of nine pig-tailed macaques studied by Fornalé, Spiezio, and Grassi (2008) failed to use Reaching tools to get food outside of their enclosure, and a 3-year-old Japanese macaque made no attempt to establish contact between a Reaching tool and a reward (Natale, Potì, and Spinozzi 1988).

Although it took more than eight hours for Beck's (1976) pig-tailed macaque to learn to use a Reaching tool by trial and error, other group members did show some socially mediated learning (stimulus enhancement and social facilitation), and one group member imitated the user's behavior.

There is an extensive body of neurophysiological work associated with the use of Reaching tools by macaques. In general, these studies have been conducted in highly controlled settings, almost always involving physical restraint of the monkey during tool use, as well as the recording of brain activity. Typically, the tool-use task requires the monkey to rake in an out-of-reach food reward. Some tasks involve Reaching another tool that is then used to obtain a food reward (Sequential Tool Use), or to pick up food with a pincerlike tool. Japanese macaques have been used most often in this type of research (Hihara et al. 2003a, 2003b, 2003c, 2006; Iriki, Tanaka, and Iwamura 1996b; Ishibashi, Hihara, and Iriki 2000; Ishibashi et al. 1999, 2001, 2002; Maravita, and Iriki 2004; Mariyama et al. 2002; Obayashi et al. 2001b, 2002a, 2002b, 2002c, 2003; Yamazaki, Namba, and Iriki 2009). Rhesus macaques (Schulman 1973) and unspecified macaque species (Escola et al. 2004a, 2004b; Iriki, Tanaka, and Iwamura 1996a; Obayashi et al. 2000, 2001a) have also been studied. In general the macaques succeeded at these tasks in this setting, and in some cases the Reaching tool appeared to be represented in the brain of the user as an extension of its limb when the tool was being used. Ferrari et al. (2004) reported that pig-tailed macaques failed in these tool-using tasks.

Beck (1973a) was the first to report cooperative tool use by baboons. One partner was a young adult hamadryas male that had learned previously to use a rod to Reach food. The food incentive was placed out of his reach on a platform erected in front of his cage. The tool was placed in an adjacent cage occupied by

A pig-tailed macaque uses a rod to Reach a pan of food. Photograph by Leland LaFrance, © Chicago Zoological Society.

his parents. The two cages were connected by a small opening through which none of these three adults could pass because of their large size. However, there were five younger cage mates who were small enough to pass through the opening with access to both cages. The only way the young male could get the food was to obtain the tool somehow from the adjacent cage. Beck confessed that the experiment was designed to demonstrate that baboons are not capable of cooperation in instrumentation. At first the four youngest animals manipulated the tool ceaselessly but actually kept it away from the door and thus away from the young male. Occasionally they would get careless and the young male was able to snatch it from them and get the food. After four such solutions, spanning an accumulated time of eleven hours, there was clearly no evidence of cooperation. One of the five young baboons in the group was a young adult female that had a youngster sired by the tool user. At one point in the study, she was grooming the young male in his enclosure. She intermittently glanced at the food and into the adjacent cage where the other four young animals were playing with the tool. When the youngsters momentarily left the tool unattended on the floor, the female immediately stopped grooming, went into the other cage, got the tool, and brought it directly to the young male's cage. She held the tool for a few moments while the male watched her raptly. He made no attempt to snatch the tool, and she soon put the tool down in front of him. Twenty-one seconds later he had obtained the food and both were eating. Prior to this sequence, the female had not even touched the tool during the experiment. During the next two months, these two animals cooperated in the same way in 193 trials.

Insert and Probe

A yellow baboon in Kenya's Tsavo National Park was observed by J. Broda (pers. comm.) Inserting a slender branch or stem into what appeared to be a subterranean termite nest. The baboon Probed momentarily with the tool, withdrew it, and then ate insects directly from it. A captive chacma baboon Inserted and Probed with a blade of grass in a pipe stem, and then ate the oils that adhered to the grass (Marais 1969).

A captive mandrill Inserted sticks and stems into its infected ear (Vincent 1973). The animal quickly learned to choose tools of the most suitable length, thickness, and stiffness. It developed considerable dexterity in this tool use and persisted for forty minutes.

A long-tailed macaque flosses its teeth with a strand of human hair. Photograph courtesy of Kunio Watanabe.

Apparently, the treatment was successful. Bayart (1982) and Bayart and Anderson (1985) described an adult male Tonkean macaque that used tools such as plant stems to Probe his nostrils. An adult female Tonkean macaque exhibited the same behavior (Thierry 1991, as cited in Thierry et al. 1994). Wild long-tailed macaques also used sticks as nasal probes (Fuentes et al. 2001).

Nonaka (2000) reported that two long-tailed macaques living near a Buddhist shrine in Thailand used human hair as dental floss to clean between their teeth. In a follow-up study at the same site conducted by Watanabe, Urasopon, and Malaivijitnond (2007), many males and females of all ages (except infants) were found to use hair as dental floss. The authors concluded that this form of tool use was widespread among the monkeys living near the shrine and could be considered a cultural behavior. Masataka et al. (2009) documented adult females that exaggerated this tool-using behavior in the presence of their infants, and they concluded that this was a means of actively facilitating social learning. A survey found that 6 percent of 135 women living near the shrine had had hair taken from them by the macaques. The authors pointed out that "the people worshipped these monkeys as servants of God and respected and tolerated them even when the monkeys rode on their heads and pulled hair from people's heads" (Watanabe, Urasopon, and Malaivijitnond 2007, 943). The monkeys were also observed using coconut-shell fibers to floss their teeth.

In addition, Watanabe, Urasopon and Malaivijitnond reported that several wild long-tailed macaques in central Java and elsewhere in Indonesia, and one in Vietnam, used twigs to clean their teeth. Matheson et al. (2008) offered human hair, dental floss, ribbon, and coconut husk to eight ex-pet long-tailed macaques living in a sanctuary. Six of the monkeys flossed with hair or dental floss during the study, although hair was preferred. Matheson et al. (2008) suggested that flossing with hair might be widespread in this species, rather than being a site-specific cultural trait. In a study of a free-ranging group of 145 Japanese macaques, Leca, Gunst, and Huffman (2009) documented dental flossing by only one adult female. She used either her own hair or the hair of a conspecific.

Lion-tailed macaques living in a zoo were provided with a log containing drilled holes that were routinely filled with highly preferred foods such as peanut butter. The monkeys Inserted sticks into the holes to remove the incentives (Pietsch 1989). Beck (1980) reported that a young captive female pig-tailed macaque Probed with wire in a cage crevice into which a mouse had run.

Watson's (1908) rhesus macaque subject failed to solve the *tube task*, which requires an animal to Insert a stick or rod to push or pull out a food incentive. One of four of Veino and Novak's (2004) rhesus macaques successfully employed a tool within eighty seconds to solve the tube task on his first attempt. Despite several hours of exposure, as well as the opportunity to watch successful solutions by this cage mate, none of the other individuals solved the task. Free-ranging Japanese macaques studied by I. Tanaka et al. (2001) and Tokida et al. (1994) were shaped to elicit Inserting and Probing in the tube task. For example, a proper tool was initially provided and placed in the tube, then near the tube, and gradually moved to 5 meters away. Ultimately, the tool was completely removed so that the monkeys were required to find and utilize a tool on their own, which they did successfully. Laidre (2008) presented a group of sixty-two wild savannah baboons (*Papio anubis*) with horizontal and vertical tubes containing high-quality foods. On the second through the fifth days of the study, the experimenter placed tools within the apparatus to facilitate solution by the monkeys. None of the baboons solved either of these tasks during the five-day experiment.

A number of different Old World monkeys have been documented using dipping tools to extract desirable edibles from tubes with restricted openings that contained a sticky food reward. Access to the food requires Probing with a tool. Westergaard (1992, 1993b) reported success by two cohorts of baboons (*Papio cynocephalus*) aged 24 to 28 and 25 to 33 weeks. Of seven Sulawesi macaques studied by Babitz (2000), one subadult male successfully utilized tools in a dipping task. Lion-tailed macaques succeeded and mandrills failed in this task (Westergaard 1988, 1991).

A long-tailed macaque Detaches a piece of human hair as a dental flossing tool. Photograph courtesy of Kunio Watanabe.

A free-ranging adult female Japanese macaque used her infant as a tool for the tube task. I. Tanaka et al. (2001) and Tokida et al. (1994) stated specifically that while several adult females took food from their babies when they emerged from the tube, only one female, Tokei, specifically brought her babies to the tube and "actively pushed them in" (I. Tanaka et al. 2001, 513). Over several years, she performed this behavior with four of her infants. If the babies would not go into the tube, Tokei obtained the food by using another reaching tool, such as a stick.

Scratch, Rub

Galat-Luong (1984) reported that four different types of Old World monkey—grivets, Campbell's mona monkeys (*Cercopithecus campbelli*), drills, and sooty mangabeys (*Cercocebus atys*)—Scratched themselves with tools (usually sticks) for personal comfort. A wild adult female bonnet macaque (*Macaca radiata*) was observed on numerous occasions to insert tools such as a "short twig, dry stick, stiff leaf or grass blade, or a leaf midrib into her vagina and scratch vigorously" (Sinha

1997, 23). The author concluded that this behavior was in response to persistent irritation.

There are two reported cases of stone-grooming. In the first, a captive adult female Japanese macaque routinely rubbed a stone in a circular motion around her infant's eye. Occasionally, a stick or piece of hard food was substituted for the stone (Weinberg and Candland 1981). In the second case, a captive sooty mangabey performed the same behavior, which the author described as "repetitive and stereotypic" (Kyes 1988, 172). The mangabey sometimes substituted a twig, a hard piece of food, or a metal rivet for the stone. At times the grooming motion was made in the air near to, but not touching, the infant. Also, multiple stones were used during a grooming session, and they appeared to be chosen very carefully. At times a new stone was chosen after each stroke. Unlike the macaque, which only performed this behavior with her infant, the mangabey stone-groomed another adult female. The behavior was also performed by two other adult females from the same mangabey group with their offspring, including the mother of the stone-groomer that was the subject of this study. The behavior could be habitual and cultural in this group. Aside from the obvious similarity in the topography of this behavior, there is no other relationship between these two reports.

Prop and Climb, Balance and Climb, Bridge, Reposition

A primiparous western red colobus monkey and her 29-day-old male offspring were separated by a gap of approximately 1 meter in the forest canopy, and the infant could not cross. The female stood bipedally, surveyed the area, and then Repositioned a branch across the gap, allowing her son to join her (D. Starin, pers. comm.).

Bolwig's (1961, 1963) chacma baboon Balanced and Climbed sticks to get suspended food and in play. A chacma baboon studied by Hall (1961) spontaneously Repositioned a rectangular box that was lying flat on the ground. The monkey placed the box with the long end standing vertically and then climbed on the box to reach a suspended food reward.

Immature members of a troop of Sulawesi macaques and a troop of pig-tailed macaques carried fallen branches and pieces of lumber and inserted them in the mesh of the chain-link fence that enclosed their spacious outdoor corral. They then sat or displayed on the projecting ends (I. Bernstein, pers. comm.). Lion-tailed macaques observed by Westergaard and Lindquist (1987) performed the same behavior with branches; some of the resulting "shelves" were 2 meters long. Lion-tailed macaques also Propped tree branches or bamboo poles against vertical elements in their enclosure and Climbed them. Some of these tools were as long as 3.5 meters, and the monkeys carried them as many as 6.5 meters from the site where they were selected to the site where they were used as tools. Westergaard and Lindquist also observed two individuals pushing simultaneously on the same pole to Prop it. On two occasions, one of the macaques Balanced a pole and Climbed it. As he ascended, the monkey and tool both fell over. A juvenile female Japanese macaque studied by Machida (1990) repeatedly Propped a 2-meter-long log against the wall of her enclosure and Climbed it. Several months after witnessing this behavior, two other juvenile group members also Propped and Climbed a log. The author speculated that social learning was involved. Ducoing and Thierry (2005) observed several males in a group of semi-free-ranging Tonkean macaques spontaneously Propping and Climbing poles against a fence. The same article reported a subsequent experimental study designed to assess the impact of social learning on the transmission of this behavior, in which none of the subjects exposed to a skilled conspecific demonstrator made any attempt to perform Propping and Climbing.

Hang

Captive Japanese macaques Hung a metal rod and chains from an elevated feature in their enclosure and swung or displayed on them, and they suspended other provided items, such as a chain with hooks and a chain with bars on either end, and swung on them (Candland, French, and Johnson 1978).

Contain

A subadult male captive green monkey (*Chlorocebus sabaeus*) began to use halves of peanut shells as water containers, and two group mates did likewise on the day following this "discovery" (J. Lombardi, pers. comm.). A captive juvenile male long-tailed macaque placed a stone and then a piece of metal into a tumbler and shook it to make a ringing noise. The same individual used pieces of cloth to carry items, such as a favorite toy or a live toad (Artaud and Bertrand 1984).

Captive rhesus macaques used a cap, a bowl, and a hollow rubber toy as tools to collect and drink water (Parks and Novak 1993). An adult male lion-tailed macaque living in a zoo used multiple items, such as a piece of plastic, a coconut shell, and an egg shell, to scoop and drink water from a pool. This behavior spread among the group, and its transmission seemed best explained by social learning (Kumar, Kumara, and Singh 2008).

In an experimental setting, Westergaard (1992, 1993a) provided infant baboons with a variety of objects. Infants as young as 4 months old placed objects inside of other objects. We agree with Westergaard's (1993a) characterization of this behavior as combinatorial object manipulation, rather than tool use, given that no other purpose was apparent. However, in a separate study, Westergaard (1992) documented tool use in a 33-week-old infant that spontaneously held a cup under flowing water, filled it, and then drank from the container. She repeated this behavior several times. This same female, and a 25-week-old male infant, both used cups to scoop water from a tub on multiple occasions.

Absorb

Free-ranging lion-tailed macaques dipped leaves into a tree hole to Reach and Absorb standing water. Users sometimes left the sponge in the cavity, and the tool was subsequently employed by a different individual (Fitch-Snyder and Carter 1993). Wild grivets in some areas heavily utilize pods from the *Acacia tortilis* tree for both moisture and the viscous exudate they produce. Hauser (1988) described one population in which individuals placed dry pods in narrow *A. tortilis* cavities that contained exudate from the tree itself. Two possible explanations for this behavior were offered. The monkeys might have been using the dry pods to acquire and consume the important exudate from the tree cavity, or they might have been softening the pods in the exudate so that the seeds could be more readily eaten. Of course, these benefits are not mutually exclusive. However, it should be noted that tool use would not be involved in the softening explanation, since the monkeys were not holding or manipulating a tool (the exudate).

A rhesus macaque (Reinhardt 1991) and a lion-tailed macaque (Pietsch 1989) were observed using their tails as sponges by dipping them into out-of-reach liquids, which were then consumed. While the behavior is notable for resourcefulness, it does not conform to our definition of tool use, since body parts are not "environmental objects."

Westergaard (1992, 1993a, 1993c) provided infant baboons (*Papio cynocephalus*) with absorptive materials and an apparatus containing liquids. Infants in each of two age cohorts (21 to 33 weeks, and 24 to 28 weeks), used paper towels repeatedly to Absorb and drink water or juice. Four infants, aged 12 to 24 months, were tested in a subsequent, similar study. A single occurrence of sponging to obtain juice was noted in one of the infants at 16 weeks of age. By 24 weeks of age, all subjects were repeatedly using sponges to Absorb and drink juice.

Wipe

A wild olive baboon Wiped sticky fluid from its face with a stone, and another Wiped blood from its lip with a corn kernel (van Lawick–Goodall, van Lawick, and Packer 1973). Worch (2001) observed a female red-tailed monkey (*Cercopithecus ascanius*) that had handled and consumed a sticky fruit. After eating, she selected a mature *Celtis durandii* leaf and placed it rough side up on a slightly inclined branch. She then Wiped each of her sticky fingers on both hands against the leaf. The entire sequence lasted approximately fifty seconds.

Crab-eating macaques inhabiting the Singapore Botanical Gardens used leaves to Wipe sticky or noxious material such as dirt, fungus, or ants from potential food items (Chiang 1967). The items were wrapped in the leaves and the mass was rubbed vigorously with the hands until the leaves absorbed the unwanted material and disintegrated. Some, but not all, of the members of each of the four troops that inhabited the Gardens exhibited the behavior, but Chiang's sampling did not allow evaluation of intra- and intertroop variability. Wheatley (1988) also observed crab-eating macaques in Ubud, Bali, Wiping food or objects with leaves or paper in what appeared to be cleaning. Although not tool use, the monkeys also wiped items in water or on other substrates. Similar behaviors such as sweet potato washing (Hirata, Watanabe, and Kawai 2001; Kawamura 1954) and washing grass roots (Nakamichi et al. 1998) by Japanese macaques also do not conform to our definition of tool use, because the water is not held or directly manipulated (see chapter 7).

Affix, Apply, Drape

Hamilton, Buskirk, and Buskirk (1978, 211) described an unusual case of baboon tool use: "[on] one occasion an adult male . . . picked up a fallen palm frond and carried it to a tree stump, placed the frond on the stump, then sat on the seat, faced into the morning sun, and sunned himself."

Wild crab-eating macaques placed leaves over caterpillars and worms and rubbed them to kill or immobilize them before they were eaten (Wheatley 1988). Hohmann (1988) relayed observations of lion-tailed macaques that placed a leaf over chrysalises and then rolled them to remove guard hairs. After preparation, the chrysalises were eaten.

A captive rhesus macaque Draped pieces of cloth over its head and face in what appeared to be play (Watson 1908). A juvenile male long-tailed macaque placed a piece of cloth over his teeth before making contact with hard items such as a walnut or a pipe (Artaud and Bertrand 1984).

Captive drills wiped fresh branches and oranges on their mouths, chins, and sternal glands in behavior that appeared to be analogous to self-anointing by *Cebus* (Fiedler 1957).

TOOL MANUFACTURE: PROSIMIANS

There are no reports of tool manufacture among prosimians.

TOOL MANUFACTURE: NEW WORLD MONKEYS

Detach

Drop

All of the New World monkeys that Dropped limbs and branches on human observers were also reported to Detach the missiles beforehand. Howler monkeys wrenched off dead limbs and branches with their hands (C. Carpenter 1934; Dampier in C. Carpenter 1934), spider monkeys broke off dead limbs and green branches (C. Carpenter 1935; Wagner 1956), brown capuchin monkeys did likewise (Klein 1974), and white-faced capuchin monkeys (*Cebus capucinus*) chewed through green branches and broke off dead branches with one or both hands, feet, or the tail (Oppenheimer 1973, 1977). Defler (1979b) reported that an adult male *Cebus albifrons* broke off and Dropped a dead branch that the author believed the monkey had previously selected and planned to use during a display.

Throw

Capuchin monkeys successfully aimed and Threw stones weighing between 20 and 60 grams (Westergaard and Suomi 1995c). When these stones were replaced with rocks that were too heavy (175 to 305 grams) for these monkeys to Throw, they struck the rocks against hard surfaces to Detach smaller pieces that they could Throw.

Drag, Roll, Kick, Slap, Push Over

Chevalier-Skolnikoff (1990) described aimed snag crashing in her direction on two occasions when a capuchin monkey Detached 2- to 3-meter-tall dead trees by pushing them over.

Club, Beat

A wild subadult male capuchin monkey broke off a thin 50-centimeter-long branch that it then used to Club a venomous snake (Boinski 1988). A 1-year-old capuchin monkey broke off a half-meter-long branch and a 20-centimeter-long branch and used them to strike its mother (Chevalier-Skolnikoff 1990).

Jab, Stab, Penetrate

During a play bout between two young capuchin monkeys, the older individual broke a dead branch from a tree and poked the younger (Chevalier-Skolnikoff 1990).

Reach

Golden lion tamarins broke twigs off larger branches and then used these to Probe under pieces of tree bark for insect prey or during allogrooming. The foraging tools were 8 to 10 centimeters long and always less than 0.5 centimeters in thickness (Stoinski and Beck 2001). The tamarins sometimes shortened a stick while foraging, but it was unclear if this improved the functionality of the tool.

The capuchin monkey studied by Klüver (1933, 1937) Detached a frond of a fern plant, lengths of tethered and anchored sticks, and pieces of anchored newspaper prior to using them as Reaching tools.

A captive capuchin monkey was exposed to an apparatus that contained a model of a snake with its head protruding from a hole. The monkey broke off small sticks from a larger branch and Reached under the apparatus and into the cylinder with them (Vitale, Visalberghi, and DeLillo 1991).

Insert and Probe

Capuchin monkeys Detached pieces of vegetation, such as a stick from a larger branch, to use these as Probing tools (Costello 1987; Lavallee 1999; Mannu and Ottoni 2009; Westergaard and Fragaszy 1987a; 1987b; Westergaard and Suomi 1995a; Westergaard et al. 1997).

Cut

A capuchin monkey Pounded stones on an experimental apparatus that contained sweet syrup, producing flakes that it then used to Cut an acetate cover over the incentive. The monkey also struck the stone with other stones to produce usable flakes, a form of Secondary Tool Use (Westergaard and Suomi 1994d).

Westergaard and colleagues conducted a series of experiments in which capuchin monkeys Detached stone flakes from larger stones but had no opportunity to use the flakes for Cutting. Although these flakes can't be called tools, because they were not used, the experiments do provide information useful for understanding tool manufacture in general and stone reduction in particular. In the first experiment (Westergaard and Suomi 1994d, 403), six of eleven capuchin monkeys manufactured stone flakes, employing three different techniques that the authors termed *stationary anvil*, *portable anvil*, and *hard-hammer percussion*. In the first, the monkey holds a stone and strikes it against a hard surface. In the second, the monkey holds a stone and strikes another stone set on a substrate. In the third, the monkey holds a stone in each hand and strikes them together. Westergaard and Suomi (1995b) presented seventeen capuchin monkeys with quartz and quartzite stones. The monkeys Threw stones against the hard floor, resulting in the production of (unused) stone flakes. Westergaard and Suomi (1995d, 1019) later refined their descriptions of stone-reduction techniques used by the capuchin monkeys. In addition to hard-hammer percussion described above, four new categories were introduced. These are *anvil*, *bipolar*, *throw*, and *soft-hammer percussion*. Anvil refers to hitting a stone against any hard surface and appears identical to the term stationary anvil from Westergaard and Suomi (1994c). Bipolar is defined as "striking a stone against another stone positioned on a flat surface." This appears identical to portable anvil from Westergaard and Suomi (1994c). Throw is the same as discussed earlier in Westergaard and Suomi (1995b). In soft-hammer percussion, the monkeys strike the stone with wood. There was a group right-hand bias (72%) among fourteen capuchin monkeys performing a one-handed stone reduction task. There were no relationships between lateral preferences and age and gender (Westergaard and Suomi 1996).

Affix, Apply, Drape

Spider monkeys observed by A. Richard (1970) picked limes from a tree and then rubbed them over their chests.

Subtract

Dig

Four of ten capuchin monkey subjects presented with containers that held buried peanuts used sticks to Dig in the soil and extract the food (Westergaard and Suomi 1995b, 4). During use, three of the monkeys modified the sticks by removing leaves and bark. Some sticks were also modified by breaking, which resulted in sharp points. These modifications made the tools more efficient, but the authors noted that "we cannot be certain that this was the animals' intended consequence."

Reach

Captive, free-ranging golden lion tamarins used the antennas from the radio collars they wore as tools while they groomed themselves and others. Stoinski and Beck (2001) noted that the monkeys removed the plastic tip of the antenna, exposing the narrower interior wire. The tamarins also removed the plastic casing approximately 4 cm from the end, allowing the tips to be more freely manipulated. However, given that individuals who did not use their antennas as tools also removed similar bits of the plastic casing, these modifications may not have been intended to make a better tool.

Spider monkeys were observed using unattached sticks to Scratch themselves, and sometimes chewing the tips of the sticks between bouts. The authors characterized this as modification, although no specific details were provided (Lindshield and Rodrigues 2009; Rodrigues and Lindshield 2007).

A captive female capuchin monkey broke a long stick into two pieces and used one of them to Reach food that was in an adjacent enclosure (Giudice and Pavé 2007).

Insert and Probe

Wild capuchin monkeys used sticks, pieces of branches, vines, and other plant parts as Probing tools (Moura 2002; Moura and Lee 2004a, 2004b, 2004c). Moura and Lee (2004a) reported that 43 percent of these tools were modified by removing leaves or stems before use. Captive studies revealed consistencies in the way in which capuchin monkeys modified tools prior to use as probes. Pieces of vegetation were first Detached, such as a stick from a larger branch. Leaves, bark, or other projections were then removed before use, and limp ends were removed in some cases. Additionally, larger tools might be broken, split, or chewed into smaller tools (in length and/or diameter) for ease of use (J. R. Anderson and Henneman 1994; Costello 1987; Lavallee 1999; Westergaard and Fragaszy 1987a, 1987b; Westergaard and Suomi 1995a; Westergaard et al. 1997). In a study requiring use of a Probing tool to extract live ants from an apparatus, the six capuchin monkeys that used tools modified their probes by Subtraction in 43 percent of their attempts (Westergaard et al. 1997).

Capuchin monkeys were tested on their ability to push a food reward out of a narrow clear tube with two potential tools requiring modification. The first was a taped bundle of three sticks that was too large to fit into the opening of the tube. The solution required disassembly and use of a single stick. The second was a single stick with a smaller transverse stick inserted into each end that prevented insertion into the tube. This solution required Subtraction of at least one transverse stick. Visalberghi and Trinca (1989) reported that each of their subjects succeeded at this task in some trials and did not succeed in others, leading the authors to conclude that they never fully understood the nature of the task. Klüver (1933, 1937) employed a similar methodology; his capuchin monkey subject failed to Subtract corks from the ends of sticks when the cork prevented use of the stick to push food from a tube.

When using stick pestles to extract fiber and sap from sugarcane, capuchin monkeys broke sticks into smaller pieces, resulting in shorter tools. They also removed features such as twigs, leaves, or bark to remove any obstructions to extraction (Westergaard et al. 1995). Breaking the sticks sometimes resulted in sharply pointed ends but "it was unclear whether [the monkeys] intended to produce specific forms or tool characteristics or merely reduce the size of the pestles" (Westergaard et al. 1995, 646).

Cut

Eighteen capuchin monkeys were presented with pieces of bamboo and an experimental apparatus containing a food reward covered with a piece of acetate. Five of the subjects manufactured tools from the bamboo to Cut through the acetate. Using their hands and teeth, they broke the bamboo apart, split or partially split pieces, and removed shoots, leaves, and stubs (Westergaard and Suomi 1995a). One subject modified 48 percent of the tools it used, and across all subjects the number of modifications to a single tool ranged from one to six.

A capuchin monkey studied by Westergaard and Suomi (1994d) manufactured stone tools in association with a Cutting task. In this instance, the capuchin monkey Pounded the stone against the experimental apparatus, and it also struck the stone with other stones, resulting in removal of flakes. In addition to using the flakes (described above under Detach), the monkey also Cut with the sharply edged core stone.

Affix, Apply, Drape

Capuchin monkeys that engaged in self-anointing with plants specifically extracted the pulp and juice. The fruit was Pounded on a hard substrate and, once opened, was held tightly against the chest while the soft contents were vigorously rubbed onto the body (Baker 1996).

Add, Combine

Reach

A male golden lion tamarin used the tip of his radio-collar antenna to groom himself or his mate. He regularly licked the antenna during grooming. Stoinski and Beck (2001) noted that this might have been a way to remove debris from the tool, or to make the antenna more functional for parting hair by adding saliva, a form of modification through Addition.

Westergaard and Fragaszy (1987b) observed a female capuchin monkey using tools that she coated with sweet syrup to touch and groom her genital area and treat wounds that she received from conspecifics.

J. R. Anderson and Henneman (1994) presented two capuchin monkeys with plastic tubes, sections of a tent pole, or pieces of a fishing rod that could be fitted together and used as a Reaching tool. One of the two demonstrated intense interest in Combining and separating the tools, but never attempted to use the finished product for Reaching.

Reshape

Reach

The capuchin monkey studied by Klüver (1933, 1937) Reshaped pieces of newspaper by crushing and rolling them to make Reaching tools of sufficient firmness and length to retrieve an incentive, but it failed to uncoil a piece of wire for use as a Reaching tool. J. R. Anderson and Henneman's (1994) male capuchin monkey subject extended a telescopic aerial and used it as a Reaching tool.

When using tools to treat a wound, an adult female capuchin monkey bit the end of the tool, creating a brushlike tip. She used both the modified and unmodified ends of the tool to explore the wound. Westergaard and Fragaszy (1987b) suggested that the brushlike end provided a broader surface for applying pressure. A mother capuchin monkey observed by Ritchie and Fragaszy (1988) modified several sticks and used them to treat her infant's head wound. The female first bit the end off a stick, resulting in a tool that measured about 11 centimeters long (Subtract). She then chewed one end of the stick, producing a frayed, brushlike tip that she used repeatedly to treat the wound (Reshape). When that tool was taken by another monkey, the mother fashioned an equivalent tool in the same way from another stick and used that to continue treating the wound.

Insert and Probe

Bortolini and Bicca-Marques (2007) noted one case in which a capuchin monkey held a twig against a large stone and then Pounded the end with a smaller stone. The tool was then used as a Probe.

A male capuchin monkey studied by J. R. Anderson and Henneman (1994) unbent a piece of wire and then used it to Probe for honey.

Scratch, Rub

Spider monkeys used sticks to Scratch themselves. The monkeys chewed the ends of the sticks that they applied to their bodies, either prior to or between Scratching bouts (Lindshield and Rodrigues 2009). The degree to which the tools may have been modified was not specifically reported.

Absorb

Captive capuchin monkeys observed by Westergaard and Fragaszy (1987a) Reshaped paper towels, sticks, leaves, and straw into sponges that they used to Absorb juice from an experimental apparatus. Paper towels were the most commonly used tool. Typically, the monkeys Reshaped the paper with their hands and mouth before using it, and frequently remodified it between uses.

Affix, Apply, Drape

Wild howler monkeys were said to "get" or "take" leaves which they then "chewed" and Applied to their own wounds or those of conspecifics (Azara in C. Carpenter 1934; Buffon in Boulenger 1936 and in C. Carpenter 1934). It can be inferred that the monkeys first Detached the leaves and then Reshaped them to increase absorbency. Wild spider monkeys self-anointed with various forms of vegetation that were chewed into a pulp, producing excessive salivation while chewing. C. Campbell (2000) reported observing this behavior with Rutaceae species, and Laska, Bauer, and Hernandez Salazar (2007) relayed similar observations with *Brongniartia*, *Cecropia*, and *Apium* leaves.

Capuchin monkeys that self-anointed with plant parts have been observed to Reshape those items prior to use. Plant pods, stems, and leaves were rubbed between the hands or manipulated orally until broken into tiny pieces. These small parts were mixed with saliva, presumably to assist with application, and then Applied energetically on the pelage (Baker 1996). A female capuchin monkey (see also Reshape: Reach) chewed bits of bark from a stick and Applied these to a wound on her offspring's head (Ritchie and Fragaszy 1988).

TOOL MANUFACTURE: OLD WORLD MONKEYS

Detach

Drop

Male red colobus monkeys Detached dead branches from trees prior to dropping them toward humans (Struhsaker 1975).

Throw

Macaques that Threw down pine cones (Kinnaman in Hall 1963) and roof tiles (Boulenger 1936) at humans presumably first Detached the missiles.

Drag, Roll, Kick, Slap, Push Over

Wild chacma baboons pried stones from the substrate with their hands before Rolling them toward humans (Hamilton, Buskirk, and Buskirk 1975). Since the stones were apparently tightly embedded in the soil, this behavior probably qualifies as Detachment.

Brandish, Wave, Shake

Silvery leaf monkeys Detached dead branches during displays (Lydekker 1910). Western red colobus monkeys Detached long leaves prior to incorporating them into bouts of social play (Starin 1990).

Bait, Entice

Western red colobus monkeys Detached branches before manipulating them to invite play from conspecifics. Adult male red colobus Detached branches and used them to challenge other males (Starin 1990).

Reach

Hooton (1942) noted that the macaque (species not identified) studied by Verlaine and Gallis manufactured tools. Spence (1937) indicated that this animal Detached branches for use as Reaching tools.

Insert and Probe

A mandrill Detached straws from a broom, and possibly grass stems and twigs from the surrounding vegetation, and Inserted and Probed with the objects in its infected ear (Vincent 1973). Lion-tailed macaques observed by Westergaard (1988, 1991) Detached twigs from large branches to use as dipping tools. Babitz (2000) noted that a Sulawesi macaque modified branches to create a dipping tool, but she provided no further details.

Long-tailed macaques removed hair from women and used it as dental floss (Watanabe, Urasopon, and Malaivijitnond 2007).

Scratch, Rub

A wild adult female bonnet macaque often Inserted tools into her vagina and Scratched or Rubbed with them (Sinha 1997). These tools were manufactured in several ways. She Subtracted the laminae of dry leaves, leaving the midrib that she then shortened to 1 to 2 centimeters long before use. She split dry leaves longitudinally and used one of the halves as a tool. She Detached short sticks from longer pieces of vegetation, and she broke dry sticks into pieces and selected a short tool from these.

Wipe

Crab-eating macaques studied by Wheatley (1988) Detached leaves from surrounding vegetation before using them to clean foods and other objects.

Subtract

Dig

A young male long-tailed macaque Detached a branch and then Subtracted the leaves and bark before using it to Dig in sand (Artaud and Bertrand 1984).

Insert and Probe

Free-ranging Japanese macaques studied by I. Tanaka et al. (2001) and Tokida et al. (1994) used sticks to extract food from a tube. One female shortened long sticks by repeatedly biting them. Once, she Detached a shrub from the ground and Subtracted the leaves and roots before using it as a tool.

Lion-tailed macaques observed by Westergaard (1988, 1991) Subtracted projections such as twigs and leaves that impeded the use of tools; modified tools were used for significantly longer periods than nonmodified tools.

A captive adult female Tonkean macaque manufactured probes that she Inserted into her nostrils to reach mucus. After selecting a stem, she sometimes stripped or shortened it before use (Thierry 1991 in Thierry et al. 1994).

Add, Combine

Insert and Probe

An adult male rhesus macaque, proficient in extracting food from a tube, was reported to combine tools "to achieve sufficient length" (Veino and Novak 2004, 118). No further details were reported.

Reshape

Hang

Candland et al. (1978) reported that an unattached metal rod was bent into an S shape by a Japanese macaque and then Hung and used as a swing. The authors subsequently gave the macaques another unbent bar, a chain with hooks attached, and a chain with unbent bars attached at each end. The macaques are reported to have used all of these as swings, after making unspecified modifications.

ASSOCIATIVE TOOL USE: PROSIMIANS

We could find no reports of Associative Tool Use by prosimians.

ASSOCIATIVE TOOL USE: NEW WORLD MONKEYS

Sequential Tool Use

On three occasions, wild capuchin monkeys used small hammer stones to loosen larger, embedded stones. These larger tools were then used to Hammer and pulverize pebbles, and the monkeys licked the powdery residue. During one of these instances, a monkey transported the first hammer stone from at least 10 meters away (Mannu and Ottoni 2009).

Two captive capuchin monkeys studied by Warden, Koch, and Fjeld (1940) used a series of up to eight tools to Reach other tools, ultimately acquiring one that they could use to get a piece of food.

Tool Set

Mannu and Ottoni (2009) reported six cases of Tool Sets. In two of the cases, a monkey Pounded on a cavity in a tree with a stone and then Probed the opening with a stick tool. In both instances, bees were living in the tree cavity. Two events involved monkeys Pounding on a crack in a rock and then Probing inside with a stick. The authors suspected that one of the monkeys was searching for a lizard or invertebrate. A pair of rocks was used in two other cases. One monkey used the stones to Dig a tuber out of the ground, and the other used the tools to Cut or crack a tuber.

Captive capuchin monkeys have used Tool Sets to Hammer open nuts and then Pry out the contents, as well as to Cut open an apparatus and then Insert a tool to obtain live ants (Westergaard 1993b; Westergaard et al. 1997).

Tool Composite

While standing on a box it had Repositioned below a food reward, a monkey used a stick as a Reaching tool to obtain the incentive (Klüver 1933, 1937). Westergaard and Suomi (1994d, 1994e) and Westergaard et al. (1996) described the use of stone hammers and chisels by capuchin monkeys to either remove the casing surrounding a bone or to break through the lid of a container holding a sweet liquid. The authors described this Tool Composite as additionally involving a Metatool, presumably because the hammer allowed the chisel to be used with additional force. Jalles-Filho (1995) and Jalles-Filho, Teixeira de Cunha, and Salm (2001) provided their subjects with a closed box containing a desirable edible. The monkeys wedged sticks into the seam of the lid and then Hammered the sticks in an attempt to access the contents. The authors considered this to be analogous to the hammer and chisel Tool Composites of Westergaard and Suomi (1994d, 1994e) and Westergaard et al. (1996), but they make no assertions about use of a Metatool.

Multi-Function Tools

Mannu and Ottoni (2009) reported two instances in which wild capuchin monkeys used a single tool in two different modes. In the first, a male Dug with a stone and then transported the tool 1 meter and Hammered open an edible seed. In the second, a female also Dug with a stone and then used the tool to Pound at four different sites. During the four minutes in which this sequence occurred, she transported the tool from 1 to 8 meters between foraging locations.

Secondary Tool Use

As described in detail under Tool Manufacture, there are examples of captive capuchin monkeys striking one stone against another to produce flakes (Westergaard and Suomi 1994d, 1995b), but in some cases the flakes were not used as tools. Even when the monkeys used the flakes to Cut something, it was unclear if their production was purposeful or incidental to object manipulation. Bortolini and Bicca-Marques (2007) described a less ambiguous case of Secondary Tool Use in which a capuchin monkey held a twig against a large stone and then Pounded the end with a smaller stone. After modification, the twig was used as a Probing tool.

OLD WORLD MONKEYS: ASSOCIATIVE TOOL USE

Sequential Tool Use

Schulman (1973) trained rhesus macaques to use a short stick to Reach a longer one that was used to Reach food. Hihara et al. (2003a) employed a similar methodology in which Japanese macaques used a short rake to Reach a longer rake, and the latter was used to obtain a food reward. Reports by Schulman (1973) and Hihara et al. (2003a) involved neurophysiological studies and were conducted in highly controlled circumstances. Obviously, the experimental design was highly influential in the production of these behaviors.

Hobhouse (1926) reported that a rhesus macaque learned to use a short stick to secure another that was long enough to Reach food. Warden's (1940) rhesus macaques mastered a sequential Reaching task involving successively longer rakes. One individual was able to solve a four-rake problem, while another mastered a seven-rake problem. Working with fifteen rhesus macaques, Metevier, Stonemetz, and Novak (2006) presented a task that required the subject to use a rake to Reach a rod, which could then be Inserted into a transparent tube to obtain a food reward. Only two individuals succeeded.

Captive chacma baboons studied by Bolwig (1961, 1963) used short sticks to get sticks and rods that were long enough to rake in food.

Obayashi et al. (2002a, 2002b, 2002c) presented Japanese macaques with a task in which two identical tools (rakes) were used differently to solve a task. The monkey was required to pull on a fixed rake to dislodge a food reward from a tube, and then use an unattached rake to reach the reward. Since the monkey was not responsible for the proper and effective orientation of the rake in the tube, we do not consider this to be tool use. As a result, the task does not qualify as a form of Associative Tool Use. Even if it did, we know now that these authors were incorrect when they stated that "only humans and chimpanzees could spontaneously utilize tools sequentially on a single object" (Obayashi et al. 2002c, 2349).

Multi-Function Tools

An adult male olive baboon used a stick to sort small stones lying on the soil (a form of Reaching), and also inserted the same tool into the substrate to Pry out pebbles (Oyen 1978).

6 Apes

THIS CHAPTER IS ORGANIZED BY GENUS: *HYLOBATES* FOR GIBBONS AND SIAMANGS, *Pongo* for all species and subspecies of orangutans, *Gorilla* for all species and subspecies of gorillas, and *Pan* for bonobos and all subspecies of chimpanzees. Generally, we did not identify species or subspecies, since there does not appear to be meaningful interspecific and intersubspecific differences in tool behavior that are independent of study site, and because species is rarely identified reliably in captive studies.

TOOL USE: GIBBONS

The hylobatid or lesser apes are members of the genus *Hylobates* (Groves 2001). They are geographically limited to Southeast Asia and are highly arboreal. Glickman and Sroges (1966) found that they are unlikely to manipulate objects on a substrate, and they experience difficulty when they try to do so (Beck 1967). C. Parker (1973, 1974) found their manipulative propensities to be roughly equal to those of cercopithecine monkeys but deficient in comparison with the great apes. Tool use is rare, perhaps because of limited manipulative dexterity (Abordo 1976).

Drop

C. Carpenter (1940) frequently saw wild white-handed gibbons (*Hylobates lar*) agonistically Drop dead branches in the direction of human observers. Ellefson (in L. Baldwin and Teleki 1976) saw adult males of the same species break off and Drop branches during intraspecies conflict. Geissmann (2009) stated that gibbons routinely break off and Drop branches as they produce their characteristic songs.

Throw

A captive young male white-handed gibbon aimed and Threw a banana peel at a cage mate in what appeared to be play (D. Judge, pers. comm.). An anonymous source cited by Kortlandt and Kooij (1963) noted Throwing by a captive gibbon.

Brandish, Wave, Shake

Chivers (1974) and Geissmann (2000, 2009) noted that gibbons routinely Shake branches at specific moments during their vocal displays. Geissmann (2009) described a captive adult female white-handed gibbon that augmented a specific portion of her song by forcefully sliding open the door of her sleeping box, producing a loud bang. This unusual behavior was documented during 53 percent of her vocalizations and consistently occurred at the climax of her great calls. This behavior did not occur in other contexts, except for a single instance in response to the recorded duet of another gibbon pair.

Reach

Boulenger (1936) cited the results of anonymous experimenters who found that captive gibbons learned to Reach toward and secure food with rakes. C. Cunningham (2006), C. Cunningham and Anderson (2004a, 2004b), C. Cunningham, Anderson, and Mootnick (2006), Drescher and Trendelenburg (1927), and Inoue and Inoue (2002) all found that captive gibbons are capable of drawing a rake towards themselves to Reach a desired food. However, in all of these cases, the food reward was placed between the rake and the ape by the experimenter; since the subjects did not establish the proper and effective orientation of the tool, the behavior does not qualify as tool use.

In exploring the ability of gibbons to properly orient a tool, C. Cunningham (2006) presented her subjects with a modified version of the rake task. After the apes were given considerable experience with a prepositioned rake, the rake was presented in such a way that the gibbons had to adjust the distal end to properly align it with the food reward. There were two such conditions. In one, the rake head was either 5 centimeters away from the reward on either the left or right and between the food and the gibbon but closer to the food, or 30 centimeters on the left or right and between the food and the gibbon but closer to the gibbon. In the 5-centimeter condition, the six subjects were able to acquire the food on 20 to 55 percent of the trials. A few successes were recorded in the 30-centimeter condition, and the author suggested that these might have been the result of chance manipulation.

Insert and Probe

C. Cunningham (2006) presented eleven gibbons with a transparent apparatus that included a rod prepositioned to rest in a sticky, sweet reward. The apes learned to extract the rod and consume the treat. After a number of these training trials, as well as demonstrations by a human who inserted the rod into the apparatus, the gibbons were tested on their ability to Insert the tool. The only successful insertion by a gibbon was attributed to a chance manipulation.

Scratch, Rub

S. Wich and T. Geurts (pers. comms.) observed a captive gibbon (*H. agilis*) hold a branch and use it to stimulate her genitals with repeated Rubbing and touching motions.

Hang

Rumbaugh (1970, 2006) reported that a white-handed gibbon fashioned swings by Hanging rope and hose on the mesh and bars of her cage. A 2.5-year-old siamang (*H. symphalangus*) was reported (Anonymous 1971) to use pieces of rope or a blanket to construct a swing. The swing was wrapped around an overhead bar of the enclosure and the loose ends were held by the ape to spin and wind up the tool, and then unwind it.

Absorb

Rumbaugh (1970, 2006) described the tool use of a white-handed gibbon that was provided with an automatic drinking fixture from which the water gushed forcibly. The gibbon learned to hold pieces of cloth in front of the dispenser until they became soaked, and then either sucked the Absorbed water from the cloth or let it drip into a puddle from which the animal drank directly. A young female white-handed gibbon observed by L. Baldwin and Teleki (1976) repeatedly picked up a leaf, Dropped it into a pool of water, retrieved it, and then sucked out the water. As the leaf crumpled during use, the gibbon began to dip her fingers and the leaf into the water, and then lick and suck water from her fingers and the leaf sponge.

TOOL USE: GREAT APES

There is an anecdote that circulates among zoo folk describing what happens when a screwdriver is inadvertently left in the enclosure of an adult gorilla, chimpanzee, bonobo, or orangutan. The gorilla would not discover the screwdriver for an hour until accidentally stepping on it. Shrinking in fear, the ape would approach the tool only after a considerable interval, venturing a cautious, tentative touch with the back of

A white-handed gibbon swings on a rope she had Hung over a bar in her cage. Photograph courtesy of Duane M. Rumbaugh.

the hand. Finding the screwdriver harmless, the gorilla would smell it and try to eat it. Upon discovering that the screwdriver was inedible, the gorilla would discard it indefinitely. The chimpanzee would notice the tool at once, seize it immediately, and try it out as a club, a spear, a lever, a hammer, a probe, a missile, a toothpick, and everything else, except as a screwdriver. The tool would be guarded jealously, taken by dominant group members, manipulated incessantly, and discarded from boredom after several days. The bonobo would also notice the tool immediately and begin vocalizing with great intensity. The tool would be quickly picked up, inspected, and then passed from bonobo to bonobo in a flurry of excitement. The good fortune of finding such a novel item would result in copulations, genital rubbing, and an enthusiastic chorus of screams. In the midst of all of the erotic intensity, the screwdriver would be dropped and forgotten. Like the chimpanzee and the bonobo, the orangutan would notice the tool at once but ignore it, lest a keeper discover the oversight. If a keeper did notice, the ape would rush to the tool and surrender it only in trade for a preferred food. If a keeper did not notice, the ape would wait until night and then proceed to use the screwdriver to pick the locks or dismantle the cage and escape. While this anecdotal comparison is scientifically inadmissible, it is a reasonably accurate portrayal of the disposition, curiosity, manipulative ability, and propensity for tool use of the four types of great apes.

TOOL USE: ORANGUTANS

The terminology in this section merits a brief explanation. Orangutans are exceptional among mammals, in that two forms of adult males exist. *Flanged* males are fertile and have fully developed secondary sexual characteristics. *Unflanged* males are fertile yet remain somewhat androgynous in their appearance. Flange refers to fatty, platelike structures that protrude from each side of the male's face. The terms flanged and unflanged appear relatively recently in the literature, and they are applied in our catalog when specifically mentioned in an original source. Otherwise, the term *adult* is used when appropriate. For a full discussion of the phenomenon of bimaturism in orangutans, see Utami-Atmoko et al. (2009).

A more regrettable distinction for orangutans is that their wild populations have been devastated, in part by infant poaching for illegal trade throughout Southeast Asia. Fortunate infants are confiscated by

The same gibbon shown in the previous photograph uses a cloth as a sponge to Absorb water from a gushing water fountain. Photograph courtesy of Duane M. Rumbaugh.

authorities, rehabilitated, and reintroduced to life in the forest. A number of terms are used to refer to these orangutans, such as "ex-captives," "rehabilitants," and "ex-captive rehabilitants." To avoid confusion, the term *rehabilitants* will be used exclusively in our catalog. Rehabilitants living in the wild will be referred to as *reintroduced*. Both of these terms will be used consistently with the definitions by Beck et al. (2007) of rehabilitation and reintroduction. See Russon (2009) for an informed discussion of orangutan rehabilitation and reintroduction.

We have not noted which orangutan tool behaviors are thought to be customary, habitual, and cultural in specific wild populations and absent in others (see chapter 1 for definitions). Readers are referred to van Schaik et al. (2003) for these designations.

Drop

A. R. Wallace (1869) provided one of the earliest descriptions of agonistic Dropping of branches and other objects by wild orangutans. In each of the three cases he observed, the ape was a female and the target was Wallace himself as he tried to shoot them. Yerkes and Yerkes (1929) cited a similar, second-hand report by Radermacher, another collector. Harrisson (1962a) quoted an identical observation by Schneider under the same circumstances, and she herself was a target while trying to observe a large male. Harrisson (1963) noted that a captive being reintroduced to the wild Dropped branches when disturbed by humans or monkeys. Attenborough, Hoogerwerf, and Luitjes (in Kortlandt and Kooij 1963), de Silva (1970), Galdikas (1978a, 1978b, 1982a), Galdikas-Brindamour (1975), Horr (1975, 1977), Rijksen (1974, 1978), P. Rodman (pers. comm.), and A. Schultz (1961) mentioned branch Dropping by wild orangutans, but the most complete descriptions were supplied by Davenport (1967), MacKinnon (1971, 1974a, 1974b), and Schaller (1961). Adults and adolescents of both sexes break off and Drop branches either toward humans on the ground or when encountering groups of noisy monkeys. The accompanying vocalizations indicate threat or agonism. The branches range from only a few centimeters to more than 3 meters in length. The apes expend considerable energy in this behavior and persist for periods of up to fifteen minutes. If the branches are merely Dropped, the ape may position itself directly over the target. Disturbed orangutans also Drop down bark and other debris from the canopy.

In some orangutan populations, individuals modify specific vocalizations by holding leaves against their mouths during production of the sound, termed a "kiss-squeak with leaves" (van Schaik et al. 2003). The sound is most commonly made in agonistic contexts, and the leaves are then Dropped towards a threatening animal or unfamiliar human (Peters 2001). The behavior is thought to augment the display.

Zookeepers J. Roberts and D. Hastings (pers. comm.) described an incident that occurred as they were repairing a water faucet at the base of an enclosure housing a group of orangutans. A juvenile female seized a sizeable stone, climbed silently to the top of the enclosure, and extended her arm through the bars so that the stone was directly above the men's heads. They looked up just in time to avoid the mischievously dropped stone.

In a novel study of reaching, N. Mendes, Hanus, and Call (2007) presented five orangutans with a narrow vertical tube containing a single peanut floating in a small amount of water. The nut could not be reached directly, and no tools were provided for extracting the peanut from the tube. The orangutans were tested individually to prevent social learning. On the first trial, all of the subjects spontaneously collected mouthfuls of water from their mechanical drinker and added them to the tube, successfully raising the floating peanut within their reach. Mendes, Hanus, and Call attributed the use of water as a reaching tool in this scenario to insight learning on the part of the apes. We view the water as being Dropped by the orangutans into the tube to bring the peanut within reach. This behavior is analogous to rooks Dropping stones in a similar experiment to raise a water level to obtain a worm (Bird and Emery 2009b; see chapter 3). Walkup (2009) conducted an experiment with three captive orangutans, who successfully learned to drop a stone down a tube to push a nut off a platform to bring it within reach.

Throw

Wild adult and adolescent orangutans of both sexes break off branches and fruits and Throw them toward humans and noisy monkeys, either with an underhand scooping motion or with an overhand continuation of the motion used to break them off (Davenport 1967; MacKinnon 1971, 1974a, 1974b; Peters 2001; and Schaller 1961). Schaller (1961) saw a female lift branches to her chest and head before Throwing them. The orangutans do not appear to specifically aim the missiles, but they are big enough and there are enough of them to be an effective deterrent to human pursuit. Galdikas (1982a) documented what she judged to be aimed Throwing of live and dead branches at human observers by wild orangutans. Harrisson (1962a) cited the observations by Beeckman, a mariner, of wild orangutans Throwing sticks, stones, and other objects at humans.

Rehabilitants display unaimed Throwing with a variety of objects in numerous settings (Galdikas 1982a), but sometimes they seem to aim. Rijksen (1978) observed a young male rehabilitant using an underhand motion to Throw a stick in the direction of a snake he encountered in the forest. A different rehabilitant used an overhand motion to Throw a handful of herbs and vegetation at a dying monitor lizard (*Varanus salvator*) he found in the forest. Aimed Throwing of mud by a rehabilitant was observed by Galdikas (1982a). Bard (1995) documented aimed Throwing by young orangutans in unspecified contexts.

Schmidt (1878 in Yerkes and Yerkes 1929) relayed the reaction of a captive orangutan when first presented with his reflection in a mirror. The agitated ape Threw a wooden hammer and bread at the mirror. During a threat display, a captive female broke off and Threw the crown of a small tree in the direction of a dog 5 meters away. In a separate agonistic incident, she dislodged a large flat stone, picked it up, and Threw it in the direction of a caretaker who was about 1.5 meters away (Koehler 1993; note that this is a posthumously published paper by Wolfgang Köhler using an Anglicized spelling of his name).

Drag, Roll, Kick, Slap, Push Over

Members of at least one population of wild orangutans perform displays that involve Dragging branches on the ground (van Schaik et al. 2009). Rijksen (1978) saw a rehabilitant orangutan Drag a branch during agitated display in the presence of the corpse of a conspecific. Rehabilitants living on Kaja Island, an important orangutan rehabilitation and reintroduction site, splashed water to draw the attention of conspecifics (Russon et al. 2010). We three authors have each observed Azy, a captive adult male orangutan, regularly Roll large plastic drums on the floor during displays while making long calls.

During agonistic or dominance displays between wild orangutans, they sometimes Push Over large dead trees in a behavior labeled "snag crashing" (Galdikas 1983; van Schaik, van Noordwijk, and Wich 2006). Although this is primarily an activity of flanged males, it has also been performed by unflanged males and adult and adolescent females. Galdikas (1983) reported that coming into contact with other orangutans was the most common reason for adult females and immatures to snag crash. On two separate occasions, an adult female orangutan crashed snags in the direction of human observers. The dramatic snag crashing of an adult male named TP was also described in detail. TP did not simply push snags over until they fell; he balanced and rocked each snag, causing it to crash in a desired direction. Remarkably, TP would also ride the snag on its way down, grasping a branch on a nearby tree with a hand and a foot once the crash was well underway. Snag crashing serves at least two purposes: as a form of aimed Throwing, it directly repels and endangers another animal or human, and it significantly augments display behavior (Galdikas 1982a). According to S. Wich (pers. comm.), snag crashing may also occur when no other orangutan is nearby, or in response to a flanged male's long call heard at a distance.

Brandish, Wave, Shake

Both wild and captive orangutans Shake, Brandish, or Wave branches as part of agonistic displays toward humans, other orangutans, and other species (Galdikas 1982a; Koehler 1993; Kortlandt 1972; Peters 2001). Galdikas-Brindamour (1975) and Rijksen (1978) provided photographs of rehabilitants Brandishing sticks at dogs. Galdikas (1978b, 1989) saw a wild juvenile break off a leafy branch and Wave it at wasps that were buzzing around him. Wild individuals Wave twigs to ward off stinging insects such as bees or wasps (Rijksen 1978; van Schaik et al. 2003). In addition to driving away insects, wild orangutans frequently fan themselves with branches for cooling (Kaplan and Rogers 1994; Rogers and Kaplan 1993). On rare occasions, orangutans have been observed carrying and hiding behind Detached branches to conceal themselves from predators or humans (van Schaik et al. 2003, 2009).

While being teased with favored foods by a caretaker, a young orangutan Shook a crate at him, picked up and Threw a large stone in his direction, then picked up the stone, Shook it, and Dropped it (Koehler 1993). In an example involving initial exposure and gradual habituation to a mirror, a male playfully Waved a piece of paper at his reflection (Schmidt 1878 in Yerkes and Yerkes 1929).

Russon (2000) described a rehabilitant female orangutan who attempted to start a cooking fire. She first picked up a smoldering stick from an old fire, used a cup to dip kerosene from a container, and then dropped the stick into the cup. Russon (2000, 78) noted that "fortunately for us, plunging the stick into the kerosene put the fire out." Undeterred, the orangutan attempted to reignite the stick: fanning it by Waving a lid, blowing on it, and touching it to another burning stick. As mentioned in chapter 1, this orangutan was a noted imitator of human behavior.

Bait, Entice

While studying rehabilitants at Camp Leakey in Tanjung Puting National Park in Borneo, Russon (2000) described orangutans who developed a taste for coffee. Upon seeing a human with coffee, the ape would find and then proffer an empty mug, or attempt to trade a leafy branch in exchange for the beverage.

An orangutan Waving a leafy branch to repel flying insects. Drawing courtesy of Perry van Duijnhoven.

In two experimental settings, orangutans traded tokens with human social partners for food rewards (Bräuer, Call, and Tomasello 2009; Dufour et al. 2009). Token exchange is regarded as tool use in the Bait, Entice mode, since the exchange alters the behavior of the receiver. We recognize only some token use as tool use, specifically when the user is responsible for the proper and effective orientation of the token to the primary reinforcer. Captive orangutans readily trade illicitly acquired objects—such as stones and light bulbs—with caretakers for food (present authors, pers. obs.), but in these cases it is the caretakers who are using the food as tools, to Entice the orangutans to surrender the objects.

Club, Beat

Orangutans have been documented using tools for hitting during agonistic interactions (Lethmate 1982). Galdikas (1978a, 1978b, 1982a) reported that a wild orangutan hit a conspecific with a piece of bark held in its hand. An orangutan in the process of being reintroduced to the wild Clubbed a snake with a stick (Harrisson 1963). Another rehabilitant hit a dying lizard with a sapling (Rijksen 1978), while others at the Tanjung Puting National Park commonly hit other orangutans, humans, dogs, and monkeys with sticks (Galdikas 1982a; Russon 2000). The Kaja Island orangutans used a branch to Club fish in shallow water and then caught and consumed them (Schuster, Smits, and Ullal 2008).

Captive orangutans have been seen using sticks to Club or hit humans (G. Rabb, pers. comm.). One individual struck a young male cage mate with a shoe held by the laces (Herzfeld and Lestel 2005). The authors also reported that this female tied a double knot in a rubber hose to make it more rigid and used it to hit the cage mate. J. Ellis (1975, 1977) observed a captive hitting at a human and at a snake with sticks and branches, and using a strip of rubber removed from a tire as a whip to hit anything within range. Dewsbury (2009) reported that a 1- to 2-year-old captive male orangutan used a block to hit at a bee that flew near him.

Pound, Hammer

Wild orangutans used tools they manufactured from branches to Hammer open invertebrate nests in tree hollows or to open tough-skinned durian (*Durio* spp.) fruits (E. Fox, Sitompul, and van Schaik 1999; van Schaik and Fox 1994; van Schaik, Fox, and Sitompul 1996; van Schaik et al. 2009). Kaplan and Rogers (1994) reported that orangutans in Sepilok broke into mounds of earth using large pieces of wood. Galdikas (1982a) noted that rehabilitant orangutans routinely used sticks to Pound on solid objects to open them. One rehabilitant Hammered on a durian fruit with a pointed stick, thus blunting the spines on the exterior of the fruit and allowing him to handle it more easily (Rijksen 1978).

A rehabilitant imitatively Hammered nails (Russon 2000). Other rehabilitants routinely cracked open hard items, such as pieces of a termite nest, by pounding them on available surfaces. Two individuals tried to open various things by pounding them on another individual's head, using it as an anvil (Russon 2003). These cases are considered proto-tool use, since there is no indication that the tool users held or manipulated the hard surfaces or their conspecific.

Rijksen (1974) mentioned captive orangutans using stones to "force" locks, presumably by Pounding on them. R. Wright's (1972) subject and a captive male (B. Beck, pers. obs.) Pounded with stones on cage walls and floors, both in play and in what seemed to be frustration. R. Horwich (pers. comm.) observed a captive male Pounding one stone with another. When the anvil broke, the individual carefully examined the flaked edge but did not use it. A juvenile female at the Brookfield Zoo twice cracked a large window in her enclosure with a stone. Since the damage occurred at night, it is not known if she threw the stone or used it as a hammer. However, she was seen Pounding on the glass with stones during the day (B. Beck, pers. obs.). C. Parker (1969a) described how a male orangutan at the San Diego Zoo dismantled an experimental apparatus and then used the pieces to Hammer bolt heads and gouge the mortar in the walls of his enclosure. K. Walkup (pers. obs.) documented captive orangutans at Great Ape Trust pounding on a drain cover with PVC pipes and with copper tubing. A male orangutan at the Como Zoo removed two drinking faucets from the wall of his enclosure and used them as a hammer and chisel while attempting to break through an adjoining wall (J. Ellis 1977). Boulenger (1936) reported that an orangutan used a large potted plant to batter through a skylight in the process of escaping from the ape house at the London Zoo. Jantschke (1972) saw captives strike cage floors and walls with sacks and other objects in anger.

An adult male at the Denver Zoo arranged a pile of sunflower seeds that he smashed with a round, hard rubber toy (L. Miller and Quiatt 1983). He was then observed carefully picking out and consuming the shelled nutmeat. Chantek, a sign-language-competent individual, demonstrated proficiency in using hammers (Miles 1993). Koehler (1993) described the behavior of a young wild-caught female orangutan who proficiently used a stone as a hammer for disassembling the 50-centimeter-thick wall of her enclosure. Eventually, the wall became so badly damaged that it had to be replaced to contain the industrious ape.

Pry, Apply Leverage

E. Fox, Sitompul, and van Schaik (1999) provided a detailed description of the way in which wild orangutans use tools to obtain seeds from the *Neesia* fruit. Although these lipid-rich seeds are generally extracted by scraping, seeds occasionally become stuck in this tough-skinned fruit. When that occurs, the apes reorient the fruit and use their tools to Pry out the individual seeds (also E. Fox, pers. comm.).

Rijksen (1978) noted that rehabilitants used sticks as levers in excavating termite mounds. Rehabilitants were observed by Galdikas (1982a) using sticks to Pry objects loose and also to propel themselves in dugout canoes by paddling. Lethmate (1982) reported that rehabilitants used sticks to Pry up boards in their enclosures, and also to force padlocks. Vosmaer (1778 in Yerkes and Yerkes 1929) described a captive individual using a nail as a lever. Camacho (1907), Hornaday

An orangutan using a stick to extract seeds from a *Neesia* fruit. Orangutans often hold their tools in their mouth during use. Drawing courtesy of Perry van Duijnhoven.

(1922), and Bradford and Blume (1992) all relayed engaging accounts of a captive male orangutan systematically dismantling his enclosure at the Bronx Zoo. He first pulled some wooden exercise bars from the wall of his cage and then used a broken portion of one board as a lever to Pry off the remaining bars, as well as some of the thick boards that made up the wall itself. The damage was repaired using stronger materials, but the orangutan then used his trapeze bar to Pry off the bars. Two more rounds of repair and destruction followed before the zoo carpenters finally stymied the lever-wielding ape. At one point, the orangutan was reported to have received cooperative assistance in his destructive efforts from a chimpanzee cage mate. He also used the lever to widen the gap between the iron bars of his cage front so that he could stick his head out and expand his field of view. Benchley (1940), C. Darwin (1871), J. Ellis (1975), Vosmaer (in Yerkes and Yerkes 1929), and Yerkes (1916) reported that captive orangutans used sticks, metal bars, keys, and nails to Pry or Apply Leverage to open boxes containing food or to dismantle their cages. A captive male at the San Diego Zoo used broken pieces from an experimental apparatus to Pry more pieces off, and also to Pry at the door and lock of the enclosure (C. Parker 1969a). Walkup, Shumaker, and Pruetz (2009, 2010) documented captive orangutans who selected rigid tools when presented with both rigid and flimsy tools, and then used them to Pry open an out-of-reach apparatus containing a food reward. Döhl and Podolczak's (1973) young subject Pried with a trowel inserted into the seam of a wooden experimental apparatus. Chantek Pried two boards apart using a screwdriver (Miles 1994).

Dig

Russon (2000) described one rehabilitant who Dug with a shovel and another who chopped weeds out of a path using a half-meter-long stick. Galdikas (1982a), Galdikas-Brindamour (1975), Harrison (1962a, 1963), Lethmate (1982), and Rogers and Kaplan (1993) noted that captive orangutans Dug with sticks in what appeared to be playful investigation. An adult male living at the Adelaide Zoo commonly saved the paper cups in which treats were presented and later used them to Dig in sand (Harper 2001). J. Ellis (1977) described an adult male at the Como Zoo who reportedly used a nail to Dig, hiding the nail in his hair when he was not using it.

Jab, Stab, Penetrate

Wild orangutans were reported to "poke and prompt" ants or bees with probing tools so that they would emerge from their nests (van Schaik, Fox, and Sitompul 1996, 186). Rijksen (1978) observed a rehabilitant aggressively Jabbing with a long stick at a caged clouded leopard (*Neofelis nebulosa*). Another rehabilitant repeatedly Jabbed at tough-skinned fruit with a stick. Galdikas (1982a) reported that rehabilitant orangutans used sticks to Jab and poke other orangutans, humans, a caged leopard, a snake, and carcasses. After Jabbing a carcass, the rehabilitant usually smelled the end of the stick that had made contact.

O'Malley and McGrew (2006) described a captive male who removed embedded raisins from an enrichment device, forcefully Penetrating each raisin by jamming a tool into it and then extracting the raisin. The Penetrating motion was performed with the tool held in the mouth, and extraction was accomplished by holding the tool manually. Walkup (2009) conducted an experimental study involving an out-of-reach apple placed in a tube that could be retrieved most easily by Stabbing (Penetrating) it with a stick. Of the three orangutans studied, one carefully pushed until the tool impaled the apple and then carefully removed the tool and the apple from the tube. The other two apes poked repeatedly and forcefully at the apple, obtaining the apple either by impaling it by chance or by raking out the small pieces that resulted from the repeated poking.

Reach

A wild flanged male was observed by E. Fox and bin'Muhammad (2002) as he attempted to cross between two trees. After both reaching with his hand and tree swaying failed to place him close enough to the destination tree, he broke off a straight branch approximately 1 meter long and attempted to use it as a hook to span the 3- to 4-meter distance. After two unsuccessful attempts, he discarded the tool and broke off a second branch that was approximately 2.5 meters long. He used this tool successfully to hook the nearby tree and then transferred into it. The use of a branch hook has been observed in one other individual at a separate study site (van Schaik, van Noordwijk and Wich 2006; van Schaik et al. 2009). This use of a sway tree and the hook can be considered a Tool Composite. Note, however, that our use of the word "hook" does not imply that the tip of the tool was shaped or recurved, although Russon et al. (2009) state that it was in some cases.

Blomberg reported to Kortlandt and Kooij (1963) that a wild orangutan used a Detached branch to Reach an incentive that may have been fruit. Rijksen (1978) observed a wild orangutan using a long stick to push a spiny durian fruit into a crevice, thus protecting the ape's hands.

In experimental and non-experimental settings, captives and rehabilitants used sticks, branches, twigs, splinters, sacks, blankets, straw, leaves, wire, PVC pipes, paper, strips of rubber, pieces of carrot, spines from lettuce and cabbage leaves, and peanut shells to rake in food and other desired objects (Drescher and Trendelenburg 1927; J. Ellis 1975, 1977; Galdikas 1982a; Galdikas-Brindamour 1975; Haggerty 1910; Herrmann et al. 2007; Jantschke 1972; Koehler 1993; Lethmate 1976a, 1976b, 1976d, 1977a, 1977b, 1977d, 1977e, 1979; Martin-Ordas, Call, and Colmenares 2008; Mulcahy and Call 2006b; C. Parker 1968, 1969a; Reuvens in Yerkes and Yerkes 1929; Rijksen 1974, 1978; Sheak 1922; Walkup 2009; Yerkes 1916). Kaja Island orangutans were observed using a Reaching tool to dredge the bottom of a pond for sunken items, to rake in and acquire floating items, to assess water depth, and as a hook to Reach and pull in leafy branches that enabled travel over water (Russon et al. 2010). Rehabilitant orangutans have been observed using sticks to stir hot beverages (Galdikas 1982a) or to pull burning pieces of wood from a fire (Galdikas-Brindamour 1975). Rijksen (1978) saw a rehabilitant touch a dying monitor lizard with a stick and then smell the end of the stick. A rehabilitant used a stick to pull in a branch that was out of reach (Rijksen 1974, 1978). A Kaja Island orangutan used a branch to extract a baited fishing line from the water, together with a caught fish that had been left by a human. The ape removed and ate the fish (Schuster, Smits, and Ullal 2008).

In one of the earliest existing references, Reuvens (1889 in Yerkes and Yerkes 1929) related the details of a captive orangutan who used two sacks as a Tool Composite to Reach an orange outside of its cage (see Associative Tools for details). Wilkie and Osborn (1912 in R. Mitchell 1999) described an unusual scenario involving a captive orangutan who was allowed to smoke. She used lit cigarettes to set bags on fire, and then moved the burning bags with a stick to set other bags on fire. Sheak's (1922) subject removed the sweater it was wearing to use it as a Reaching tool. Sanyal (1902) described the behavior of a young male orangutan living in the Alipore Zoological Gardens in Calcutta. In order to obtain a nearby pan of food, this individual collected straw, twisted it into a rope, and used it as a Reaching tool. A similar observation was made of a female orangutan living in a rehabilitation and reintroduction center (Schuster, Smits, and Ullal 2008). She was seen plucking hairs from her arm to twist into a rope, tying a banana skin to the end (presumably as weight for throwing), and then using the rope to obtain food that was out of reach from her cage. An orangutan being studied by Osvath and Osvath (2008) used a piece of rubber hose as a straw to suck up fruit soup contained in a plastic bottle.

In a reported case of planning (Lethmate 1982), a 4- or 5-year-old captive orangutan carried an elongated stick to a stand of distant trees, stood bipedally, and used the tool to hook an out-of-reach branch that was hanging overhead. Russon (2003) described a male adolescent who used a very skinny female as a Reaching tool. The male led the female by the arm to an area of their enclosure where food had fallen through the elevated barred floor. The male's arm was too large to Reach through the openings, but he held the female in place until she retrieved the food, which he promptly took from her.

In one of the earliest reported experimental studies with orangutans in captivity, Haggerty (1913) documented that his pair of 4-year-old females spontaneously learned to use a hook for Reaching food on a table adjacent to their barred enclosure. They employed the tool for both Reaching and impaling to secure the treats. Shepherd (1923) reported that an orangutan failed to use a rake to Reach food, even after demonstration by the experimenter. However, the ape was allowed only two trials of two minutes each to get the food. Guillaume and Meyerson (1934) blamed disinterest for their adult male orangutan subject's failure to use a Reaching tool.

In an experimental setting, captives used sticks to Reach toward and knock down suspended food (Lethmate 1976c, 1976d, 1977a, 1977c, 1979, 1982) and unspecified objects (anonymous in Kortlandt and Kooij 1963). G. Rabb (pers. comm.) saw an orangutan Reach with a stick to knock fruit from a person's hand. Lethmate (1979) observed an orangutan raking a baited container from a box and an edible from a hollow tube. Döhl and Lethmate (1986) described a 6-year-old male

An orangutan Combines a stick and a tube to manufacture a tool long enough to knock down suspended food. Note the strips of wood that the orangutan had previously Subtracted from the stick so that it could fit into the tube. Photographs courtesy of Jürgen Lethmate.

who used tools to Reach food placed outside of his enclosure. Call and Tomasello (1994) reported that their orangutan subjects successfully used rakes to obtain out-of-reach food. However, they noted that their subjects devised their own strategies for this, rather than imitating the solutions offered by both human and orangutan demonstrators. Mulcahy, Call, and Dunbar (2005) offered orangutans a piece of out-of-reach food and Reaching tools of different lengths. The apes made useful tools, selected tools of the optimal length from those provided, refused tools that were too short to Reach the food, and also demonstrated Sequential Tool Use by using a tool to Reach a second tool that was used to obtain the food. In a subsequent study, Mulcahy and Call (2006a) presented two orangutans with an out-of-reach bottle of juice hanging from a string, a hook, and several other tools that were not usable for the task. The apes spontaneously used the hook to Reach the string and retrieve the juice.

Swartz, Himmanen, and Shumaker's (2007) orangutan subjects used wooden styluses to Reach through the enclosure mesh and touch illuminated panels in a computerized test of memory. One individual held the stylus in her mouth.

Insert and Probe

Van Schaik, Fox, and Sitompul (1996) detailed tool use by wild orangutans at the Suaq Balimbing study site. The orangutans at this site were observed foraging for insects or their honey and processing *Neesia* fruits to remove the seeds. All recorded tool use took place in the trees.

When these orangutans Probed in insect nests, they removed the residents, broke off sections of termite nests, or extracted honey (E. Fox, Sitompul, and van Schaik 1999; Sitompul, Fox, and van Schaik 1998; van Schaik, Fox and Sitompul 1996). The orangutans held the tools with their mouths during use in 83 percent of the occurrences. Van Noordwijk and van Schaik (2005) reported that juveniles as young as 4 to 6 years of age used sticks left by their mother to Insert and Probe into insect nests. The 6- to 7-year-olds made their own Probing tools and sometimes engaged in this activity independently.

Neesia fruits contain calorie-rich seeds protected by razor-sharp hairs. The orangutans used Probing tools to extract these seeds by first scraping them toward the apex of the fruit, blowing to remove any stinging hairs that were dislodged, wiping with a fingernail, and then scooping out the seeds with a finger or the tool (E. Fox, Sitompul, and van Schaik 1999; Sitompul, Fox, and van Schaik 1998; van Schaik, Fox, and Sitompul 1996). In all of the observations, the tool was orally manipulated as the seeds were scraped and then transferred to the hand when they scooped. The authors concluded from these observations (tool use

An adult female orangutan prepares to Insert and Probe with a stick in an arboreal insect nest. Photograph courtesy of Perry van Duijnhoven.

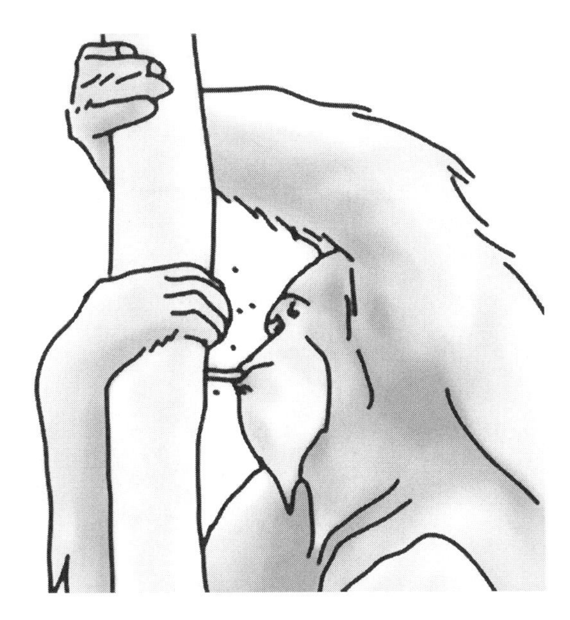

An orangutan Probing in an arboreal insect nest. The tool is held in his teeth. Drawing courtesy of Perry van Duijnhoven.

during feeding on insects and on fruit) that orangutans possess a Tool Kit in the wild, and Sitompul, Fox, and van Schaik (1998, 202) labeled these behaviors as "intelligent tool use." Further, geographic variation in the occurrence of tool use to process *Neesia* fruit supports the interpretation that this is a learned skill that persists as a cultural tradition only in specific populations (van Schaik and Knott 2001; see van Schaik et al. 2009 for an analysis of the geographic technique to document cultural variation). Use of tools to extract *Neesia* seeds was not seen in orangutans younger than about 7 years of age.

Harrisson (1963) reported that a rehabilitant being introduced to a forest habitat Inserted a twig and Probed into an insect nest attached to a fallen branch, holding the twig first in its hand and then with its teeth. Rijksen (1978) observed rehabilitants Inserting and Probing with sticks in subterranean termite nests and rat burrows. J. Ellis (1975) observed an orangutan Inserting a stick into a deep hole, withdrawing it, and then smelling and tasting the distal end. Russon (2003) reported that an adolescent female Inserted a Probing tool into the "eye hole" of a coconut to extract the jelly contained within. Another female attempted to siphon fuel from a drum using a hose (Russon 2000). An orangutan studied by Furness (1916) used keys to open locks and padlocks in her living quarters. She was reported to know the correct key for each lock and could correctly select them from an array with up to twelve other keys. Rehabilitant orangutans used sticks to Probe in holes in walls, tree trunks, and the ground, as well as in keyholes and locks (Galdikas 1982a; Russon 2000). When probing keyholes and locks, the orangutans employed a twisting motion, presumably in response to having seen humans use keys. Vosmaer (1778 in Yerkes and Yerkes 1929) relayed a virtually identical incident with a lock. Döhl and Podolczak (1973), Lethmate (1977a, 1977d, 1978), and Rensch and Dücker (1966) also described orangutans opening locks and mechanically operated boxes by Inserting keys and hand tools, such as screwdrivers. Linden (1999) relayed the story of an adult male living in a zoo who was observed inserting a piece of wire, which he had kept in his mouth, through the thin gap in a door frame to flip open the latch and exit his enclosure.

Hirata and Ohashi (2003) presented young free-ranging rehabilitants with a bottle containing honey that could be accessed through a 1-centimeter hole. They installed the bottle in the forest where these individuals were being reintroduced. Of the twenty-three orangutans who approached the apparatus, fifteen spontaneously Inserted probes into the bottle. The youngest tool user was 3 years old.

A rehabilitant orangutan prepares to Insert a stick into a lock (the small hole at right) while her infant observes. Photograph courtesy of Anne E. Russon.

An orangutan tries to Insert a key into a locked box. Photograph courtesy of Jürgen Lethmate.

Lethmate's (1982) review of tool use by orangutans reported Inserting and Probing by rehabilitants and captives during autogrooming and allogrooming, as well as for acquiring bits of food. Nogge (1984) reported that a captive orangutan used a stick to pick his teeth, and Kaplan and Rogers (1994) observed rehabilitants using small sticks to clean their teeth or ears. R. Shumaker (pers. obs.) routinely saw a flanged male, Azy, carefully select and use a rigid piece of straw as a toothpick just after he had eaten popcorn. Wild orangutans used toothpicks as well as small sticks to clean under their fingernails (van Schaik et al. 2009).

A number of orangutans have been experimentally presented with the tube task. Haggerty (1913) provided his two young subjects with a horizontally secured hollow pipe containing preferred food. Both learned to solve this task with proficiency by using a stick to push the food away from them and out of the tube. Bourne (1971), Haggerty (1910), Lethmate (1977a, 1977b, 1977d), Martin-Ordas, Call, and Colmenares (2008), Mulcahy and Call (2006b), Visalberghi, Fragaszy, and Savage-Rumbaugh (1995), Walkup (2009), and Yerkes (1916) reported successful retrieval of a reward involving Insertion of a stick into a tube by raking or pushing it out. Martin-Ordas, Call, and Colmenares (2008) and Mulcahy and Call (2006b) reported that given a choice, raking was preferred over pushing. However, Walkup's (2009) replication of the same task found no significant preference for raking in, rather than pushing out, a food reward. Yerkes' (1916) subjects carried their tools several meters to the point of use, and Lethmate's (1977a, 1977b, 1977d) combined as many as five sticks to construct a tool of sufficient length. Lethmate's subjects also used sticks to retrieve saturated leaf sponges from a narrow cylinder that contained juice.

Mulcahy and Call (2006a) presented five captive orangutans with hanging bunches of grapes that could only be accessed by Inserting a tube into a hole and breaking the support for the grapes so that they fell within reach. The apes were given the tube, along with several other tools that were not usable for the task. All of the apes spontaneously solved the task and correctly selected the tube from the array of tools. They were also able to plan for future tool-using opportunities by transporting and saving the correct tool during delays lasting from one to fourteen hours.

Captive orangutans have used various enrichment devices providing an opportunity to Insert and Probe with a tool (C. Becker 1984; C. Buckley 2003; Harper 1992; B. King 1986; Nakamichi 2004; Nogge 1984, 1989; O'Malley and McGrew 2000, 2006; B. Wright 1994). When presented with "grape sugar" in narrow tubes, the apes observed by Nogge (1984) licked their tools before inserting them. O'Malley and McGrew (2000) provided five adult orangutans and three juveniles of varying ages with wooden blocks containing raisins in small holes, as well as bamboo stalks that the orangutans modified into appropriate tools. Overall, the apes held and used their tools with their lips or teeth in about 75 percent of all bouts. The three adult males and two adult females employed oral tool use in 92 percent of the bouts, although the authors stated that this might be correlated with hand size or individual preference rather than age, sex, or body weight. O'Malley and McGrew also reported that oral tool use by captive orangutans was customary, consistent with reports by van Schaik, Fox and Sitompul (1996) on wild orangutans at Suaq Balimbing.

Scratch, Rub

Orangutans routinely Scratch themselves with branches and twigs in the wild, and with a variety of implements in captive settings (Galdikas 1978b, 1982a, 1982b; Jantschke 1972; R. Shumaker, pers. obs.; van Schaik, van Noordwijk and Wich 2006; van Schaik et al. 2003). Given the length of orangutans' hands and arms, and their ability to touch all parts of their body with ease, scratching with a stick is not fully explained simply as an extension of their reach. Rather, these tools probably assist them in targeting a specific location on the body and intensifying the sensation as they Scratch. A similar explanation is likely for male and female orangutans who utilize sticks for masturbation (C. Becker 1984; Rijksen 1978; van Schaik 2004; van Schaik, van Noordwijk, and Wich 2006; van Schaik et al. 2003).

Russon (2000) described rehabilitants who used combs or brushed their teeth using a toothbrush and toothpaste. Other captives also combed or brushed themselves and, occasionally, other orangutans (R. Shumaker, pers. obs.).

Cut

Russon (2000) described a rehabilitant who chopped wood. Another female learned to use a saw after observing humans perform construction work. On one occasion, and on her own initiative, she used a saw to Cut deeply into a log that was 10 centimeters thick. Another individual used the saw from a Swiss Army knife to Cut into a small tree (A. Russon, pers. comm.).

Using a social learning paradigm with a human demonstrator, R. Wright (1972) detailed how a young orangutan learned to use stone flakes as Cutting tools. The orangutan subsequently learned how to create flakes without demonstration. The ape was able to use the flakes to Cut cords, thereby releasing the lid of a box that contained food. An adult female living at Great Ape Trust rapidly and confidently learned to Cut with stone flakes made by a human, but she was not given the opportunity to create her own flakes (R. Shumaker, pers. obs.).

Block

Russon (2003) described two young orangutans chasing each other through the forest. The first swayed a "vehicle tree" to cross a gap, and then held on to the tree after crossing, preventing his pursuer from following (Russon 2003, 283). During a visit to a zoo, Galdikas (1982a) observed an adult male using a pile of leaves as a dam to interrupt the flow of water on the floor of his enclosure. Orangutans observed by J. Ellis (1977) routinely placed tires over drains to stop the movement of water or waste. One individual did

A rehabilitant orangutan uses a saw to Cut wood. Photograph courtesy of Anne E. Russon.

this repeatedly, even when moved among different enclosures. C. Parker (1969a) described how a male living at the San Diego Zoo destroyed an experimental apparatus and then used the broken pieces to jam a sliding door.

Caretakers and managers at seven captive institutions reported that orangutans were observed inserting a variety of items—such as sticks, straw, paper, and vegetables—as wedges into spring-loaded water dispensers, creating a continuously running fountain (T. Geurts, unpub. data).

Debris jammed by an orangutan into a spring-loaded water fountain to maintain a constant flow of water. Photograph by Tine Geurts, courtesy of Great Ape Trust.

Prop and Climb, Balance and Climb, Bridge, Reposition

Several wild orangutans have bitten through vines to free them from their lower connections and then used the vines to swing across a gap in the forest, while other individuals have bitten through a vine to free a vehicle tree that is then swayed in order to reach a second tree (van Schaik, van Noordwijk, and Wich 2006).

Russon et al. (2010) observed rehabilitants who bent small trees over water, thus allowing them to cross. Bard (1993, 1995) and Russon (2003) described young orangutans who bent or swayed trees during travel in order to cross gaps in the forest. In one case, Russon reported that a young male repeatedly swayed a tree but could not reach his intended target. He stopped and cracked a branch so that it was loose but still attached to the vehicle tree. He then used this "handle" to extend his reach, but failed. Undeterred, he made another, longer handle that allowed him to succeed; Russon described this as coming "close to a meta-tool" (Russon 2003, 285). We think that in this case the handle is indeed a Metatool and with the vehicle tree is part of a Tool Composite. In a similar case, a rehabilitant used a hooked stick to Reach and pull in a branch that it then used as a bridge to cross water (Russon et al. 2010).

Russon (2002, 2003) described several wild orangutans who Repositioned vegetation to create a "work seat" as they spent time extracting a palm heart. An adolescent female pulled two small trees together, forming an X. She then "sat on the outer side of one lower leg of the X," using it as a chair (Russon 2003, 284). MacKinnon (1974a) mentioned orangutans bending branches to make a screen that obscured them from view when they were disturbed by people.

A rehabilitant Propped a stick against Galdikas's house at Camp Leakey and Climbed it to gain access to a window (Galdikas-Brindamour 1975). Rehabilitants were often observed dragging logs and vines to the riverbank and placing them as bridges to cross the water (Galdikas 1982a). Although orangutans may wade, play, and even submerge most of their bodies in water, there are no documented cases of orangutans swimming to cross rivers and streams. Rehabilitants, however, may employ tools to travel in or across bodies of water. Lethmate (1982) reported the use of boats, rafts, or pieces of floating wood for transporta-

A rehabilitant orangutan bends over a small tree and uses it as a Bridge to cross water. Photograph courtesy of Anne E. Russon.

tion. Galdikas (1982a) and Russon (pers. comm.) provided detailed accounts of rehabilitants using dugout canoes, rafts, or floating wood. Acquiring a canoe frequently involved untying complex knots in the rope that fastened the vessel to the shore. Some orangutans used the canoes to travel around a lake, propelling the vessel by grasping vegetation growing just below the water's surface or along the bank. Others routinely crossed a river in a canoe. Although no orangutans were seen to use paddles, they did propel themselves with sticks, boards, or dippers (Galdikas 1982a). After crossing the river, at least one unflanged male kept his canoe by holding the rope while he travelled and foraged along the bank, allowing him to return when he desired. This same male also used the dugout as a place to sleep during the night. Galdikas noted that in addition to using canoes, the apes untied a floating dock constructed of logs and use it to raft across the river.

Yerkes' (1916) orangutan Propped and Climbed sticks to secure suspended food. A captive female rolled a tire below the feeder, leaned it against the mesh wall, and then climbed on top of it to reach an elevated enrichment device (C. Buckley 2003).

Yerkes (1916) reported that a captive orangutan Balanced and Climbed on a stick to reach suspended food. Lethmate (1976d) and Rice (in Harrisson 1962a) observed similar behavior in contexts not involving food; in these cases, the apes seemed to be motivated to devise a new means of locomotion. Lethmate's (1976d) subject combined several short sticks to make a longer one for Balancing and Climbing.

Propping objects may not be necessary to climb them; Repositioning them may suffice. Dewsbury (2009) described a captive 1- to 2-year-old male orangutan moving a stool beneath a suspended towel and climbing on the stool to get it. Lethmate (1982) described a young orangutan moving a box approximately 12 meters to position it beneath a tree and then climbing on the box to grasp the branches overhead. The same individual moved the box 7 meters to a different location, while carrying a Reaching tool, to Reach an elevated nest box. In this case, the ape transported and then used two tools as a Tool Composite. Both male and female adults used large barrels as elevated seats. The apes rolled or pushed the barrels into position, placed them upright, and sat on the top (R. Pietsch, pers. comm.). Once seated, the apes sometimes traveled on the barrels by grasping nearby affordances with their hands and rocking and gently moving the barrel along the floor to the desired location, never leaving their seats. In a study of orangutan locomotion, K. Walkup, R. Shumaker, and C. Pitman (unpub. data) termed this "barrel travel."

Lethmate (1976c, 1976d, 1977a, 1977b) observed a captive orangutan stacking up to four boxes to build a tower high enough to reach suspended food. In our definitional scheme (see chapters 1 and 7), each of the boxes is a tool, but the tower is a Construction because it is not held or manipulated in its entirety.

A rehabilitant orangutan unties a boat prior to using it to cross a body of water. Photograph courtesy of Anne E. Russon.

An orangutan Repositions and stacks boxes to make a climbing tower to reach suspended food. Photographs courtesy of Jürgen Lethmate.

Rensenbrink (1960) reported that a captive orangutan stacked chairs on a table and playfully climbed to the top of the structure. Yerkes' (1916) subject was initially unable to stack boxes. After extensive human tuition and demonstration, he was able to stack, awkwardly at first, but finally confidently and efficiently. Captive orangutans at the Smithsonian National Zoo competently stacked two large plastic barrels on top of each other and used the structure to escape from and subsequently reenter their outdoor, dry-moated enclosure (R. Shumaker, pers. obs.). According to zoo visitors who witnessed the event, the apes did this several times over the course of a morning before staff members discovered the unfolding drama. The apes returned to their enclosures, and barrels were subsequently banned in the outdoor yard.

Bresard (1993) observed a 2-year-old orangutan using a 6-month-old chimpanzee with whom he lived as a "social tool." The orangutan picked up and placed the chimpanzee directly beneath a door handle that was out of reach and then climbed on the chimpanzee to reach the handle. The use of a human experimenter as a ladder for accessing an out-of-reach item was also reported by Yerkes and Yerkes (1929).

Hang

A young female rehabilitant observed by Russon (2000) took an empty rice sack high into a tree, hooked it over a branch, and then climbed in head first. Russon (1999b, 2000) reported that rehabilitant orangutans Hung and used hammocks, most likely in imitation of human demonstrators. R. Shumaker (pers. obs.) watched an adolescent male orangutan, Kiko, at the Smithsonian National Zoo as he constructed a hammock from a sheet. Kiko wound one end of the sheet securely around the branch of an artificial tree, then stretched the sheet out until it reached a nearby branch, over which he looped the other end. While continuing to hold the looped end, he climbed into the hammock and rested. By holding the looped end, Kiko kept both himself and the hammock from falling to the floor. Hornaday (1922) described a captive orangutan twisting straw into a rope more than 2 meters long and Hanging the rope over a bar in the cage to use as a swing. Gewalt (1975), Jantschke (1972), and MacKinnon (1978) later published observations of identical behavior by zoo-housed orangutans. Cage mates of Gewalt's subjects did not exhibit the behavior. Beck (1980) observed a captive female weaving an impressive rope from straw, but she did not swing from it. Koehler (1993) reported that a young female placed a blanket, a sack, or a rope over the horizontal bars of her cage and Hung on and slid while holding the free ends with her hands. Captive orangutans Hang pieces of cloth, hose, or chain over the bars at the top of their cages and swing from them (B. Beck, pers. obs.; Russon 2003). Jantschke (1972) reported numerous cases of similar behavior. A juvenile male at the Smithsonian National Zoo placed a small strip of burlap over a cable approximately 15 meters from the ground and, either fearlessly or foolishly, swung from the loose ends (R. Shumaker, pers. obs.).

Contain

Rehabilitants used coconut shells or husks as dippers or containers for fluids, and also employed sticks as utensils for feeding themselves (Galdikas 1982a). Rehabilitant orangutans commonly used utensils with food or drink, particularly when the comestibles were too hot to touch directly (Russon 2000). Russon et al. (2010) observed individuals living on an island who used containers such as cups or shells to scoop and drink water. A rehabilitant broke eggs into a cup, added flour, and then mixed the batter, as she had seen a human cook do when preparing pancakes (Russon 2000). Galdikas (1982a) described an occasion when one individual placed a mound of rice on a piece of bark and handed it to her. Rogers and Kaplan (1993, 1994) provided detailed accounts of rehabilitants using leaves to Contain food. After acquiring a large mouthful of provisioned food, the apes moved away from the feeding site and collected several large leaves. As a result of being partially chewed, the food was thin in consistency. Leaves appeared to be selected based on size and sufficient strength to hold and Contain the semisolid food. The leaves were arranged in a fan shape, and the food was deposited on the leaves and eaten at a leisurely pace. Rogers and Kaplan (1993) suggested that this behavior might be a strategy to decrease food competition at a feeding site or to assist with digestion. The use of containers is, however, not limited to captives or rehabilitants. Knott (1999) observed wild orangutans using leaves as containers to dip drinking water from puddles.

There are numerous examples of captive orangutans using objects such as cups, plastic construction hats, bags, bowls, or tubs to Contain food, drink, or other items (B. Beck, pers. obs.; J. Ellis 1975, 1977; L. Miller and Quiatt 1983; R. Shumaker pers. obs.). The apes drank from their vessels, soaked hard food in them, or used them to transport items around their enclosures. J. O'Connor (pers. comm.) observed a captive female at the Brookfield Zoo filling a large tub with straw and carrying it to an outside enclosure in one efficient trip to build a nest. A young female observed by Koehler (1993) spontaneously offered an empty bowl when a can of milk was presented to her. C. Becker (1984) described zoo-living orangutans who drank from tubs, and who also urinated into tubs. R. Wright (1972) mentions that his young male subject placed a hammer stone into a tire and rolled it across his enclosure.

Absorb

Examples of sponging by wild orangutans are relatively rare. Orangutans at the Ketambe field site in Sumatra and the Sabangu field site in Borneo have been reported to use crumpled leaves to Absorb water (van Schaik et al. 2003, 2009). The use of leafy branches (termed a "branch scoop") to Reach into deep tree holes that contained water has also been documented, although at different locations (van Schaik, van Noordwijk, and Wich 2006; van Schaik et al. 2003, 2009).

Kaja Island rehabilitants Absorbed and drank water with various items, such as sponges, cloth, husks, or plastic bags. These individuals also dipped leafy twigs into water and then allowed the fluid to drip into their mouths (Russon et al. 2010). Numerous items, such as pieces of cloth, plastic bags, and vegetation, were used by other rehabilitants as sponges (Galdikas 1982a).

An orangutan wears a rain hat made of leafy branches. Photograph courtesy of Anne E. Russon.

The use of absorbent materials as sponges has been most frequently observed in zoos or other captive facilities (B. Beck, pers. obs.; J. Ellis 1975; Jantschke 1972; Lethmate 1976a, 1976d, 1977a; MacKinnon 1974b; C. Parker 1968, 1969a; V. Thompson, pers. comm.; Walkup 2009). In most cases, the behavior allowed access to fluids that could not otherwise be drunk, either because the containers were fixed and nearly empty or had narrow openings. Thompson's (pers. comm.) subject saturated rope by placing it into a gushing waterfall and then sucked out the water. Lethmate (1976a, 1976d, 1977a, 1977b) observed a captive Inserting leaves into a narrow cylinder containing juice. When the leaves had Absorbed the fluid, the individual withdrew and sucked them. C. Parker (1969a) reported similar behavior, with the ape using a piece of rope instead of leaves. Walkup (2009) and B. Beck and R. Shumaker (pers. obs.) each observed an orangutan use a stick and paper towels to retrieve an out-of-reach liquid (see Tool Composite).

Wipe

Rijksen (1978) reported that wild and rehabilitant orangutans Rubbed or Wiped their own genitalia with inanimate and animate objects, for example, a cat. MacKinnon (1974b) twice saw wild female orangutans using leaves to Wipe feces from their own pelage. Rijksen (1978) observed a wild adolescent male Wiping saliva from his mouth with leaves. Galdikas (1978) saw wild orangutans at Tanjung Puting Wiping their faces with crumpled leaves, but she felt the behavior resulted from generalized frustration rather than hygienic concern. Once the orangutans finished Wiping their faces, they Dropped or Threw down the used leaves at the observers, who likely stimulated the Wiping behavior by their presence (Galdikas 1982a; Russon and Galdikas 1993). This behavior was termed "leaf wipe" by van Schaik et al. (2003). Several geographically separate populations of wild orangutans used leaves or moss to Wipe or clean various parts of their bodies (van Schaik et al. 2003, 2009). It was also reported that wild orangutans cleaned their teeth, possibly as a result of eating sticky fruit, by chewing leaves and spitting them out (van Schaik et al. 2009).

The use of tools for Wiping the body, objects, and surfaces is common and widespread for rehabilitants and captives. Russon (2000) observed a rehabilitant Wiping her face with a tissue, and another cleaning a porch with a broom. Felce (1948) reported that a young male used a towel to dry his ears and other body parts when he was wet. Harrisson (1962b) saw captive males Rubbing or Wiping their own and their companions' penises with items such as orange peels. A captive male Wiped topical medication from his face and eyes with straw and pieces of cloth (B. Beck, pers. obs.). Jantschke (1972) also reported that captives Wiped their bodies, as well as cage floors and walls, with sacks and towels.

Affix, Apply, Drape

The tendency of orangutans in various settings to place a variety of objects on their heads and bodies appears universal, although the functional significance may not always be apparent (Koehler 1993; Lethmate 1982). Similar behavior incorporating cloth, leaves, branches, rope, chains, food, straw, and paper is frequently seen among captives during solitary or social play (Alcock 1972; B. Beck pers. obs.; C. Becker 1984; Freeman and Alcock 1973; Harrisson 1962a, 1962b, 1963; Jantschke 1972; Nogge 1989; R. Shumaker, pers. obs.). Wild young males sometimes Draped vegetation on their heads or around their necks when approaching conspecifics (Rijksen 1978). MacKinnon (1971, 1974a, 1974b), and Rijksen (1978) observed Draping by wild and rehabilitant youngsters in play contexts. Russon, Vasey, and Gauthier (2002) described a game played by two young individuals who took turns placing a shirt over the other's head and eyes. The sighted orangutan would charge the unsighted one, resulting in a vigorous bout of wrestling. The authors reported that the game lasted for more than fifteen minutes.

A. R. Wallace (1869) cited reports from local indigenous peoples that wild orangutans Draped large leaves over themselves during rain. Galdikas (1978), Galdikas-Brindamour (1975), MacKinnon (1971, 1974a, 1974b), and Rijksen (1978) later confirmed that wild orangutans held or Draped leaves and other vegetation over themselves in heavy rain or intense sun. MacKinnon (1971, 1974a, 1974b) saw them behave similarly in the presence of humans and suspected that they were trying to hide. Van Noordwijk and van Schaik (2005) reported that infants as young as 2 years old independently covered their heads with large leaves during rain, whether their mother was doing so or not. A female rehabilitant used a parasol to shade herself from the sun (Russon 2000).

Orangutans have been observed Draping objects for protection or comfort in a number of situations other than rain or intense sun. C. Darwin (1871) saw a captive Draping a cloth over itself before being

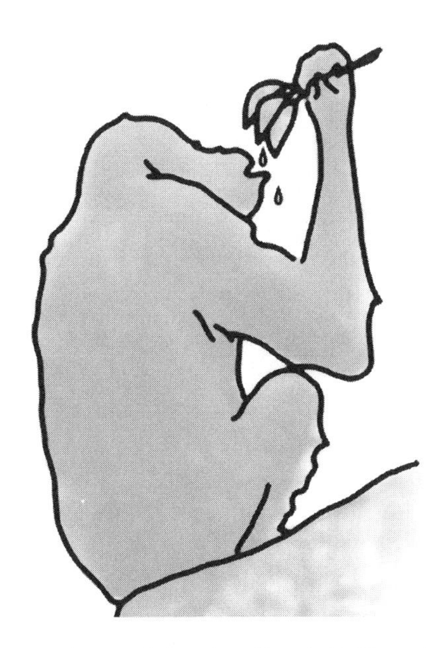

An orangutan using a leafy branch to Absorb drinking water. Drawing courtesy of Perry van Duijnhoven.

whipped. Rijksen (1978) reported that a wild female piled twigs on herself to escape pursuing bees. Western (1994) described a male constructing a "hat" from small branches as he approached a bee nest. The leafy brim allowed him to avoid stings on his face while he consumed the honey.

Wild orangutans were reported to line their nests with leaves or small branches, presumably for comfort (Chevalier-Skolnikoff, Galdikas, and Skolnikoff 1982; MacKinnon 1974a). They may be very selective in choosing material to use for nest lining (Russon et al. 2007). Captives routinely use blankets, sheets, paper, cardboard, straw, and other materials to line their nests and to cover themselves when sleeping or resting (B. Beck, pers. obs.; Bourne 1971; Harrisson 1962a, 1963; R. Shumaker, pers. obs.; K. Walkup, pers. obs.). Both in the wild and in captivity, individuals arranged nesting materials into pillows on which they rested their heads while sleeping (R. Shumaker, pers. obs.; van Schaik et al. 2003). Wild orangutans have also been seen Draping a small number of leaves or twigs on a much larger branch, creating a "branch cushion" to lie down on. The cushions were a loose assemblage, unlike the more elaborate night nests (van Schaik, van Noordwijk, and Wich 2006). Seat cushions made in a similar manner were also used by wild orangutans when in trees with spines (van Schaik et al. 2003). Wild orangutans were observed Applying poultices of chewed leaves to wounds on their bodies (van Schaik et al. 2009). Morrogh-Bernard (2008) observed a wild adult female who Detached leaves from a *Commelina* plant, chewed them until a lather was produced, Applied this to the elbow area of one arm, and then repeated the process for the same area of her other arm. Since this plant is regularly applied by local people to relieve muscular pain, Morrogh-Bernard hypothesized that this could be a case of medicinal plant use by an orangutan. Russon et al. (2010) noted that Kaja Island rehabilitants routinely rubbed water over their bodies, and a female rehabilitant rubbed insect repellant on herself (Russon 2000).

A wild orangutan wrapped an ant nest with leaves in what seemed to be an attempt to prevent bites, and rehabilitants were seen to wrap spiny fruits with paper, leaves, and a sack prior to handling them (Rijksen 1978). Protecting the hands with "leaf gloves" occurs in other contexts as well (van Schaik et al. 2003). A wild unflanged male observed by E. Fox and bin'Muhammad (2002) stacked a pad of five to ten leaves together and then placed the pad in the palm of his hand as he climbed a thorny tree trunk. As he moved throughout the tree, he transferred the pad from hand to foot, preceding shifts in his body weight. Several transfers over several minutes occurred before a new pad was constructed and used. The use of these leaf gloves was reported to be habitual at the Ketambe field site in Sumatra, while occurring rarely at the Tanjung Puting site in Borneo (van Schaik et al. 2003).

An orangutan using a leaf glove. Drawing courtesy of Perry van Duijnhoven.

In an example from captivity, an individual was observed to hold clumps of grass that functioned as an "insulating mitt," allowing him to climb over an electrical wire without being shocked (Linden 1999).

Orangutans are known to Apply leaves to their mouths in order to modify specific vocalizations. Termed a "kiss-squeak with leaves" by van Schaik et al. (2003), this vocalization occurs in agonistic contexts, such as the presence of a threatening animal or the first encounter with a human (Peters 2001). Leaves are held to the mouth as the kiss-squeak is produced, and then Dropped in the direction of the intended recipient. According to Peters, who witnessed the production of the call, the leaves served to increase the volume of the sound. He found that a sound spectrograph of a human simulation confirmed this impression.

Orangutans also utilize Draping for personal adornment. A captive female constructed a "necklace" with string, small tubes, and beads and then placed it around her neck (Herzfeld and Lestel 2005). In a similar case, after viewing his reflection in a large mirror, an adult male left and returned with a burlap bag. He Draped the bag over his head, leaving his face fully exposed, and then adjusted the cloth while looking into the mirror (D. Shillito, pers. comm.).

Symbolize

A female rehabilitant observed a field assistant using a Swiss Army knife to cut and shape a walking stick for himself. The orangutan collected two sticks and rubbed one against the other, imitating the Cutting and scraping motions she had just observed. Noting her interest in the knife, the assistant opened the scissors and pretended to snip a lock of her hair. In response, the orangutan held a fistful of her hair and

A zoo orangutan adorns her head with a piece of cloth. This was the cover photo on the original edition of *Animal Tool Behavior*. Photograph by Leland LaFrance, © Chicago Zoological Society.

made a cutting motion across it with her sticks. The assistant interpreted this as a request from the ape, and cut her hair at the spot she had indicated with her tools (Russon 2000). The orangutan used the sticks Symbolically to represent observed behavior and to communicate.

Russon (2003) observed a rehabilitant female named Siti, who attempted to open a coconut in the forest. Encountering difficulty, she handed the coconut to a human research assistant who was nearby, who then handed it back to her. Siti handed it back, and he returned it again. The female then took a stick and made repeated chopping motions on the coconut. This series of events was recorded on video and shown to people with considerable experience interpreting the behavior of orangutans. Russon (2003, 286) noted that all viewers interpreted it "as Siti's twice asking the assistant to chop her coconut open with his parang (Indonesian machete-like knives), his refusing twice, then her reiterating and clarifying her request by acting out what she wanted done." We interpret the stick Siti used as a Symbolic tool that represented Cutting with the parang.

At two geographically separated sites in Borneo and one in Sumatra, a small number of young wild female orangutans have been observed constructing leaf bundles, treating them like infants, and holding them while they slept in their nests at night. Field researchers have dubbed these "dolls" (van Schaik et al. 2003). Without question, these dolls are held and carried by the user, but since the doll did not have a head or feet, proper and effective orientation appeared to be subjective to the ape.

Chantek, the young male orangutan with proficiency in sign language, used metal washers as currency with his human social partners. He earned "money" by responding to requests from researchers and could spend it on favored foods or activities (R. Shumaker, pers. obs.).

Indah, an adult female orangutan, and Beck used to provide a public demonstration of tool behavior at the Think Tank exhibit at the Smithsonian National Zoo. Beck gave Indah several stalks of freshly cut green bamboo from which she could make tools and use them to Reach incentives. After the demonstration, Beck wanted to remove all the long, stout bamboo stalks lest Indah use them later to get hoses and dismantle light fixtures outside of her enclosure. Indah would surrender the bamboo in exchange for food rewards, and she quickly learned to retain the longest, thickest stalks for really preferred foods. When Beck offered a single raisin, Indah would exchange only a short leafy twig. When Beck offered a piece of mango, Indah would trade a prime stalk. She would decline a raisin offered for a prime stalk and halt further exchanges. There were also one or two intermediate categories, with a resultant three- or four-step ordinal scale matching the size and flexibility of the tool with the desirability of the food. Indah used the pieces of bamboo, and Beck used the foods, as Symbolic tools. Zoo visitors were so impressed with Indah's grasp of value-matching that this exchange became part of the tool demonstration. But soon, after finishing the Reaching tool part of the demonstration, Indah began to gather every scrap of bamboo into a pile and offer it unconditionally, trusting Beck to reciprocate with all of the preferred foods. The only way that the value-matching could be preserved for the demonstration was for Beck to "cheat" by taking the bamboo and providing no food. He reports that he could not do that because it would not be fair (B. Beck, pers. comm.).

TOOL USE: GORILLAS

Drop

A wild gorilla trapped in a tree Dropped down branches toward its pursuers (Merfield 1956). Kortlandt and Kooij (1963) relayed Baumgartel's observation of a wild gorilla Dropping a branch toward an unspecified target.

Throw

A common mode of gorilla tool use is unaimed Throwing, which occurs as a component of agonistic display. Wild gorillas Threw branches, twigs, leaves, and herbs while displaying in the presence of conspecifics and humans (Emlen 1962; Fontaine, Moisson, and Wickings 1995; Geddes, Merfield, Rahm, and Rollais in Kortlandt and Kooij 1963; Groves 1970; Kortlandt 1972; Schaller 1963). Chevalier-Skolnikoff (1977) relayed Fossey's observation of underhand aimed Throwing by wild mountain gorillas during aggression. A population of non-habituated wild gorillas living on Kagwene Mountain in Cameroon have been observed employing aimed Throwing toward humans. These gorillas typically Threw clumps of grass with roots and mud still attached, but a Detached branch was used on at least one occasion (Wittiger and Sunderland-Groves 2007). A human was observed throwing rocks at the gorillas, who responded in turn by Throwing clumps of grass at the man. The gorillas bit into some of these clumps before Throwing them. The authors suggested that the use of these tools might be a local behavioral tradition.

In similar contexts, captives are known to Throw straw, sand, and water without aiming, and they may also Throw objects during play (anonymous in Kortlandt and Kooij 1963; B. Beck, pers. obs.; Shafer 1988; R. Shumaker, pers. obs.).

Aimed Throwing by captive gorillas has been reported frequently. In a survey by S. Parker et al. (1999) concerning tool use by zoo gorillas, 60 percent of fifty-six individuals exhibited aimed Throwing. J. Gómez (1999) described a 2.5-year-old gorilla who imitated a human companion after seeing aimed Throwing of an apple at a glass wall. A male gorilla was documented Throwing sand on his caretaker (Henschen 1926 in L. Harris 1993). A female gorilla at the Brookfield Zoo Threw objects such as tires, sand, straw, and feces at a male when he was displaying toward and pursuing her. The male was not deterred, even when hit by a missile, but began to Throw objects at her in retaliation. Similar observations of aimed Throwing were reported for a male gorilla living with a female at the Zoo in Gulf Breeze, Florida (J. E. Gould and Snyder 1990), and in a variety of social settings in different captive facilities (Vancatova 2008).

Most aimed Throwing by captive gorillas is directed toward humans. Sand, water, feces, stones, branches, and vegetables are among the objects Thrown, and some individuals attain remarkable velocity and accuracy (B. Beck, pers. obs.; Böer 1990; Fontaine, Moisson, and Wickings 1995; J. E. Gould and Snyder 1990; Groves 1970; R. Shumaker, pers. obs.; Smith in Kortlandt and Kooij 1963; S. Woods 1991, 1995). A female gorilla was frequently observed stripping side branches from a larger branch, sharpening one end with her teeth, and then Throwing the "spear" at a disliked caretaker (S. Woods 1992). This behavior was observed regularly over eighteen months. Kortlandt and Kooij (1963) also mentioned a case of aimed Throwing in play. In the context of foraging, gorillas living at the San Diego Zoo were documented Throwing sticks into trees in order to dislodge leaves and seeds to consume (Nakamichi 1998, 1999). The apes were selective in both their tools and targets. They chose long and thick sticks, aimed only at fig (*Ficus* spp.) trees, and scanned the targets prior to throwing. The timing of their efforts also appeared to be strategic, since they only performed this behavior when certain adult members of the group were not present. J. Gómez (1999) described a juvenile gorilla who skillfully Threw bundles of straw, used cardboard, pieces of plastic, and rope to Reach and retrieve desired items outside her enclosure. Gómez also noted that sticks, branches, and similar items were not available to her, and the use of less suitable items required significant skill. In an experimental reaching task, a juvenile female gorilla Threw a stick at an out-of-reach food reward (Natale 1989).

Drag, Roll, Kick, Slap, Push Over

Wild and captive gorillas splash or Slap water as a means of augmenting their displays. Wild gorillas in the Republic of the Congo create "splash displays" in at least ten different ways, although three techniques account for 90 percent of the observations (Parnell and Buchanan-Smith 2001): charging into standing water with the whole body, using one hand to forcibly Slap the water at an angle, and using two arms simultaneously in the same manner. Each of these creates large and impressive sprays. The authors reported that most of these displays were made by adult males in agonistic contexts, most likely to intimidate rivals and acquire potential mates. Parnell and Buchanan-Smith suggested that the open forest and swampland at this site might favor the evolution of visual displays that are visible at long distances.

Captive gorillas also produce "water displays," which might be influenced by visitor behavior (S. Brown, Dunlap, and Maple 1982). Water display has also been seen as part of play behavior by young males (Quick 1976), who enthusiastically drop into shallow pools, creating attention-getting splashes.

Brandish, Wave, Shake

Gorillas Brandish or Wave objects during display (Kortlandt 1972). Both Groves (1970) and Kortlandt and Kooij (1963) cited Cordier's observation of a wild gorilla Brandishing a pole toward a human. A series of photographs presented by Joines (1976) depicted a gorilla carrying a bough while displaying. Zenker (in Matschie 1904), subsequently cited by Armbruster (1921) and Yerkes and Yerkes (1929), described gorillas Waving bunches of twigs to disperse flies. J. E. Gould and Snyder (1990) observed an adult male named Colossus using a stick to swat a fly. Patterson and Linden (1981) described how Koko, a sign-language-competent gorilla, Brandished a toy alligator at humans in a playful attempt to elicit a "fearful" reaction.

Bait, Entice

An adult female gorilla at the San Francisco Zoo used branches as lures to encourage her infant to crawl, and also to come to her (S. Woods 1992). Several gorillas tested by Chalmeau and Peignot (1998) learned to exchange one item for another with an experimenter, such as a stick for a nut.

Club, Beat

A wild female mountain gorilla was observed to repeatedly hit the ground with a branch near a hyrax (Procaviidae). It was unclear if she was attempting to hit the animal directly, with poor aim, or if she was simply trying to scare it off (B. Blaine, unpub. data).

Kortlandt and Kooij (1963) noted that captive gorillas Clubbed or hit unspecified targets, but they provided no other details on objects or contexts. In their survey of zoo gorillas, S. Parker et al. (1999) reported that 29 percent of fifty-six gorillas were observed hitting with weapons. Fontaine, Moisson, and Wickings (1995) stated that one of their subjects hit a human with a stick. S. Woods (1992) noted that a gorilla used a tool to squash insects.

Pound, Hammer

Of the fifty-six gorillas in the survey by S. Parker et al. (1999), 17 percent of the subjects were observed Hammering. J. Gómez (1999) reported that 12-month-old gorillas hit floors and walls with sticks, although the intention of these actions was not clear. Published reports of Hammering by gorillas often involve individuals who have lived in close association with humans. A pet gorilla raised as a human child attempted to open a coconut with a literal hammer (A. Cunningham 1921). Patterson (1985) reported that when Michael was presented with a log with drilled holes in which food was embedded, he Pounded on it with PVC pipe in an attempt to dislodge the edibles. Pettit (1997) reported that Michael regularly Pounded two pieces of wood together rhythmically to imitate the sound of human hammering, and Michael was also seen to Hammer wooden "nails" into the ground outdoors. A zoo-living gorilla tapped on a window with a stick in order to attract the attention of visitors (J. E. Gould and Snyder 1990). A 2.5-year-old female gorilla living in a sanctuary in the Democratic Republic of the Congo was documented using a hammer-and-anvil technique to open nuts (Owen 2005). This behavior is well known among capuchin monkeys (*Cebus apella*) and chimpanzees (*Pan troglodytes*), but was never before seen in gorillas, particularly in such a young individual. Owen reported that this gorilla devised the behavior on her own, without any tutelage.

In the contexts of play or displaying, gorillas routinely Pound objects such as bowls or plastic boxes on hard surfaces or against each other to create noise (J. Gómez 1999; Nierentz 2007; R. Shumaker, pers. obs.; Vancatova 2008; S. Woods 1995). This behavior has been performed by infants as young as 12 months old (J. Gómez 1999).

During an experimental reaching task, a juvenile female gorilla repeatedly hit the testing apparatus with a stick, which sometimes caused the food reward to come within reach (Natale 1989). In addition to Throwing objects into trees to get edibles, one of the San Diego Zoo gorillas hit an overhead branch with a tool to knock down leaves (Nakamichi 1999).

Pry, Apply Leverage

In their survey, S. Parker et al. (1999) recorded that 15 percent of the subjects were observed "levering," although no further details were provided. Pettit (1997) reported that two adult gorillas, Koko and Michael, used toys, sticks, nylon bones, carrots, and a corn stalk as Prying tools, often in attempts to open windows.

Dig

S. Parker et al. (1999) reported that only 4 percent of the fifty-six gorillas in their survey engaged in "digging with a stick." S. Woods (1995) observed an adult female Digging with a stick in a planter shortly after she watched a caretaker performing the same behavior. Pettit (1997) reported that Koko routinely Dug outdoors with sticks, filling bowls with the dirt she excavated.

Jab, Stab, Penetrate

Koko used chopsticks, combs, toys, and a variety of other tools to prod or Jab at screens, windows, and virtually any other part of her living space that appeared vulnerable (Patterson and Linden 1981; Pettit 1997). B. Blaine (unpub. data) described the behavior of a captive female, Mandara, who routinely attempted to Stab unfamiliar or disliked humans working near her enclosure. Mandara favored the use of bamboo, and she consistently oriented the cut end, usually pointed, toward her intended victim.

Reach

T. Phillips (1950) and Pitman (in Schaller 1963) reported that wild gorillas used sticks to Reach toward and secure fruit. In a personal communication to Schaller, however, Phillips later expressed uncertainty about the observation, and neither Schaller nor other field workers subsequently documented such behavior in gorillas.

Breuer, Ndoundou-Hockemba, and Fishlock (2005) documented a unique case of Reaching by a wild gorilla. An adult female used a branch as a walking stick as she attempted to move across a pool of water. Upon entering the waist-deep water bipedally, she inserted the stick, approximately a meter long, into the pool to assess the water depth and/or the stability of the bottom. The authors reported that she did this a number of times while in the pool, also relying on the stick for support when moving. Eventually she abandoned the stick and returned to her crying offspring on the shore.

S. Parker et al. (1999) found that 27 percent of zoo gorillas exhibited "raking" to obtain items beyond their reach. J. Gómez (1999) described 22-month-old gorillas using sticks to manipulate things such as feces. At 28 months of age, these same gorillas used sticks, cardboard, and straw for the "manipulation of spiders and other obnoxious targets, picking up food from a container, Reaching the latch of a door, trying to poke objects out of a hole, etc." (J. Gómez 1999, 166). These individuals also used cardboard or wood to touch ants.

There are numerous other reports, such as one by R. Wood (1984), of gorillas who routinely and spontaneously used items such as sticks, branches, burlap bags, blankets, and flexible hose to Reach desired items outside of their enclosures. While the use of Reaching tools is common in captivity, there is variation within social groups. R. Wood (1984) documented the use of Reaching tools by only eight individuals of her group of twenty-one. Böer (1990) described a gorilla who stood on the shoulders of a group mate, then used a stick to Reach for a plant. There is no indication that the tool user manipulated the group mate in any way, suggesting that the only tool was the stick. A detailed observation provided by D. deGraffenreid (pers. comm.) documented how Sylvia, a female gorilla then living at the Smithsonian National Zoo, used a tool to gain the attention of a favored caretaker. The caretaker and a colleague were talking with another gorilla and did not know that Sylvia had approached in an enclosure behind them. Sylvia first tried to get their atten-

A female gorilla using a branch to test the depth and bottom stability of a pond. Photograph © Thomas Breuer / Wildlife Conservation Society.

tion by pushing her hand through the mesh, but she could not reach them. She left the area and returned with a 15-centimeter-long stick, and pushed it in the direction of the caretaker. Since she made no sound, he remained unaware of her efforts. She left the area for a second time and returned with a stick that was at least 1 meter long. Sylvia stuck this through the mesh and gently poked the caretaker in the back, finally succeeding in gaining his attention.

Foods are by far the most frequently cited rewards that gorillas work to obtain. However, gorillas also work to procure a variety of non-food items such as rakes, trash bags, latex gloves, office curtains, items of clothing, soft drinks, and a wall-mounted telephone (B. Blaine, unpub. data; Fontaine, Moisson, and Wickings 1995; J. E. Gould and Snyder 1990; Nakamichi 1999; Pettit 1997; Prince-Hughes 2001; Vancatova 2008; S. Woods 1991, 1995). Gorillas also use tools to Reach and examine non-food items. Teleki (in van Lawick–Goodall 1970) observed a captive gorilla Reaching with a piece of straw to touch urine beyond the cage bars. The animal then put the soaked end of the straw to its lips. An adult male used small sticks to investigate scabs or irregularities on his skin. Typically, this male would use the tool to move the hair to expose the skin and then Probe the area directly with the tool (B. Blaine, unpub. data). Patterson and Linden (1981) reported that Koko used a straw to drink water.

Fontaine, Moisson, and Wickings (1995) described the near-daily use of tools by gorillas to Reach foods outside of their enclosure. This included using a flat stick to scoop up porridge. These apes were also seen to chew on the end of a stick until it resembled a brush, which was then used like a mop to clean up porridge. S. Parker et al. (1999) reported that 4 percent of the fifty-six gorilla subjects included in their survey used sticks as utensils with food.

Gorillas also employed Reaching tools to get to tree branches that were otherwise inaccessible. Nakamichi (1999) described an 8-year-old male who stood bipedally while Reaching with a long stick to successfully pull down a branch in order to forage on the leaves.

A gorilla Reaches with a stick. Photograph courtesy of The Revealed Project and Marina Vancatova.

Enrichment devices provided by caretakers readily stimulate tool use. Oliver (1996) and R. Wood (1984) both documented gorillas using sticks confidently and frequently to obtain preferred foods from a dispensing apparatus.

Captive gorillas used rakes and sticks to Reach food in formal experimental settings (Chevalier-Skolnikoff 1977; Martin-Ordas, Call, and Colmenares 2008; Mulcahy and Call 2006b; C. Parker 1968, 1969b; Redshaw 1975). Yerkes (1927a) found that a 5-year-old captive gorilla did not learn independently to use a Reaching tool and did so awkwardly, even after tuition. At 6 and 7 years, however, the same gorilla skillfully used sticks to Reach food and other desired objects (Yerkes 1927b, 1928–1929). Yerkes attributed the improvement to maturation. At 7 years, the subject retrieved sticks that were not in the same visual field as the incentive and skillfully used Reaching tools in novel situations. Antinucci, Spinozzi, and Natale (1986) reported that a 5-year-old gorilla failed to use a stick properly as a Reaching tool in an experimental setting. A 4- to 5-year-old tested by J. Gómez (1999) also initially showed poor skill with Reaching tools. However, Gómez suggested that lack of previous exposure, as well as having to manipulate the tool through restrictive caging, might have affected performance. When tested three days later in a less restrictive setting, the same individual performed perfectly.

In contrast to these reports, Redshaw (1978) documented the successful use of a Reaching tool by a 26-month-old gorilla, although earlier trials produced negative results (Hughes and Redshaw 1974). Natale, Potì, and Spinozzi (1988) reported that a 2-year-old gorilla was able to use a long stick successfully, but not consistently, as a Reaching tool. By 3 years of age, this gorilla consistently used a blanket successfully as a preferred Reaching tool. Further, if the blanket were not nearby when an out-of-reach food was provided, the youngster would search for it, suggesting the presence of a mental image of the blanket as a tool. Mulcahy, Call, and Dunbar (2005) presented their gorilla subjects with an experimental setting designed to assess both the use and selection of a proper tool. Their subjects successfully used straight tools to Reach out-of-reach food items. They were able to select tools of appropriate lengths, rejecting tools that were too short. The gorillas also used tools sequentially to obtain other tools so they could successfully complete the task.

Insert and Probe

In the survey by S. Parker et al. (1999), 46 percent of the zoo gorillas were observed Probing. The vast majority of these reports arose from similar scenarios in which caretakers or researchers provided enrichment devices requiring the use of a thin tool to extract preferred foods (B. Blaine, unpub. data; Böer 1990; Boysen and Frisch 1987; Boysen et al. 1999; M. Cole 1987; Downman 2000; J. E. Gould and Snyder 1990; Jarvis and Gould 2007; Lonsdorf et al. 2009; Mentz and Perret 1999; Oliver 1996; Patterson 1985; Pouydebat et al. 2005; Prince-Hughes 2001; S. Woods 1991, 1992; B. Wright 1994, 1995). These devices usually contain sticky, pasty, or buttery foods that adhere to tools that are Inserted and withdrawn. In a maze apparatus, the apes had to move small foods such as peanuts along a series of pathways until the treat dropped out of a small hole (Mentz and Perret 1999). Gorillas at times transported their tools to a probing site (S. Woods 1992) or carried portable devices as they searched for proper tools to use on them (B. Wright 1995).

Captive gorillas Probe with tools in personal grooming. An adult female gorilla chewed small sticks to create a pointed end and then used the tool to clean the nostrils and navel of her offspring while he slept (Fontaine, Moisson, and Wickings 1995). A 15-month-old used either tissue or straw to remove mucus from her nose (J. Gómez 1999). Shortly after receiving a deep bite from a conspecific, an adult female was seen to Insert and slide a piece of straw in and out of the site, perhaps assessing the depth of the wound (B. Blaine, unpub. data). After removing the tool, she carefully inspected the blood on the tool. S. Parker et al. (1999) reported that 35 percent of zoo gorillas groomed themselves with tools; however, it is not clear if this "grooming self" category included behaviors other than Inserting and Probing.

Scratch, Rub

J. E. Gould and Snyder (1990) reported that a captive male Rubbed his teeth with a stick. Patterson and Linden (1981) described Koko as performing a similar behavior with a toothbrush. Fossey (1981) observed an older male using a long-stemmed flower to tickle a youngster. B. Blaine (unpub. data) reported that a captive adult male used a piece of straw to tickle his own foot. S. Woods (1992, 1995) described zoo-living gorillas who tickled themselves with branches.

Block

Gorillas inserted sticks as wedges into spring-loaded water dispensers to cause the water to run continuously (S. Woods 1992).

Prop and Climb, Balance and Climb, Bridge, Reposition

In a unique example, a wild female gorilla living in a swampy habitat picked up and pushed a dead and detached trunk into the wet ground, and then held it with one hand to stabilize herself while she foraged with the other. After foraging, she removed the trunk, placed it on the swampy ground, and used it as a Bridge to cross, walking bipedally (Breuer, Ndoundou-Hockemba, and Fishlock 2005).

S. Parker et al. (1999) provide two related categories, "using ladder/stool" and "using bridge," in the results of their survey; 29 percent of the gorillas used ladders or stools, and 17 percent used bridges. Of the twelve responding institutions, 75 percent had one or more gorillas who used ladders or stools.

Some published reports of tool use in these modes result from escapes or attempted escapes by gorillas living in moated enclosures. A gorilla at the Houston Zoo scaled the wall of a moat after Propping a ladder left within reach (Quick 1976). A male at the Woodland Park Zoo Dropped a log into his moat, moved down into the moat, Propped the log against the wall, and Climbed out (Hancocks 1983).

Gorillas have been documented using a variety of items as ladders for purposes other than escape. A captive gorilla Balanced and Climbed sticks and poles to reach suspended food (Yerkes 1927a, 1927b). Others have Repositioned and stood on plastic barrels, tubs, rolling tables, boxes, and logs, using them as ladders to access elevated incentives (Fontaine, Moisson, and Wickings 1995; J. Gómez 1988; A. Henry, unpub. data; S. Woods 1991). An adult female living in the Apeldoorn Zoo Propped a branch against the wall of her outdoor enclosure, giving her access to a nearby tree where she foraged on leaves and bark (Vancatova 2008). The use of Bridges has been documented to occur in individuals as young as 10 months old (J. Gómez 1999). Muni, an infant female gorilla being raised in a zoo nursery, leaned a pole against a wall and then Climbed it in order to reach the latch on a door (J. Gómez 1990). She used a tricycle, boxes, crates, bins, poles, and brooms in similar fashion (J. Gómez 1990, 1999). Koko Propped a PVC pipe against the wall of her outdoor enclosure, Climbed it, and foraged on otherwise out-of-reach plants (Pettit 1997).

A. Henry (unpub. data) described an adult female who routinely rolled a large plastic ball to a specific place in her enclosure and then sat on it. Her elevated position gave her an unobstructed view of the kitchen where caretakers prepared her food. Vancatova (2008) reported that female gorillas at the Prague Zoo Repositioned and used plastic boxes as elevated seats. S. Woods (1995, 386) also noted that gorillas positioned objects and then used them to "sit, lay, or stand on."

A female gorilla at the Los Angeles Zoo used a young male twice as a ladder by climbing onto his shoulders and escaping her enclosure. It is not known how she convinced him to stand against the wall while she ascended the barrier (anonymous 1987 in S. Woods 1992). S. Woods also mentioned that at the San Francisco Zoo, an adult female's infant stood on her back below an elevated window in order to play with human children through the glass. The impression of the reporter was that the adult female provided this opportunity for the infant, in essence allowing herself to be used as a tool. A 1-year-old infant at the Smithsonian National Zoo used her older brother as a ladder to reach her mother, who was seated on an elevated platform. Her brother appeared to be a willing participant (A. Henry, unpub. data). In an analogous example from the wild, infants routinely used their mother as slides (Fossey 1970). These examples are not definitive acts of tool use, since it is unclear if the "users" pushed or moved the other individuals prior to use.

A gorilla collects and stacks baskets on which to stand to look out of a window. Photograph courtesy The Revealed Project and Marie Farova.

The infant Muni climbed on human companions to access out-of-reach items or places. In some cases, she directly manipulated the person's body by lifting and positioning their legs against a wall and then using the adult as a Bridge to reach a window (J. Gómez 1990). Alternatively, she would Reposition them below a desired object or location and then gesture to be picked up. Once in her companion's arms, Muni would immediately reach for what she wanted (J. Gómez 1988). At other times, she simply climbed on a person without reorienting them in any fashion, a behavior not interpreted as tool use (J. Gómez 1990). Although no climbing was involved, a gorilla used a stick to prop up a sore limb (S. Parker et al. 1999), and another Repositioned a chair so that she could use it to prop up her feet (Pettit 1997).

Pettit (1997) described an adult female who routinely stacked several objects, such as chairs or tubs, to create an elevated resting spot for herself. This was reported to occur as a way of avoiding the cool floor and accessing the warmer air near the ceiling. All of the females at the Prague Zoo were reported to frequently stack and stand on boxes to use more three-dimensional space or to look through an elevated window (Vancatova 2008).

An individual gorilla stacked as many as four boxes to acquire food in an experimental setting, but human tuition may have contributed to the acquisition and perfection of stacking (Yerkes 1927a, 1927b, 1928–1929; Yerkes and Yerkes 1929). In one trial, the gorilla held a box in its hand and used it to Reach up toward the food.

Contain

S. Parker et al. (1999) reported that 8 percent of zoo gorillas were observed using a "container for water." J. Gómez (1999) described gorillas approximately 42 months old holding open a water dispenser with one hand while simultaneously filling a container with the water using the other hand.

Other captive gorillas used food boxes and other hollow objects to hold drinking water (C. Carpenter 1937). J. E. Gould and Snyder (1990) observed a gorilla dipping a cup into a pool to acquire drinking water. S. Woods (1992) noted that gorillas used a variety of objects as vessels for obtaining and drinking water. A number of gorillas at the Smithsonian National Zoo regularly used plastic construction helmets as containers (B. Blaine, unpub. data). Individuals in this group were observed to dip their hardhats into a waterfall and then drink the water. At least one other individual was reported to regurgitate food into a hardhat and carry it with him as he reingested the regurgitant in small portions (B. Blaine, unpub. data). An adult female urinated into a hardhat and then drank the fluid. On some occasions, she was observed to leave the vicinity of her group, retrieve a hardhat, bring it back to the where her group was located, and then urinate into the hat and drink the contents (A. Henry, unpub. data). An adult male held and filled a hardhat from the trickle of a hose offered by a caretaker. He became quite skilled at holding the hat level to maximize the quantity of acquired water (B. Blaine, unpub. data).

Pettit (1997) reported that an adult female used a chair to hold food she was given by a caretaker. A similar observation was made by Vancatova (2008), who saw an adult male use a box as a "table or tray" for his food. He walked bipedally while carrying the box that held his food.

Absorb

S. Parker et al. (1999) reported that 48 percent of fifty-six zoo gorillas used tools for sponging. One of three gorilla subjects studied by S. Parker (1968, 1969b) used rope to Absorb sweet liquid from a concavity. The rope was subsequently sucked or licked. Some tuition was involved in acquisition of the response. J. E. Gould and Snyder (1990) observed an adult female frequently using a towel to Absorb water from a small pool or puddle and then drinking the liquid. A young female extracted water from a small hole by using coconut fibers as a sponge and squeezing the Absorbed water from the fibers into her mouth (Fontaine, Moisson, and Wickings 1995). Böer (1990) described a gorilla's use of leaves, grass, and straw that she formed into a clump and dipped into water, functioning as a sponge.

Wipe

J. E. Gould and Snyder (1990) provided an account of a captive female gorilla who used towels to Wipe various surfaces in her surroundings, such as windows, as well as Wiping her own body. Based on his observations, S. Woods (1995, 386) made the general statement that captive gorillas will "wipe an object with another object." An adult male used water to wash his loose stool from a rock (Hancocks 1983). Juvenile gorillas estimated to be 2 to 2.5 years of age who were studied by J. Gómez (1999) used straw or tissue to clean themselves or other people, and to touch things such as feces when moving them. Fontaine, Moisson, and Wickings (1995) observed an adult female who soaked coconut fibers in water and used them to clean her infant's body. When the infant first began to move about on his own, she cleaned his feces-soiled hands with coconut fibers.

Affix, Apply, Drape

Draping various materials on the body, a behavior universal in orangutans, is also common in gorillas. Wild gorillas observed by Schaller (1963) Draped *Lobelia* leaves and moss on their heads in a play context. C. Carpenter (1937) reported that captive gorillas placed straw under their chins in an unspecified context. Captives have been documented covering themselves with blankets, fabric, pieces of rubber tubs, burlap, cardboard, straw, pieces of wood, branches, feces, sand, and other objects (Böer 1990; Pettit 1997; S. Woods 1995; B. Wright 1995). Some Draping occurred for warmth (Pettit 1997), during play (B. Beck, pers. obs.), and for personal adornment (Böer 1990), and some had no obvious function. S. Woods (1992) cited the reports of Emlen (1962) and Schaller (1963), describing mountain gorillas who placed vegetation between their lips during a display.

Yerkes (1928–1929) observed his gorilla subject collecting and placing straw in a puddle to provide a dry cushion on which to sit. T. Phillips (1950) noted similar behavior by wild gorillas, who constructed a "bamboo drainage seat" when it rained. Prince-Hughes (2001) observed a young gorilla flattening a rectangle of paper and then sitting on it. An adult male regularly manipulated and sat in a rubber tub (Pettit 1997). Adult females at the Prague Zoo used excelsior to make cushions when sitting on hard surfaces (Vancatova 2008).

Another behavior that qualifies as tool use was observed in conjunction with chest beating by captive males. Two zoo-housed males were observed to collect water in their cupped palms immediately before chest beating. The water seemed to increase intensity and resonance (B. Beck, pers. obs.; V. Thompson, pers. comm.).

Upon encountering snow, gorillas at the Prague Zoo made "primitive shoes" (Vancatova 2008). Vancatova stated that similarly, an adult female held "excelsior pillows" with her toes as she walked in the snow. This same individual duplicated the behavior months later when walking on wet ground. A group of gorillas at the Zurich Zoo performed the same behavior when walking on snow, holding straw in their hands and feet and keeping them warm and dry (Nierentz 2007).

Symbolize

There are a number of reports of female gorillas using objects as dolls (Fontaine, Moisson, and Wickings 1995; J. Gómez and Martin-Andrade 2002; Matevia, Patterson, and Hillix 2002). Koko routinely treats dolls maternally. She has been reported to place a doll's face on her nipple, kiss dolls, make sign language gestures on appropriate places on their bodies, move their hands as if they are gesturing in sign language, cradle them, carry them, and make dolls interact with each other (Matevia, Patterson, and Hillix 2002).

J. Gómez and Martin-Andrade (2002, 259) described two juvenile gorillas providing maternal behaviors to "stones, shoes, balls, pieces of cloth, and even a rubber doll." They placed the objects between their arms and sides, or on their back and neck area, while standing or walking quadrupedally. If necessary, the apes repositioned the objects with their hands while walking to keep them in place. The gorillas were also seen holding the objects against their chests with one arm while walking on the other arm and legs. These authors stressed that no behaviors or facial expressions indicating playfulness were witnessed in association with these incidents.

Fontaine, Moisson, and Wickings (1995) relayed the details of a case involving an adult female whose 1-month-old infant had died. Two weeks after the death, she formed vegetation into a ball by pressing it against her abdomen and then placed the ball on her back and neck area, rolling it forward and catching it three times in a row. She had performed the exact behavior previously with her infant. Additionally, she cradled the ball in her arm, carrying it in this way while ascending a platform, reminiscent of the way in which she handled her infant.

TOOL USE: BONOBOS

In a study of handedness that included twenty-two bonobos in three different captive populations, Harrison and Nystrom (2008) noted that twelve individuals used tools, but no further details were provided.

Drop

A wild male bonobo Dropped pieces of vegetation as part of a courtship display (Kano 1997). The male, who had an erection, positioned himself in a tree approximately 4 meters above a female who was on the ground below. The male quickly and repeatedly broke off several small pieces of the tree, Dropping each of them successively within 1 meter of the female. The female then climbed the tree, where she approached, presented, and copulated with the male.

In a separate context, Ingmanson (1996) reported that bonobos commonly Dropped objects toward researchers observing them from below. Young bonobos, 3 or 4 years of age, have been observed Dropping an object from above, retrieving it, Dropping it again, and then repeating the cycle several times in succession. This may be considered play (Ingmanson 1996) or rehearsal of a behavior. Captive bonobos were reported Dropping leaves during play (Walraven, van Elsacker, and Verheyen 1993).

Throw

Wild bonobos have Thrown vegetation at human observers, as well as other conspecifics. Juvenile bonobos were observed Throwing unripe fruit at a tortoise (Hohmann and Fruth 2003).

S. Savage (1976) observed unaimed Throwing by a captive bonobo in an unspecified context. Individuals living in a sanctuary Threw various objects without aiming as part of agonistic displays (C. Andre, B. Hare, S. Hirata, and V. Woods, joint pers. comm.; Jordan 1982) or in play (Jordan 1982).

Bonobos living in a sanctuary performed aimed Throwing with various objects directed at other bonobos, humans, and birds (C. Andre, B. Hare, S. Hirata, and V. Woods, joint pers. comm.; Jordan 1982). An individual at a rehabilitation center Threw woodchips and feces at a domestic cat (Messinger 2007). The author reported that this same bonobo demonstrated extreme accuracy in Throwing feces at human visitors who peered at him through openings in a wall of his enclosure. S. Savage (1976) reported that a captive male Threw a pail at a female. Aimed Throwing may also serve as an invitation for play (Jordan 1982; Walraven, van Elsacker, and Verheyen 1993). An adolescent male Threw water playfully at an adolescent female (Jordan 1982), and a subadult female repeatedly Threw a nut and a yellow cap to a human social partner as a game (Pika and Zuberbühler 2008). Kanzi, a language-competent bonobo, accurately responded to requests such as "throw the [toy] dog to Kelly" (Savage-Rumbaugh et al. 1993, 116).

A lower-ranking female Threw tennis balls at a higher-ranking female who was afraid of them (Jordan 1982). Bonobos Threw objects at groups involved in grooming, which Walraven, van Elsacker, and Verheyen (1993) interpreted as a means of causing disruption and distraction. These authors noted that objects may also be Thrown at an individual who is displaying, or at a conspecific pursuer.

In an experimental study of stone tool making and tool using, Kanzi utilized aimed Throwing as a means of making flakes for cutting. This technique is further described under Tool Manufacture.

Drag, Roll, Kick, Slap, Push Over

Wild bonobos broke off and Dragged small trees 1.5 to 2 meters long as part of display behavior. Trees with multiple small branches were most commonly used, presumably for the noise they created when Dragged through the underbrush (Kuroda 1980). Hohmann and Fruth (2003) noted that vines were sometimes Dragged.

In anticipation of group travel, male bonobos were observed breaking off small trees and engaging in coordinated Dragging activity. The carefully selected trees were typically about 2 meters long, with enough branching to create significant noise when moved through the underbrush (Ingmanson 1989, 1996). Once a tree was selected and broken off, but before Dragging began, the tree might be held by an individual for up to thirty minutes while he sat, fed, or groomed. As the Dragging behavior commenced, the trees were pulled in the direction of travel and functioned as a means of communicating directional information to the other group members. Ingmanson (1992, 1996) and Ingmanson and Kano (1993) also reported that these small trees were Dragged during travel to encourage group members to stay near each other. According to Ingmanson (1996), 23 percent of the cases involving branch Dragging in this study occurred when the apes were "excited," such as when entering a feeding site; 3 percent occurred during dominance or agonistic situations; and 64 percent in contexts involving travel. Context for the remaining 10 percent could not be determined. However, Ingmanson interpreted all occurrences of branch Dragging, regardless of context, as an augmentation of display or the communication of information.

Captive bonobos Dragged, hit, pushed, or Kicked various items in their environments—such as woodwool, branches, or plastic canisters—during agonistic displays (C. Andre, B. Hare, S. Hirata, and V. Woods, joint pers. comm.; Jordan 1982; S. Savage 1976; Walraven, van Elsacker, and Verheyen 1993). One variant of this behavior was Dragging tree boughs along the

A bonobo Drags a small leafy tree in anticipation of travel. The direction in which the tree is Dragged may influence the direction of the group's travel. Photograph by Ellen J. Ingmanson, courtesy of Bridgewater State College.

mesh of enclosures to make noise, augmenting their displays (Jordan 1982). During social play, bonobos Dragged wood-wool and a variety of other items (Jordan 1982; Walraven, van Elsacker, and Verheyen 1993), varying the behavior by Dragging an object held by another bonobo, resulting in the playmate being Dragged as well (Walraven, van Elsacker, and Verheyen 1993). During solitary play, Walraven, van Elsacker, and Verheyen reported that various objects were Dragged while being held in the mouth or hands, and Jordan (1982) observed a tennis ball being Kicked.

Captives Rolled a canister to other individuals as an invitation for play (Walraven, van Elsacker, and Verheyen 1993). As a game, an adolescent male repeatedly Rolled a grapefruit back and forth with a human social partner (Pika and Zuberbühler 2008).

Brandish, Wave, Shake

Wild bonobos have been seen using vegetation as a flyswatter (Ingmanson 1992). An adult female also used a leafy stick to repel bees for about twenty minutes and then discarded it as she left the area (Ingmanson 1996). Hohmann and Fruth (2003) reported that female bonobos used small leafy twigs as flyswatters to repel flies from their sexual swellings. A male performed a similar behavior to chase flies away from his injured hand.

Captives Waved objects such as wood-wool and Detached tree branches during displays or play (Jordan 1982; Walraven, van Elsacker, and Verheyen 1993). These actions were directed at other bonobos, humans, and other species of primates (Jordan 1982). A male Shook a plastic bucket, perhaps as a result of observing chimpanzees performing the same behavior (S. Savage 1976).

When Panbanisha, a language-competent female bonobo, was asked to extinguish candles, she Waved a towel over the flames to successfully put them out (Niio 2000).

Bait, Entice

Kano (1992) reported that young bonobos used a small branch as Bait during games of chase. One individual would pick up a small branch and run away with it, as a signal to a playmate. Each chased the other in turn, trying to snatch the branch.

Adults also used vegetation as tools for social signaling to attract, or attract the attention of, other bonobos. Adult females ripped leaves from trees or vegetation growing on the ground in the same manner as they would when foraging, except that the leaves were not eaten. This behavior was interpreted by Hohmann and Fruth (2003) as a means of attracting the attention of a male for copulation. The authors also noted that to invite play, bonobos clipped leaves and, while looking at another bonobo, held the leaves in their lips.

In an experimental setting, bonobos traded tokens with human social partners for food rewards (Bräuer, Call, and Tomasello 2009).

Savage-Rumbaugh et al. (1996) provided a detailed discussion of the possibility that bonobos used various forms of vegetation as directional markers for other bonobos to follow as they traveled. Large and small branches were laid on the trail, smaller pieces of vegetation were broken off and left on the ground, and both small and large branches were left standing upright on the route used by the bonobos (Savage-Rumbaugh et al. 1996). The authors concluded that this assortment of markers could not have accumulated accidentally, randomly, or through typical behaviors such as foraging or walking. Although the evidence is not definitive, these tools do provide the basis for a compelling discussion on the origins of Symbolic communication and the potential complexity of tool use among these great apes.

Club, Beat

Kano (1979) relayed observations by local people that wild bonobos cornered a small duiker (*Cephalophus* sp.) and hit it with branches. The bonobos played with the duiker but apparently did not kill it or eat it.

Gold (2002) described the behavior of a wild-born female bonobo living in a zoo who used a 1.5-meter branch as a Club. A peahen (*Pavo cristatus*) had nested and hatched chicks in the bonobo enclosure of the Apenheul Primate Park. In an attempt to protect her chicks, the hen threatened and attacked the curious bonobos. The female bonobo left the scene and went to the indoor housing area, returned with a branch, and used it to Beat the hen to death. Gold reported that this young female delivered more than forty "over-arm blows" to the bird in over eight minutes. Jordan (1982) described a captive juvenile who used leaves to Beat an insect. Walraven, van Elsacker, and Verheyen (1993) reported playful Clubbing of conspecifics with objects such as branches, wood-wool, and paper.

During studies designed to assess receptive language competence, the male bonobo Kanzi accurately responded to requests such as "hit the [toy] dog with the stick" (Savage-Rumbaugh et al. 1993, 116).

A sanctuary-living bonobo uses a stone to Hammer open nuts set on a fixed stone anvil. Photograph courtesy of Friends of Bonobos.

Pound, Hammer

Several bonobos living in the Lola ya Bonobo sanctuary in the Democratic Republic of the Congo used hammers on a fixed anvil to open oil palm (*Elaeis guineensis*) nuts (C. Andre, B. Hare, S. Hirata, and V. Woods, joint pers. comm.). This did not involve tutelage by humans, but observational learning by the bonobos from humans cannot be discounted. At the same facility, individuals also hit tough-skinned fruits with tools to open them. The objects used as the tools were not specified (C. Andre, B. Hare, S. Hirata, and V. Woods, joint pers. comm.).

Kanzi was involved in a longitudinal study of stone tool using and tool making. Details of Kanzi's technique, which involved Pounding and Hammering, are included under Tool Manufacture. In a different context, Kanzi played a xylophone with a xylophone hammer held in each hand (Niio 2000).

Pry, Apply Leverage

Captive bonobos used levers such as sticks for breaking the mesh of their enclosures (Ingmanson 1988; Jordan 1982).

In an unusual example noted during play, an individual bent a branch and inserted both ends into his mouth simultaneously, causing his jaws to open wide (Walraven, van Elsacker, and Verheyen 1993).

Dig

Sanctuary-living bonobos were observed Digging with a stick. The context was unclear, although this behavior was not associated with foraging (C. Andre, B. Hare, S. Hirata, and V. Woods, joint pers. comm.).

Jab, Stab, Penetrate

During playful interactions, individuals Jabbed each other with sticks (Walraven, van Elsacker, and Verheyen 1993). These authors also reported that when trying to open canisters, zoo-living bonobos prodding them with sticks.

Reach

Captive bonobos used sticks as Reaching tools to obtain items outside their enclosures (Jordan 1982; Walraven, van Elsacker, and Verheyen 1993). They also used Reaching tools to touch things that frightened them, including other bonobos, scary objects, or stinging insects (Jordan 1982). Panbanisha used a spoon to stir hot liquids (Ikeo 1993). A sanctuary-living female used a stick as a utensil for eating papaya, while another used a piece of grass to pick ants off the ground to play with but not eat. Others used a stick to steal someone's camera (C. Andre, B. Hare, S. Hirata, and V. Woods, joint pers. comm.). Kanzi used a hairbrush on the hair of a human social partner (Niio 2000). When asked to snuff out lighted candles, Kanzi blew some out and Reached with the tip of a knife to extinguish others (Niio 2000).

Mulcahy and Call (2006a) presented two bonobos

with an out-of-reach bottle of juice hanging from a string, a hook, and several other tools that were inappropriate for the task. The apes spontaneously demonstrated the ability to retrieve the juice by Reaching the string with the hook, correctly selecting the appropriate Reaching tool from the array of objects presented to them.

Insert and Probe

Researchers tracking wild bonobos in the Lomako Forest found broken termite mounds with "ant harvesting tools" nearby (Badrian, Badrian, and Susman 1981). The tools had termite-mound mud stuck on them, and it was inferred that the resident bonobos had used these sticks and stems for acquiring termites. Similar circumstantial observations were made by F. White et al. (2008) in the same forest, and by Bila-Isa (2003) in the Salonga National Park.

Ingmanson (1992, 1996) described a wild male bonobo using a small twig as a toothpick to clean between his teeth after consuming provisioned sugarcane.

Captive bonobos were seen Inserting items such as straw, twigs, or paper to Probe features of their environment, such as a screw hole (Jordan 1982; Walraven, van Elsacker, and Verheyen 1993), or to extract food from small spaces (Ingmanson 1988). Some bonobos have used keys to open locks (Niio 2000; Segerdahl, Fields, and Savage-Rumbaugh 2005). They also used various objects as probes to explore their own ears or nose (Jordan 1982; Walraven, van Elsacker, and Verheyen 1993). Kanzi used a small twig to clean under a human's fingernails (Ikeo 1993).

Like other zoo-living apes, bonobos have regularly been provided with an artificial termite mound as an enrichment device. In research by Parish (1994), the mound contained favored foods that could be extracted via a small hole by using long, thin tools. The seven individuals studied by Parish all fashioned tools from tree foliage to successfully obtain the food rewards. Neary (1997) mentioned captive adolescent bonobos spontaneously using Probing tools, but the author did not provide additional specifics.

A number of bonobos have experimentally been presented with the tube task. Martin-Ordas, Call, and Colmenares (2008), Mulcahy and Call (2006b), and Visalberghi, Fragaszy, and Savage-Rumbaugh (1995) all reported successful solutions involving the Insertion of a stick and raking or pushing to retrieve a reward.

Mulcahy and Call (2006a) supplied five bonobos with hanging bunches of grapes that could only be retrieved by Inserting a tube into a hole and breaking the support for the grapes, which caused them to fall within reach. The apes were provided with the tube and several other possible tools that were not functional for the task. All of the apes spontaneously solved the task, correctly selecting the tube from the array of tools. They were also able to plan for future tool-using opportunities by both transporting and saving the correct tool during delays lasting from one to fourteen hours.

Scratch, Rub

Kuroda (1982) observed a wild male bonobo masturbating with leaves. Masturbation in captivity with other items, such as wood-wool, branches, or plastic balls, has been reported (K. Walkup, pers. obs.; Walraven, van Elsacker, and Verheyen 1993). Individuals were also reported to Rub their bodies with items such as rope, which Walraven, van Elsacker, and Verheyen (1993) interpreted as Scratching itchy skin.

Both wild and captive bonobos used sticks to Scratch their own backs or other parts of their body (Hohmann and Fruth 2003; Ingmanson 1996; Walraven, van Elsacker, and Verheyen 1993). Savage-Rumbaugh et al. (1993) reported that Kanzi used a toothbrush to brush his own teeth. When requested to do so, he properly used a toothbrush on the teeth of a human social partner, on masks with teeth, and on toys (such as a dog) with teeth.

Cut

In the stone tool-making and tool-using studies described above, Kanzi was initially presented with human-made stone flakes that could be used for Cutting, as well as a closed box that contained a desired treat. The box could only be opened by Cutting a cord that held the door to the box shut. Kanzi rapidly learned to Cut the cord with a stone flake by observing a human demonstrator. He subsequently Cut with flakes that he himself made (Savage-Rumbaugh et al. 2007; Schick et al. 1999; Toth et al. 1993).

In a separate context, Panbanisha used scissors to Cut hair but showed little interest in stone flakes for Cutting (Niio 2000).

Prop and Climb, Balance and Climb, Bridge, Reposition

Wild bonobos used conspecifics as tools by Repositioning them and then climbing on their bodies so that a branch was within reach (Ingmanson 1996).

A male living in a sanctuary broke off a tree near a pond and put the trunk in the water. He broke off another tree, attempted to put the trunks together, and used the resulting bridge to try to cross the pond (C. Andre, B. Hare, S. Hirata, and V. Woods, joint pers. comm.). Jordan (1982) reported that captive bonobos Balanced and Climbed sticks to obtain items above them, but she did not provide specific examples. To jump over a shallow, muddy pool, a bonobo utilized a pole (Jordan 1982). A sanctuary-living bonobo used a stick to jump over an electric fence (C. Andre, B. Hare, S. Hirata, and V. Woods, joint pers. comm.).

Jordan (1982) and Ingmanson (1988) stated that captives used ladders, they but did not provide any additional information. Savage (1976) reported that a male used a plastic bucket as a portable seat. Jordan (1982) noted that her subjects used wooden blocks as seats to improve their view. Walraven, van Elsacker, and Verheyen (1993) reported that bonobos used canisters instead of blocks for the same purpose. In a separate context described by Jordan (1982), a female bonobo with a maximally tumescent sexual swelling was observed using a block as a seat. She positioned her swelling so that it hung off of the block while she was seated, presumably increasing her comfort. Jordan noted that the bonobos also used objects such as blocks and balls in play. They seated themselves on the items and used their legs to push themselves across the floor.

Neary (1997) reported that bonobos Propped and Climbed logs to explore their enclosures. These "ladders" allowed the bonobos to access areas that were two-and-one-half times their height when standing bipedally. A captive adult female was documented Propping a 4-meter branch against a 5-meter-high metal cylinder used to protect live trees in her enclosure (Gold 2002). She climbed the branch, leaped to the top of the protective cylinder, and then climbed into the tree. Gold reported that while the female was dangling from a low branch, she "hoisted up four other bonobos into the tree" (Gold 2002, 609). This female used the same branch-propping technique to gain access to the trees on several other occasions, until caretakers removed all long branches from the ground inside the enclosure. She then attempted unsuccessfully to use the shorter branches that remained. Gold added that on at least one subsequent occasion, the bonobo successfully got back into the trees by using a long branch that had fallen to the ground.

Walraven, van Elsacker, and Verheyen (1993) described an adult male who commonly positioned objects such as balls, canisters, or branches against the back of a copulation partner. The use of these tools facilitated ventro-ventral intromission by thrusting the partner's pelvis forward. Repositioning such a tool validates and perpetuates the reputation for sexual creativity earned by bonobos.

Hang

Captive bonobos placed long branches and a rubber rope through the overhead bars of their enclosures and used them as swings (Jordan 1982).

Contain

S. Savage (1976) noted that captive bonobos used a pail as a repository for urine. Walraven, van Elsacker, and Verheyen (1993) reported the use of containers for holding urine, saliva, and water. Jordan (1982) observed individuals using half of a red pepper as a container for drinking water. The same bonobos also used a plastic container, which they transported about 3 meters to a water sprinkler, filled, and then carried to a dry spot where they drank the water. After scooping out and consuming the flesh from a whole pineapple, a female observed by van Elsacker and Walraven (1994) held the cup-shaped rind under a water dispenser, filled it, and drank the liquid. On a subsequent occasion, the same female used a "cuplike remainder" of the rind to scoop out pineapple juice that had spilled into a small bucket as the fruit was consumed. Several sanctuary-living bonobos were observed using the empty half-rind from a piece of fruit as a container from which they drank water (C. Andre, B. Hare, S. Hirata, and V. Woods, joint pers. comm.).

Bonobos scooped water out of a moat with a bucket, and they also scooped water and mud into buckets with their hands (Walraven, van Elsacker, and Verheyen 1993). An individual living in a sanctuary rolled a large leaf into a cup and used it to take and drink water from a river (C. Andre, B. Hare, S. Hirata, and V. Woods, joint pers. comm.).

Walraven, van Elsacker, and Verheyen (1993) reported objects such as balls or wood-wool being placed into containers. To move a large amount of food, one female dragged wood-wool on which she had placed the food.

Bonobos involved in a longitudinal language-acquisition project routinely used a large array of different objects to Contain things. These included boxes, foil, bowls, plates, and cups, as well as bags and backpacks for carrying a variety of foods, liquids, and objects (Ikeo 1993; Niio 2000; Savage-Rumbaugh 1993; Savage-Rumbaugh et al. 1993).

Absorb

Wild bonobos at the Lomako site used mosses as sponges to Absorb water from tree holes (Hohmann and Fruth 2003).

Zoo-living bonobos were observed using long leafy branches to Reach and Absorb soapy water outside their enclosure (Jordan 1982). These individuals also used wood-wool and leaves to Absorb water from the floor or from a container, after which they sucked the water out of the sponging tool. In some instances, the apes collected dry leaves and transported them approximately 4 meters to the water source. Jordan reported that tennis balls were also used for Absorption by one zoo population. These bonobos soaked the balls under a water spigot and then sucked out the water. Bonobos at a different zoo used both wood-wool and chewed banana skins for sponging. They placed either of these in their mouths, which they then filled with water. The bonobos then removed the soaked wood-wool or banana skins and sucked out the water (Walraven, van Elsacker, and Verheyen 1993). These bonobos also used wood-wool to Absorb fluids that collected on surfaces.

Wipe

Two wild adult male bonobos used leaves to Wipe off "feces that had fallen on them or that they had stepped into," and a female "wiped her infant's urine off her thigh" (Ingmanson 1996, 195).

Captive bonobos Wiped their bodies with material such as wood-wool, leaves, or paper in order to remove water, feces, urine, mud, or food remains (C. Andre, B. Hare, S. Hirata, and V. Woods, joint pers. comm.; Jordan 1982; Walraven, van Elsacker, and Verheyen 1993). A female was observed Wiping a glass window with a moist towelette (Niio 2000). Rubbing floors and windows with paper or wood-wool has also been documented and is hypothesized to result from observing caretakers as they performed cleaning chores (Walraven, van Elsacker, and Verheyen 1993).

Affix, Apply, Drape

Wild bonobos Draped leafy branches as "rain hats" over their heads, necks, and shoulders during rain showers (Ingmanson 1992, 1994, 1996; Ingmanson and Kano 1993; Kano 1982) and held large leaves over their heads when it rained (Ingmanson and Kano 1993). During rainy weather, individuals covered themselves with leafy twigs after entering their night nests (Hohmann and Fruth 2003). The authors interpreted this as a means of regulating temperature for personal comfort.

Captive bonobos were observed covering their bodies, and sometimes their faces, with items such as paper or branches during play or in unspecified contexts (Walraven, van Elsacker, and Verheyen 1993). The "blindfolded" individuals then swung and ran around their enclosure playfully. A male was observed placing a plastic bucket on his head in play (S. Savage 1976). Bonobos in a language-acquisition project spontaneously donned and wore masks in play with humans and other bonobos (Ikeo 1993; Niio 2000). Walraven, van Elsacker, and Verheyen (1993) noted that captives sometimes hid from playmates by covering themselves completely in a game similar to hide-and-seek.

When nervous, zoo-living bonobos pressed wood-wool against their bodies (Walraven, van Elsacker, and Verheyen 1993). They also protected their hands with wood-wool while attempting to open coconuts by rolling or hitting them against a hard surface. The same bonobos, when trying to unscrew caps from jars, held the caps with wood-wool, interpreted by Walraven, van Elsacker, and Verheyen as a means of improving their grip.

Captive bonobos were reported to protect themselves from the elements with wood-wool when outside, but the specifics of this behavior were not presented (Walraven, van Elsacker, and Verheyen 1993). During hot weather, an adult male bonobo repeatedly sprinkled his head and torso with water scooped by hand from a small pool (R. Shumaker, pers. obs.).

Symbolize

Wild bonobos may use pieces of vegetation to Symbolically convey information (Savage-Rumbaugh et al. 1996). Many different forms of vegetation, such as broken branches, bent branches, or flattened plants, were noted along trails used by bonobos. Additionally, Savage-Rumbaugh et al. (1996, 183) noted that male bonobos "were seen to carefully select certain branches, break them off slowly and deliberately at the base, carry them to another location, drop them and continue to travel." The authors suggested that these items may be conveying information, such as the direction of travel, to other bonobos, but they also cautioned that more information is required before reaching a definite conclusion.

TOOL USE: CHIMPANZEES

In the first edition of this book, Beck (1980, 79) wrote that "chimpanzees show the greatest frequency and diversity of tool behavior of any non-human animal and, were it not for them, this book would be a slim volume." This is not as true as it was 30 years ago, as chapters 2 through 6 of this edition demonstrate. Other non-humans—such as orangutans, capuchin monkeys, New Caledonian crows (*Corvus moneduloides*), hermit crabs (*Diogenes edwardsi*), and woodpecker finches (*Camarhynchus pallidus*)—frequently use tools and, like chimpanzees, employ diverse and complex Tool Kits. It is still true, however, that the literature on chimpanzee tool behavior far exceeds that found for any other non-human animal; over 475 publications are listed for this taxon alone. In part, the focus on chimpanzees may be a result of "chimpocentrism," a term coined by Beck (1982) to describe the disproportionate focus on chimpanzees as a result of their close genetic, morphological, and behavioral similarities with humans. Because of these similarities, chimpanzees have been studied more intensely in the wild and in their natural habitats than most other species, with much emphasis placed on behaviors considered pivotal to hominin evolution, such as tool use.

The first detailed studies of chimpanzee tool behavior were with captives, as reported in the works of Köhler (1925), Kohts (1923), Yerkes and Learned (1925), and Yerkes and Yerkes (1929). These studies motivated numerous researchers, who have continued to investigate chimpanzee behavior. In the wild, long-term field sites for the study of chimpanzees began in 1960 in Tanzania, with Jane Goodall's establishment of a site in what became Gombe National Park, and with the founding of the Mahale site in 1966 by Junichero Itani and Toshisada Nishida. The work of these pioneers was pivotal to our understanding of chimpanzees and of tool behavior. Goodall was not the first to discover that wild chimpanzees use tools, as others often assert. Chimpanzee tool use was first reported by T. Savage and Wyman (1843–1844). But Goodall did document an unexpected diversity and frequency of tool behavior among the Gombe chimpanzees. She discovered that wild chimpanzees manufacture tools, documenting the creation of tools for termite fishing. Today, many long-term and short-term research sites are in place across the chimpanzees' range in Africa (table 6.1), contributing to the substantial literature on this topic. In the first edition of this book, Beck (1980) described tool behavior as occurring in twenty different wild and released populations of chimpanzees. In a later review by McGrew (1994), forty-one populations were reported as using tools. We report on sixty tool-using chimpanzee populations (see table 6.1). We have not noted which chimpanzee tool behaviors are thought to be customary, habitual, and cultural in specific wild populations, and absent in others (see chapter 1 for definitions). Readers are referred to Whiten et al. (1999, 2001) for these designations.

Drop

Wild chimpanzees Dropped branches, twigs, and vines on their pursuers (Izawa and Itani 1966; Nishida 1970; Nishida et al. 1999; Owen, and also Reynolds, in Kortlandt and Kooij 1963; Sabater-Pi 1972, 1974; Sugiyama 1969). Albrecht and Dunnett (1971) saw wild chimpanzees Drop branches from trees in the presence of leopard models placed on the ground, and Nishida (1968) reported that a wild chimpanzee Dropped branches and bits of vine while looking down from a tree toward a wild leopard (*Panthera pardus*). Takenoshita, Yamagiwa, and Nishida (1998) documented a branch-dropping display by an adult female, who Dropped a branch in the presence of humans and a wild elephant (*Loxodonta* sp.). When harassing older individuals, typically females, male adolescents at Mahale engaged in a behavior termed "thumping" (Nishida 2003a). While standing bipedally, they lifted a heavy log from the ground, and then Dropped it to produce a thumping sound. Wild chimpanzees also Dropped branches and leaves on conspecifics during social play (Albrecht and Dunnett 1971; J. Goodall 1963a, 1964; van Lawick-Goodall 1968, 1970).

Köhler (1925) once saw a captive chimpanzee Drop a stone at a human. Kellogg and Kellogg (1933) reported that the home-reared infant Gua repeatedly Dropped objects from his chair onto the floor, much as a human child would, apparently to hear the clatter. This is probably best considered as object play, not tool use.

Table 6.1
Research sites where chimpanzee tool behavior has been documented. Expanded from McGrew 1994. Sites containing released populations are indicated by an asterisk (*).

Site	Country	Site	Country
Campo	Cameroon	Bassa*	Liberia
Dja Biosphere	Cameroon	Kanton	Liberia
Ebo	Cameroon	Sapo	Liberia
Ntale	Cameroon	Cape Palmas	Liberia / Côte d'Ivoire
West Cameroon	Cameroon	Gashaka	Nigeria
Bai Hokou	Central African Republic	Goualougo	Republic of the Congo
Ndakan	Central African Republic	Guga	Republic of the Congo
Ngoubunga	Central African Republic	Lokoué Bai	Republic of the Congo
Banco	Côte d'Ivoire	Lossi	Republic of the Congo
Taï	Côte d'Ivoire	Mbomo	Republic of the Congo
Yealé	Côte d'Ivoire	Ndoki	Republic of the Congo
Ituri	Democratic Republic of the Congo	Odzala	Republic of the Congo
Kahuzi-Biega	Democratic Republic of the Congo	Assirik*	Senegal
Tongo	Democratic Republic of the Congo	Fongoli	Senegal
Ayamiken	Equatorial Guinea	Niokolo-Koba	Senegal
Dipikar	Equatorial Guinea	Tenkere	Sierra Leone
Mayang	Equatorial Guinea	Tiwai	Sierra Leone
Okorobikó	Equatorial Guinea	Filabanga	Tanzania
Belinga	Gabon	Gombe	Tanzania
Ipassa*	Gabon	Kabogo	Tanzania
Loango	Gabon	Kasakati	Tanzania
Lopé	Gabon	Kasoje	Tanzania
Moukalaba	Gabon	Mahale	Tanzania
Abuko*	Gambia	Rubondo Island	Tanzania
Baboon Island*	Gambia	Ugalla	Tanzania
Bossou	Guinea	Budongo	Uganda
Diécké	Guinea	Bwindi	Uganda
Kanka Sili	Guinea	Kalinzu	Uganda
Seringbara	Guinea	Kibale	Uganda
Nimba	Guinea / Côte d'Ivoire	Semliki	Uganda

Throw

Throwing has been found in all chimpanzee populations (McGrew et al. 2003). Chimpanzees Throw in an overhand, sidearm, or underhand manner, with and without aiming (J. Goodall 1986).

Unaimed Throwing often occurs during displays by male chimpanzees. Wild males have been observed Throwing (without aiming) branches, sticks, stones, sand, handfuls of grass and leaves, sugarcane, grapefruits, a palm nut, cans, kettles, and other small objects during intraspecific agonistic charging displays; at the onset of a thunderstorm; in the presence of a leopard model, tape-recorded chimpanzee vocalizations, baboons (*Papio* spp.), a goat (*Capra hircus*), and a mongoose (Herpestidae, genus not reported); and when frustrated by unsuccessful attempts to gain access to incentives such as estrus females (Albrecht and Dunnett 1971; Boesch and Boesch 1990; J. Goodall 1964, 1979; Kortlandt 1962, 1966; McGrew 1992; McGrew et al. 2003; Nishida 1968, 1970, 1977; Nishida et al. 1999; Reynolds and Reynolds 1965; Rollais in Kortlandt and

Kooij 1963; Riss and Goodall 1977; van Lawick–Goodall 1965, 1968, 1970, 1971, 1973). Albrecht and Dunnett (1971), Nishida (1968, 1970, 1977), and Reynolds and Reynolds (1965) did not consistently distinguish aimed from unaimed Throwing, so it is assumed that Throwing in these cases was unaimed. Chimpanzees Throw a variety of objects without aiming during play (van Lawick–Goodall 1968). Youngsters may Throw a stone a short distance and then retrieve it (J. Goodall 1986). In one instance, Goodall reported that a young Gombe male Threw a *Strychnos* fruit approximately 1 meter into the air and caught it when it fell. Nishida et al. (1999) noted that males might also Throw branches in courtship in the presence of estrus females.

Captive males incorporated unaimed Throwing into agonistic charging displays in the presence of conspecifics and humans. The Thrown objects included sticks, stones, sand, straw, water, cans, a box, and a pail (R. Eaton 1978; Köhler 1925; Kortlandt and Kooij 1963; S. Savage 1976; Takeshita and van Hooff 1996, 2001). It is again assumed that Throwing was unaimed when Kortlandt and Kooij did not specify aiming. Jackson (1942) noted that captives frustrated in attempts to Reach food with sticks often Threw the sticks about their enclosures. Bernstein (1962) and Fletcher and Weghorst (2005) mentioned captives Throwing objects, but these authors provided no further details. Similarly, N. King, Stevens, and Mellen (1980) described captives as occasionally Throwing hay in play or in display, but they did not provide details regarding aiming. Pfeiffer and Koebner (1978) noted that captives attacked hippopotamuses (*Hippopotamus amphibius*) with rocks; Pfeiffer (pers. comm.) reported that the rocks were Thrown with no clear evidence of aiming. K. Hayes and Nissen (1971) mentioned that the female Viki threw rocks, but they did not provide further information.

With regard to aimed Throwing, wild chimpanzees threw sticks, rocks and stones, banana skins, grass, and other bits of vegetation at conspecifics during aggressive interactions (Albrecht and Dunnett 1971; Boesch and Boesch 1990; J. Goodall 1963a, 1964; Matsuzawa 1999; McGrew et al. 2003; Sanz and Morgan 2007; Sugiyama and Koman 1979; van Lawick–Goodall 1968, 1970; Watty in Kortlandt and Kooij 1963). Several authors noted that the missiles were of sufficient size and Thrown with sufficient velocity and accuracy to cause injury or at least discomfort. Indeed, captive chimpanzees learned to Throw stones at targets with considerable accuracy (Kats 1972a, 1972b). Immature males at Mahale Threw branches at others, usually adult females, in what was considered harassment (Nishida 2003a; Nishida et al. 1999). Stones may also be Thrown into water in display (Bygott 1979; Nishida et al. 1999). Nishida (2003a) described a pattern of rock Throwing, termed "throw splash," among Mahale chimpanzees. This was typically seen when the group crossed canyons; adult males stood bipedally, lifted heavy rocks with both hands, aimed, and Threw them into the canyon, splashing in the water below.

Chimpanzees also engage in interspecific aimed Throwing. Wild chimpanzees were seen Throwing sticks, rocks and stones, bananas and banana skins, and leaves at baboons (Albrecht and Dunnett 1971; J. Goodall 1964, 1986; de Leest in Kortlandt and Kooij 1963; van Lawick–Goodall 1968, 1970, 1971); sticks, saplings, stones, and grapefruits at leopard models (Albrecht and Dunnett 1971; Kortlandt 1965, 1967a); sticks, branches, dead leaves, bark, stones, and banana skins at humans (J. Goodall 1964; McGrew et al. 2003; Millot in Kortlandt and Kooij 1963; Nishida, Matsusaka, and McGrew 2009; J. Pruetz, pers. comm.; Sugiyama and Koman 1979; van Lawick–Goodall 1968, 1970, 1971; Watts 2008); rocks at a monitor lizard (*Varanus* sp.) (van Lawick–Goodall 1968, 1971); a rock at a bush pig (*Potamochoerus larvatus*) (Plooij 1978); and rocks at baboons, cows, and spotted hyenas (*Crocuta crocuta*) (J. Pruetz, pers. comm.). At Gombe, Mahale, and Taï, chimpanzees Threw objects at leopards (*Panthera pardus*) (Boesch 1991a; J. Goodall 1986; Nishida 1968). J. Goodall (1986) reported a description from a local Tanzanian of sticks used as missiles to threaten lions (*Panthera leo*). Boesch and Boesch (1989) documented an adult male who Threw a branch at two colobus monkeys (Colobinae, genus not specified) while hunting them, hitting them in the process. Kortlandt (1967a) emphasized that wild females with dependent offspring used tools more frequently than males when attacking a leopard model. Kasokwa males carried stones when returning from crop raiding, perhaps for use in defense if attacked by humans (Kyamanywa 2001 in Reynolds 2005). Nishida et al. (1999) noted that Mahale males also Threw sticks or rocks against the wall of a metal house to create a loud noise.

Captive chimpanzees aim and Throw sticks, rocks and stones, sand, tufts of grass, and locks during intraspecific agonism (Cowper 1971; Hopkins et al. 1993, 2001, 2005; Kollar 1972; Kortlandt and Kooij 1963; Menzel, Davenport, and Rogers 1970; Okano et al. 1973; Savage-Rumbaugh in de Waal 1997; Takeshita and van Hooff 1996, 2001; van Hooff 1973; Wilson and Wilson 1968). Males again accounted for most intraspecific aimed Throwing by captives, but Cowper (1971), Okano et al. (1973), Savage-Rumbaugh (in de Waal 1997), and van Hooff (1973) saw females throw objects. Ladygina-Kohts (2002) described a captive male Throwing a stone at a mirror that showed his own reflection. Captive infants and juveniles have been documented Throwing sand or sticks at other apes, in what was interpreted as "teasing" and "annoyance" (de Waal and Hoekstra 1980).

Captive chimpanzees Throw sticks, rocks and stones, sand, mud, water, potatoes, apples, feces, rolls of wire netting, cans, cups, and wooden blocks at humans and other animals, such as a tiger (*Panthera tigris*), birds, squirrels (Sciuridae), and dogs (*Canis lupus familiaris*) (B. Beck, pers. comm.; Bettinger and Carter 1991; Hopkins et al. 2001; Hostetter, Cantero, and Hopkins 2001; Köhler 1925; Kortlandt 1962, 1965; Kortlandt and Kooij 1963; Ladygina-Kohts 2002; Merfield 1956; Mottershead 1963; Osvath 2009; Poulsen 1974–1975; Schiller 1957; A. Schultz 1961; Wilson and Wilson 1968). They also Threw sticks at fear-producing stimuli, such as reptile models (Menzel 1971, 1973b). A female in a mixed-sex captive group Threw most of the fifty stones that broke the thick plate-glass window of an observation booth (Menzel 1972); another female observed by Merfield (1956) learned to annoy her captors by Throwing stones on a tin roof. Osvath (2009) reported the spontaneous planning of missile Throwing at humans by a zoo-living male. The chimpanzee was observed to gather and cache stones or concrete blocks from the water moat before the zoo opened each day. He retrieved the stones or blocks later in the day to hurl at visitors. Caretakers removed hundreds of these caches from the enclosure. Osvath argued that the calm behavior of the chimpanzee during the retrieval and caching indicated a different mental state than during the display and thus was indicative of planning.

Captive chimpanzees also spit water at others, which can be considered a form of Throwing. R. Wood (1984) documented a zoo-living chimpanzee spitting water at gorillas through the mesh barrier that separated them. J. Pruetz (pers. comm.) reported that captive chimpanzees at one institution sometimes spat on chimpanzees in a neighboring cage, greatly enraging their neighbors. Captive chimpanzees also spit water at humans. Brewer (1978) described a rehabilitant male with a hot cup of coffee, spontaneously retrieving a mouthful of cold water from a jug and spitting it into his cup to cool the coffee before drinking it.

Not all aimed Throwing occurs as agonism. Wild juveniles have been observed using an aimed underarm swing to Throw dead sticks at each other in play (van Lawick–Goodall 1970). Captives Throw stones of various sizes, sticks, and clods of dirt at playmates (Cowper 1971; Köhler 1925; Wilson and Wilson 1968). Takeshita and van Hooff (1996, 2001) observed captive chimpanzees Throwing sticks into the foliage of a tree in order to break leaves away for consumption. Brink (1957) described a cigarette-smoking captive Throwing used cigarettes out of his cage. If they were still burning, he sometimes tossed them onto wet areas on the floor. Aimed Throwing in one captive population was interpreted as a signal for mating (Takeshita and van Hooff 1996, 2001). Several chimpanzees in this group, mainly adult and juvenile males, Threw sand and leaves at others to attract attention or as a precursor to mating or presenting. Captive chimpanzees also Threw objects to capture the attention of an inattentive chimpanzee or a human (Hostetter, Cantero, and Hopkins 2001). K. Hayes and Hayes (1952) conducted a "ball-throwing problem" in which a food lure was suspended out of reach but could be knocked down by Throwing a ball. Viki solved this problem without a demonstration of the solution. Washoe, the first chimpanzee to use sign language, made alarm calls and retreated from a new doormat. She then Threw one of her dolls at the mat (R. Fouts 1997).

Drag, Roll, Kick, Slap, Push Over

Wild male chimpanzees Dragged, Slapped, or Rolled branches, vegetation, fallen trees, stones, cans, and camp furniture during agonistic display in the presence of conspecifics (Albrecht and Dunnett 1971; Boesch and Boesch 1990; J. Goodall 1963a, 1964, 1965, 1979; Kummer and Goodall 1985; Matsuzawa 1999; Nishida 1970; Nishida et al. 1999; Riss and Goodall 1977; Sanz and Morgan 2007; van Lawick–Goodall 1965, 1968, 1970, 1973). Whiten et al. (2001) described branch Dragging as habitual at Bossou, Budongo, and Kibale and customary at Gombe, Mahale, and Taï. Mike, an adult male at Gombe, is the most publicized user of tools of this sort. Many of the Gombe males occasionally Dragged or Kicked discarded kerosene cans to enhance their agonistic charging displays, but Mike's incorporation of cans became frequent. He learned to Drag and Roll as many as three of the cans simultaneously while charging rapidly over distances exceeding 50 meters. Although his original use of cans was said to have occurred by chance, van Lawick–Goodall (1971) emphasized that he soon learned to seek out and use cans deliberately. The resultant noise and clamor effectively intimidated other adult males, even high-ranking ones, and was said to have contributed to Mike's meteoric rise to the alpha rank. Infants typically avoided or fled from males during charging displays, but van Lawick–Goodall also described how two youngsters approached displaying males, one doing so repeatedly. Both were Dragged by the charging males as if they were inanimate objects. Teleki (1973a) observed a wild adult male chimpanzee charging across a clearing while Dragging a large branch in reaction to the noisy approach of a baboon troop. Young males at Mahale Slapped or hit branches to harass or annoy adult females (Nishida et al. 1999). A Bossou alpha male Dragged a dead infant chimpanzee body in his display (Matsuzawa 1997). Chimpanzees at Fongoli were reported to Roll boulders at human observers (J. Pruetz, pers. comm.). Captive male chimpanzees also Dragged, Rolled, and Kicked wooden boxes, metal drums, pails, and branches while displaying toward conspecifics (I. Bernstein, pers. comm.; Köhler 1925; S. Savage 1976).

Males also displayed with tools at the onset of a thunderstorm. While the patterns exhibited in such "rain dances" are similar to those of charging displays, the behavior appears to be an expression of general excitement rather than of social agonism. Nishida (1980b) and van Lawick–Goodall (1971) saw wild males Dragging branches during thunderstorms, and van Hooff (1973) saw identical behavior by captives. Nishida (1980b) also saw a wild female Dragging branches during a rain dance. Whiten et al. (2001) characterized rain dances as habitual at Budongo and Taï; customary at Gombe, Kibale, and Mahale; but absent at Bossou (C. Boesch, unpub. data; van Lawick–Goodall 1971; R. Wrangham, unpub. data).

Chimpanzees also use or move objects in the context of courtship. "Branch-slap," in which a chimpanzee sitting on a branch Slaps the branch with its hand, is presumed to occur to attract attention, possibly in courtship (V. Reynolds, pers. comm., in Whiten et al. 1999). Whiten et al. (2001) described this behavior as customary at Bossou, Budongo, and Taï and present in the Mahale M group. "Shrub-bend," bending or squashing a stem underfoot repeatedly, is described as a male courtship behavior used to attract the notice of a mating partner (Nishida 1987, 1997). Whiten et al. (2001) characterized this as habitual at Bossou, and customary at Budongo and Mahale M group.

Chimpanzees may Drag, Kick, Slap, or bend objects in play. Immature wild chimpanzees Dragged sticks during social play (van Lawick–Goodall 1970, 1971). Boesch and Boesch (1989) documented an instance of subadults playing with the corpse of a blue duiker (*Cephalophus monticola*). The chimpanzees attempted to carry the corpse, as well as Kicked and pulled on it. Boesch and Boesch noted that the adolescents were more interested in the body as a toy rather than as food. "Leaf-pile pulling" was documented at Mahale and occasionally at Gombe (Nishida and Wallauer 2003). The authors described the behavior as sometimes occurring when a chimpanzee party traveled down a slope; a young individual would turn and walk backward while using his or her hands to rake and accumulate dry leaves. Dragging the leaves produced sound. "Leaf-pile pushing" is a similar behavior, in which adult or young males pushed dead leaves on the forest floor (Nishida, Matsusaka, and McGrew 2009). Adult males were presumed to do so for intimidation, while young males did so in play.

Brandish, Wave, Shake

Chimpanzees universally Wave or Shake both attached and unattached branches, a behavior termed "branching" or "branch-shake" (J. Goodall 1986; Whiten et al. 1999). Branching attracts attention in threat and intimidation displays, or in solicitations for another to approach in courtship (J. Goodall 1986; Kortlandt 1965; Reynolds and Reynolds 1965; van Lawick–Goodall 1968; Whiten et al. 1999, 2001). As an element of intimidation displays or in harassment, wild chimpanzees Brandished or Waved uprooted saplings, attached and unattached branches, sticks, palm fronds, pieces of sugarcane, and even a bushbuck (*Tragelaphus* sp.) carcass at or in the presence of conspecifics, human intruders, baboons or colobus monkeys, leopard models, a mongoose, a porcupine (*Hystrix cristata*), and their own mirror images (Albrecht and Dunnett 1971; Boesch and Boesch 1989; J. Goodall 1965, 1979; Kortlandt 1962, 1965, 1967a, 1972; Matsusaka 2007; W. McGrew, pers. comm.; K. Morris and Goodall 1977; Nishida 1970; Nishida et al. 1999; van Lawick–Goodall 1965, 1968, 1970, 1971; Watts 2008). J. Goodall (1963b, 1965) reported that a wild chimpanzee Brandished an axe while pursuing a human. A 3-year-old male once Brandished twigs toward a young female in a miniaturized version of adult male display (van Lawick–Goodall 1971). Nishida (2003a) and Nishida et al. (1999) reported that a male at Mahale Brandished and hit the ground with long sticks to threaten older females. Boesch and Boesch (1989) described a male Waving a branch at a group of colobus monkeys as a threat or as a weapon while hunting the monkeys.

Displaying captive males Brandished or Waved sticks at conspecifics, humans, a leopard model, a hippopotamus, and cameras (Köhler 1925; Kortlandt 1965, 1967a; Kortlandt and Kooij 1963; Ladygina-Kohts 2002; Pfeiffer and Koebner 1978; Schiller 1957). Schiller also saw captives Waving sticks when frustrated by their inability to reach food. Captives, like wild chimpanzees, Brandished sticks during an explosive episode of excitement and animated locomotion, triggered by a rainstorm (van Hooff 1973).

Branch shaking is also done to attract attention in a non-agonistic context. Branch shaking may occur in courtship, for example at Assirik (P. Baldwin 1979; McGrew et al. 2003), Fongoli (J. Pruetz, pers. comm.), Gombe (van Lawick–Goodall 1968), Mahale (Nishida 1997), and in the Ngogo population (Watts 2008). Nishida et al. (1999) described chimpanzees Shaking both attached and detached branches in courtship. J. Goodall (1986) and J. Pruetz (pers. comm.) stated that branch shaking might sometimes be a "follow-me signal," used in courtship or in travel to persuade others to follow. In the context of attracting attention, Tutin (unpub. data in Whiten et al. 1999) described "branch-din" at Lopé in which vegetation was pulled down and released, resulting in noise. Whiten et al. (2001) described this as a habitual behavior that is done before moving into areas of dense vegetation or before entering savannah landscapes. Wild chimpanzees Brandished sticks in play (van Lawick–Goodall 1970), as did captives (Köhler 1925; Watty in Kortlandt and Kooij 1963).

Objects may also be Waved as "fly-whisks" to disperse insects (Whiten et al. 1999). Sugiyama (1969) twice saw wild Budongo female chimpanzees Waving boughs to disperse swarming flies. Boesch (1995) documented a Taï female who swayed a sapling, causing it to hit a nearby sapling containing a tsetse fly (*Glossina* sp.). Watts (2008) documented three cases of the use of a "bee whisk," in which a chimpanzee Waved leafy branches to displace bees from the hive entrance. Watts classified this behavior as a comfort tool. Sanz and Morgan (2007, 2009) also documented bee whisking or swatting at Goualougo. Matsuzawa et al. (2004) reported a mother Brandishing a stick to chase flies from her dead infant's body.

Bait, Entice

Chimpanzees used objects to attract attention for play, behaviors termed "play-start" (J. Goodall 1986; McGrew et al. 2003; Nishida et al. 1999; Reynolds 2005; Sanz and Morgan 2007; Whiten et al. 1999). Whiten et al. (1999, 2001) considered this to be a universal chimpanzee behavior found at all sites where chimpanzees have been studied. To initiate a play session, a young chimpanzee approaching or retreating from a conspecific detached or took a twig or other object (a scrap of monkey skin in one observation) and carried it in the mouth, hand, or foot. The object might also be Waved and flailed to display it to the other individual. Nishida et al. (1999, 175) described a case of a Mahale adolescent female using a detached monkey tail in an attempt to "fish" for a juvenile male, in what they labeled as "tease." They reported that she climbed into a tree and dangled the monkey tail at the male, withdrawing it from reach when he jumped at it. After several repetitions, the male got the tail, "and the female grimaced, suggesting she was trifling with the male" (Nishida et al. 1999, 175). Chimpanzees in captivity also used tools to Bait or attract others' attention for play (M. Vancatova, pers. comm.). Captives Baited other types of animals, too. Köhler (1925) was the first to document this, as his captive chimpanzees Baited chickens with bread in play, but also to draw them near enough to their cage to Stab them with sticks or wire. J. Pruetz (pers. comm.) reported that two adult males lured a dog close to their cage using monkey chow, and then hit the dog when it was in range, once with a stick and once with fists.

What Nishida (1980a) termed "leaf-clipping" occurs when a chimpanzee repeatedly pulls a leaf through its lips or teeth, or uses its fingers to remove the leaf blade to produce a ripping noise. He noted that the sound thus produced may be used as a communicative signal. At Mahale, the leaf was nibbled in small pieces from the blade, while at Taï the blade was removed in one movement (Boesch 1995). It is considered a culturally acquired behavior, which varies by population in its form (use of mouth or fingers), frequency, and function (Whiten et al. 1999, 2001). The function of leaf clipping is somewhat ambiguous, since it is associated with many different behaviors (Boesch 1995; Matsumoto-Oda and Tomonaga 2005). Leaf clipping has been observed in the context of courtship, with males using it to attract the attention of receptive females at Bossou (Matsuzawa 1999), Budongo (Quiatt 2006; Reynolds 2005), Mahale (Matsumoto-Oda and Tomonaga 2005; Nishida 1980a; Nishida et al. 1999), and Ngogo (Watts 2008). McGrew (1989, 461) refers to leaf clipping in courtship as a "billet-doux." Leaf clipping also may occur in the context of loud pounding or slapping with hands on tree buttresses in display, as at Ngogo (Watts 2008) and Taï (Boesch 1995). It was done to attract the attention of a playmate at Bossou (Sugiyama 1981), Mahale (Hayaki 1985), and Ngogo (Watts 2008). It was used to augment displays at Fongoli (J. Pruetz, pers. comm.) and Mahale (Nishida 1980a). It has been observed and interpreted as an expression of frustration, impatience, or tension release, such as while waiting for humans to distribute provisioned food or waiting under a group of monkeys prior to commencing a hunt (Boesch 1995). Leaf clipping in a context suggesting frustration has been documented at Bossou (Sugiyama 1981), Gombe (J. Goodall, pers. comm., in Nishida 1987), Mahale (Nishida 1980a), Ngogo (Watts 2008), and Taï (Boesch 1995). J. Pruetz (pers. comm.) reported that Fongoli chimpanzees also leaf clip when they have the hiccoughs or when they are about to regurgitate, which she interpreted as a signal of frustration or anxiety. Whiten et al. (1999) reported that leaf clipping habitually occurred at Kibale as well, though neither the context nor the specific Kibale population were noted. At Taï, a new context for leaf clipping was observed in 1990, when multiple individuals began to clip while resting on the ground, with no evident cause of frustration or solicitation (Boesch 1995). In a few instances, the behavior deviated from the original form, in that the leaves were ripped using the fingers while the leaves were still attached to the sapling. As the function of this behavior is still unclear, Bait may not be an appropriate classification of Boesch's observations. Indirect evidence supporting an inference of leaf clipping, a pile of clipped leaves, was documented at Mahale in an area used by an unhabituated group (Uehara 2002). McGrew (2004) noted that leaf clipping may be Symbolic.

"Leaf-strip," a display in which the leaves are torn from a stem by the thumb and fingers, is habitual at Gombe and Kibale, and has been observed at Bossou and Mahale M (R. Wrangham, unpub. data, in Whiten et al. 1999). The behavior of "stem pull-through," in which a chimpanzee pulls a leafy branch or grass clump through the hand without tearing off the leaves, is used to produce a sound presumed for courtship (Nishida 1997). This behavior is customary at Bossou, habitual at Kibale and Mahale M, and has been observed at Gombe (Whiten et al. 1999, 2001).

Wild chimpanzees engaged in "leaf-grooming" in several populations (J. Goodall 1986; Whiten et al. 1999, 2001). Goodall first observed this behavior, in which a chimpanzee held leaves, inspected them visually, and groomed the leaves, sometimes while lip smacking (J. Goodall 1986; van Lawick–Goodall 1968, 1973). The function of leaf-grooming remains unclear, despite repeated observations (van Lawick–Goodall 1973; Nishida et al. 1999). McGrew (1992) included it as tool use and summarized some of its possible functions, such as Plooij's (1978) suggestion that it was used to get attention, and Wrangham's observation (1980, unpub. ms., in McGrew 1992, 188) that it might "perk up" the grooming bouts of other chimpanzees.

J. Goodall (1986) reported what she termed "social tool use" in Gombe chimpanzees. She observed multiple occurrences in Flo's family line of an older offspring, usually a male, carrying off an infant sibling to persuade his mother to follow him. This occurred while the mother was resting or feeding, after the youngster had already failed to persuade his mother to move. This was also seen at Fongoli (J. Pruetz, pers. comm.).

Agonistic buffering was documented in one group of captive chimpanzees (Casanova, Mondragon-Ceballos, and Lee 2008). However, unlike what has been observed in monkeys, this behavior was interpreted as a means of appeasing group members, perhaps through "forced reconciliation" (Casanova, Mondragon-Ceballos, and Lee 2008, 54). These researchers documented thirty-three episodes in which a dominant male showed agonistic behavior toward a community member and was then chased by members of the group. In response to the uproar, the dominant male took an infant (typically the alpha female's), halting the chase.

Meat shared between individuals is sometimes viewed as a form of social tool use (Mitani and Watts 2001; Nishida and Hosaka 1996; Nishida et al. 1992). Teleki (1973c) proposed that a male might share meat with females in exchange for sex. C. Stanford (1996, 1998) and C. Stanford et al. (1994) found support for this proposal at Gombe, where the hunting of red colobus (*Piliocolobus* sp.) monkeys was best predicted by the number of females in estrus in the group. However, Mitani and Watts (2001) noted that only scant data supported this hypothesis. They argued instead that Ngogo chimpanzees shared meat as a social tool to develop and maintain alliances between males, as Nishida and colleagues asserted regarding Mahale chimpanzees (Nishida and Hosaka 1996; Nishida et al. 1992). Hockings et al. (2007, e886) documented the sharing of plant food by chimpanzee at Bossou, arguing the "'food-for-sex-and-grooming' and 'showing-off' strategies" may explain the behavior. Captive chimpanzees also exchanged objects between each other for food or services (de Waal 1989; Savage-Rumbaugh, Rumbaugh, and Boysen 1978). Savage-Rumbaugh, Rumbaugh, and Boysen (1978) documented the exchange of tools between two chimpanzees in a cooperative task. The first chimpanzee had the tools, but no access to the task. He provided a tool to the second chimpanzee, who solved the task and then shared the resulting food. De Waal (1989) found that food sharing between captive chimpanzees was correlated to grooming between the sharers, which suggests that the food may be used as a tool to influence behavior.

Captive chimpanzees are known to exchange objects with humans (Hyatt and Hopkins 1998; Lefebvre and Hewitt 1986). As reviewed by Hyatt and Hopkins (1998), objects exchanged with humans for food included empty peanut shells (Premack 1976) or human tools, such as a screwdriver (de Waal 1996). Tokens or other synthetic objects were also exchanged for food in other studies of chimpanzee behavior and cognition (Bräuer, Call, and Tomasello 2009; Brosnan and Beran 2009; Brosnan and de Waal 2005; Lefebvre 1982). Hyatt and Hopkins (1998) found that chimpanzees did not push PVC tubes out of the enclosure without human solicitation, but instead would barter the tubes with humans for bananas after human solicitation. Chimpanzees were quicker to exchange the PVC tubes for preferred foods (bananas) than less desirable ones (monkey chow). In some studies (Cowles 1937; Sousa and Matsuzawa 2001, 2006), tokens were inserted in a vending machine to get food. We do not classify the use of tokens in vending machines as tool use (see chapter 1).

Club, Beat

Clubbing conspecifics has been observed in several chimpanzee populations, including those at Taï, Gombe, Mahale, and Ngogo (Boesch and Boesch 1990; J. Goodall 1986; Nishida et al. 1999; Watts 2008). Wild chimpanzees sometimes Clubbed or hit conspecifics with sticks during agonistic episodes (Itani and Suzuki 1967; van Lawick–Goodall 1968, 1970). Linden (2002) reported Wrangham and Hooven's observations of a Kibale male using a stick to Beat a female of his community. Kortlandt (1965) and Kortlandt and Kooij (1963) cited unnamed sources who observed wild chimpanzees Clubbing the ground with sticks while pursuing conspecifics. Nishida (1970) actually observed such behavior. Nishida (2003a) and Nishida et al. (1999) reported that immature males Clubbed the ground to threaten adult females, who either retaliated or ignored them. Nishida et al. (1999) described Clubbing the ground at Mahale in the context of courtship. Captive chimpanzees also Clubbed or hit conspecifics with sticks and objects such as branches, burlap bags, clods of earth or grass, and pails during agonistic interactions (Fletcher and Weghorst 2005; Köhler 1925; Kollar 1972; S. Savage 1976; Takeshita and van Hooff 1996, 2001; van Hooff 1973; Vancatova 2008; Wilson and Wilson 1968). Wilson and Wilson (1968) noted that one animal suffered many lacerations as a result of being Clubbed.

Clubbing and Beating were also directed at other animals or humans. An elderly villager's recollection of four chimpanzees chasing a lion with sticks was skeptically reported by van Lawick–Goodall (1971). However, van Lawick–Goodall (1968, 1970) observed a wild chimpanzee hitting a baboon forcefully with a palm frond, and Albrecht and Dunnett (1971) saw wild chimpanzees Club an animated leopard model with sticks. Kortlandt (1965, 1967a, 1967b) provided vivid descriptions of the use of sticks by captive chimpanzees to Club leopard models. Albrecht and Dunnett (1971) also saw a chimpanzee hitting the ground with a stick as it attacked the model, while W. McGrew (pers. comm.) saw a chimpanzee do the same thing in the presence of a mongoose. A repetition of Kortlandt and Kooij's (1963) leopard experiment at Kibale failed to elicit Clubbing (Whiten et al. 2001). Boesch (1991a) documented seven instances in which Taï chimpanzees attempted to hit a leopard in a hole. This was interpreted as an attempt to scare the leopard. Kummer and Goodall (1985) observed a juvenile Clubbing an insect with a stick. Boesch and Boesch (1990) described chimpanzees killing wood-boring bees (*Xylocopa* sp.) at Taï, both to consume them and to gain access to their nest. On six occasions, they observed a chimpanzee Probing into a hole to determine if bees were present, as the disturbed bees tend to block the entrance with their bodies. The chimpanzees then killed them, a behavior described as "disables them with the stick" (Boesch and Boesch 1990, 89). Whiten et al. (1999) called this "bee-probe" and noted that this behavior was present at Mahale (K community) as well. Since the function appears to be to avoid stings, we classify these behaviors under Club, Beat. Hirata et al. (2001) observed an adolescent male who bent a sapling and used it to hit a western tree hyrax (*Dendrohyrax dorsalis*). Whiten et al. (2001) mentioned chimpanzees Clubbing snakes at Gombe. Captives have been observed Clubbing humans with whips or sticks (Kortlandt 1965; Sheak 1924), Beating a gorilla with a whip (Steinbacher, and also Weinert, in Kortlandt and Kooij 1963), and Clubbing a variety of other mammals and reptiles with sticks (Köhler 1925; Kortlandt and Kooij 1963). Kummer and Goodall (1985) reported a juvenile smashing an insect with a stone. Matsusaka (2007) described a chimpanzee using a stick to hit the ground near a porcupine.

Clubbing also occurs in play. Wild chimpanzees are reported to Club each other with sticks and with a grass tuft during play (Albrecht and Dunnett 1971; van Lawick–Goodall 1970). A wild chimpanzee Clubbed an insect with a stick in play (van Lawick–Goodall 1970). Köhler (1925) reported playful Clubbing with sticks among captives. Elsewhere, a captive Clubbed the ground with a stick in what may have been play (Wilson and Wilson 1968).

Pound, Hammer

Chimpanzee nut cracking (opening nuts and seeds by Pounding or Hammering them on hard objects) is common in some areas of western Africa, but there are important differences between sites in terms of the nut species cracked and in the properties of the tool and anvil materials (Sugiyama 1989b). Nut cracking is absent in eastern Africa, despite the presence of suitable nuts and stones (McGrew 1992).

Sept and Brooks (1994) recounted Manual Alvares's report, from about 1615, of chimpanzees in Sierra Leone cracking palm nuts using a hammer stone. Whitesides (1985) reported circumstantial evidence of nut cracking of the sweet detar tree (*Detarium senegalense*) in Sierra Leone after hearing Hammering and chimpanzee vocalizations, as well as finding nut pods that were scattered around tree roots that exhibited fresh scarring consistent with Hammering.

A medical missionary working near what is now the border between Liberia and Côte d'Ivoire also provided one of the earliest records of nut cracking in wild chimpanzees (T. Savage and Wyman 1843–1844). Beatty (1951) saw a wild Liberian chimpanzee place an oil palm nut on a rock and smash it with another rock. Since he found heaps of broken nuts at the site, it is likely that other group members also opened nuts with rock hammers. At Mt. Kanon, also in Liberia, wild chimpanzees cracked open *Coula* nuts (*Coula edulis*) (Kortlandt and Holzhaus 1987). In the Sapo Forest, in eastern Liberia, stone hammers were also employed to open the nuts of four tree species, the Gabon nut or African walnut, *C. edulis*; *Panda oleosa* (*Panda* nuts); the Guinea plum, *Parinari excelsa*; and the bitter bark tree, *Sacoglottis gabonensis* (J. R. Anderson, Williamson, and Carter 1983). Most of the cracking at this site occurred on anvil stones (71%), although tree roots were also utilized as anvils. Hannah and McGrew (1987) reported on wild-born Liberian chimpanzees, released from captivity onto an island in Liberia, who were observed cracking open oil palm nuts (*Elaeis guineensis*). The behavior originated with one of the released females and eventually spread to others in the group and to other locations on the island. The authors stated that both hammer stones and nuts were transported by the apes.

Rahm (1971) heard Pounding sounds as he explored the Taï Forest of Côte d'Ivoire. When he approached the source, he saw chimpanzees moving rapidly away from a clearing where he found smashed *Panda* nuts wedged into crevices formed by the shallow exposed roots of the tree. Nearby were two sizable quartz rocks that could otherwise be found only in a small stream about 50 meters away. Rahm inferred that the chimpanzees had placed the nuts in the root crevices for stability and then opened them with the rocks carried from the stream. Struhsaker and Hunkeler (1971) made similar observations independently, about one year earlier, in the same forest. Responding to Pounding sounds, they found piles of pericarps and nutshells of *Coula* nuts near an exposed root that bore a depression. Four sticks lay nearby, apparently abandoned by three chimpanzees who had fled at the authors' approach. Two of the sticks had indentations, indicating that they might have been used as hammers. Struhsaker and Hunkeler had other similar encounters, including one in which a female left the scene with a rock. Fifteen specific Pounding sites were located, some bearing the remains of *Panda* nuts. Struhsaker and Hunkeler inferred that the chimpanzees cracked open the nuts of both species with rocks and sticks in order to gain access to the edible seeds. They also concluded that the chimpanzees carried rocks to the Pounding sites, where nuts and sticks were already present. The sticks were variable in length, though more consistent in diameter. The stones, however, were strikingly similar in shape and size. Four of six weighed stones were between 5 and 6 kilograms. One weighed 3.5 kilograms, and an unusually large one weighed about 16 kilograms. Stones were always found in association with *Panda* remains. The nutshell of this species is considerably thicker than that of the *Coula* nuts and would be difficult for a chimpanzee to break with a stick.

Also in the Taï Forest, Boesch (1978, 1991b, 1991c), Boesch and Boesch (1981, 1983, 1984a, 1984b, 1990, 1993a, 1993b, 1994), Boesch and Boesch-Achermann (1991a, 1991b, 1994, 2000), Boesch et al. (1994), Günther and Boesch (1993), Joulian (1995, 1996, 2000), and Kortlandt (1986) have observed chimpanzees using branches and stones to Hammer nuts open. Nut species cracked at Taï include *Detarium senegalense*, *Coula edulis*, *Parinari excelsa*, *Sacoglottis gabonensis*, and *Panda oleosa*. Boesch and Boesch (1983) found that using a wooden hammer to crack the nut required greater force than using a stone hammer. Taï chimpanzees cracked *Coula* nuts with hammers either on an anvil on the ground or on an attached branch anvil in the *Coula* nut tree. *Panda* nuts were cracked only on ground anvils. Ground anvils consisted of either surface roots or rock outcrops, neither of which are considered tools, because they are not held or directly manipulated. The chimpanzees were observed bringing as many nuts as they could carry to the anvil location. Tools to open the nuts consisted of either wooden or stone hammers; harder stones and heavier hammers were transported to open the harder *Panda* nuts (Boesch and Boesch 1983). Boesch and Boesch (1983, 1984a) and Boesch-Achermann and Boesch (1993) argued that the chimpanzees used a mental map in transporting the rare hammer stones for opening *Panda* nuts; 40 percent of hammer transports occurred between nut trees that were not within sight of each other, an average distance of 120 meters. This contrasts with the transportation of tools for opening *Coula* nuts, where wooden clubs were moved an average distance of 20 meters, and stones up to 50 meters.

Archeological evidence supports the observation that chimpanzees at Taï have been using stones for nut cracking for at least 4,300 years (Mercader, Panger, and Boesch 2002; Mercader et al. 2007).

Chimpanzees at Bossou in Guinea crack the nut of the oil palm (Biro et al. 2003; Fushimi et al. 1991; Inoue-Nakamura and Matsuzawa 1997; Inoue-Nakamura, Tonooka, and Matsuzawa 1996; Kortlandt 1986; Kortlandt and Holzhaus 1987; Matsuzawa 1994, 1996, 1999; Matsuzawa et al. 2001; Sakura and Matsuzawa 1991; Sugiyama 1981, 1993; Sugiyama and Kawai 1990; Sugiyama and Koman 1979, 1987; Sugiyama et al. 1993; Yamakoshi 1998). Fushimi et al. (1991) and Sakura and Matsuzawa (1991) studied the nut-cracking process in what they called an outdoor laboratory, containing marked and documented stones and oil palm seeds. To crack nuts, the stone anvil is placed on the ground

A Bossou chimpanzee uses a stone hammer and a portable stone anvil to open a nut. Note that the stones are numbered by the researchers for precise recording of their use and transport. Note also the rapt attention of the two younger animals. Photograph courtesy of Tetsuro Matsuzawa.

and the nut is placed centrally on the anvil, with the hammer in the other hand. The hammer is then raised 10 to 40 centimeters to strike the nut. Nuts, anvils, and hammers were transported, though tools were not as frequently transported as at Taï, and transport generally occurred over a shorter distance (Sakura and Matsuzawa 1991). Hammer selection varied by age class, with adults tending to choose heavier hammers than juveniles (Fushimi et al. 1991). In comparison with nut cracking at Taï, hammers at Bossou were smaller, and only stone hammers were employed. Sakura and Matsuzawa (1991) explained these differences by noting that the palm seeds cracked at Bossou were much softer and easier to crack than *Coula* or *Panda* nuts. In addition, stones were easier to find at Bossou, while locations for appropriate root anvils were rare. Nearly all anvils at Bossou were stone. The use of wooden anvils was rare, occurring only when stone was unavailable (Matsuzawa 1994; Sakura and Matsuzawa 1991).

Nuts were also cracked at the Diécké Forest research site in Guinea, 50 kilometers west of Bossou (Carvalho, Sousa, and Matsuzawa 2007; Carvalho et al. 2008; Humle and Matsuzawa 2001; Matsuzawa et al. 1999). Diécké chimpanzees cracked *Panda* and *Coula* nuts with stone hammers on stone and root anvils. In addition, chimpanzees in the Nimba Mountains of Guinea have been documented occasionally cracking oil palm and *Coula* nuts with stones (Humle and Matsuzawa 2001, 2004; Matsuzawa and Yamakoshi 1996). Matsuzawa and Yamakoshi (1996) also reported finding three locations at Nimba where chimpanzees had used a pair of stones as hammer and anvil to crack open a fruit.

Brewer (1976, 1978) reported that the chimpanzees she was rehabilitating in Gambia placed hard-shelled fruits and pods on stone anvils and Pounded them with repeated blows of a rock. Some tuition by humans may have contributed to behavior acquisition in this case.

Cracking *Coula* nuts with hammer stones has also been documented in the Ebo Forest in Cameroon (B. Morgan and Abwe 2006). These are the first observations of nut cracking in *Pan troglodytes vellerosus*, occurring more than 1700 kilometers east of the previously believed western limit of this behavior in Côte d'Ivoire.

Itani and Suzuki (1967) relayed the conclusion by Azuma and Toyoshima that chimpanzees crack hard fruits with stones in the Kabogo region of Tanzania. However, Izawa and Itani (1966) stated that the evidence indicated that the fruits (*Strychnos* sp.) were pounded against rocks, making this proto-tool use rather than true tool use (see chapter 1). It appears that true nut cracking, Pounding *with* stones or sticks, occurs only in western chimpanzee populations.

Koops, McGrew, and Matsuzawa (2010, 175) described preliminary evidence that might indicate that Nimba Mountain chimpanzees use "cleavers" to fracture the volleyball-sized *Treculia africana* fruit. They found stone and wooden tools in association with the fruit, with indications of tool use on both the tools (fruit remnants) and the fruit (halved with a indentation of the tool). They argued that these tools are a type of percussive technology, used to reduce the size of the fruit into "manageable bite-sized pieces."

Chimpanzees at Bossou habitually use "pestle-pounding" to extract fibrous vegetation from the oil palm tree for consumption (Humle and Matsuzawa 2004, 2009; Ohashi 2006; Sugiyama 1994; Yamakoshi 1998; Yamakoshi and Sugiyama 1995). From the crown of the tree, the apes first spread apart the mature leaves with their hands to reveal the young, centrally placed shoots. Then they pull out the shoots, consuming the white vegetation at each extracted shoot's base. Finally, they use a tool, a hard leaf petiole, pounding it into the hole where the shoots were removed. Through Pounding, the chimpanzees deepen the hole, allowing them to insert an arm into the hole to extract juicy and fibrous vegetation, most likely the apical meristem or apical bud, parts of the oil palm that are otherwise inaccessible. Researchers first observed this behavior in 1990 (Yamakoshi and Sugiyama 1995), and, as of 1995, pestle-pounding had spread to nearly half of the group members. Yamakoshi and Sugiyama suggested that the behavior had been invented recently, just before the first 1990 observation, and had become habitual in the group.

Sugiyama and Koman (1979) reported Pounding with a stick into a tree hole by two Bossou adults to Reach, mash, and consume termites (*Macrotermes* spp.). T. Hicks, Fouts, and Fouts (2005) found a large stick swarming with insects near an ant nest and in-

A Bossou chimpanzee "pestle-pounding" to access juicy vegetation from the top of an oil palm. The chimpanzee first spreads the palm leaves (A), uses the stem of one leaf to Pound the area within the palm from which the leaves grow (B), withdraws the stem tool (C), and uses its hand to excavate the juicy fibrous material that has been produced (D). Drawings courtesy of Gen Yamakoshi.

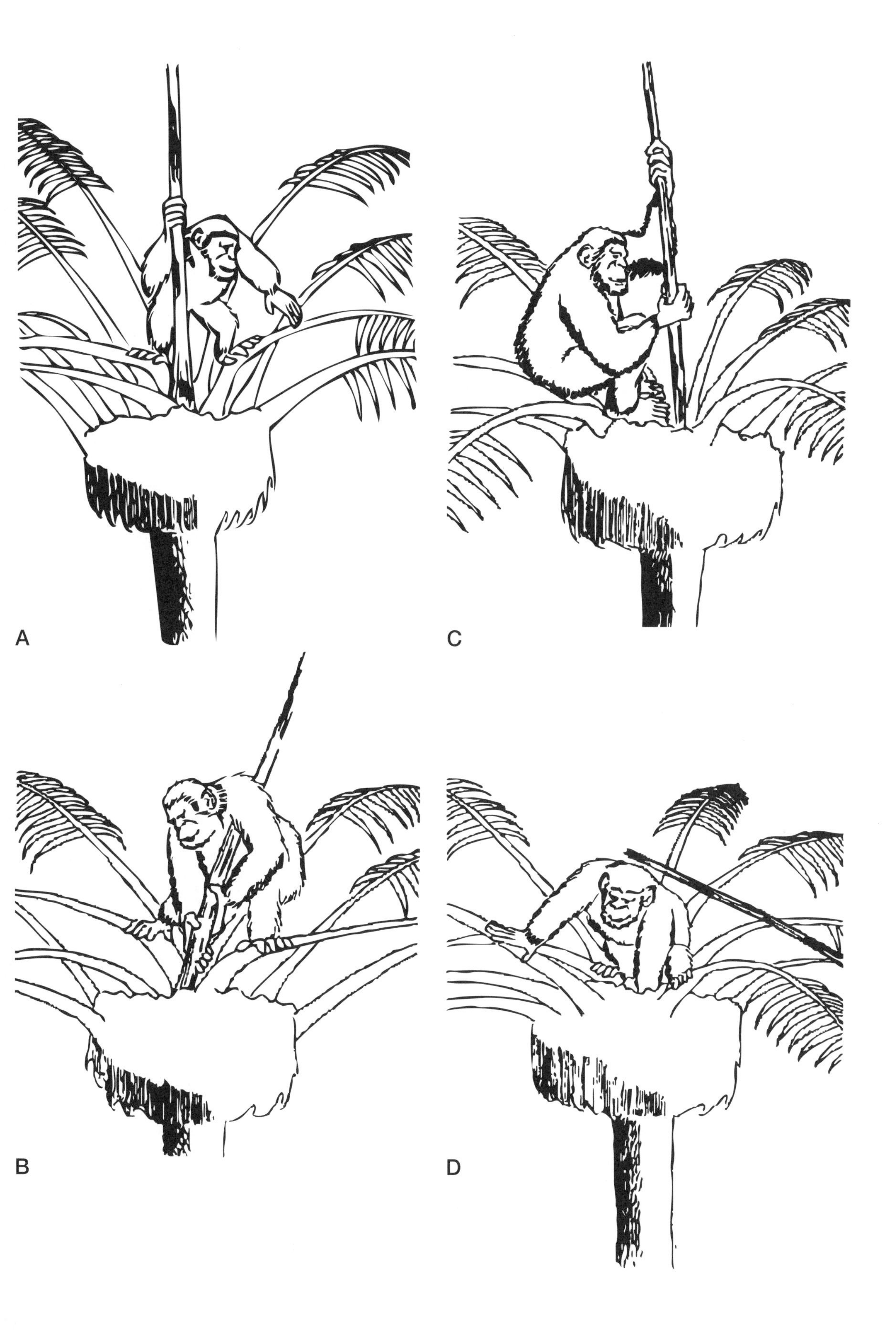
A
C
B
D

ferred that it had been used to punch the deep holes that were observed in the nest. They named the tool an "ant club."

Thick, stout sticks—termed "pounders," "pounding stick," "pounding tool," "pounding club," "honey hammer/club," or "stout chisel"—were used to break the exterior of beehives for access to honey (Bermejo and Illera 1999; Boesch, Head, and Robbins 2009; Fay and Carroll 1994; T. Hicks, Fouts, and Fouts 2005; Sanz and Morgan 2009; Sanz, Morgan, and Gulick 2004). Sanz and Morgan (2009) noted that the behavior is often referred to as "honey pound," "honey hammer," or "hive pound." It has been documented at several locations, including the Goualougo Triangle in the Republic of Congo (Sanz and Morgan 2007) and the Ituri Forest in the Democratic Republic of Congo (Hart, pers. comm., in McGrew 1992), and at Loango in Gabon (Boesch, Head, and Robbins 2009). This sort of Pounding tool creates a hole or enlarges an already-present hole. Enlarging the hive entrance may also fall into the mode of Pry, Apply Leverage (Brewer and McGrew 1990; Fay and Carroll 1994; J. Goodall 1986; van Lawick–Goodall 1970). After Pounding, these tools may have one blunt end (Boesch, Head, and Robbins 2009). Sanz and Morgan (2009, 420) also documented tools they described under "flail/club." Sticks used for this purpose were Pounded on and around the hive, not necessarily on the hive entrance, employing the side of the tool. The authors speculated that these blows served to weaken the hive structure.

During agonistic display, West African chimpanzees are reported to Pound on tree buttresses with sticks (anonymous in T. Savage and Wyman 1843–1844; Robillard in Rahm 1971). Wrangham (1974) reported that wild chimpanzees used sticks to hit locked banana boxes. "Resin-pound" was observed at Bossou, where chimpanzees used a stick, noted as longer than that used for insect pounding, to Pound into a tree hollow to obtain resin (Sugiyama and Koman 1979; Whiten et al. 1999).

Brewer's (1978) work contained descriptions and photographs of rehabilitants using rocks to Hammer open hard fruits.

Jackson (1942) noted that captive chimpanzees who failed to use sticks to Reach food sometimes used the sticks to hit the cage mesh and the platform on which the problem was displayed. Menzel, Davenport, and Rogers (1970) noted that captives were seen to Pound objects on walls, and stones on stones, in what was apparently play or frustration. Gaspar (1993) observed a captive male using a plastic brick to Pound fruit and cooked potatoes prior to consumption. K. Hayes and Nissen (1971) reported that the home-reared chimpanzee Viki used rocks and bricks to Hammer objects. C. Hayes (1951) described Viki's imitation of Hammering a stake after observing a human companion performing this behavior. She also had access to a toy workbench and was able to use a claw hammer (on nails) and a screwdriver. Kitahara-Frisch, Norikoshi, and Hara (1987) conducted an experimental study of Hammering on long bones. They demonstrated the use of a stone hammer to crack open long bones that had been filled with chocolate. After observing the human demonstrators, three captive females acquired the Pounding behavior. In addition, many researchers have studied captive chimpanzees' use of stone hammers to Pound open nuts, in order to explore the apes' cognition, manual skills, acquisition, and social learning (Bril et. al. 2009; Foucart et al. 2005; Funk 1985 in Boesch 1996; Hayashi, Mizuno, and Matsuzawa 2005; Hirata, Morimura, and Houki 2009; Marshall-Pescini and Whiten 2004; Sumita, Kitahara-Frisch, and Norikoshi 1985; Yoshihara 1985).

Pry, Apply Leverage

Wild chimpanzees Pry or Apply Leverage with sticks in attempts to open or widen spaces, for example, to gain access to arboreal insect or bird nests or to separate the nests from branches (J. Goodall 1986; Teleki 1974; van Lawick–Goodall 1968, 1970, 1973; Watts 2008). This behavior is termed "lever open" (Whiten et al. 1999). After inserting the tool in the opening, they moved it vigorously back and forth to enlarge the hole (J. Goodall 1986; van Lawick–Goodall 1971). These tools have been termed "enlargers" (Boesch, Head, and Robbins 2009), "prying sticks" (Fay and Carroll 1994), "bodkins" (in the sense of a chisel-like punch) (Bermejo and Illera 1999), or "dip sticks" (Fay and Carroll 1994; Sanz and Morgan 2007). Fay and Carroll (1994) did not observe the use of a Prying stick to access a beehive at Bai Hokou, but they inferred it. Bermejo and Illera (1999) described the use of the bodkin as following the use of a Pounding tool to create an entrance into the hive. The Pounding tool was turned around and the other, sharper end of the stick (sharper, perhaps, due to its previous Detachment from the substrate or in its not having been used to Pound) was used to Penetrate the nest wall, in combination with pushing and Levering actions to enlarge the hole. Sanz and Morgan (2007) described these lever-open tools as medium-sized twigs with the secondary vegetation removed. They noted that chimpanzees often made them after using Pounding tools, traveling to adjacent trees to Detach the tool. The lever-open tools were then used to widen the hole that had been created with the Pounding tool, with the lever-open stick moved rapidly back and forth in the hole. Using a lever to widen a nest entrance was documented at Gombe (J. Goodall 1986), Goualougo (Sanz and Morgan 2007, 2009), Loango (Boesch, Head, and Robbins 2009), Lopé (Tutin, Ham, and Wrogemann 1995), the Lossi Forest (Bermejo and Illera 1999); Ngogo (Watts 2008), Taï, (Boesch and Boesch 1990) and by a rehabilitant in Gambia (Brewer and McGrew 1990). Leverage tools have been used to open sturdy boxes containing provisioned bananas (J. Goodall 1964, 1986; Kummer and Goodall 1985; van Lawick–Goodall 1968, 1970, 1971; Wrangham 1974).

Captive chimpanzees used sticks and iron bars as levers to widen holes in cage mesh and open a wooden septic tank (Köhler 1925), to open a fruit feeder (Kollar 1972), to open a box (Hobhouse 1926), and to open the door of a food storage room (McGrew, Tutin, and Midgett 1975). Captives Pried with sticks in crevices in playful exploration (Schiller 1957) or when frustrated by Reaching-tool experiments (Jackson 1942; Menzel, Davenport, and Rogers 1970). Takeshita and van Hooff (1996, 2001) described captives Applying Leverage between two fallen trees by putting a stick in the crack of the contact point between the trees and pushing and pulling the stick. They noted that this appeared to be an attempt to move the upper tree to access the space between the trees. Ladygina-Kohts (2002) described a captive using a knife to remove plaster from the cage walls, in what may be considered Prying behavior.

Dig

Digging tools are used by wild chimpanzee to access insect nests, underground plant storage organs, or water. To access invertebrate prey, chimpanzees used the stick in the manner of a spade to Dig away the soil or nest (McGrew, Tutin, and Baldwin 1979; S. Suzuki, Kuroda, and Nishihara 1995). There is ambiguity regarding the use of Digging tools by chimpanzees, for several reasons. Descriptions of possible Digging behaviors are imprecise, and Digging is conflated with inserting and Prying with a tool. Indeed, if the use of a spade shovel by a human is used as a model, Digging might better be called Inserting and Prying. Teleki (1974) actually distinguished between Prying and Digging with sticks by Gombe chimpanzees feeding on ants, but the author did not elaborate on this. The use of sticks to enlarge the openings of subterranean bee nests, described by van Lawick–Goodall (1970, 1971), is listed above as Prying, but it might also be designated as Digging. In addition, behaviors are sometimes referred to as both Digging and Pounding (Sugiyama and Koman 1979) or Digging and perforating (Penetrating) (Fay and Carroll 1994). These distinctions are especially challenging in cases where the tool and some evidence of use are found without behavioral observations.

Whiten and McGrew (2001a; 2001b, 413) published what they consider the first "realistic depiction of tool use," a 1906 Liberian postage stamp illustrating what they considered to be a chimpanzee using a stout stick to Dig into a termite mound. They noted that this finding was surprising to them, as termite Digging has not been documented among Liberian chimpanzees. We question whether this image illustrates Digging, but we still include it here, based on Whiten and McGrew's labeling. Nonetheless, as artwork, it is subject to interpretation. Kattmann (2001) illuminated the discussion by noting that the original image upon which the stamp was based came from an 1887 German text and was drawn by Gustav Mützel, based on a zoo chimpanzee. Kattmann suggested that the ape was probably using the stick to probe into a hole in a tree trunk. However, Whiten and McGrew (2001b) contended that the stamp designer might have been more informed about wild chimpanzees than Mützel, as what appeared as a tree in Mützel's work appeared

A 1906 Liberian postage stamp depicting a tool-using chimpanzee. The mode of tool use is subject to debate.

as a mound in the stamp, and the twig was shorter in the stamp.

Sugiyama (1985) found sticks in Cameroon presumed to have been used by chimpanzees for Digging into termite mounds. These tools had been stripped of leaves and side twigs, and some had one brushlike end. As evidence that the tools had been used for Digging, he described five bowl-shaped depressions (2 to 5 centimeters deep and 10 to 25 centimeters wide) in the termite mounds. One or both sides of the tools were encrusted in mud, though the brush ends were seldom mud covered. Sugiyama sorted the Digging sticks into two forms, those containing blunt ends and those containing at least one brush-shaped end (see discussion of brushed sticks under Tool Manufacture). The mud-free brushed end was presumably not used for Digging.

Sugiyama, Koman, and Sow (1988) reported on an army ant (*Dorylus* spp.) nest that was Dug out by a chimpanzee. They observed the chimpanzee from a distance as she focused on a spot on the ground. Once she and her infant left the spot, the observers discovered an ant nest that had been excavated by Digging. The hole in the nest was approximately 40 centimeters wide and 20 centimeters deep. The authors speculated that the hole had been excavated both by hand and, possibly, with stems of an oregano-like plant (*Lippia* sp.) that were found at the site. Sugiyama (1995b) reported another observation of a Bossou chimpanzee using a stick to Dig into an army ant nest. The Digging stick was a straight, hard branch, with one end encrusted in mud.

Fay and Carroll (1994) reported that Bai Hokou chimpanzees used Digging tools to access arboreal and subterranean bee nests. They observed two chimpanzees from a distance, one of whom was looking intently at the ground. After the apes left the scene, a freshly dug hole and Digging stick remained, along with fragments of honey and hive. It was presumed that the chimpanzees had dug up the nest. In addition, Fay and Carroll described Digging sticks found protruding from termite mounds at Bai Hokou and Ndakan. Although the authors referred to these as perforating tools, their inferred use most closely places them in the Dig mode. Takenoshita (1996) reported the use of Digging sticks in the Ndoki Forest. Chimpanzees at the Dja Biosphere Reserve in Cameroon were presumed to use sticks to extract honey from both tree and subterranean nests of African honeybees (*Apis* sp.) and stingless bees (Meliponini) (Isra 2006). The sticks, described as sturdy and straight, were thought to be used to "dig for honey" (Isra 2006, 2). These later examples, while certainly indicative of tool use, do not allow us to clearly differentiate between Digging, Penetrating, or perhaps a combination of these modes.

A. Goodall (1979), Mwanza et al. (1992), and Yamagiwa et al. (1988) found tools used by chimpanzees in Digging for subterranean bees (*Meliplebeia tanganyikae*), larvae, or honey in the Democratic Republic of the Congo. Yamagiwa et al. (1988) found two sticks, free of leaves and branches, near the entrance of a nest, along with chimpanzee footprints and feces. Mwanza et al. (1992) noted that these sticks appeared similar in size to those used for Digging termite mounds in Cameroon and Equatorial Guinea. A. Fowler and Sommer (2007), in a study of the Gashaka chimpanzees in Nigeria, found stout sticks that had been used to Dig into bee nests, through leverage and spading. The tools were found at the entrance of the subterranean bee nests and were usually mud encrusted, with frayed ends. Though the researchers did not observe the behavior, they presumed that the tools were used for Digging, noting that other behavioral patterns (perforate, open and probe, lever open, and expel/stir, from Whiten et al. 2001) might be involved.

Chimpanzees in the Niokolo-Koba National Park area in Senegal sometimes used a wooden stick to Dig water wells (Galat-Luong and Galat 2000). This was also documented at Semliki in Uganda (K. Hunt and McGrew 2002; McGrew et al. 2003). Nishida, Matsusaka, and McGrew (2009) reported two observations of Mahale females Digging into the ground with a stick for water.

Hernandez-Aguilar, Moore, and Pickerings (2007) described indirect evidence of chimpanzees Digging for tubers at Ugalla in Tanzania. The tools consisted of stick tools, with wear suggestive of Digging with the larger, proximal end; bark tools, comprising long fragments from the outer bark of a tree; and wood tools, such as fragments of tree trunk extracted from a fallen log.

Captive chimpanzees also use tools for Digging. Cowper (1971) observed a captive chimpanzee Digging up grass with a stone, and Köhler (1925) saw captives Dig up roots with sticks. N. King, Stevens, and Mellen (1980) observed a captive female Digging into cracks in the floor to dislodge dirt and seeds that had became embedded in the cracks. The excavated material was eaten. Captives also used sticks, pieces of wood, metal rods, nails, spoons, shovels, and wire to Dig in the ground in what was apparently exploratory manipulation (K. Hayes and Hayes 1954; Köhler 1925; Kollar 1972; Ladygina-Kohts 2002; Menzel 1972, 1973a; Menzel, Davenport, and Rogers 1970; van Hooff 1973).

Jab, Stab, Penetrate

Chimpanzees at several locations used a stout stick to create access holes in insect nests through perforation or puncturing, what we call Penetration in the mode Jab, Stab, Penetrate in table 1.1 (Boesch, Head, and Robbins 2009; Deblauwe et al. 2006; C. Jones and Sabater-Pi 1969; Sabater-Pi 1972, 1974; Sanz and Morgan 2007, 2009; Sanz, Morgan, and Gulick 2004; Suzuki, Kuroda, and Nishihara 1995). After creating the access holes, chimpanzees either used their hands to extract insects or Inserted fishing tools into the newly made hole (see Tool Set, under Associative Tool Use: Chimpanzees, below). Sanz, Morgan, and Gulick (2004) used the term "puncture" when describing tool use to gain access to subterranean nests, and the term "perforate" for reopening entrance/exit holes in epigeal (above ground) nests.

C. Jones and Sabater-Pi (1969) found sticks protruding from termite mounds in Equatorial Guinea. Gonzalez-Kirchner and Sainz de la Maza (1992) saw similar sticks at Rio Campo, also in Equatorial Guinea. Struhsaker and Hunkeler (1971), Muroyama (1991), and Sugiyama (1985) reported the discovery of similar sticks in Cameroon. Fay and Carroll (1994) also found such tools at Ndakan and Bai Hokou in the Central African Republic, and Abwe and Morgan (2008) saw tools of this type in the Ebo Forest. Penetrating tools have also been found at Ndoki-Nouabale (Kuroda 1992; S. Suzuki, Kuroda, and Nishihara 1995). McGrew and Rogers (1983) found indirect evidence of perforation to access termites in Gabon. In none of these cases was any actual use of the sticks observed, although there was circumstantial evidence that the chimpanzees had abandoned them.

Bermejo and Illera (1999) discussed two types of termite-mound perforating sticks found at Lossi. One kind consisted of blunt-ended sticks; the other had brush-shaped ends, which they inferred had been either chewed or frayed, but their manufacture was not observed. However, they did see a juvenile repeatedly pushing and pulling a stick from a mound in the manner of Penetration. After extraction, the chimpanzee turned the same stick around and Inserted the opposite, brush-shaped end into the hole for fishing. This may have been a Multi-Function Tool (see below). Sanz and colleagues, using video monitoring, provided the most detailed description of Penetration by chimpanzees. Sanz, Morgan, and Gulick (2004) reported that the chimpanzees of the Goualougo Triangle used stout sticks to puncture subterranean termite nests and twigs to perforate and reopen entrance/exit holes in epigeal termite nests. The termite gallery of subterranean *Macrotermes nobilis* nests is approximately 30 centimeters below the ground and is not visible from the surface. To puncture these nests, leaf litter is raked by hand from the soil surface, and then the end of the stout stick is pushed into the soil of the cleared area. Both hands are used to grasp the tool as the chimpanzee pushes, frequently using one foot to push the tool as well. This allows the body weight to shift over the tool to increase force. The tool is extracted and the end of the tool is smelled and visually examined. If a nest is successfully located, puncturing is followed by Inserting termite-fishing probes. Sanz, Morgan, and Gulick (2004) reported that chimpanzees were observed on several occasions arriving at the nest with their puncturing and fishing tools in hand. They also reused tools left at the nest by other chimpanzees (Sanz and Morgan 2007). Sanz and Morgan further stated that the chimpanzees preferred to use sticks from a small tree (*Thomandersia hensii*) to access the subterranean nests. To enter epigeal nests, Goualougo chimpanzees perforated the exit and entrance holes with twigs, pressing the tip of the tool into the mound to create a hole through which the termite-fishing probes could be Inserted (Sanz, Morgan, and Gulick 2004). Deblauwe et al. (2006) found sticks presumed to be used for puncturing subterranean nests and sticks for perforating epigeal nests at the Dja Biosphere Reserve.

Goualougo chimpanzees also used perforating sticks to gain access to driver ant (*Dorylus* spp.) nests (Sanz, Schöning, and Morgan 2009). To create an opening into the nest, a chimpanzee would grasp a Detached woody sapling by its distal leafy branches and Insert the entire length of the tool (average length of 92.3 centimeters) into the nest. The tool could be partially withdrawn and reinserted several times to create a hole into which a dipping probe could be Inserted.

Boesch, Head, and Robbins (2009) documented the use of Penetrating tools to reach the inner chamber of a beehive for honey extraction. The tools were used

A Goualougo chimpanzee using a twig to perforate an above-ground termite nest. Drawing courtesy of David Morgan, Goualougo Triangle Ape Project.

both to locate the honey or insect chamber and to create a hole into which a probe could be inserted. After use, the tools often had one blunt end, encrusted with soil. Bermejo and Illera (1999) reported the use of a stick as a bodkin to puncture a beehive, in combination with Pounding and Prying to gain entry. Sanz and Morgan (2009) also reported the use of a Penetrating tool at Goualougo to enter a honey hive. The chimpanzee aligned the stick tool with the nest and then put his foot on the opposite end of the tool, pushing the tool into the nest to create an opening. Sanz and Morgan (2009) also documented two more new types of tool behavior related to gaining access to the honey hive, termed "press-and-hold," and "rotate/twist." Press-and-hold involves a chimpanzee holding and pushing a tool to (but not into) the hive entrance for several seconds. The authors described this as possibly providing information regarding the structure of the hive and/or possibly weakening the hive structure. Chimpanzees at Goualougo also used rotate/twist to Penetrate the hive, placing the tool at the hive entrance and rotating it as if drilling into the hive.

Chimpanzees may use tools to assist in retrieving or extracting insect or mammalian prey, or in intimidating other animals through prodding, Jabbing, or Stabbing with tools. This mode of tool use may overlap with Insert and Probe, when prodding takes place inside a cavity. For consistency, all examples of attempting to expel an animal are included under Jab, Stab. J. Goodall (1986) termed this "rousing the occupants of holes." Nishida (1973) first described the use of an "expelling stick" by Mahale females. The chimpanzees forcefully and repeatedly inserted a stick into a tree cavity. The behavior was quite distinct from Inserting and Probing, and Nishida presumed it was probably an attempt to drive ants or other insects from the hole. However, M. Nakamura and Itoh (2008) included a personal communication from Nishida describing other observations of this behavior by different females. These additional observations led Nishida (in Nakamura and Itoh 2008) to assume that this expelling behavior was aimed at small mammals.

At Gombe, Jabbing was documented in the context of investigating the contents of a hole; the tool was inserted into the hole and vigorously moved back and

A Goualougo chimpanzee using a stick to puncture an underground termite nest. Drawing courtesy of David Morgan, Goualougo Triangle Ape Project.

forth (J. Goodall 1986). Often nothing came out of the hole, but she reported two cases of ants, one case of termites, and three cases of bees swarming out. She stated that in play, Gombe infants also prodded at ant trails with sticks. Ohashi (2006) reported a chimpanzee using a stick to probe unidentified animal species in holes at Bossou.

Boesch (1991a) noted that some Taï chimpanzees attempted to hit a leopard in a hole and tried unsuccessfully to Stab the leopard with the branch. At Mahale, Huffman and Kalunde (1993, 94) observed a 12-year-old wild female "forcefully probe into the hole at least once and rouse the squirrel [*Protoxerus stangeri* or *Heliosciurus rufobrachium*] which she grabbed." This was the first recorded observation of tool-assisted mammalian predation by a chimpanzee. However, the authors did not use Jab or Stab in their description of the behavior. M. Nakamura and Itoh (2008) suggested that hunting with tools might be habitual at Mahale. They documented an 8-year-old female poking a stick into a tree hollow and extracting a dying or dead squirrel (*H. rufobrachium*). Nakamura and Itoh assumed the chimpanzee had wounded or killed the squirrel. Of interest, this individual was the adopted daughter of the chimpanzee observed by Huffman and Kalunde (1993), described above. Nakamura and Itoh (2008) also reported one incident of a presumed hyrax (*Heterohyrax brucei*) hunt in which three males used sticks, vines, and stems to Jab, Club, Beat, and whip into a cave where the authors believed a hyrax lived.

Chimpanzees at Fongoli in Senegal habitually hunt bushbabies (*Galago senegalensis*) by using modified sticks "in the manner of a spear" to Stab at the bushbabies in their nesting tree holes (Pruetz and Bertolani 2007, 412). After manufacturing tools, the chimpanzees forcefully Jabbed branches into the bushbaby's hollow, licking and smelling the end of the tool upon withdrawal. Pruetz and Bertolani reported that the chimpanzees actually acquired the prey only once in the twenty-two observed cases of this behavior. J. Pruetz (pers. comm.) stated that this behavior has now been observed ninety-two times, occurring in about half of the Fongoli community, with a 10 to 15 percent success rate at acquiring the prey. Tumbo, an adult female, was the most prolific and successful hunter.

Captive chimpanzees also Jab and Stab with tools. Köhler (1925), Kollar (1972), and Sheak (1924) saw captives Jabbing conspecifics with sticks. Birch (1945), A. Jones (1986), and Köhler (1925) noted prodding or Jabbing at humans with sticks. Köhler (1925) and Sheak (1924) mentioned prodding or Jabbing at dogs, chickens, and a baboon with sticks, wire, and a whip. In all cases, the behavior appeared to be agonistic. Birch (1945) also reported prodding with sticks during "play-attacks." N. King, Stevens, and Mellen (1980) saw captives spearing insects with lengths of straw. The Tenerife chimpanzees poked or Stabbed at free-ranging chickens that came near their enclosure (Köhler 1925). They did not capture and eat the chickens, but rather poked or Stabbed at them with sharp sticks and wire in what appeared to be a game. A captive chimpanzee was observed poking cockroaches with a stick, in what was interpreted as amusement (Kohts 1935; Ladygina-Kohts 2002).

Reach

An adult male chimpanzee at Gombe used a stick to Reach toward bananas held by a human and then hit the bananas with the stick (van Lawick–Goodall 1968). Sugiyama and Koman (1979) documented "branch hauling" at Bossou, in which an individual modified several branches, using them to attempt to pull in an out-of-reach branch on a fruiting fig tree. Whiten et al. (1999) later referred to this behavior as "branch-hook," to avoid confusion with Dragging tools. "Algae-scoop" is a customary behavior of Bossou chimpanzees (Matsuzawa 1999; Matsuzawa, Yamakoshi, and Humle 1996; Ohashi 2006; Whiten et al. 1999; Yamakoshi 1998). This form of Reaching involves using a stripped stem to acquire algae and other vegetation from the surface of a water source. Devos, Gatti, and Levréro (2002) documented this behavior by a male at Lokoué Bai in Odzala National Park in the Republic of the Congo. He used a sedge stem to scoop the algae from pools, carrying the tool in his mouth as he walked between pools.

Chimpanzees obtain army ants through ant dipping by using a long and rigid twig, stem, stalk, or stick as a wand to get ants from terrestrial trails or nests (McGrew et al. 2003; Sugiyama, Koman, and Sow 1988). Ant dipping from terrestrial trails of traveling ants is best described under Reach, while ant dipping from nests is best classified under Insert and Probe. We describe this behavior in depth under Insert and Probe.

Dipping for ants was observed by Köhler (1925) among the Tenerife chimpanzees. He noted that the chimpanzees, "armed with straw or twigs like anglers on a river's bank," Reached into processions of ants moving outside their cage (Köhler 1925, 80). When the tools were covered with ants, the chimpanzees drew them in and feasted.

Through observations, experiments, and problem-solving situations, captive chimpanzees also have been found to use Reaching tools (sticks, branches, twigs, long poles, PVC pipes, rakes, straws, wood-wool, a rug, and a weighted rope) to obtain a variety of items (Bania et al. 2009; Birch 1945; Bourne 1971; Brent, Bloomsmith, and Fisher 1995; Chamove 1989; de Waal 2007; Denisov 1958 in Windholz 1984; Döhl 1966, 1969; Dufour and Sterck 2008; Firsov 1972, 1982; Fletcher and Weghorst 2005; Girndt, Meier, and Call 2008; Goustard 1986; Guillaume and Meyerson 1930, 1934; C. Hayes 1951; K. Hayes and Hayes 1952; Hobhouse 1926; Hrubesch, Preuschoft, and van Schaik 2009; Jackson 1942; Jacobsen, Wolfe, and Jackson 1935; Jennison 1927; A. Jones 1986; Kats 1972b; N. King, Stevens, and Mellen 1980; Köhler 1925; Kollar 1972; Ladygina-Kohts 2002; Levykina 1959 in Ladygina-Kohts and Dembovskii 1969; Mathieu et al. 1980; Menzel, Davenport, and Rogers 1970; Nagell, Olguin, and Tomasello 1993; Nissen 1956; Novoselova 1959 in Ladygina-Kohts and Dembovskii 1969; Okano et al. 1973; Poulsen 1974–1975; Povinelli et al. 2000; Prince et al. 2009; Roginskii 1945 in Ladygina-Kohts and Dembovskii 1969; Rothmann and Teuber in Bierens de Haan 1931; Savage-Rumbaugh 1986; Schiller 1952, 1957; Takeshita and van Hooff 1996, 2001; Temerlin 1975; Tomasello et al. 1987; van Hooff 1973).

Köhler (1925) presented food beyond the reach of a hungry chimpanzee and provided a stick with which the subject could rake in the food. Köhler's subjects spontaneously used sticks, plant stalks, straw, wire, rags, blankets, pieces of cardboard, and a variety of other objects to Reach food and other incentives that were either placed outside their cage or suspended from above. Experiments requiring raking out-of-reach objects from water have also been conducted with semi-free-living chimpanzees; the chimpanzees were able to select tools of appropriate length without simultaneously looking at the food reward (Goustard 1986). Kollar (1972) and van Hooff (1973) also reported that captive chimpanzees used sticks and twigs to retrieve objects from a moat. Fletcher and Weghorst (2005) similarly observed captives fishing with a stick to obtain a sunken object from water and paddling with the tool in the water to draw in a floating object. Lucy, a home-reared chimpanzee, used a garden hose to spray birds, in what may be considered a form of Reach to irritate the birds (Temerlin 1975). Nissen (1956) reported a captive's selecting a stick of an appropriate length, strength, and thickness and then using it to flip the light switch outside her enclosure on and off.

In some of Guillaume and Meyerson's (1930, 1934) and Köhler's (1925) tests, the subjects first had to push the food away with the stick, maneuver it around a detour, and only then rake it in. Döhl (1969) presented a captive with different types of tools, requiring the subject to examine a complete problem and then choose the most appropriate tool to solve the task. She correctly solved the problem in 238 of 250 trials. Captive chimpanzees used screwdrivers correctly to extract and remove screws (Fernberger in Dewsbury 2009; R. Fouts 1997; Rensch 1973; Rensch and Döhl 1967; Temerlin 1975).

The chimpanzees' understanding of the physical properties of tool use have been investigated as well. For example, Furlong, Boose, and Boysen (2008) and Povinelli et al. (2000) examined the chimpanzees' comprehension of tool properties through the flimsy-tool paradigm, in which the apes had to select Reaching tools of sufficient rigidity. However, in these studies the subjects were not responsible for establishing the proper and effective orientation of the tool, and therefore it does not conform to our definition of tool use. This is true of some trap-table studies as well, where two rakes are placed on a partitioned table, one side of which contains a trap into which the food will fall and be lost if the chimpanzee pulls the food over it. The task requires the chimpanzees to pull the rakes from the non-trap side to obtain a reward (Martin-Ordas, Call, and Colmenares 2008; Povinelli et al. 2000). Chimpanzees performed poorly on trap-table studies, not succeeding at greater-than-chance levels (Povinelli et al. 2000). However, Girndt, Meier, and Call (2008) conducted a trap-table study that does conform to our definition of tool use. They presented chimpanzees with trap-tables with only one rake, which was not pre-positioned by the researcher on the table, thus allowing the chimpanzee to select where to use the tool. Approximately 80 percent of the subjects solved the task correctly on the first trial when they were allowed to establish the effective orientation and position of the tool.

Contrasting with the many impressive successes with Reaching tools is the failure of the subjects studied by Birch (1945), Haggerty (1910), Mathieu and Bergeron (1977), C. Parker (1968, 1969b), Povinelli et al. (2000), Schiller (1957) and Shepherd (1923) to use tools successfully to obtain an out-of-reach reward. Haggerty (1910) provided no information about the age of his subject or the time allowed to solve the problem. Mathieu and Bergeron's (1977) two subjects were only about 1 year old and had been tested monthly since they were 3 months of age. C. Parker's (1968, 1969b) three chimpanzees were probably adults who were each tested in three forty-five-minute sessions. Two of the subjects did eventually use Reaching tools successfully to obtain an out-of-reach reward, after some tuition and shaping by the experimenter. Shepherd's (1923) subject was 7 years old and was allowed only four trials of two minutes each. Age, social history, and richness of the rearing environment might have been significant factors in such negative results. Schiller (1957) showed that the opportunity for captives to play freely with sticks greatly facilitated the chimpanzees' later use of sticks as Reaching tools. McGrew (1992, 21) noted that among chimpanzees living alone or in small barren enclosures, "any performance of tool-use in experimental testing is therefore remarkable, and is probably positively correlated with the socio-ecological validity of the captive environment." McGrew contrasted the spontaneous tool use by chimpanzees in an enriched facility (Nash 1982) to the long training required for the acquisition of tool use by a deprived chimpanzee (Birch 1945). Brent, Bloomsmith, and Fischer (1995) examined this empirically by studying the effect of rearing on tool use. They used a basic raking experiment involving the use of a PVC pipe to obtain an out-of-reach reward and concluded that wild-born chimpanzees excelled at the task, because they raked in more food rewards in a trial than either captive mother-reared or captive nursery-reared subjects, with the captives all performing similarly. The authors did not find that sex or age influenced tool-using performance.

Wild chimpanzees use tools such as stems, grass blades, twigs, or sticks to Reach toward and touch objects that they are reluctant to approach or touch directly (J. Goodall 1986). Albrecht and Dunnett (1971) saw a wild chimpanzee use a stick to touch a suspended chimpanzee doll; others used sticks to touch a leopard model. A Gombe youngster repeatedly used a twig to touch an infant sibling while the baby was cradled protectively by their apprehensive mother (van Lawick–Goodall 1970, 1971). Two infants touched the genitals of a female with a twig (J. Goodall 1986; van Lawick–Goodall 1968). A young adult touched the bloody head of a dead python with a palm frond (van Lawick–Goodall 1968, 1970, 1971, 1973). This same individual was once observed to poke a grass stem into Goodall's pocket to investigate a piece of fruit within (J. Goodall 1986). Albrecht and Dunnett (1971) saw wild chimpanzees smelling the ends of sticks after using them to touch a leopard model. In many other of these "investigatory probe" cases, the chimpanzee also smelled the end of the tool after Reaching with it (J. Goodall 1986). As in the cases where chimpanzees smelled the ends of tools after Inserting them in insect nest holes and pants pockets, the apes, according to Goodall, gained additional sensory information about the objects they touched. A similar behavior was observed in Brewer's Gambia group (Brewer and Brewer in van Lawick–Goodall 1973). Chimpanzees at Bossou also poked with sticks at traces of animals or humans on the ground and then withdrew to sniff the sticks (Ohashi 2006). Boesch and Boesch (1990) listed Taï chimpanzees as using leaf stems or twigs to probe at wood-boring bee nests, corpses, wounds, and tree cavities. Chimpanzees may use tools to stir water in play, as has been documented for a young male using a leaf sponge to stir water at Mahale after he splashed and stirred it with his hand (Matsusaka et al. 2006). Gombe youngsters also investigated water and feces with sticks, in what may have been play or exploration (J. Goodall 1986).

In addition, captives use tools to avoid touching objects. Köhler (1925) observed that his chimpanzees were sometimes ambivalent about approaching fear-producing, potentially dangerous, or unpleasant objects. In such cases, the apes often used tools to Reach toward the object from a safe distance, even though they could have approached and explored the object directly with their hands, mouths, or noses. For example, Köhler's chimpanzees used sticks to Reach toward and touch a strange chimpanzee, sticky banana skins, fire, mice, lizards, and an electrically charged basket of fruit. Brink (1957) described a cigarette-smoking chimpanzee as sometimes holding other objects (banana or orange peels) to push a used-up but still burning cigarette into a wet patch on the floor of his enclosure until the flame was extinguished. Mottershead (1959, 1963) saw a captive touch a shock wire with a twig, and Wilson and Wilson (in McGrew, Tutin, and Midgett 1975) reported that captives used objects to short-circuit the electrical fence designed to contain them. Menzel (1971) observed captives Reaching toward hidden reptile models with sticks or using sticks to rake food from piles on which such models had been placed. H. van Lawick (in van Lawick–Goodall 1970) saw a captive touch a pangolin (*Manis* sp.) with a stick. Kortlandt and Kooij (1963) reported this same incident and inferred that the chimpanzee was Clubbing the pangolin as well as touching it. Raphael, a chimpanzee studied by Pavlov, was able to use objects to extinguish a flame (Denisov 1958, and also Orbeli 1949, in Windholz 1984).

Reaching tools may also be useful in grooming or examining hard-to-reach portions of the body, such as areas in the mouth. At Gombe, van Lawick–Goodall (1970, 1971) saw a wild female chimpanzee use a twig in what appeared to be dental autogrooming, and she saw an infant pick its nose with a piece of straw. Kummer and Goodall (1985) described a young chimpanzee using a termite-fishing technique to probe the hair on his mother's leg. Both wild and captive apes have been documented to use a type of comb to groom their hair. While absent at most chimpanzees sites, the use of leaf stems or twigs as combs has been documented at Budongo (V. Reynolds, and also C. Assersohn, unpub. data in Whiten et al. 1999) and the Goualougo Triangle (Sanz and Morgan 2007).

Nishida (2002) observed a Mahale (M group) female using a small twig to probe under her toenail. The author presumed she was attempting to remove a sand flea. Nishida, Matsusaka, and McGrew (2009) also reported this behavior in another Mahale female.

A captive used cotton swabs to probe in its mouth, bamboo splinters to probe between the toes of humans, and straw to probe under human fingernails (Merfield 1956). Home-reared Viki used pliers to extract a deciduous molar from her own mouth (K. Hayes and Hayes 1954). Mignault (1985) reported that a human-reared chimpanzee used a hairbrush to brush his doll after observing a human use a hairbrush. Captive chimpanzees used human-made combs for grooming their hair (M. Vancatova, pers. comm.). A captive chimpanzee, Loulis, created a tool to use to probe a sore on a conspecific (D. Fouts 1983). The home-reared chimpanzee Lucy used human tools to explore her body, including the use of pliers and a mirror to examine her labia (Temerlin 1975).

Insert and Probe

Hayashi and Matsuzawa (2003, 231) postulated that chimpanzees have a "strong tendency to insert objects into a hole or hollow." This hypothesis is well supported by numerous observations in the wild and in captivity. There are many examples of chimpanzees Probing into cavities, such as tree trunks, insect nests, or fallen logs or crevices, for both extractive foraging and in exploration. Termite fishing and ant dipping or fishing are subsumed under this mode of tool use; but as McGrew (1974) has emphasized, there are important distinctions between these activities, and, as Teleki (1974) has noted, further distinction should be made between procuring arboreal and terrestrial ants. Probing for other types of insects (e.g., larvae, beetles) and mushrooms also falls under this category, as does fluid dipping to obtain honey or water from cavities.

Many wild populations of chimpanzees Insert and Probe for termites into subterranean mounds or aboveground epigeal nests, using fishing probes consisting of flexible stalks of grass, herb stems, twigs, lengths of vine, pieces of bark, palm frondlets, leaf petioles, or leaf midribs. Termite fishing by the chimpanzee David Greybeard was the first form of tool use observed by Jane Goodall at Gombe (J. Goodall

Flo, a chimpanzee of the Gombe National Park, fishes for termites. Photograph by Hugo van Lawick, courtesy of Jane Goodall, Ph.D., DBE, Founder, the Jane Goodall Institute, and UN Messenger of Peace (www.janegoodall.org).

Fongoli chimpanzees fish for termites. Photograph courtesy of Stephanie L. Bogart.

1963a), and this can fairly be said to have revolutionized our understanding of apes and humans. *Macrotermes* spp. is the termite most commonly consumed by chimpanzees in Africa, but *Pseudacanthotermes* spp. are consumed at Mahale K, where *Macrotermes* seems to be absent (Collins and McGrew 1985; McGrew 1992; Uehara 1982). During termite fishing, the probe is Inserted into the nest, often after the chimpanzee has created a hole with a finger or a Penetrating tool. After the termite soldiers seize the probe with their mandibles, the chimpanzee withdraws the tool to eat the attached termites (J. Goodall 1963a; McGrew and Collins 1985; van Lawick–Goodall 1968, 1970) As in ant dipping (see above), termites are consumed by "direct mouthing" (also called "nibbling") (J. Goodall 1986; Humle 1999; McGrew and Collins 1985; Sanz, Morgan, and Gulick 2004) or "pull-through" (Sanz, Morgan, and Gulick 2004). Direct mouthing, the most prevalent behavior, consists of bringing the wand end directly to the mouth to nibble the insects from the wand. Pull-through consists of holding the tool in one hand and swiftly drawing the tool through the other hand to gather the attached termites into a bundle, which is then consumed. One chimpanzee might use both techniques. J. Goodall (1986) reported Gombe chimpanzees using the lips and teeth to nibble soldier termites from the tool, followed by a pull-through to gather the workers.

Termite fishing has been recorded from direct and indirect observations at Assirik (P. Baldwin 1979; Bermejo, Illera, and Sabater-Pi 1989; McBeath and McGrew 1982; McGrew, Tutin, and Baldwin 1979; McGrew et al. 2003), Bossou (Humle 1999), Campo (Muroyama 1991; Sugiyama 1985; Takemoto 2004), the Dja Biosphere Reserve (Deblauwe et al. 2006), Ebo (Abwe and Morgan 2008), Fongoli (Bertolani et al. 2007; Bogart 2009, Bogart and Pruetz 2008; Bogart, Pruetz, and McGrew 2005; McGrew, Pruetz and Fulton 2005; Pruetz 2006), Gabon (McGrew and Rogers 1983), Gombe (J. Goodall 1963a, 1963b, 1964, 1965, 1986; Greenfield et al. 2000; Lonsdorf 2005, 2006; Lonsdorf, Eberly, and Pusey 2004; Lonsdorf and Hopkins 2005; McGrew 1977, 1979; McGrew and Marchant 1992, 1999; McGrew, Tutin, and Baldwin 1979; Pandolfi, van Schaik, and Pusey 2003; Teleki 1974; van Lawick–Goodall 1968, 1970, 1971, 1973), the Goualougo Triangle (Sanz, Call, and Morgan 2009; Sanz, Morgan, and Gulick 2004); the Kasakati Basin (A. Suzuki 1966, 1969), Lossi (Bermejo and Illera 1999), Mahale K and B (Collins and McGrew 1987; McGrew and Collins 1985; Nishida and Uehara 1980; Norikoshi 1994; Takahata 1982; Uehara 1982), Ndoki (S. Suzuki, Kuroda, and Nishihara 1995), Ngogo (Watts 2008), and among a group of captives reintroduced in Gambia and later moved temporarily to Senegal (Brewer 1976, 1978, 1982). The reintroduced chimpanzees frequently fished for termites, although they were not trained by humans to do so (McGrew, Tutin, and Baldwin 1979). Termite fishing is noted as absent in the Mahale M community (Uehara 1999). At Bossou, termite fishing is not habitual. Humle (1999) first observed this rare behavior at Bossou, seeing two chimpanzees fish with stalks taken from herbaceous plants.

Differences in the size of the tool and the tool material have been documented between sites (McGrew, Tutin, and Baldwin 1979). McGrew, Tutin, and Baldwin (1979), A. Suzuki (1966), and Teleki (1974) all noted that probes must be slender enough to be Inserted, strong enough to prevent collapse, flexible enough to be snaked through twisting passages, soft enough for the termites to grip, and resilient enough to retain the original shape for reuse. J. Goodall (1986) found that tools might be selected and discarded prior to use, presumably for unsuitability. Teleki (1974) emphasized that considerable knowledge and skill are required to locate and expose workable tunnels, select and manufacture tools, and utilize the tools effectively. Chimpanzees at times reused tools left at the termite mound by other chimpanzees (J. Goodall 1986).

Chimpanzees Insert and Probe with elongate objects in contexts that may appear non-functional and are best characterized as playful. Wild Gombe infants Inserted twigs and grass stems into termite mounds in what might be either play or unpracticed attempts at termite fishing (van Lawick–Goodall 1968, 1970, 1973).

Ant dipping consists of using a long and rigid twig, stem, stalk, or stick as an ant-catching wand to extract *Dorylus* spp. (army ants, also known as safari or driver ants) from terrestrial trails or nests (McGrew et al. 2003; Sugiyama, Koman, and Sow 1988). The ants live in very large colonies in subterranean nests, from which they emerge in large columns to forage. Their bites are very painful, and the soldier castes swarm onto and over objects and organisms that they encounter or that disturb them. Ant dipping was discovered and initially described by J. Goodall (1963a, 1964, 1965) and van Lawick–Goodall (1968, 1970, 1973). A more detailed description was provided by McGrew (1974, 1977, 1979). After locating a nest, a chimpanzee typically pulled and scraped out handfuls of soil, which stimulated massed active aggression by the soldiers. The ape then selected and/or modified a branch and Inserted it into the nest. The sticks used to dip for ants averaged 1 centimeter in diameter and 66 centimeters in length. They were straight, without side branches. McGrew (1974) argued that the chimpanzees selected and modified ant-dipping sticks for these characteristics, so as to provide tools that were sturdy, long, and not cumbersome. J. Goodall (1986) noted that dipping wands were more uniform in appearance and size than the ones McGrew (1974) described that were used for termite fishing. These tools must be long enough to prevent ants from quickly swarming up the stick to bite the user, but they cannot be so long that their use becomes awkward (J. Goodall 1986). Boehm (1992) summarized an observation of a Gombe female trying to obtain army ants using a short tool. He interpreted her retreat and fashioning of a new, longer tool for dipping as the result of being "too close for comfort" using the short tool (Boehm 1992, 10). J. Goodall (1986) reported that these tools might also be reused after being left by a previous user. T. Hicks, Fouts, and Fouts (2005) found a vine tool projecting from a driver ant nest, in one of the few reported cases of a vine being used in ant dipping.

Direct and indirect evidence (finding wands near/in nests and seeing ants in fecal remains) indicated the presence of ant dipping at Assirik (P. Baldwin 1979; McGrew 1983), Bili (Hicks, unpub. data, in Schöning et al. 2008), Bossou (Ohashi 2006; Sugiyama 1989b, 1995b; Sugiyama, Koman, and Sow 1988; Yamakoshi 1998), Fongoli (McGrew, Pruetz, and Fulton 2005), Gashaka (A. Fowler and Sommer 2007; Schöning et al. 2007), Goera (of the Nimba mountains) (Sugiyama 1995b), Gombe (J. Goodall 1963a; McGrew 1979), Goualougo (Sanz and Morgan 2007; Sanz, Schöning, and Morgan 2009), Kalinzu (Hashimoto 1998; Hashimoto, Furuichi, and Tashiro 2000), Ndakan (Fay and Carroll 1994), Ngotto (T. Hicks, Fouts, and Fouts 2005), Ntale (Ingmanson 1997), Seringbara (Humle 2003; Humle and Matsuzawa 2001), Taï (Boesch and Boesch 1990, 1993b), Tenkere (Alp 1993), and Yealé (Humle and Matsuzawa 2001; Shimada 2000).

Ant-dipping chimpanzees frequently stood bipedally as far from the nest or trail as possible, rushed to the nest, Inserted the tool, and then withdrew it and themselves to eat the ants. Alternatively, they elevated themselves on nearby tree trunks, shrubs, vines, or branches. Each measure provided some protection against the painful bites of the aroused ants. Though dipping for terrestrial ants is subsumed under the Insert and Probe mode of tool use, McGrew's (1974) description of the behavior and his enumeration of the techniques used by ant-dipping chimpanzees to maintain a safe distance from the disturbed insects indicated that there may also be an element of the Reaching mode. Additionally, dipping ants from terrestrial trails is also referred to as ant dipping, but it does not involve Insertion into a nest. Both van Lawick–Goodall (1970) and McGrew (1974) saw chimpanzees Reach into the ant processions with sticks, even though the insects were completely exposed. Ant dipping thus includes not only acquiring ants from inaccessible subterranean nests by Inserting and Probing, but also Reaching for them from a distance to minimize exposure to their painful bites. A. Fowler and Sommer (2007) reported that the tools used for ant dipping were the longest tools in the Tool Kit of the Nigerian chimpanzees they observed. T. Hicks, Fouts, and Fouts (2005) also reported that the tools used for ant dipping by Ngotto chimpanzees were longer than those used for honey dipping.

As with termites, once ants are acquired on the dipping tools, there are two observed methods for consuming them, known as the "direct mouthing" and "pull-through" techniques. Direct mouthing—also called "ant-dip-single" (Whiten et al. 1999), "nibbling" (Boesch and Boesch 1990), or the "one-hand method" (Yamakoshi and Myowa-Yamakoshi 2004)—occurred at Bossou (Humle and Matsuzawa 2002; Sugiyama 1995b; Sugiyama, Koman, and Sow 1988), Taï (Boesch and Boesch 1990) and, occasionally, at Gombe (McGrew 1974). This method involves bringing the tool directly to the mouth and either sweeping the tool through the lips or nibbling off the ants. The "pull-through" technique, also called "ant-dip wipe" (Whiten et al. 1999), or the "two-hand" (Yamakoshi and Myowa-Yamakoshi 2004), is typical of Fongoli and Gombe chimpanzees (McGrew 1974; J. Pruetz, pers. comm.). This method involves holding the tool with one hand, dipping it into the nest hole until the tool swarms with ants, and then swiftly drawing the tool through the other hand to gather the ants into a bundle. This technique requires the tool to be free of side branches that would hinder the movement (J. Goodall 1986). With the pull-through technique, a chimpanzee can sweep together a mass of up to 300 swarming ants with a rapid movement of its hand up the stick and into its mouth. The mass of ants is then chewed frantically to avoid the ants' bite. McGrew et al. (2003) documented that the pull-through wand was longer (50 to 100 centimeters) than the wand used for direct mouthing (25 to 50 centimeters). When a tool is recovered without direct observations of use, these differences in length assist in making inferences regarding the form of dipping that occurred. For example, at Assirik, ant-dipping tools were found but the behavior was not observed (P. Baldwin 1979). Based on the length of the tools, McGrew et al. (2003) inferred that a pull-through method was probably used. It is not known if the correlation of length and technique results from a chimpanzee choosing a tool to match the technique it plans to use, or if the ape matches the technique to the tool in hand.

Differences in the length of the dipping wands among research sites have been documented. As Humle and Matsuzawa (2002) summarized, dipping wands from Gombe were significantly longer than those from the Taï population (Boesch and Boesch 1990; McGrew 1974), while tools at Bossou were intermediate in length (Sugiyama 1995b; Yamakoshi and Myowa-Yamakoshi, unpub. data, in Humle and Matsuzawa 2002). Such differences may be due to differences in ant-dipping techniques among the three sites, with pull-through postulated to be the most efficient form of this tool behavior (Boesch and Boesch 1990; Hashimoto, Furuichi, and Tashiro 2000; Humle and Matsuzawa 2002). Humle and Matsuzawa (2002) found a correlation between tool length, technique, and the type of army ant. Tools for obtaining less aggressive red ants (*Dorylus kohli*, *D. lamottei*, *D. militaris*) were shorter and were associated with direct mouthing; longer tools were associated with more aggressive black ants (*D. nigricans*). Furthermore, Humle (2006) noted that ant-dipping chimpanzees demonstrated behavioral plasticity by minimizing their risk of being bitten as well as by increasing their ant-dipping efficiency. Schöning et al. (2008) examined army ant eating across fourteen research sites, and found that the characteristics of the tool and the tool-use technique are only partly explained by the aggressiveness of the ant prey, arguing for the important role of culture in the consumption of army ants. Möbius et al. (2008, 37) similarly concluded that the "interaction of cultural and ecological factors shapes the differences in army ant predation between Taï and Bossou chimpanzees."

Ant fishing refers to the extraction of arboreal ants from nests on tree branches or trunks and includes the genera *Camponotus* (carpenter ants), *Paltothyreus*, and *Oecophylla* (weaver ants; recall from chapter 2 that weaver ants are themselves tool users). Teleki (1974) noted that the materials and techniques used to fish for termites and to dip for fossorial (living on or under the ground) ants are very different. However, tools used to acquire arboreal ants are similar to those used in termite fishing. The chimpanzees Insert limber tools such as sticks, twigs, lengths of vine, strips of bark, grass stems, and the midribs of leaves into the ant nest holes in trunks and branches, and into the nests constructed on tree branches. The ants swarm up the probe and are eaten directly from it. Ant-fishing tools may vary in size in relation to the size of the ant nest opening (J. Goodall 1986). At Mahale K, the chimpanzees sometime shook the tool violently while it was in the nest hole (Nishida and Hiraiwa 1982).

Fishing for arboreal ants has been documented by direct and indirect evidence at Assirik (McGrew 1983), Bossou (Sugiyama 1995b; Yamakoshi 2001; Yamamoto et al. 2008), Gashaka (A. Fowler and Sommer 2007), Gombe (J. Goodall 1963a, 1964, 1965, 1986; Teleki 1974; van Lawick–Goodall 1968, 1970, 1971, 1973), Lopé (Tutin and Fernandez 1992; Tutin, Ham, and Wrogemann 1995), and the Mahale M and K groups (Kawanaka 1990; Marchant and McGrew 2007; Nishida 1973, 1977; Nishida and Hiraiwa 1982; Nishida and Uehara 1980; Nishie, Itoh, and Nishida 2006; Takasaki 1992). Ant fishing has also been documented in a free-ranging group introduced to an island in the Ipassa Reserve in Gabon (Hladik 1973, 1975, 1977; Hladik and Viroben 1974). Brewer (1976) mentioned that a chimpanzee in the Gambia group once Probed for ants with a branch. McGrew (1974) cited Kade's observation of a chimpanzee feeding on ants after translocation to an island in Lake Victoria, Tanzania. Hladik and Viroben (1974) noted Grasse's observation of the use of a "fishing pole" to capture insects by chimpanzees in Côte d'Ivoire.

Nishida (1973) stated that arboreal ant-fishing tools averaged 33.2 centimeters in length and were 1 to 8 millimeters in diameter, suggesting that much of the variation resulted from the chimpanzees' accommodating to varying nest hole depths and diameters as they selected and/or modified materials to produce suitable tools. Nishida saw a chimpanzee carry an arboreal ant-fishing tool about 70 meters to the point of use; J. Goodall (1963a) saw such a tool transported nearly 20 meters. Mahale K chimpanzees transported their ant-fishing tools, after making the tools as much as seventeen minutes prior to finding the ant nest (Nishida and Hiraiwa 1982). The authors noted that tools were also reused, including tools made by other individuals.

Several chimpanzees of the Ngogo community of Kibale National Park were documented making and using tools to forage for insects (possibly wood-boring beetles, genus not specified) from a fallen dead tree (Sherrow 2005). After creating a tool with a brush end, the chimpanzee Inserted the tool into the log and moved it circularly before removing it to eat the insects that adhered to it. Nishimura et al. (2003) observed four chimpanzees using sticks at the Moukalaba Reserve, Gabon, to extract what may have been insect larvae from a fallen log. Ohashi (2006) also reported the use of a stick to extract larvae from a nest tunnel at Bossou.

Boesch (1995) documented one observation of grub eating at Taï, involving a male making and using a twig to Probe into a cavity to extract otherwise unreachable prey. In addition, Boesch documented twig use to extract larvae from a fallen log. Finally, he reported one observation in which a male used two twigs to extract and eat mushrooms from inside a termite mound.

Chimpanzees also Insert and Probe with tools to get honey from beehives. As summarized by Boesch, Head, and Robbins (2009, 562), tools used for the extraction of honey from hives are referred to by several names, including "collector" (Boesch, Head, and Robbins 2009), "bee probe" (A. Fowler and Sommer 2007), and "dip-stick" (Bermejo and Illera 1999; Sanz and Morgan 2007). Bark strips were used as a "swabber" at Loango Park, Gabon (Boesch, Head, and Robbins 2009). The process of extracting honey is labeled as "fluid-dip" (Sanz and Morgan 2007, 2009; Whiten et al. 1999), and/or "honey fish" (Bermejo, Illera, and Sabater-Pi 1989). Sanz and Morgan (2009) summarized the tool behaviors associated with honey gathering, including behaviors we review under other modes of tool use. Honey-fishing tools are easily identified by the presence or scent of honey. The behavior is also confirmed by the presence of bees in feces. Honey extraction was first documented by Merfield (1956). He recorded wild chimpanzees in Cameroon Inserting twigs into the subterranean nest of a "small black bee" (Merfield 1956, 43). The chimpanzees Probed briefly with the twigs, then withdrew them and licked the honey off the tool. Izawa and Itani (1966), who founded the long-term field study of wild chimpanzees in the Kasakati Basin, were the first to provide a scientific description of this behavior. They observed a female Inserting and Probing with a twig in the nest of a stingless bee of the genus *Trigona*. The twig was about 22 centimeters long, and the nest opening was about 10 meters above the ground in a large tree. The chimpanzee worked for at least fifteen minutes, Inserting and withdrawing the tool at least thirty times, and she could clearly be observed licking copious amounts of pure yellow honey from the twig after each withdrawal.

Honey dipping into the nests of arboreal African honeybees and the arboreal and subterranean nests of large stingless (*Melipona* spp.) and small stingless bees (*Trigona* spp.) has been recorded from Assirik (Bermejo, Illera, and Sabater-Pi 1989), Bossou (Ohashi 2006), the Budongo-Kasokwa group (Reynolds 2005), Bwindi (C. Stanford et al. 2000), the Gambia Complex (Takenoshita 2002), Gashaka (A. Fowler and Sommer 2007), the Goualougo Triangle (Sanz and Morgan 2007), the Kasakati Basin (Izawa and Itani 1966), Kibale (Wrangham, pers. comm., in McGrew 1992), Lopé (Tutin and Fernandez 1992; Tutin, Ham, and Wrogemann 1995; Tutin et al. 1991), the Lossi Forest (Bermejo and Illera 1999), Mahale K (Nishida and Hiraiwa 1982), Ndoki-Nouabale (Kuroda 1991), Ngogo (Watts 2008), Ngotto (T. Hicks, Fouts, and Fouts 2005), and Taï (Boesch and Boesch 1990). J. Pruetz (pers. comm.) reported that an adult female chimpanzee at Fongoli dipped a stick into a beehive but abandoned the stick without eating any honey. Brewer and McGrew (1990) observed a female rehabilitant in Gambia honey dipping.

Chimpanzees in the Bwindi-Impenetrable National Park, Uganda, used tools of different sizes to extract honey, depending on the type of bee (C. Stanford et al. 2000). Smaller sticks (with an average length of 27 centimeters and an average diameter of 0.5 centimeter) were used for foraging in tree or ground nests of stingless bees. In comparison, larger tools (with an average length of 60 centimeters and an average diameter of 1.6 centimeters) were used in honey extraction from the nests of the aggressive African honeybees. The authors argued that these findings suggested task specificity, with the Bwindi chimpanzees selecting longer tools for the more aggressive species.

Chimpanzees also use tools to extract water from cavities. Young chimpanzees at Gombe sometimes Inserted blades of grass, twigs, and other elongate objects into cavities containing water and then sucked or licked the moisture from the tip (Kummer and Goodall 1985; McGrew 1977; van Lawick–Goodall 1968). Albrecht and Dunnett (1971) saw wild youngsters Insert twigs into knotholes and suck the ends, but they felt that the behavior was playful. To obtain water from a tree hollow, Bossou chimpanzees Inserted a hard leaf into the hole, without crumpling, and then extracted it to suck and lick the water from the leaf (Matsuzawa 1991; Ohashi 2006; Sugiyama 1995a; Sugiyama and Koman 1979; Tonooka 2001). Captives observed by Hobhouse (1926), Köhler (1925), Kollar (1972), Schiller (1957), and van Lawick–Goodall (1970) sucked the wet ends of sticks, twigs, and straw after Inserting them in water and other fluids.

Just as chimpanzees use Reaching tools as olfactory aids, they also Insert and Probe with tools to get olfactory information. Chimpanzees approaching a termite mound often Inserted a blade of grass, withdrew it, and smelled the end (J. Pruetz, pers. comm.; van Lawick–Goodall 1968). Van Lawick–Goodall (1968) inferred that the ape acquired information about the presence or absence of termites. The chimpanzees were also seen to Insert sticks and twigs into small holes in dead branches and then smell the withdrawn probe. Subsequently, they sometimes broke the wood apart and ate insect larvae they found within (van Lawick–Goodall 1968, 1970, 1971, 1973). On three separate occasions, a juvenile female used a grass stalk as an olfactory probe by Inserting it into Goodall's pocket where a banana was hidden (van Lawick–Goodall, 1968, 1970, 1971, 1973). Kuroda, Suzuki, and Nishihara (1996) also documented this type of investigatory Probe in Tschego chimpanzees at Ndoki-Nouabale, where a female poked a branch into a cavity but withdrew it with nothing gained. Investigatory Probing with tools occurred at Bossou, Gombe, the Goualougo Triangle, Kibale, Mahale M and K, and Taï (Matsuzawa 1999; Sanz and Morgan 2007; Whiten et al. 1999).

Taï chimpanzees used leaf stems or twigs to extract marrow from the bones of colobus monkeys. They also used tools to Probe the eye orbits, as well as the brain case through the foramen magnum (Boesch and Boesch 1989, 1990). This behavior was also documented in the Goualougo Triangle, where one individual was observed using a twig to extract marrow from a duiker bone (Sanz and Morgan 2007).

At Mahale (M group), an adult male with a flu-like illness was observed on four occasions using nasal probes of dry twigs, grass stems, or leaf midribs (Nishida and Nakamura 1993). The tool (10 to 18 centimeters long) was Inserted into a nostril, often eliciting a sneeze. The authors felt this behavior was performed specifically to encourage the sneeze, in an attempt to clear mucus from his nasal passage. They noted the significance of this, stating, "a chimpanzee can manipulate even its own involuntary body response to relieve an unpleasant body condition with the aid of tools" (Nishida and Nakamura 1993, 220). Marchant and McGrew (1999) also documented a Mahale individual Probing into her nostril with a twig. She extracted mucous with the tool and induced a sneeze.

Brewer (1978) described a rehabilitant male using a twig as a toothpick. McGrew and Tutin (1972, 1973) observed the use of sticks and twigs by a captive female chimpanzee during long sessions of dental allogrooming. The tool user was one of a group of adolescent chimpanzees living in a large outdoor enclosure at the Delta Regional Primate Research Center. She applied the tools in all cases to a young male shedding his deciduous molars. The sticks were used to scrape at or probe the male's dentition and, in one case, may have been used to remove a loose molar. Later, several colony members began to use twigs and sticks to groom their own teeth, and several cases of tool-aided removal were suspected. One animal also inserted a strip of cloth around a loose tooth and used it like dental floss to remove the tooth. N. King, Stevens, and Mellen (1980) and M. Vancatova (pers. comm.) reported similar observations of dental auto- and allogrooming.

Captive chimpanzees used a plastic string, chopsticks, rubber tubing, branches, vines, leaves, grass, sticks, straw, stems, or a metal rod to acquire jelly, honey, juice, or foods such as raisins from containers (Cantalupo et al. 2008; Celli, Hirata, and Tomonaga 1999, 2004; Celli et al. 2001; Gaspar and Reis 1993; M. Harvey 1998; Hirata 2006; Hirata and Celli 2003; Hirata and Morimura 2000; Hopkins and Rabinowitz 1997; Hopkins, Russell, and Cantalupo 2007; Ladygina-Kohts 1959; Morimura 2003; Morimura et al. 2002; Ochiai and Matsuzawa 1998; Perret, Buechner, and Adler 1998; Tonooka 1994). Pfeiffer and Koebner (1978) mentioned that members of a captive group used tools to acquire ants, but they provided no further details. Chimpanzees at the San Francisco Zoo accessed sunflower seeds and chopped fruits and vegetables by Inserting a length of hose into a perforated cylinder (Murphy 1976), and captives at the Copenhagen Zoo got porridge by inserting twigs into small holes in boxes (Poulsen 1974–1975). Several captive facilities have installed artificial termite mounds or puzzle feeders for enrichment purposes to encourage species-typical use of sticks, hay, grass stems, or twigs (Besch 1981; Bettinger and Carter 1991; Bodamer 1990; Brandibas, Chalmeau, and Gallo 1995; Fletcher and Weghorst 2005; Lonsdorf et al. 2009; Maki et al. 1989; Nash 1981, 1982; Paquette 1992, 1994; Terdal 2005; Vancatova 2008; Wehnelt, Bird, and Lenihan 2006). Such mounds elicit Probing behavior, though the food reward is typically a semiliquid or sticky substance such as mustard, honey, baby food, or yogurt. Perhaps "artificial beehive" would better characterize the structure. Members of Köhler's colony on Tenerife had no such opportunities, but they did Probe with elongated objects in the cracks of a wooden waste tank. Köhler (1925, 68) described how "it became a perfect mania with the apes to squat beside the tank armed with straws and sticks which they dipped in the foul liquid and then licked." Ladygina-Kohts (2002, 180) described the chimpanzee Joni as using sticks or straws to "scare cockroaches from cracks in his cage." He also used these tools to extract liquid foods from small vials. Nissen (1956) reported a captive using a stick to clean her ears.

Captives of all ages Probed with sticks, twigs, grass stems, straw, and nails in holes and crevices, playfully or in exploration (Bingham 1929; Köhler 1925; Kollar 1972; Menzel, Davenport, and Robers 1970; Schiller 1957; van Hooff 1973). Captives also picked locks by inserting sticks, twigs, and nails into them (Bingham 1929; Fletcher and Weghorst 2005; Jennison 1927; McGrew, Tutin, and Midgett 1975; Menzel, Davenport, and Rogers 1970). A chimpanzee studied by Furness (1916) used keys to open locks and padlocks in her living quarters. She was reported to know the correct key for each lock, and she could correctly select it from an array with as many as twelve other keys. The home-reared chimpanzee Lucy also excelled at using (and hiding) keys to open doors (Temerlin 1975). Döhl (1968) found that a captive chimpanzee could solve a task requiring the sequential use of keys to open locked cases. Schastnyi (1963, 1972) conducted a similar experiment with multiple locked, baited boxes, finding the chimpanzees able to use the appropriate keys to open the boxes.

A standard laboratory analogue of the Insert and Probe mode is what Yerkes called the "box-and-stick" problem (Yerkes 1943; Yerkes and Learned 1925; Yerkes and Yerkes 1929), now, as noted above, known as the tube task. A choice bit of food is placed at the midpoint of a tube or tunnel, out of the subject's reach. A stick or pole is provided for the ape to push the incentive through the tube in order to obtain it. Khroustov (1964) found that at least one chimpanzee, whose age was not specified, could solve such a problem. Haggerty's (1910) subject failed, but the author noted neither the length of exposure to the problem nor the age of the animal. A young male studied by Yerkes and Learned (1925) failed to get the food from the tunnel during four sessions of fifteen to thirty minutes each. A 6-year-old female solved the problem in her

A captive chimpanzee Inserts and Probes with a stick in a simulated termite mound. Sequential photograph courtesy of Noel Rowe (www.alltheworldsprimates.org).

thirteenth fifteen-minute trial (Yerkes 1943). Three of six chimpanzees between 4 and 5 years of age tested by Birch (1945) solved the problem in a one-hour session. An adult studied by Hobhouse (1926) did likewise, with little delay. Ladygina-Kohts (1959) and Ladygina-Kohts and Dembovskii (1969) found that their chimpanzee was able to select appropriate tools or modify tools, such as a leafy branch, bent wires, and short sticks, to solve the task. The subject's age and length of exposure to the tube task are important variables, and a lack of specification or failure to control the age and length of exposure make a comparison among these reports difficult. In a study of imitation, K. Hayes and Hayes (1952) presented this task to Viki when she was between 28 and 36 months old, using a wire-mesh rectangular box attached to a wooden base as the container, calling it the "stick-and-tunnel" problem. She solved the task and retrieved the reward after two demonstrations of the solution. K. Hayes and Nissen (1971) presented Viki with a baited box that required Insertion of a crank to open the box to reveal the reward. After Viki attempted to Pry the box open with the crank, the experimenter brought her attention to the hole through which the crank should be inserted. Viki proceeded to Insert the tool and crank the box open.

As already described in other chapters and other sections of this chapter, variations of the box-and-stick problem and the trap-tube task have been presented to many different species to investigate causal understanding in relation to tools. There have been many such studies with chimpanzees (Bard, Fragaszy, and Visalberghi 1995; Horner and Whiten 2007; Limongelli, Boysen, and Visalberghi 1995; Martin-Ordas, Call, and Colmenares 2008; Mulcahy and Call 2006b; Povinelli et al. 2000; Seed et al. 2009; Visalberghi, Fragaszy, and Savage-Rumbaugh 1995). Insertion and Probing tasks have also been used to study future planning (Osvath and Osvath 2008), tool modification (Bania et al. 2009; Povinelli et al. 2000; Visalberghi, Fragaszy, and Savage-Rumbaugh 1995), tool-material selection and tool acquisition (Tonooka, Tomonaga, and Matsuzawa 1997); cultural transmission (Hopper et al. 2007; Whiten, Horner, and de Waal 2005; Whiten et al. 2007), cumulative culture (Marshall-Pescini and Whiten 2008), social learning (Horner and Whiten 2005; Pescetta et al. 2008), cooperation (Savage-Rumbaugh, Rumbaugh, and Boysen 1978), and language use and acquisition (Greenfield and Savage-Rumbaugh 1984; Savage-Rumbaugh 1986; Savage-Rumbaugh, Rumbaugh, and Boysen 1978).

Scratch, Rub

Wild chimpanzees at Gombe and the Goualougo Triangle used objects such as stones, twigs, sticks, fruit, or seedpods to tickle themselves, a behavior called "self-tickle" (Goodall 1986; Sanz and Morgan 2007). J. Goodall (1986) reported examples of a chimpanzee Scratching herself with a stick. A Gombe chimpanzee was also observed bending a vine to Rub her back and shoulder.

Captive chimpanzees Scratched or groomed themselves with sticks, straw, or wood splints (Gaspar and Reis 1993; N. King, Stevens, and Mellen 1980; Köhler 1925; Kummer 1971). Gaspar and Reis (1993) observed that whenever fresh wood-chip bedding was introduced, the chimpanzees would search for Scratching tools of hard wood splinters. One individual was observed picking up several splinters and sharing them with a conspecific. The home-reared chimpanzee Lucy used a pencil to Rub her clitoris (Temerlin 1975).

Cut

Cutting behavior has been elicited only in captive settings with enculturated chimpanzees. K. Hayes and Nissen (1971) documented Viki using scissors to Cut a string, enabling her to open a box. Viki also used scissors to Cut her own hair (K. Hayes and Hayes 1952). Hayes and Hayes also described a social learning experiment, the string-and-candle problem, involving a box that contained a reward and was tied shut with a string. Viki was given a lighted candle that could be used to burn through the string. After four demonstrations, Viki imitated this behavior to obtain the reward. Kitahara-Frisch, Norikoshi, and Hara (1987) found captive chimpanzees able to use bone fragments to Cut through animal hide covering the openings of milk bottles.

Block

Chimpanzee sometimes left Penetrating sticks (see Jab, Stab, Penetrate) protruding from termite mounds after use (Fay and Carroll 1994; S. Suzuki, Kuroda, and Nishihara 1995). As described above, chimpanzees first used the Penetrating stick to gain access to the termite mound, and then Inserted a fishing tool to extract the termites (see also Tool Set). Suzuki, Kuroda, and Nishihara (1995, 227) speculated that the "protruding [penetrating] sticks may function as a stopper for these holes to reuse them for fishing later, though there was no evidence to show the chimpanzees left the protruding stick as a stopper on purpose."

Prop and Climb, Balance and Climb, Bridge, Reposition

Köhler (1925) reported that captive chimpanzees Propped and Climbed branches, poles, ladders, iron bars, and planks to secure suspended food lures. McGrew, Tutin, and Midgett (1975) and Menzel (1972, 1973a) provided an entertaining account of the spontaneous development of Propping and Climbing from the former Delta Primate Center chimpanzee colony. Using branches that had fallen from trees into their enclosure, the apes first Propped them against the wall and then Climbed into a glass-fronted observation booth built above. Once inside the booth, the chimpanzees wreaked havoc, effectively preventing its use for observation. Since clearing the corral of trees was impractical, the booth was boarded up and abandoned. The chimpanzees then turned their attention to the trees themselves. The trees had been wrapped with electrified wire to a height of about 5 meters. Planks on sunken wood posts interconnected the trunks of the trees at a height of about 2.5 meters. These features were intended to provide an elevated structure for climbing and resting, while preserving the trees for shade by keeping the animals from the crowns. Once the observation booth became unavailable, the chimpanzees began to take dead branches to the elevated walkway, Prop them on the planks with the tips leaning against a tree trunk above the shock wires, and Climb them. This maneuver provided unimpeded access to the tree crown. The trees were soon denuded and killed. One fell and damaged the walkway, and a ladder user was injured in a fall from a dead branch. It was decided that clearing the corral of trees was not only practical, but also necessary. Some long-dead branches were left in the corral, however, and the chimpanzees soon began to Prop these against the corral wall and escape. These developments all occurred within one year. To prevent further escapes, the corral was cleared of all branches suitable as tools. However, the stumps of some of the cut trees and some of the planks and wooden posts of the elevated walkway remained. The stumps and posts were all about 4 meters high and were firmly embedded in the ground. The chimpanzees began to break off or uproot them to continue their escape attempts. As the stumps and poles were confiscated, the chimpanzees began to use short sticks as pitons to scale the fence. The walls of the enclosure were 5.5 meters high. The lower half was made of chain link fencing and the upper half of galvanized sheet metal. The chimpanzees learned to insert branches into the mesh or into crevices between the two halves of the fence. Then they stood on these pitons while gaining fingerholds in small seams between the galvanized sheets and climbed or jumped to the top of the fence, where escape was simple. The authors noted that the chimpanzees were selective in choosing their pitons. Dead sticks might break, sticks longer than 100 centimeters were difficult to carry and insert, and sticks shorter than 30 centimeters did not provide adequate standing space. Thick sticks could not be inserted, and thin ones wobbled or fell out. The authors noted that the apes rarely chose inappropriate sticks. This prolonged and remarkable matching of wits and will power was terminated by the translocation of the colony to an island.

McGrew, Tutin, and Midgett (1975) and Menzel (1972, 1973a) described additional aspects of this behavior that demonstrated the sophistication involved in the Propping and Climbing phase. The chimpanzees typically broke off or selected only the poles or branches from their enclosure that were of sufficient dimensions to be effective. To be useful as ladders, the sticks had to be at least 3.5 meters long. Further, they had to be at least 5 centimeters in diameter to support the climber's weight. Although such tools might weigh 15 kilograms, they were carried or dragged 20 meters or more from areas not in the same visual field as the point of use. They were generally positioned with the thick end as the base, at an angle that provided maximum length and stability. Since most of the branches were curved, the concave side had to face outward to provide a secure ladder. Most branches were erected accordingly. When using branches to climb into the trees, the chimpanzees sometimes Propped the tool against the walkway from the ground, Climbed to the walkway, hauled the branch up to the walkway, and then Propped it against the tree trunk from the walkway. Suitable positioning of the base of the branch on the walkway was a delicate matter that required considerable skill and practice. Sometimes one animal steadied the branch while others Climbed. Such cooperation was gesturally solicited on some occasions. Menzel (1972) suspended toys and food on the corral wall of the enclosure or in the trees. The chimpanzees readily Propped and Climbed branches to secure these lures. Takeshita and van Hooff (1996, 2001) also described chimpanzees Propping and Climbing poles or sticks to reach inaccessible locations in a captive enclosure. Similarly, de Waal (2007) described and provided illustrations of captive chimpanzees Detaching long branches from trees and Propping them against other trees, thus bypassing electric fencing at the base of the tree in order to reach the fresh leaves in the canopy. In addition to tool manufacture and use, cooperation was demonstrated when one chimpanzee held the Propped stick in place while a conspecific Climbed. Brewer (1982) reported the behavior of William, a rehabilitant, who dragged a dead tree trunk next to a baobab (*Adansonia digitata*) tree and attempted to situate it as a ladder.

Captive chimpanzees also Balanced and Climbed sticks or poles to secure suspended food (K. Hayes and Nissen 1971; Köhler 1925; Yerkes and Yerkes 1929) and in play (Menzel 1972, 1973a; Menzel, Davenport, and Rogers 1970). The upper end of a stick that is Balanced and Climbed does not touch a vertical surface, as does the end of a stick that is Propped and Climbed (see table 1.1). But sometimes, as a chimpanzee Climbed a Balanced tool, it held on to a nearby vertical surface with an extended hand or foot. Poles as much as 3 meters long were Balanced and Climbed skillfully, and even 4.5-meter-long poles were used by larger chimpanzees (Menzel 1972, 1973a).

J. Goodall (1986) described examples of Bridging at Gombe. While moving through the canopy, infants were sometime unable to follow their mother across large gaps between the trees. Goodall observed mothers returning to the infant, grasping and pulling a branch in the infant's tree, and then holding it in place until the infant had crossed to her.

Gua, a home-reared infant chimpanzee studied by Kellogg and Kellogg (1933) was presented with a cookie hung from the ceiling. Gua was more skilled than Donald, the Kellogg's own human infant, at Repositioning a chair to retrieve the cookie. In addition, the Kelloggs documented that by 10 months of age, Gua could spontaneously push a child's walker under a doorknob, bringing it within reach for grasping and swinging. She also dragged cardboard boxes from adjacent rooms to the base of a chair or table, allowing her to climb onto the furniture. Viki, another home-reared chimpanzee, also moved a chair to reach a suspended lure (C. Hayes 1951; K. Hayes and Hayes 1952).

Chimpanzees in captivity have been documented Repositioning and climbing another individual to reach elevated areas and incentives (Köhler 1925; Takeshita and van Hooff 1996, 2001). Köhler was the first to document one chimpanzee pulling another beneath a lure. Takeshita and van Hooff (1996, 2001) described how one chimpanzee maneuvered another under a window in a crouch, and then climbed onto its back to reach the window. They stated that "the supporting individuals adjusted to this role reluctantly in most cases" (Takeshita and van Hooff 2001, 524). Captives also Repositioned boxes, tires, or straw as steps to reach objects overhead (Takeshita and van Hooff 1996, 2001). McGrew (1992) documented a female using a tire as a support for her body, while another chimpanzee inspected her sexual swelling.

Denisov (in Windholz 1984), Pavlov (1949, 1955, 1957), Shtodin (in Razran 1961), Vatsuro (in Razran 1961), and Vatsuro and Shtodin (in Razran 1961) conducted several studies of problem solving with a young male chimpanzee named Raphael. In 1948, Vatsuro (in Razran 1961, 366) provided an English summary of "The Experiment on Water." In this experiment, a pole was provided to enable travel between two floating barges on a lake. Raphael positioned the pole between the barges, so that he could travel across the pole to the barges.

Captive chimpanzees Reposition metal drums and other objects and stack them to secure suspended food (Bingham 1929; Denisov 1958 in Windholz 1984; K. Hayes and Nissen 1971; Köhler 1925; Orbeli 1949 in Windholz 1984; Pavlov 1975 in Windholz 1984; Schiller 1957; Wazuro in Döhl 1966; Yerkes 1943; Yerkes and Learned 1925). One individual stacked two tires tread to tread to escape from his zoo enclosure (Lorenz in McGrew, Tutin, and Midgett 1975), and another stacked a stool on a table to procure a balloon (von Buttel-Reepen in Bierens de Haan 1931). Each of Bingham's (1929) four subjects was able to construct towers of three boxes, and two individuals stacked four boxes. In some cases, the chimpanzees Repositioned the longest dimension of the box vertically to maximize the height of the tower. Wazuro's (in Döhl 1966) subjects stacked four boxes, Schiller's (1957) stacked up to three, and the chimpanzees observed by Yerkes (1943) and by Yerkes and Learned (1925) stacked two. Some of Köhler's (1925) apes stacked as many as four boxes after considerable experience with stacking fewer. They brought boxes from a distance to a point beneath the lure, emptying boxes of sand and stones so that they could be transported. On one trial, three colony members participated in moving an especially heavy box. In some cases they stood on the boxes and used a stick to Reach or hit at the suspended fruit. Pavlov (in Windholz 1984) concluded that box stacking in chimpanzees was acquired by trial and error. In these cases, we consider each stacked object to be a tool, but the completed tower is not a tool, because it is not held or manipulated in its entirety. We call the finished tower a Construction (see chapter 7).

The ability of chimpanzees and other apes to Prop and Climb, Balance and Climb, and Reposition objects is of more than academic interest. These behaviors facilitate escape from open outdoor enclosures. Countless millions have been spent designing and building deeper enclosures with higher walls and wider moats to counter these forms of tool use. Additionally, caretakers have to be constantly vigilant to confiscate objects that might make good climbing tools. The apes appear to enjoy the competition, and in many cases they have been known to win.

Hang

Gaspar (1993) described captives suspending folded clothes from a bar in the ceiling of their enclosure to create a rope swing.

Contain

The use of a leaf "spoon" involves dipping an unmodified leaf into water, followed by sucking or licking the water clinging to the leaf's surface or pouring the clinging water into the mouth (van Lawick–Goodall 1968). Matsusaka and Kutsukake (2002) and Matsusaka et al. (2006) reported spooning by two juveniles, who picked up leaves from the ground, dipped them into a stream, and transferred the water to their mouths. The authors speculated that the purpose of this behavior might have been play, as the water could have been drunk without the leaves. Materials such as sticks, twigs, vines, grass stems, leaf midribs, and stones are also used to pick up water (Matsusaka et al. 2006).

Wild chimpanzees at Bossou used containers manufactured through leaf folding to hold and drink water. The technique of leaf folding is further described under Tool Manufacture (Biro, Sousa, and Matsuzawa 2006; Tonooka 1995, 2001; Tonooka and Inoue-Nakamura 1993; Tonooka, Inoue-Nakamura, and Matsuzawa 1994).

Chimpanzees in the wild use containers for holding or catching non-food objects. At Gombe, a male chimpanzee defecated into leaves while holding them (J. Goodall 1986). The chimpanzee then used his lips to remove pieces of undigested meat from the feces prior to discarding the leaves and feces. Halperin (in McGrew 1979) reported an identical observation.

Ectoparasites were regularly placed on leaves for visual inspection, termed "leaf-inspect," before being discarded or consumed at Budongo (Assersohn et al. 2004; Whiten et al. 1999). Brewer (1978) described the use of cups and spoons for feeding and drinking in rehabilitant chimpanzees. Chimpanzees in captivity used environmental objects—such as orange peels, toys (including a deflated ball), cups, spoons, cans, lids, bowls, unchewed leaves, eggshells, and empty coconut shells—to Contain or scoop liquids and solids for consumption (Bettinger and Carter 1991; Fletcher and Weghorst 2005; R. Fouts 1997; Furness 1916; Gaspar 1993; C. Hayes 1951; Ladygina-Kohts 2002; McGrew and Tutin, unpub. data, in McGrew 1992; Takeshita and van Hooff 1996, 2001; Temerlin 1975; Tonooka,

Tomonaga, and Matsuzawa 1997). Van Hooff (1973) reported that captive chimpanzees used suitable objects to Contain water scooped from a moat. The water was sometimes drunk, but at other times the behavior appeared to be playful. These chimpanzees were also observed to carry water in containers to other locations in their enclosure (Takeshita and van Hooff 1996, 2001). In each case the behavior appeared to be play. N. King, Stevens, and Mellen (1980) observed a captive making a plate for carrying food by shuffling dry leaves together.

Absorb

Leaf sponging is documented at most wild chimpanzee research sites (Whiten et al. 1999, 2001). The behavior consists of placing chewed or crumpled vegetation in water, then squeezing or sucking the water into the mouth (J. Goodall 1964). Documented observations include reports from Bossou (Biro, Sousa, and Matsuzawa 2006; Sousa and Matsuzawa 2004; Sugiyama 1989a, 1995a; Tonooka 2001; Tonooka, Tomonaga, and Matsuzawa 1997), Gombe (J. Goodall 1964, 1986; McGrew 1977; van Lawick–Goodall 1965, 1968, 1970, 1971, 1973; Wrangham 1975), the Goualougo Triangle (Sanz and Morgan 2007), Kasokwa (Reynolds 2005), Lopé (Tutin, Ham, and Wrogemann 1995), Mahale (Matsusaka and Kutsukake 2002; Matsusaka et al. 2006; Takasaki 1992; Takasaki and Tsukahara, unpub. data, in Nishida 1990), Ngogo (Watts 2008), Semliki (K. Hunt and McGrew 2002; Marchant, McGrew, and Hunt 2007), Sonso (Quiatt 2001, 2006), Taï (Boesch and Boesch 1990), and Tongo (Lanjouw 2002). The use of sponge drinking tools is not common at Mahale, but it has been documented (Matsusaka and Kutsukake 2002; Takasaki and Tsukahara, unpub. data, in Nishida 1990). Rehabilitant chimpanzees reintroduced to Gambia and later to Senegal also used leaf sponges to obtain water (Brewer 1978).

Chimpanzees used leaves to Absorb water from holes or hollows in trees, doing so even when other sources of water were readily available. J. Goodall (1964) compared the amount of water she could secure by dipping her fingers with that attained by the use of a leaf sponge, concluding that the sponging yielded eight times as much water.

Chimpanzees sometimes even used tools to sponge readily available water from springs or rivers (K. Hunt and McGrew 2002; Matsusaka et al. 2006; Quiatt 2006). For example, Gombe chimpanzees were documented using a sponge to drink from a stream (J. Goodall 1986; Wrangham 1975).

When chimpanzees chew a leaf or stem before inserting it in a hole to extract water, it is known as a form of "fluid-dip" (Whiten et al. 1999). Matsusaka and Kutsukake (2002) reported several observations of juveniles using a leaf sponge to drink water. On

A chimpanzee uses a deflated ball to scoop water from a moat. Photograph courtesy of Hideko Takeshita (in Burger's Zoo, Arnhem, the Netherlands).

these occasions, individuals pushed a leaf (or leaves) into a tree hollow, using the same tools repeatedly in a "soak and suck" action to drink water from the hollow (Matsusaka and Kutsukake 2002). Unlike chimpanzees at Bossou and Gombe, these juveniles did not first put the leaf into the mouth to crumple it, but used an uncrumpled leaf.

There does not seem to be a vegetation preference for the leaf material at some sites, such as Gombe (McGrew 1977), but a preference for a particular species of vegetation has been found at others. For example, Sonso chimpanzees of the Budongo Forest who prefer to drink water with sponges choose the hairy leaves of the chenille plant (*Acalypha* spp.) for their sponging (Reynolds 2005).

Though most wild chimpanzees use leaves for sponging to dip for fluids, they also employ stems, moss, fruit, and wadges (masticated wads of plant material). Not all chimpanzee communities make their sponges from leaves. The Tongo population regularly used moss sponges, preparing the tools from the bark of a tree (Lanjouw 2002).

Chimpanzees at Sonso and one individual at Mahale have been observed dipping a portion of fruit into water, presumably for its absorptive properties (Matsusaka et al. 2006; Quiatt 2006). A similar behavior, in the form of wadge dipping, was recorded at Taï (Boesch 1991b). The use of stem sponges was documented at Kibale, where the stem was chewed before use (Wrangham, de Waal, and McGrew 1994).

Leaf sponges are also used to extract liquids other than water. Palm wine is collected by humans by placing bottles in the crown of the *Raphia* tree. Ohashi (2006) documented several chimpanzees at Bossou finding these bottles and using crumpled leaves to get the wine from the bottles.

After pestle-pounding in an oil palm tree, chimpanzees at Bossou extracted juice and fibrous vegetation from the interior of the crown of the tree (Sugiyama 1994; Yamakoshi and Sugiyama 1995). After chewing the fibrous vegetation, the chimpanzees dipped the mass into the hole to sponge up additional juice. This behavior was referred to as "fibre-sponge sucking" (Sugiyama 1994). Leaves were used once as sponges at Gombe to Absorb residual fluids and brain tissue from the skull of a monkey (Teleki 1973b, 1973c). The behavior occurred because the skull was insufficiently broken to allow the material to be licked. The leaves were undoubtedly used to Reach and Wipe the residual material as well, but Teleki compared the behavior with leaf sponging for water. Wrangham (1977) reported a single case in which leaves were used similarly to sponge out the inside of a *Strychnos* fruit.

Captive chimpanzees used straw, rags, bread, paper, cardboard, and toys to Absorb water that was then sucked (Cowper 1971; Fletcher and Weghorst 2005; Gaspar 1993; Hobhouse 1926; A. Jones 1986; Köhler 1925; Takeshita and van Hooff 1996, 2001). In an experimental version of leaf sponging, one of two captive chimpanzees learned, without behavioral shaping, to Insert a length of rope into a narrow tube containing banana mash (C. Parker 1968, 1969b). When the rope had Absorbed the mash, the ape withdrew the rope and sucked on it. A second subject removed and sucked the rope after it was inserted by the experimenter but did not reinsert it. Captive chimpanzees also sponged with sticks and leaves modified through chewing to Absorb water or juice from out-of-reach containers in experimental situations (Bettinger and Carter 1991; Norikoshi and Kitahara 1979). Captive chimpanzees also used materials such as leaves, paper, and cardboard as sponges (M. Vancatova, pers. comm.)

Wipe

Chimpanzees use objects such as leaves, termed "leaf-napkins," to Wipe or mop material—such as feces, urine, ejaculate, blood from wounds or menstruation, sticky food residues and juices, water, and mud—from their bodies, from other chimpanzee's bodies, or from other surfaces. This may have been done either for removal or for consumption (P. Baldwin 1979; Boesch 1995; J. Goodall 1964, 1986; Kuroda 1991; Nishida 1973, 1977; Nishida, Matsusaka, and McGrew 2009; Nishida et al. 1999; Sanz and Morgan 2007; van Lawick–Goodall 1965, 1968, 1970, 1971, 1973). J. Goodall (1986, 545) described chimpanzees as "quite fastidious" in this regard, particularly concerning feces. The behavior varied from careful dabbing with the leaf to more vigorous Wiping (van Lawick–Goodall 1968; McGrew et al. 2003). J. Goodall (1986) documented nineteen cases of one chimpanzee Wiping another. All of these cases involved family members, such as mothers or offspring Wiping feces, mud, blood, mucus, or urine from each other. Two cases of Wiping to remove a "stranger's touch" were recorded from Gombe (J. Goodall 1986, 546). The first of these involved a new, strange female touching the arm of a resident male, who promptly moved and used leaves to Wipe the place where she touched his arm. The second example occurred after an infant stamped on a human visitor's head and then used leaves to Wipe the bottom of her foot. Nishida et al. (1999) noted that Wiping behavior occurred rarely at Mahale, citing only a handful of "rump wiping with a leaf" observations. Nishida (2003b) and Corp et al. (2009) described several observations of chimpanzees Wiping their faces with attached leaves while feeding in a lemon tree.

Penis Wiping is a frequent behavior at several sites (J. Goodall 1986). This behavior was documented in the Sonso community of Budongo (Quiatt 2006), Gombe (J. Goodall 1986), the Goualougo Triangle (Sanz and Morgan 2007), Kibale (R. Wrangham, pers. comm., in O'Hara and Lee 2006), Ngogo (Watts 2008), and Taï (Boesch and Boesch-Achermann 2000). O'Hara and Lee (2006) documented the use of leaf napkins for postcoital penis cleaning by males in the Sonso community at Budongo. After Detaching the leaf or leaves from a nearby source, they Wiped or scraped fluids from their own penises with the leaf blades or petioles. O'Hara and Lee reported that this behavior occurred three to four times more often at this site than at either Gombe or Kibale. Ngogo chimpanzees of the Kibale National Park reportedly used tools more often for hygienic purposes than for extractive foraging (Watts 2008). J. Goodall (1986) noted that female posterior wiping after copulation was only rarely observed. She explained the lack of such Wiping by females as related their inability to see their own posteriors, while the males' fastidiousness was related to both their ability to see their penises and to contamination from the female's feces during copulation. Goodall described two examples of males Wiping a female's posterior with leaves after she presented; in one case they subsequently copulated.

Captive chimpanzees used twigs, straw, rags, and paper to Wipe blood, feces, pus, water, and oil from their bodies (Fletcher and Weghorst 2005; N. King, Stevens, and Mellen 1980; Köhler 1925). The Tenerife females Wiped menstrual blood as well as blood from wounds. J. Goodall (1986) noted a contradiction observed in Köhler's chimpanzees: they were willing to pick up feces with their hands, but they Wiped fastidiously with rags, straw, or paper if they happened to step in feces. Bernstein (1962) mentioned that captives Wiped their faces and arms with newspapers. Kortlandt and Kooij (1963) noted but did not specifically describe chimpanzees' use of sticks or fruit as "toilet aids." Viki used a cloth to Wipe her face, as instructed by her caretakers, looking into a mirror to guide her efforts (C. Hayes 1951).

Chimpanzees may inspect a wound by touching it with leaves or holding leaves on a wound, termed "leaf dab" (Boesch 1995; J. Goodall 1986; Sanz and Morgan 2007; Watts 2008; Wrangham, unpub. data, in Whiten et al. 1999). The leaves are generally examined by the chimpanzee before they are discarded. T. Savage and Wyman (1843–1844) reported that wild chimpanzees pressed grass and leaves to bleeding wounds; Merfield (1956) relayed a native guide's account of identical behavior. We include this behavior under Wipe, as there are no modern accounts offering specific descriptions, and the chimpanzees did Wipe off blood. If the purpose is to cover the wound or stop bleeding, the more appropriate mode would be Affix, Apply, Drape.

A female at Gombe used leaves to brush bees from a hive entrance and later to brush driver ants from a sapling where she dipped for the ants (J. Goodall 1986). Whiten et al. (1999) referred to this as "leaf-brush." Nishida (1973) once saw a Mahale chimpanzee Wipe ants from a tree trunk with a clump of leafy boughs. The behavior allowed efficient collection of large numbers of the ants. The use of leaves to Wipe out the inside of skulls and fruits for food remnants was observed at Gombe and at Taï (Boesch and Boesch 1990; Teleki 1973b, 1973c; Wrangham 1977).

Bernstein (1962), Grzimek (1941), N. King, Stevens, and Mellen (1980), and Kollar (1972) saw captives Wiping their cages with paper, burlap, straw, and leaves, possibly in imitation of caretakers. Home-reared chimpanzees and those engaged in language studies also Wiped their surroundings, for example, using a mop to Wipe the floor, a cloth to dust the furniture, and a washcloth to wash and Wipe dishes (R. Fouts and Fouts 1989; C. Hayes 1951; Temerlin 1975; Terrace 1979).

Affix, Apply, Drape

Izawa and Itani (1966) saw a wild chimpanzee cover itself with a leafy branch as it settled into its nest on a rainy evening. Nishida (1980b) reported a similar observation of an older female Detaching a leafy fig branch for use as an umbrella in a rainstorm. At the Goualougo Triangle, Sanz and Morgan (2007) described the use of a rain cover by chimpanzees, where they placed leafy twigs on their backs and shoulders during heavy rains.

Albrecht and Dunnett (1971) saw a wild youngster Drape a piece of liana on its head in play. H. R. Bauer

(1977) reported that the Gombe chimpanzees Draped animal skins and purloined camp objects on themselves. Reynolds and Reynolds (1965) saw a juvenile at Budongo playfully Drape lichen on his head. Many individuals at Budongo Draped twigs over their necks, including fig-containing twigs ("pearl necklaces"). Budongo chimpanzees also tucked branches or leaves into their body folds, termed "groin-tuck" or "neck-tuck" (Reynolds 2005). McGrew and Marchant (1998) reported a case of personal adornment by a young Mahale chimpanzee, who Draped a colobus monkey skin over its neck. When the skin was later found by the researchers, it was tied in a "necklace" with a single knot. Although the tying of the knot was not observed and may have been accidental, the authors stated that it still served to keep the skin attached. Nishida et al. (1999) noted that a wild juvenile female once Draped cardboard and a cooking pan over her head and shoulder. J. Pruetz (pers. comm.) noted that Fongoli juveniles were observed Draping baboon and bushbuck skin on themselves as they played; adults did so when they traveled. In 1998, an adult female chimpanzee on Rubondo Island in Tanzania was observed stealing a blanket from a human campsite and wrapping it around her body (Pusey in Matsumoto-Oda 2000). This female may have been among the group of former zoo chimpanzees who were reintroduced to the island in the late 1960s.

Captives Draped branches, plants, leaves, hay and straw, fruit skins, stones, blankets, cloth, paper and cardboard, string, rope, hose, chains, and other objects on their heads, necks, and backs (Bernstein 1962; Cowper 1971; Gaspar 1993; K. Hayes and Hayes 1952; Kellogg and Kellogg 1933; Köhler 1925; Ladygina-Kohts 2002; Menzel, Davenport, and Rogers 1970; Merfield 1956; Takeshita and van Hooff 1996, 2001; van Hooff 1973; Yerkes 1943; Yerkes and Learned 1925). Thermoregulation, protection from rain, and enhancement of display or sexual attractiveness were mentioned as possible functions of the behavior. In many cases the behavior seemed to be playful. Arnhem Zoo chimpanzees engaged in two playful games involving Draping (Takeshita and van Hooff 1996, 2001). Young chimpanzees at the zoo also played a peek-a-boo game in which one chimpanzee covered its own face with a towel and then solicited play by holding its hands out towards a conspecific. Kellogg and Kellogg (1933) documented Gua spontaneously playing peek-a-boo by pulling bed covers over his face and back down again repeatedly. Both wild and captive chimpanzees also played blindman's bluff, Draping a towel over their faces and then groping around on the ground or in a tree (R. Fouts and Fouts 1989; Rensch 1973; van Lawick-Goodall 1971). Ladygina-Kohts (2002) documented a captive covering a bug with a cloth and then smashing the bug with his hand through the cloth.

Wild chimpanzees make and use "seat sticks" (Alp 1997), "leaf cushions" (Hirata, Myowa, and Matsuzawa 1998), "seat vegetation" (Whiten et al. 1999), and "stepping sticks" (Alp 1997) to avoid contact with uncomfortable surfaces. In using seat vegetation or a leaf cushion, the apes placed leaves on wet ground to create dry seats. This was first recorded at Bossou (Hirata, Myowa, and Matsuzawa 1998), and has now been documented at Goualougo Triangle (Sanz and Morgan 2007) and at Kibale and Taï (Whiten et al. 2001). Chimpanzees also lined the inner surface of their nests or sleeping platform with Detached leafy twigs or leaves (P. Baldwin 1979; McGrew et al. 2003). This has been reported at both Assirik and Gombe. McGrew et al. (2003) included one observation in which the

A captive chimpanzee Drapes a cloth on its head while playing with conspecifics. Photograph courtesy of Hideko Takeshita (in Burger's Zoo, Arnhem, the Netherlands).

twigs were concentrated in one area to make a pillow. Sugiyama (1998) reported a similar behavior, in which leaves were placed in the nest to create a more comfortable spot in which to lie down.

Seat sticks are used for protection while foraging in thorny trees. The behavior consists of arranging or placing vegetation (such as branches) over thorny vegetation to create a comfortable sitting location in a thorn-covered tree, such as *Ceiba pentandra*. The use of seat sticks and stepping sticks was first observed in Tenkere, Sierra Leone (Alp 1997), and this behavior has now been documented at the Goualougo Triangle (Sanz and Morgan 2007). Stepping sticks are short and smooth branches that are grasped by one or both feet during locomotion. Alp (1997) noted that these tools are akin to human footwear and protect the soles of the feet from thorns.

Bossou chimpanzees used small stones as wedges to prop up uneven anvils, in what has been termed Metatool Use (Matsuzawa 1991, 1994; see chapters 1 and 7). By propping the anvil, the chimpanzees created a flat surface on which to place a nut and then Hammered open the nut with a stone (see Pound, Hammer). We classify the use of the wedge under the anvil as being in the mode Apply, and we discuss the use of Metatools further under Tool Composites. Carvalho et al. (2008) reported the most complex variation of this behavior, involving the use of two wedges as anvil props at Bossou.

At Gombe, ectoparasites were also destroyed through "squashing ectoparasites" (Boesch 1995), termed "leaf-squash" (Whiten et al. 1999). When a chimpanzee found an ectoparasite during allo- or autogrooming, the parasite was placed on the leaf and the chimpanzee tried to squash it with his or her thumbs. Boesch (1995) documented this as a new function of leaf grooming (see Bait, Entice). Ectoparasites were most often placed on leaves for inspection or squashing, but other materials, such as eye exudate and pus, were also placed on leaves for close inspection. Zamma (2002) described a single observation of using leaves for ectoparasite squashing at Mahale.

Captives have also been documented placing objects (such as boxes or tires) on the floor to sit on, particularly if the surface is cold or wet (McGrew 1992; Takeshita and van Hooff 1996, 2001).

Lucy used a vacuum cleaner to masturbate, turning it on and Applying the suction hose on her genitals (Temerlin 1975). Viki, after observing her caretakers using sandpaper to rub a wooden stake, copied the behavior, but she also used the paper to sand the indoor furniture (C. Hayes 1951; K. Hayes and Hayes 1952).

Brink's (1957) description of the smoking habits of two zoo chimpanzees, Bango and Tyrus, suggested a sophisticated understanding of how to Apply and use fire. While holding a nearly burned-out cigarette, Bango retrieved an unlighted cigarette and used the burning tip of the other to light it. The author noted that he passed the chimpanzee a new cigarette, which Bango lit in the same way. Brink described Tyrus as a self-taught smoking chimpanzee, noting that zoo officials were firm in denying that they taught Tyrus this behavior.

Symbolize

Chimpanzees may carry objects that, in our interpretation, provide psychological security. Wrangham and Peterson (1996) reported two separate instances where a young Kibale male, Kakama, carried a log as a doll, which the authors interpreted as possible anticipation of his mother's giving birth. Kakama carried the log with him, dragging it behind him or placing it on his neck or back to carry. He even made a nest in which he placed the log, sat by it, and eventually climbed in.

Hirata et al. (2001, 93) described several instances in which Bossou adolescents were "toying" with western tree hyraxes. They reported that a young female kept a dead hyrax for fifteen hours. She carried and groomed it, and slept with it in her nest. In an online article, Matsuzawa (2008) also recounted this example, calling it "doll play." In addition, Matsuzawa described a 2.5-year-old chimpanzee at Bossou carrying a large branch with her. He noted that "she held the branch doll at her side with her arm, as if it were an imaginary infant," and continued, "doll play behavior is a good practice for the future, for bringing up your own young."

Symbolic behavior is typical for enculturated or home-reared chimpanzees, many of whom carried toys or dolls in the same manner (R. Fouts 1997; C. Hayes 1951; Jensvold and Fouts 1993; Kellogg and Kellogg 1933; Temerlin 1975). Fifi, a captive chimpanzee at the Toledo Zoo, carried a specific rubber alligator with her for at least 7.5 years and became nervous if she was separated from it (S. Husband, pers. comm.).

Experimental studies have demonstrated that chimpanzees are able to use tokens Symbolically in exchange for a food reward. When the different tokens represent different reinforcers or different quanti-

Fifi, a chimpanzee who lived at the Toledo Zoo, with her rubber alligator "companion." Photograph courtesy of Linda S. Milks.

ties, the tokens are Symbolic tools (Brosnan and Beran 2009; Brosnan and de Waal 2005). We do not recognize tokens placed into a slot machine as Symbolic tools (Sousa and Matsuzawa 2001, 2006), as these do not conform to our definition of tool use (see chapter 1).

Sarah, a captive chimpanzee, was taught to use plastic "language chips" to communicate (Premack 1976). The chips could be arranged on a magnetic board and were used Symbolically to express words and short phrases. As each chip was physically manipulated into place on the magnetic board, we can consider the use of these objects as tools, used Symbolically to communicate.

TOOL MANUFACTURE: GIBBONS

Detach

Drop

White-handed gibbons Detached dead branches from trees before Dropping them (C. Carpenter 1940; Ellefson in L. Baldwin and Teleki 1976; Geissmann 2009).

TOOL MANUFACTURE: ORANGUTANS

Detach

Drop

Orangutans Detach branches or fruit before Dropping them on human intruders (Davenport 1967; Galdikas 1978b, 1982a; Galdikas-Brindamour 1975; Harrisson 1963; MacKinnon 1971, 1974a, 1974b; Peters 2001; Rijksen 1978; Schaller 1961; A. Schultz 1961; A. R. Wallace 1869) or in agitation at hearing other orangutans (Horr 1975). Davenport (1967), MacKinnon (1971), and Rijksen (1978) added that branches, bark, lianas, epiphytes, and even an arboreal termite nest were broken off and Dropped. From Schaller's (1961) account, it appears that the branches are primarily Slapped off with the hands or Kicked off with the feet, although they may be held momentarily between Detachment and Dropping. Given the tremendous strength of orangutans, it is unlikely that an adult would have to expend much effort in Detaching even a large dead branch. Rijksen (1978) noted that objects are sometimes manipulated before being Dropped. In an imitative approximation of its mother's intimidation display, an infant broke off a tiny twig before Dropping it on a human (Horr 1977).

Throw

Galdikas (1982a) observed wild orangutans Detach live and dead branches that were then Thrown at human observers. Rehabilitant orangutans Detached herbs and other vegetation which were used for aimed Throwing (Rijksen 1978). A captive broke off and Threw the crown of a small tree at a dog, and also Detached a flat stone embedded in the ground and threw it at a caretaker (Koehler 1993).

Drag, Roll, Kick, Slap, Push Over

In a dramatic example of Detachment, wild adult and subadult males, and at least one adult female, have engaged in snag crashing. During this behavior, large dead trees are Pushed Over as part of a display (Galdikas 1983).

Brandish, Wave, Shake

A wild juvenile broke off a leafy branch and used it to deter wasps that were flying around him (Galdikas 1978b, 1982a, 1989). Other individuals Detached branches and used them to repel flying insects or to fan themselves for cooling (Kaplan and Rogers 1994).

Wild orangutans have been reported to Detach a branch and carry it in front of themselves for concealment from predators or humans (van Schaik et al. 2003, 2009).

Club, Beat

Rijksen (1978) reported that a rehabilitant pulled a small sapling out of the ground before hitting a lizard with it. J. Ellis (1975, 1977) observed a captive male who ripped a strip of rubber from a tire and then used it to whip anything within range.

Pound, Hammer

Wild orangutans presumably Detached branches that they then used to Hammer open invertebrate nests in tree hollows or to open tough-skinned durian fruits (E. Fox, Sitompul, and van Schaik 1999; van Schaik and Fox 1994; van Schaik, Fox, and Sitompul 1996; van Schaik et al. 2009).

Pry, Apply Leverage

A captive orangutan Detached a wooden bar from a wall and then used it as a lever (Camacho 1907; Hornaday 1922). A captive male studied by C. Parker (1969a) dismantled an experimental apparatus and then used some of the broken pieces as levers.

Jab, Stab, Penetrate

In an experimental setting, O'Malley and McGrew (2006) described an adult male Detaching material from a branch to form a small tool that was Inserted into tiny holes to Penetrate and extract a raisin.

Reach

Blomberg (in Kortlandt and Kooij 1963) described a wild orangutan who Detached a branch and used it to Reach what was likely a fruit. In another case, a male attempting to travel from one tree to another broke off a branch and used it as a hook to Reach the destination tree (E. Fox and bin'Muhammad 2002).

Russon et al. (2010) described Kaja Island rehabilitants making tools in unspecified ways to Reach and acquire floating items and to pull a tree branch closer.

Captives have Detached splinters from shelves and wooden climbing poles, twigs and lengths of bark from boughs, and strips from rubber tires for use as Reaching or fishing tools in a variety of experimental settings (Lethmate, 1976a, 1976d, 1977a, 1977b, 1977d). J. Ellis (1975, 1977) provided additional examples of orangutans manufacturing Reaching tools, and Lethmate (1979) offered an extensive summary. Mulcahy, Call, and Dunbar (2005, 31) stated that their subjects "fashioned tools" for Reaching, but these authors did not provide any details. Walkup (2009) reported that a captive orangutan Detached a side branch from a mulberry limb and then used the detached piece to Reach for an out-of-reach reward.

Insert and Probe

Wild orangutans Detached and then modified sticks as Probing tools for obtaining insects, eggs, larvae, and honey from tree holes (see Tool Crafting for details).

Rehabilitants studied by Hirata and Ohashi (2003) broke small branches from trees and used them to reach honey inside of a bottle. Russon (2003) reported that an adolescent female made a Probing tool to extract the contents of a coconut through an "eye hole," but she provided no further details. In an experimental study of tool use by O'Malley and McGrew (2000), orangutans Detached small branches from larger ones and Inserted the small ones into holes that contained raisins.

Scratch, Rub

Galdikas (1978b, 1982b) noted that wild orangutans Detached sticks before Scratching with them.

Cut

The young orangutan studied by R. Wright (1972) learned to Detach flakes by holding and striking one stone with another. He used the sharp flakes as Cutting tools.

Block

A zoo-living male destroyed an experimental apparatus and then used some of the broken pieces to jam a sliding door (C. Parker 1969a). Tool manufacture may have been incidental to the destruction of the device.

Contain

Rogers and Kaplan (1994) observed an unusual form of tool manufacture in which six to twelve large and oblong leaves from one plant were selected, Detached, and arranged into a fan shape to make a container for holding food.

Absorb

Some of Lethmate's (1976a, 1977a) subjects Detached leaves from boughs for use as absorbent sponges.

Wipe

Galdikas (1978b, 1982a) reported that wild orangutans Detached leaves before Wiping themselves with them.

Affix, Apply, Drape

Wild and captive orangutans doubtlessly Detached branches, leaves, and twigs prior to Draping them, although this was seldom described specifically. Wild youngsters were described breaking off branches and playfully Draping them on themselves (MacKinnon 1974a, 1974b). Koehler (1993) observed captive youngsters Detach large leaves and use them as hats or seats on the ground. Wild adults also Detached the branches they Draped over themselves in rain, sun, or in the presence of humans (MacKinnon 1974b; Rijksen 1978). Rijksen (1978) saw one male leave his nest, collect his cover 15 meters away, and return with it to the nest. The rain covers were arranged to deflect water effectively. In a case of presumed Detachment, a female piled twigs on herself as protection from bees (Rijksen 1978). Western (1994) described a male who Detached small branches and then arranged them as a hat that protected his face from bee stings. A wild orangutan Detached leaves from a branch and held them between his hand and an ant nest while feeding on the ants (Rijksen 1978).

Orangutans Detached leaves and twigs to create a specific lining for their nests, to make pillows for sleeping, to fashion covers for themselves, or to construct cushions for sitting (Bourne 1971; Chevalier-Skolnikoff, Galdikas, and Skolnikoff 1982; Harrisson 1962a, 1963; MacKinnon 1974b; Russon et al. 2007; van Schaik, van Noordwijk, and Wich 2006; van Schaik et al. 2003, 2009).

In some wild populations, individuals Detach leaves and Apply them to their mouth while making the kiss-squeak with leaves vocalization (Peters 2001).

Subtract

Club, Beat

Orangutans pulled twigs from branches, bit or split off splinters from sticks, and stripped bark from sticks so that the branches and sticks could be fitted into hollow tubes to form long tools for Beating (Lethmate 1976b, 1976c, 1976d, 1977a, 1977b, 1977c, 1977d, 1977e).

Pound, Hammer

Wild orangutans used branches as chisels to access invertebrate nests or to open tough-skinned fruits. In these cases, the branch was modified through Subtraction until only a tool of the appropriate size and shape remained (E. Fox, Sitompul, and van Schaik 1999; van Schaik and Fox 1994; van Schaik, Fox, and Sitompul 1996; van Schaik et al. 2009).

Reach

Dittmar (in J. Ellis 1975) and Lethmate (1976a, 1976d, 1977a, 1977b) saw captives Subtract portions of leaves to use the midribs as Reaching tools. Orangutans pulled twigs from branches, bit or split off splinters from sticks, and stripped bark from sticks so that the branches and sticks could be fitted into hollow tubes to form long Reaching tools (Lethmate 1976b, 1976c, 1976d, 1977a, 1977b, 1977c, 1977d, 1977e). While attempting to construct a Reaching tool from a thick branch and a tube, Döhl and Lethmate's (1986) 6-year-old subject gnawed away pieces from the end of the branch so that it would fit into the tube. Walkup's (2009) subjects were presented with an out-of-reach food reward and unmodified branches. Successful solution required Reaching with a tool through the enclosure mesh. The apes sometimes Subtracted leaves, side branches, and bark to make a tool that easily fit through the barrier.

Insert and Probe

Wild orangutans modified live branches by reducing them into relatively short, suitable tools, as well as Subtracting bark, prior to using them to remove seeds from *Neesia* fruits (van Schaik, Fox, and Sitompul 1996). Stick tools used to extract insects, eggs, larvae, and honey from tree holes also had bark removed prior to Reshaping (see Tool Crafting for details).

Rehabilitants studied by Hirata and Ohashi (2003) removed branches or reduced the length of twigs to construct a suitable tool for extracting honey from a bottle.

Harrisson (1963) relayed information provided by Schuhmacher describing a rehabilitant orangutan who removed bits of the distal end of a stick so that it could be Inserted into holes to extract insects from a rotten log. Zoo-living orangutans removed leaves and twigs from branches to manufacture a useful tool for Insertion into an artificial termite mound that contained food rewards, such as applesauce (Nakamichi 2004). They commonly left a small number of leaves on the tip of the branch, making it a more effective tool. Another adult male removed twigs from large sticks so they would fit into the holes of the mound. This same male also collected sticks that were not near the apparatus, and modified them as he travelled to the foraging site (Harper 1992). C. Becker (1984) described how orangutans manufactured similar tools for the same type of apparatus. B. Wright (1995) studied captives who were supplied with an enrichment device, termed a "tool board," that required a thin tool as a probe for removing desired foods. To manufacture a tool from branches supplied by caregivers, the apes first consumed all of the leaves, stems, and bark. The denuded branch was then used as a tool. As the tip became too wet from saliva, or worn from use as a probe, the orangutans would break it off, creating a progressively shorter tool. Harper (1992) stated that a captive adult male regularly used his teeth to sharpen sticks that were then Inserted into an enrichment device containing a food reward. Walkup (2009) reported on captive orangutans who Subtracted leaves, side branches, and bark to create appropriate Probing tools.

As in Reaching, Lethmate's (1976b, 1976c, 1976d, 1977a, 1977b, 1977c, 1977d, 1977e) orangutans pulled twigs from branches, bit or split off splinters from sticks, and stripped bark from sticks so that the branches and sticks could be Inserted into a lock, or fitted into hollow tubes and then used to pull or push incentives from the tubes.

Visalberghi, Fragaszy, and Savage-Rumbaugh (1995) presented Madu, an orangutan, with the tube task and a tool with protruding pieces on each end that prevented Insertion. The orangutan Subtracted the end pieces and used the tool successfully. However, she did make numerous errors, such as trying to Insert the tool with no modification, trying to Insert the non-modified end, and trying to Insert the Subtracted pieces, which were too short. In Walkup's (2009) replication, one of three subjects made about the same amount of errors as Madu, and two made considerably fewer.

Visalberghi, Fragaszy, and Savage-Rumbaugh (1995) also offered Madu a bundle of straight dowels that had to be separated before one dowel could be used to extract a food reward from the tube. Their subject easily solved the task. Walkup's (2009) replication yielded identical results with her three subjects. While the orangutans clearly manufactured their tool, it is not clear if they Detached the one utilized stick or Subtracted the two unutilized ones.

Add, Combine

Reach

Lethmate's (Döhl and Lethmate 1986; Lethmate 1976b, 1976c, 1976d, 1977a, 1977b, 1977c, 1977d, 1977e, 1978) orangutans Combined up to five sticks and tubes to form tools of sufficient length to Reach or knock down food. Lethmate (1977a, 1982) found orangutans able to first split sticks and then taper the ends and join them with tubes to make a five-section tool. One captive tried to use straw as a splint on a broken stick so that it would be long enough to Reach

An orangutan Combines two sticks and a tube to make a Reaching tool of sufficient length. Note that the ape had previously removed strips from the sticks so the sticks could be inserted into the tube. Photograph courtesy of Jürgen Lethmate.

a rope that controlled a skylight (Kunkel in Lethmate 1977a).

Insert and Probe

Lethmate's (1976b, 1976c, 1976d, 1977a, 1977b, 1977c, 1977d, 1977e, 1978) subjects also Combined up to five sticks to form tools of sufficient length to push or pull incentives from tubes. While using an enrichment device that contained flavored sugar, individuals studied by Nogge (1984) licked their probes prior to Inserting them. By Adding saliva, more sugar adhered to the probe.

Prop and Climb, Balance and Climb, Bridge, Reposition

Lethmate's (1976b, 1976c, 1976d, 1977a, 1977b, 1977c, 1977d, 1977e, 1978) subjects Combined sticks that they Balanced and Climbed.

Contain

Orangutans observed by Rogers and Kaplan (1994) Combined six to twelve large, oblong leaves from one plant into a fan shape to make a container for holding food.

Affix, Apply, Drape

E. Fox and bin'Muhammad (2002) reported a unique case in which individuals stacked five to ten leaves together into a protective pad that was held in the hand or foot while climbing in thorny trees.

Reshape

Club, Beat

A female tied a double knot in a rubber hose to make it more rigid and then used it to club a conspecific (Herzfeld and Lestel 2005).

Reach

A young male living in a zoo collected straw and twisted it into a rope that was subsequently used as a Reaching tool to collect food outside of his enclosure (Sanyal 1902). A female living in a rehabilitation center made a rope from hairs plucked from her body and Added a piece of banana peel to the end of the rope, presumably for weight (Schuster, Smits, and Ullal 2008).

Lethmate (1979) noted that an orangutan removed wires from cables, then straightened and used them as tools.

A young orangutan did not succeed convincingly in straightening a wire to use as a Reaching tool (Koehler (1993). Koehler called her behavior "disorganized."

Insert and Probe

E. Fox, Sitompul, and van Schaik (1999) reported that orangutans fray the ends of branch tools that are used to extract seeds from *Neesia* fruits (see Tool Crafting for details).

Nakamichi (2004) studied a captive female who was experienced in obtaining foods, such as applesauce, from an artificial termite mound. On one occasion, she Reshaped a 120-centimeter-long stick by chewing on the distal end until it had become spongy. The resulting tool was described by Nakamichi (2004, 90) as a "sponge at the end of an 80 cm handle." As food was extracted from the mound, the female sucked on the sponge for an extended period.

Walkup (2009) described a captive female who attempted to solve the tube task by straightening a piece of bent copper tube, bracing the tool under the mesh of her enclosure wall. She was successful in unbending the tool to some extent, but not sufficiently to allow her to insert the tool into the tube.

Hang

A number of captive orangutans Reshaped straw by twisting and braiding it to form large "ropes" that were Hung over cage bars and used as swings (Gewalt 1975; Hornaday 1922; Jantschke 1972; MacKinnon 1978).

Absorb

Wild orangutans at the Ketambe and Sabangau field sites have been observed to crumple leaves and use them to Absorb water (van Schaik et al. 2003, 2009). Captives chewed and crumpled leaves to form sponges and twisted cloth into a compact rope to Absorb liquids from narrow containers (Lethmate 1976a, 1977a).

Wipe

Wild orangutans first crumpled leaves that they used to Wipe their faces (Galdikas 1978b).

Affix, Apply, Drape

Wild orangutans chewed leaves into a poultice that they then Applied to wounds on their bodies (van Schaik et al. 2009). Morrogh-Bernard (2008) observed a wild adult female who Detached leaves from a *Commelina* plant, chewed them until a lather was produced, and Applied them to her arms, perhaps to ease muscle aches.

An orangutan trying to straighten a length of copper tubing by bracing the tubing under the mesh wall of the enclosure. Drawing by Heather LaPorte, from a photograph courtesy of Kristina Walkup.

TOOL MANUFACTURE: GORILLAS

S. Parker at al. (1999) reported that 93 percent of the fifty-six gorillas in their survey used tools, and 52 percent of the users engaged in some form of modification of objects that were then used as tools. Details involving the modifications were not provided.

Detach

Drop

Gorillas Detached branches, twigs, leaves, and other forms of vegetation before Dropping them on humans during agonistic display (Kortlandt and Kooij 1963; Merfield 1956; Schaller 1963).

Throw

When threatened by a human, a wild gorilla Detached clumps of grass and Threw them at the intruder (Wittiger and Sunderland-Groves 2007). A captive female gorilla on several occasions Detached side branches from a larger branch, bit away pieces of the side branches to sharpen one end (see Subtract), and then threw the "spear" at a disliked caretaker (S. Woods 1992).

Brandish, Wave, Shake

Zenker (in Matschie 1904) described a lowland gorilla who Detached bunches of twigs and Waved them to disperse flies. Vegetation and branches are Detached and then Brandished to augment a display (S. Woods 1992).

Reach

Fontaine, Moisson, and Wickings (1995) documented spontaneous tool making by captive gorillas. Three adults Detached long splinters of wood from logs, shaped them by breaking or chewing, and then carried the long splinters to where they were used as Reaching tools.

Insert and Probe

A wild adult female Detached a stick that was protruding from of a pool of water and then used it as a walking stick as she attempted to move across the pool (Breuer, Ndoundou-Hockemba, and Fishlock 2005). Many published reports detailed tool use by captive gorillas to extract food from enrichment devices and mentioned that the tools had been made (B. Blaine unpub. data; Boysen and Frisch 1987; Boysen et al. 1999; M. Cole 1987, 1989; Downman 2000; Jarvis and Gould 2007; Lonsdorf et al. 2009; Patterson 1985; S. Woods 1991, 1992). However, only a few reports provided any details of manufacture in this context. An adult male living in a zoo was observed to break a twig from a larger branch and then Insert this small tool into a PVC tube to extract peanut butter (J. E. Gould and Snyder 1990). A female presented with a tool board that required a long thin tool for Probing was observed to use one of her canine teeth to Detach a piece of wood from a platform in her enclosure. This was then used as a tool to extract food from the enrichment device (B. Wright 1995).

Prop and Climb, Balance and Climb, Bridge, Reposition

A wild female Detached a branch that she Propped to steady herself and used it as a Bridge (Breuer, Ndoundou-Hockemba, and Fishlock 2005).

Affix, Apply, Drape

Gorillas Detach vegetation and Drape it over their heads and bodies (S. Woods 1992).

Subtract

Throw

A captive female gorilla Subtracted small twigs from a branch. She then used her teeth for a second-order modification, Subtracting pieces from one end of the branch to sharpen it. Finally, she Threw the "spear" at a disliked caretaker (S. Woods 1992). This behavior was observed on several occasions.

Reach

Captive gorillas studied by Fontaine, Moisson, and Wickings (1995) routinely modified plants to create suitable Reaching tools. Large leaves of *Aframomum*, banana plants, and palms were stripped until only the long stem remained. The authors relayed one memorable incident in which a male rapidly modified a palm frond and then used it to obtain nearby video equipment, including 20 meters of cable.

R. Wood (1984) documented the various ways in which a group of zoo-living gorillas modified their Reaching tools. They were observed to remove long twigs from a larger branch with their hands or to shorten a long branch by holding an end in each hand and then pushing in the center with a foot. A less common method of modification involved biting off pieces to make a shorter Reaching tool. R. Wood suggested that the biting method could have been an expression of frustration, associated with failed attempts at tool using. More practically, modification by biting also occurred when the gorillas were climbing and needed to hang on with at least one hand. Once a stick was successfully modified, it was usually kept and transported around the enclosure by the ape. This was accomplished by holding it in the hand, foot, or mouth, or between the thigh and abdomen. R. Wood also reported that some modification appeared nonproductive. At least one individual was observed to break a Reaching tool several times during unsuccessful tool-using attempts. In this case, failure was attributed to poor manipulation on the part of the ape, rather than to excessive length of the tool.

Insert and Probe

Gorillas Subtracted leaves and stems (which they ate) from fresh branches, stripped bark from the branches, and then used the branches to extract food from a tool board (B. Wright 1995). B. Wright also noted that his gorilla subjects used relatively longer branches in this task than did orangutans in the same study. Each of three adult gorillas observed by Pouydebat et al. (2005) made and used tools that were appropriate for removing crushed figs and honey from holes drilled in wooden logs. The holes were all 1 centimeter in diameter, but were 3, 9, or 15 centimeters deep. The gorillas carried fresh branches to the foraging site and removed all leaves, stems, and bark. The gorillas started with long sticks, and then broke them into smaller pieces before selecting the most appropriate one to be used as a tool The manufactured tools were between 7.6 and 25.8 centimeters long, and the length was positively correlated with the depth of the hole in which it was used. A similar observation was made by Mentz and Perret (1999), who described a gorilla who Subtracted side branches and leaves, and then broke long pieces off, to manufacture a Probing tool approximately 20 to 30 centimeters long.

Add, Combine

Throw

A young female studied by J. Gómez (1999) Threw straw to acquire objects beyond her reach. Since a single piece of straw was inadequate, she constructed "bundles" of straw by pressing together numerous pieces of straw. The bundle was long and stiff enough to reach and acquire the incentive.

Reshape

Reach

Fontaine, Moisson, and Wickings (1995) reported that gorillas used sticks as utensils for obtaining out-of-reach porridge. The apes Reshaped the ends of the sticks by chewing on them to produce a tool that resembled a brush, which improved the effectiveness of the tool for picking up porridge.

Insert and Probe

An adult female living in a zoo was observed to chew the end of a small stick into a point. She used the pointed end of the tool to clean the nostrils and navel of her offspring while he slept (Fontaine, Moisson, and Wickings 1995).

Absorb

Böer (1990) described the use of grass, leaves, and straw that were formed into a clump and then dipped into water, acting as a sponge.

Symbolize

An adult female collected vegetation and Reshaped it into a clump by pressing it against her abdomen (Fontaine, Moisson, and Wickings 1995). The authors reported that she treated the clump in the same way that she had treated her recently deceased infant.

TOOL MANUFACTURE: BONOBOS

Detach

Drop

A wild male bonobo Detached several pieces from a tree and Dropped them successively within 1 meter of a female sitting below him in the same tree (Kano 1997).

Throw

Kanzi used Throwing as one mode to Detach flakes from stones (Toth et al. 1993). Additional details regarding this Secondary Tool Use are presented under Associative Tools.

Drag, Roll, Kick, Slap, Push Over

Wild bonobos Detached small trees or vines and Dragged them near other group members (Hohmann and Fruth 2003; Ingmanson 1989, 1992, 1996; Ingmanson and Kano 1993; Kuroda 1980).

Bait, Entice

As a means of attracting the attention of a male for copulation, wild adult females Detached leaves from trees or vegetation. Bonobos also clipped leaves and held them in their lips as a play invitation (Hohmann and Fruth 2003).

Savage-Rumbaugh et al. (1996) described wild bonobos Detaching various forms of vegetation for use as possible directional indicators.

Pound, Hammer

Kanzi Detached stone flakes to use as Cutting tools, using a variety of techniques described below under Secondary Tool Use (Savage-Rumbaugh et al. 2007; Schick et al. 1999; Toth et al. 1993).

Insert and Probe

A captive adult male Detached a small twig and used it to Probe under a human's fingernail (Ikeo 1993).

Kanzi, a language-competent adult male bonobo, Pounds one stone with another to produce sharp-edged flakes. Photograph courtesy of Great Ape Trust.

Scratch, Rub

A wild adult female broke a twig from a tree and used it to Scratch her back (Hohmann and Fruth 2003).

Prop and Climb, Balance and Climb, Bridge, Reposition

A sanctuary-living male Detached a tree located near a pond for use as a Bridge (C. Andre, B. Hare, S. Hirata, and V. Woods, joint pers. comm.).

Affix, Apply, Drape

Kano (1982) relayed multiple observations of bonobos Detaching various forms of vegetation and placing it on their bodies to provide protection from the rain. Ingmanson (1996) provided details of a specific example in which a young male Detached a large, leafy plant, Subtracted the bottom portion of the stem, and then used the modified plant as an umbrella.

Subtract

Insert and Probe

Seven captive bonobos Subtracted leaves and bark from branches to be able to use the branch to Probe in an artificial termite mound (Parish 1994). The author also noted that the apes further modified the sticks so that they would fit into the holes of the device, but she did not specifically describe those modifications. Jordan (1982) described a juvenile female who modified the end of stick by biting so that it would be thin enough to fit into a screw hole.

In an experimental study designed to assess comprehension and causal understanding associated with tool use and manufacture, Visalberghi, Fragaszy, and

Savage-Rumbaugh (1995) provided four bonobos with the tube task. When given a bundle of straight sticks held together with paper tape, all of the bonobos successfully removed one of the sticks and used it as a tool. Note that as with orangutans in the same study, the utilized stick can be seen as having been Detached from the bundle, or the unutilized sticks as having been Subtracted. Visalberghi, Fragaszy, and Savage-Rumbaugh describe the apes as "easily" solving the task. When given an H-shaped stick requiring Subtraction of at least one of the protruding pieces for solution, all of the bonobos modified the stick and eventually solved the task. However, all individuals also made errors. Examples included attempting to insert the stick without Subtracting the protruding pieces, removing a protruding piece but then Inserting the unmodified end, or use of the protruding piece itself, which was too short to be successful.

Add, Combine

Prop and Climb, Balance and Climb, Bridge, Reposition

A male living in a sanctuary Detached two trees and Combined them in a pond, attempting to manufacture a Bridge to cross the water (C. Andre, B. Hare, S. Hirata, and V. Woods, joint pers. comm.).

Affix, Apply, Drape

Kano (1982) described several situations in which wild bonobos Detached multiple pieces of vegetation and Combined them on their bodies to shield themselves from the rain. In one specific example, an adult male collected seven different twigs that were 40 to 50 centimeters in length and piled them on his body as a protective covering.

Reshape

Insert and Probe

A wild adult male Reshaped the end of a stick by biting it immediately before using it as a toothpick (Ingmanson 1996).

Hang

Jordan (1982) reported that zoo-living bonobos made a rope by winding long twigs together. The rope was then Hung over a bar and used as a swing.

Contain

A sanctuary-living bonobo rolled a large leaf into a cup and used it to drink water from a river (C. Andre, B. Hare, S. Hirata, and V. Woods, joint pers. comm.).

Absorb

Zoo-living bonobos chewed banana skins into sponges to Absorb water (Walraven, van Elsacker, and Verheyen 1993).

TOOL MANUFACTURE: CHIMPANZEES

Detach

Drop

Chimpanzees have been recorded to break off branches, twigs, and leaves from trees, and leaves from a nest, to Drop on conspecifics (Albrecht and Dunnett 1971; J. Goodall 1963a; van Lawick–Goodall 1968), in display toward human pursuers (Izawa and Itani 1966; Sugiyama 1969, Takenoshita, Yamagiwa, and Nishida 1998), and on a leopard model (Albrecht and Dunnett 1971). The subjects of both Izawa and Itani's (1966) and Takenoshita, Yamagiwa, and Nishida's (1998) reports each used one hand to break off a green bough that was 3.5 centimeters in diameter before Dropping it toward human observers.

Throw

Chimpanzees Detached branches and uprooted saplings to Throw during agonistic displays in the presence of humans, conspecifics, baboons, and a leopard model, as well as during rain dances (J. Goodall 1963a, 1963b, 1965; Kortlandt 1965, 1967a; Reynolds and Reynolds 1965; Teleki 1973a; van Lawick–Goodall 1965, 1970, 1971).

Drag, Roll, Kick, Slap, Push Over

Chimpanzees Detached branches and uprooted saplings to Brandish, Drag, and Throw without aiming during rain dances and in agonistic displays in the presence of humans, conspecifics, baboons, and a leopard model (J. Goodall 1963a, 1963b, 1965; Kortlandt 1965, 1967a; Reynolds and Reynolds 1965; Teleki 1973a; van Lawick–Goodall 1965, 1970, 1971). In one case the act itself of Detaching may have been part of the display, as the chimpanzee ran and forcefully tore at the vegetation (Kortlandt 1972).

Brandish, Wave, Shake
Two wild females Detached leafy twigs that they Waved to disperse flies (Sugiyama 1969). Boesch and Boesch (1989) reported that a young male chimpanzee Detached a branch to Brandish at a group of red colobus monkeys.

Bait, Entice
Leaves used in leaf-clipping displays were Detached from nearby vegetation and then bitten into pieces (Nishida 1980a). Young chimpanzees Detached twigs to hold or place in their mouths to initiate play (J. Goodall 1986; Sanz and Morgan 2007).

Pound, Hammer
Taï chimpanzees Detached wooden clubs to use for nut cracking by breaking a root, a fallen branch, or a tree branch (Boesch and Boesch 1983). T. Hicks, Fouts, and Fouts (2005) found a large stick swarming with ants, which they presumed had been Detached from a nearby fallen branch and used to punch holes into the nest.

Sanz and Morgan (2007) observed that tool manufacture to create tools for beehive pounding usually took place in the tree adjacent to the hive. They did not describe the manufacture, but it presumably involved Detaching a branch and Subtracting vegetation.

Matsusaka (2007) observed a chimpanzee Detaching a stick before using it to hit the ground near a porcupine.

Pry, Apply Leverage
The Gombe chimpanzees Detached branches and sticks in attempts to Pry open food boxes (J. Goodall 1964; van Lawick–Goodall 1968, 1970). Nishida (1973) saw a Kasoge chimpanzee break off a stick and use it in lever fashion to expel ants from an arboreal nest. Brewer and McGrew (1990) reported that a free-ranging Gambian chimpanzee Detached a branch that was shorter and thinner than her Penetrating tool and used it to widen a hole she had created in a beehive for honey extraction.

Jab, Stab, Penetrate
Brewer and McGrew (1990) reported that a free-ranging Gambian chimpanzee Detached a dead stout branch and used it as a chisel to Penetrate a beehive. She also Detached and trimmed a tool to use as a bodkin to Penetrate the hive.

Chimpanzees from the Goualougo Triangle Detached sticks for Penetrating subterranean termite nests. Sanz and Morgan (2007) reported that the sticks were not always in visual proximity to the subterranean termite mound that they would later be used to puncture.

Chimpanzees broke off branches to Jab or Stab at animals in holes (Huffman and Kalunde 1993; Ohashi 2006; Pruetz and Bertolani 2007).

Reach
A Gombe chimpanzee Detached a stick before using it to Reach toward and knock down bananas held by a human (van Lawick–Goodall 1968, 1970). A Bossou male Detached a long stick to touch human and animal traces on the ground (Ohashi 2006). Nishida (2002) documented a female chimpanzee who Detached a twig from a shrub to use to extract a sand flea from under her toenail.

A captive broke off a twig to Reach toward an electric wire (Mottershead 1963). A male at Bossou Detached several branches from a kapok (*Ceiba pentandra*) tree to create a Reaching tool to retrieve fig fruit (Sugiyama and Koman 1979). Ohashi (2006) documented that Bossou chimpanzees uprooted grass stems to use for algal scooping. The Tenerife apes broke branches from trees, boards from boxes and tables, pieces from a fixed boot scraper, and segments from a roll of wire to Reach for food (Köhler 1925). Other captives Detached lengths of straw to use for Scratching and Reaching (N. King, Stevens, and Mellen 1980). Mottershead (1959) documented a captive who obtained a twig from a tree and then used it to touch the electric fence.

Insert and Probe
Chimpanzees Detached grass, twigs, and branches from trees and shrubs, broke off pieces of liana, and stripped off fibers of bark to fish or Dig for termites (Brewer 1976; J. Goodall 1963a, 1963b, 1964, 1965, 1986; C. Jones and Sabater-Pi 1969; McGrew, Tutin, and Baldwin 1979; Sabater-Pi 1974; A. Suzuki 1966; Teleki 1974; van Lawick–Goodall 1968, 1970). Sabater-Pi (1974) inferred from the breakage scars that branches were torn off by hand and/or bitten off. An adult and an adolescent Detached grass stems to use as insect probes (Sherrow 2005).

Chimpanzees Detached both green and dead branches before dipping for terrestrial ants (McGrew 1974; van Lawick–Goodall 1968). Ant-dipping tools were generally Detached from the nearest appropriate shrub or sapling, but one female carried a tool 75 meters, directly to an active nest (McGrew 1974). Chimpanzees Detached grass, small branches, and pieces of liana and stripped bark fibers to Insert and Probe for arboreal ants (J. Goodall 1963a, 1964; Nishida 1973). Sanz, Schöning, and Morgan (2009) noted that dipping tools at Goualougo were sometimes gathered by uprooting the entire herbaceous stem, leaving roots on the unused end of the tool. They noted that these tools were different from those used in termite fishing, which were typically clipped at the base of the stem.

Potential honey-dipping tools were uprooted or Detached from standing vegetation (Bermejo and Illera 1999; Brewer and McGrew 1990; Ohashi 2006). Young chimpanzees have been documented picking grass and breaking off twigs before Inserting them into a water-bearing tree concavity (J. Goodall 1964; van Lawick–Goodall 1968).

Chimpanzees Detached small branches or vines to extract larvae (Ohashi 2006). Captive chimpanzees also Detached branches, twigs, vines, grass stems, or leaves from vegetation in their enclosure for fishing or for dipping into puzzle feeders, such as artificial termite mounds or other types of restricted-access containers (Celli, Hirata, and Tomonaga 2004; Lonsdorf et al. 2009; Ochiai and Matsuzawa 1998).

A captive observed by Jennison (1927) bit splinters from a stick, using the Detached splinters as keys to open the cage lock.

Prop and Climb, Balance and Climb, Bridge, Reposition

Captives in the Delta colony broke dead branches from trees, tore planks from an elevated walkway, and uprooted stumps and posts for use as ladders (McGrew, Tutin, and Midgett 1975; Menzel 1972). The stumps and posts were first loosened by vigorous and prolonged swaying.

Contain

On separate occasions, two juvenile females at Mahale were observed Detaching twigs and vines to use to drink water from a tree hole (Matsusaka and Kutsukake 2002). A captive male made a cup for Containing water by Detaching the side of a toy plastic brick (Gaspar 1993). Captives also Detached grass, bark, stems, twigs, leaves, and branches to dip into water and drink (Tonooka, Tomonaga, and Matsuzawa 1997).

Absorb

Leaves are Detached from trees for use as sponges to Absorb water from concavities (J. Goodall 1964; Matsusaka and Kutsukake 2002; McGrew 1977; Sanz and Morgan 2007; Sugiyama 1995a; van Lawick–Goodall 1968, 1970). Van Lawick–Goodall (1968, 1970) and McGrew (1977) saw infants picking leaves in unskilled attempts to Wipe themselves and to sponge water. Ohashi (2006) reported wild chimpanzees Detaching leaves prior to inserting them into a bottle of oil palm wine. Tongo chimpanzees used moss for their sponges, collecting the moss from the bark of trees (Lanjouw 2002). The moss sponges were often made well before arriving at the drinking tree.

Wipe

Wild chimpanzees Detached clumps of leaves, which they used for Wiping feces, urine, mud, ants, or sticky foods from their bodies or for dabbing at wounds (J. Goodall 1964; Nishida 1973; Sanz and Morgan 2007; van Lawick–Goodall 1970, 1973).

Affix, Apply, Drape

Chimpanzees Detached leafy branches or twigs to Drape over their backs in rainstorms (Izawa and Itani 1966; Sanz and Morgan 2007). Leaves and twigs were also Detached for use as leaf cushions (Hirata, Myowa, and Matsuzawa 1998; Sanz and Morgan 2007). To make a stepping stick, a short branch was Detached from a tree (Alp 1997). Alp mentioned that this branch was also smooth, but she did not specify Subtractive bark removal. She also observed Detachment of a small branch for use as a seat stick. Detached leafy twigs or leaves were used to line the insides of nests (P. Baldwin 1979; McGrew et al. 2003).

Subtract

Pound, Hammer

Hammers for nut cracking may be modified through both Detachment and Subtraction. If a chimpanzee (or other primate) breaks a stone or stick into two or more pieces and uses one of those *pieces* as a tool, the mode of manufacture is Detachment. But if the *remaining* stone or stick is used (the core or stock, respectively), the mode is Subtraction. In some cases, a broken (Detached) piece is refinished by further removal of material (Subtraction). These distinctions also apply to Probing tools made by birds and mammals.

At Bossou, Matsuzawa (1994) documented seven instances in which a stone hammer or an anvil broke during cracking, seemingly accidently. In four of these cases, one of the broken fragments was subsequently used as a hammer; we consider these to have been manufactured by Subtraction. At Bossou, 1.6 percent of stone hammers and 6.5 percent of wooden hammers were modified for nut cracking by breaking the tool in two and selecting one of the smaller tools for Hammering (Boesch and Boesch 1990). These, too, would be examples of manufacture by Subtraction. The authors noted that such manufacture was intentional, breaking "a branch by hitting it powerfully on a root until it breaks in two without any nut under it" (Boesch and Boesch 1990, 94). They also documented six instances at Taï of chimpanzees breaking a branch in two to create a hammer by standing with one or two feet on part of the branch and pulling on the opposite end with their hands. Note that in these cases it is hard to tell from the descriptions which broken piece was used, and thus difficult to distinguish between Detachment and Subtraction.

Two adult males at Bossou Subtracted side branches and leaves from a twig before using it to Pound into a termite nest (Sugiyama and Koman 1979). Adolescents and juveniles, also at Bossou, Subtracted leaves and twigs from branches and used the branches to Pound and stir into a tree hollow to extract resin.

Sanz and Morgan (2007) observed the manufacture of tools for beehive pounding. They did not describe the manufacture, but it presumably involved Detaching a branch and Subtracting vegetation.

Pry, Apply Leverage

The Gombe chimpanzees removed leaves from branches and then used the branches as levers on banana boxes (J. Goodall 1964; van Lawick–Goodall 1968, 1970).

Jab, Stab, Penetrate

Brewer and McGrew (1990) reported that a free-ranging Gambian chimpanzee Detached and trimmed a tool via Subtraction to use as a bodkin to Penetrate a beehive.

Goualougo chimpanzees Subtracted vegetation, such as leaves and side branches, from sticks used to Penetrate a termite mound, and they sharpened the end of the sticks to a point before using them (Sanz and Morgan 2007). Bark was also sometimes peeled from the tool (Sanz 2004). Pruetz and Bertolani (2007) noted that before hunting bushbabies, chimpanzees at Fongoli manufactured tools by (1) Detaching a branch; (2) Subtracting leaves and side branches; (3) trimming the end(s) and/or Subtracting bark; and (4) trimming the tip of the tool to sharpen it (occurring in 50% of the cases; see Tool Crafting). Leaves, side branches, and bark might be Subtracted from tools used by chimpanzees at other sites to Jab at animals in holes (Huffman and Kalunde 1993; Ohashi 2006). Captives removed material from lengths of straw to produce pointed tips used to Jab insects (N. King, Stevens, and Mellen 1980).

Reach

A male at Bossou Detached branches from a kapok tree and then Subtracted thorny bark and side branches to create a "hook-type stick" to retrieve fig fruit (Sugiyama and Koman 1979). While the individual created several such sticks, he did not succeed in acquiring the figs. A male at Lokoué Bai in Odzala National Park, Republic of the Congo, used his incisors to trim off the head from a sedge after uprooting it (Devos, Gatti, and Levréro 2002). He then used the sedge stem to scoop algae from water pools. Ohashi (2006) reported that Bossou chimpanzees Subtracted leaves from grass stems prior to using the stems for algal scooping. One of Köhler's (1925) apes picked leaves from a stick before using it as a Reaching tool. Others removed corks from hollow sticks so that a stick could be joined with another to make a sufficiently long Reaching tool; they also removed sand and stones from crates so that the crates could be moved beneath suspended food. The Tenerife chimpanzees bit pieces from boards to form pointed shafts that could be inserted into hollow sticks to form longer Reaching tools.

Insert and Probe

Prior to fishing or Digging for termites, leaves were stripped from branches, twigs, and vines (Brewer 1976; J. Goodall 1963a, 1963b, 1964, 1986; Hladik 1977; C. Jones and Sabater-Pi 1969; McGrew 1979; McGrew, Tutin, and Baldwin 1979; Sabater-Pi 1974; Sanz 2004; A. Suzuki 1966; Teleki 1974; van Lawick–Goodall 1968, 1970, 1971, 1973); narrow strips were Subtracted from broad-bladed grass (J. Goodall 1964; McGrew 1979; van Lawick–Goodall 1968, 1970, 1973); suitably short segments were Subtracted from lengthy branches, pieces of grass, and lianas (J. Goodall 1963a, 1965;

McGrew 1979; Sabater-Pi 1974; Teleki 1974; van Lawick–Goodall 1970); leaf blades were Subtracted from the midribs (van Lawick–Goodall 1968); and bark was Subtracted from branches (C. Jones and Sabater-Pi 1969). The tool might also have been shortened by clipping (Sanz 2004). These operations were done by hand and/or with the lips and teeth. In the course of termite fishing, the tools frequently got bent. The chimpanzee then either reversed the probe and used the other end or Subtracted the bent portion by biting or picking it off (J. Goodall 1963b, 1965; McGrew, Tutin, and Baldwin 1979; van Lawick–Goodall 1968, 1971). McGrew, Tutin, and Baldwin (1979) found that Subtraction of leaves and worn ends from termiting tools occurred at Gombe in Tanzania, Mont Assirik in Senegal, and Okorobikó in Equatorial Guinea. The Mont Assirik chimpanzees usually peeled all bark from twigs and vines before using them for termite fishing, although none of the termiting tools recovered at Gombe had the bark removed. The Okorobikó chimpanzees peeled bark from a high proportion of their termite-fishing tools, but they rarely peeled off all of the bark. The authors concluded that these differences in the frequency and degree of bark peeling were cultural.

Many Subtractive behaviors are also seen in conjunction with Inserting and Probing for arboreal ants and dipping for terrestrial ants. Leaves and side branches were stripped from branches and twigs, bark was stripped from branches, leaf blades were torn from the midvein, and segments were removed from branches, lianas, and grass, either to reduce them to suitable lengths or to repair worn or broken ends (A. Fowler and Sommer 2007; McGrew 1974; McGrew et al. 2003; Nishida 1973; Ohashi 2006; Sanz 2004; Tutin, Ham, and Wrogemann 1995).

Boesch (1995, 4) reported that a male "made a stick" for grub extraction, but he did not specify the modification. We presume Subtractive modification was involved. The author also noted that twig tools were modified by Subtracting the bark prior to using them for extracting larvae from a fallen tree.

Tools were also modified for honey extraction through Subtractive modification. Leaves and side branches were removed, the bark was stripped from dip sticks, and a branch might be snapped in half before being used to Probe for honey (Bermejo and Illera 1999; A. Fowler and Sommer 2007; T. Hicks, Fouts, and Fouts 2005; Sanz 2004; C. Stanford et al. 2000; Tutin, Ham, and Wrogemann 1995).

A chimpanzee modified several tools, including a dry twig and a dry leaf, before using each as a nasal probe (Nishida and Nakamura 1993). The modifications included Subtracting leaves from the twig and removing the leaf blade from the leaf, so as to leave the midrib as a probe. Watts (2008) twice found stems stripped of leaves and Inserted into a tree hole, though he was unsure of the target.

The Tenerife chimpanzees who Inserted and Probed with straw in holes and crevices also Subtracted bent portions of the straw (Köhler 1925). McGrew and Tutin (1972, 1973) saw the Delta captives strip leaves and Subtract segments from twigs to produce effective dental probes. The branches used by the same animals as pitons for escape were stripped of leaves, bark, and projecting twigs and shortened before use. McGrew, Tutin, and Midgett (1975) emphasized that since a substantial time usually elapsed between modification and use, it should not be inferred that the apes were intentionally manufacturing pitons.

In an experimental context, Ladygina-Kohts (1959) and Ladygina-Kohts and Dembovskii (1969) found that a chimpanzee appropriately modified a branchy stick by removing side branches prior to Insertion in a tube to push out a reward. M. Harvey (1998) found that captives appropriately modified tools by removing side branches in an experimental test, and they avoided doing so when it was unnecessary in an alternative protocol. In an experimental series presenting progressively more difficult tests of Subtractive modification, Khroustov (1964) found that a captive could Subtract one and two crosspieces from a stick so that the stick could be Inserted into a tube to acquire food. The animal then broke rectangular and circular pieces of wood to produce suitable tools. It broke out these tools by following the grain of the wood, and it compensated when the grain was either obscured or masked by a painted grain running perpendicular to the actual grain.

Captive chimpanzees also stripped sticks or stems of leaves or side branches to Insert them into artificial termite mounds or puzzle feeders (Lonsdorf et al. 2009; Terdal 2005).

Visalberghi, Fragaszy, and Savage-Rumbaugh (1995) presented the "bundle of sticks" and "H-shaped stick" variants of the tube task to chimpanzees. They found that the apes always broke up the bundle of three sticks to get a single suitable stick in the bundle of sticks test. As noted above, the apes may have Detached the utilized stick from the bundle or Subtracted the two unused sticks. The chimpanzees made many errors in the H-shaped stick condition, which required Subtracting a perpendicular obstruction from the tool prior to use. Although the chimpanzees showed some improvement over the course of subsequent trials, the difference in their overall performance on the first block of trials did not differ significantly from their performance on the last block of trials.

Scratch, Rub

A captive chimpanzee bit off the end of a piece of straw, resulting in a sharp point to use in her autogrooming (N. King, Stevens, and Mellen 1980).

Contain

Ohashi (2006) observed a male chimpanzee break in half a stick that had been used for pestle pounding. He dipped the stick into the hole created by the Pounding to retrieve juice that clung to the stick. J. Goodall (1964) and Sanz and Morgan (2007) noted that chimpanzees often removed leafy side branches from tools prior to inserting them in cavities to dip for fluids, such as honey or water.

Add, Combine

Reach

Chimpanzees have been seen to Add or Combine objects to produce Reaching tools only in captivity. They have been reported to join as many as three sticks or bamboo poles together to manufacture Reaching tools of sufficient length (Frolov 1937; Kats 1972b; Köhler 1925; Pechstein and Brown 1939; Schiller 1957; and Wazuro 1962 in Windholz 1984). Prince et al. (2009) found that captive chimpanzees who observed a conspecific Combining two elements (a rod and a polycarbonate pipe) to rake in a reward performed significantly better on the task than chimpanzees without a model. Ladygina-Kohts (1959) and Ladygina-Kohts and Dembovskii (1969) found that a chimpanzee was unable to Combine two short sticks to retrieve an object from a tube in an experimental context.

Bania et al. (2009) reported on the tool-manufacture abilities of captive chimpanzees when confronted with one of two tasks. One required the chimpanzees to make a tool by Adding or Combining two elements to make a probing tool. The other required them to use Subtractive manufacture to make a hook tool to obtain an out-of-reach reward. They succeeded, and their success was cited as evidence of foresight and planning.

Sultan, one of Köhler's chimpanzees, combines two sticks to make a longer Reaching tool. Reprinted from Köhler 1917.

Contain

A captive female shuffled leaves together, forming a plate on which she carried her food, an example of the Add or Combine mode (N. King, Stevens, and Mellen 1980).

Reshape

Bait, Entice

Nishida (1987) documented cushion making as a courtship display in the Mahale M group, which we classify as a form of Reshaping. In this display, to Entice a female, a male "facing an estrous female makes a crude day bed or cushion, which he sits on. Then he stamps with one foot" (Nishida 1987, 466).

Pry, Apply Leverage

The Gombe chimpanzees Reshaped splinters by chewing the end of a stick to produce a chisel-shaped edge that was used for Prying (van Lawick–Goodall 1970).

Jab, Stab, Penetrate

Brush sticks have been found at several chimpanzee research sites, presumably having been used in termite fishing and honey gathering. There is some question as to whether the brushed ends are formed intentionally by the chimpanzees by Reshaping, or whether they are the unintentional result of Detachment of the branch from a tree. It is especially important to differentiate tools used for Digging or Penetrating from those used for Inserting and Probing (see below). Here we focus on Digging and Penetrating tools. Takemoto, Hirata, and Sugiyama (2005) examined and experimented on a sample of brush-shaped sticks that had been used as Digging or Penetrating tools and then collected by Sugiyama (1985). They concluded that the brushlike shape found on such tools is often formed naturally when the tool is Detached. Sanz, Morgan, and Gulick (2004) documented brushed ends on puncturing (Penetrating) sticks, concluding that they were byproducts of Detaching the sticks from trees. Heaton and Rayne Pickering (2006) examined the brushed ends of 160 puncturing sticks used for termite predation for evidence of deliberate modification. They also concluded that the brushed end of puncturing tools was best explained as an unintended result of Detaching the tool. These results contrast with what now appears to be the intentional manufacture of brushed ends on tools used to Insert and probe into nests (see below).

The Tenerife chimpanzees Reshaped splinters through chewing to produce Probing and Jabbing tools with pointed tips (Köhler 1925).

Reach

After Detaching the end from a willow branch, Loulis, a captive chimpanzee, intently chewed the end before using it to probe at a sore on a fellow captive (D. Fouts 1983).

The Tenerife chimpanzees also Reshaped tools for Reaching. They bent clumps of straw to produce Reaching tools of sufficient firmness, as well as uncoiling wire to make a sufficiently long Reaching tool (Köhler 1925).

Insert and Probe

Tools used for termite fishing have been found with brushed ends, suggesting modification through Reshaping (Bermejo and Illera 1999; Sugiyama 1985). Approximately half of the termite-fishing sticks Sugiyama (1985) recovered had a 2- to 17-centimeter-long brushed end that was not clogged with mud, as would be expected if that end of the tool had been used to Dig or Penetrate the nest. He hypothesized that the brushed end might be used to Probe into termite mounds, and therefore might result in more efficient termite collection by creating a greater surface area for the termites to attack and cling to. Sugiyama (1985, 1997) also noted that the termites did not easily bite sticks greater than 1 centimeter in diameter, so the brush sticks might facilitate attachment and thus increase the catch. He observed that the Guérze people also make brush sticks for termiting and that, for them, brush sticks are four times more efficient for termite fishing than regular sticks.

Since these initial discoveries, termite-fishing tools containing one frayed or brushed end have been found at many sites, including Bai Hokou (Fay and Carroll 1994), Campo (Sugiyama 1985), the Goualougo Triangle (Sanz and Morgan 2007), the Lossi Forest (Bermejo and Illera 1999), Mboete (Takemoto, Hirata, and Sugiyama 2005), the Ndoki Forest (Takenoshita 1996), Okorobikó (McGrew, Tutin, and Baldwin 1979), and Tschego (Kuroda 1991). Sanz, Call, and Morgan (2009) referred to these tools as "brushed-tipped fishing tools." The tools were originally thought to have been modified (Sanz, Call, and Morgan 2009; Takenoshita 1996). Fay and Carroll (1994) had found Probing stalks for termite fishing at Bai Hokou and at Ndakan with one end frayed by chewing. One stalk was still coated in chimpanzee mucous, and small termites adhered to the brush. Bermejo and Illera (1999) inferred that the brushed ends of tools found in the Lossi Forest had been intentionally made, based on the presence of saliva and teeth markings on the brushed end, and on observations of a chimpanzee eating a termite from a brushed end. Sanz, Morgan, and Gulick (2004) documented a chimpanzee from the Goualougo Triangle chewing on a stick, presumably to create a brush stick. Sherrow (2005) documented that a stem had been modified after Detachment to create a 4- to 7-centimeter-long brushed end; the chimpanzee crushed and modified the end using its teeth. Tools were discarded and replaced when the brush-end bristle became worn or bent.

Since the behaviors of creating the tools or eating termites from the brushed ends had not been observed at all of the sites, however, it was difficult to know definitively whether all found brushed tools were created intentionally by Reshaping, or unintentionally when the branch was pulled from the tree (Bermejo and Illera 1999; Muroyama 1991; Takemoto, Hirata, and Sugiyama 2005).

There has now been confirmation of the intentional creation of brush-tipped fishing tools. Goualougo chimpanzees have been observed creating brush-tipped tools, wetting the frayed ends with their saliva and then pulling the tools through either their hands or their mouths to compact the frayed strands before

Inserting them into the narrow openings of epigeal termite nests (Sanz, Morgan, and Gulick 2004). Sanz, Call, and Morgan (2009) elaborated on how the Goualougo chimpanzees deliberately Reshaped the ends of the herbaceous stems to fashion the brush-tipped termite fishing tools. They documented five different types of modifications, including the apes using their teeth to fray the end of the tool, splitting the probe lengthwise, biting to separate the fibers, clipping the probe length to shorten it, and Subtracting irrelevant vegetation. After manufacture, chimpanzees usually (93% of the cases) pulled the brushed ends through their fists to compact the fibers before Inserting the tool into the nest. They repeated this straightening if the probe was not successfully Inserted. The chimpanzees also refined and repaired the brushes during use.

Nishida (1973) saw chimpanzees gently bite dull or bent vines to remove the ends so that they could be Inserted into arboreal ant nests. These cases might be called Subtractive manufacture, but they seem to be distinctive in that a special configuration resulted. Tutin, Ham, and Wrogemann (1995) found brushed tools used for ant fishing. Through they did not observe the manufacture, the authors believed that the tools had been deliberately frayed, most likely by chewing. Using condensed milk, they found they could extract six times more milk with a brushed tool than with a compacted or non-brushed tool. Sanz (2004) noted that the manufacture of an ant-dipping tool at Goualougo included fraying the end of the stem.

Tutin, Ham, and Wrogemann (1995) and A. Fowler and Sommer (2007) found sticks used for honey probing with frayed ends, possibly indicating biting or chewing to create a brush tool.

Boesch and Boesch (1990) described the manufacture of small sticks for extracting bone marrow or *Panda* nut kernels, tools subsequently called "marrow-picks" (Whiten et al. 1999). Boesch and Boesch (1990) noted that 83 percent of *Panda*-kernel extraction tools were modified before use. Leafy twigs were Detached from the substrate and shortened to a specific length by clipping them with either hands or teeth. The chimpanzee then further modified the length or Reshaped the end of the tool, using its teeth to sharpen the end. Boesch and Boesch stated that "in almost all the cases the chimpanzees prepare an adequate tool with all the necessary modifications beforehand, and modifications after first use occur for only 6.5 percent of the sticks used," which allowed them to conclude, "Taï chimpanzees hence seem to show an advanced understanding of the relations between objects, allowing them to make specific tools for a determined task" (Boesch and Boesch 1990, 94). Sanz (2004) also described the tool manufacture of marrow picks, including Subtraction of side branches and leaves.

In an experimental context, Ladygina-Kohts (1959) and Ladygina-Kohts and Dembovskii (1969) found that a chimpanzee straightened bent wire to Insert it into a tube and push out a reward.

Only two of seven chimpanzee subjects attempted to straighten a bent tool to solve an experimental task (Povinelli et al. 2000). Each of the two who straightened the tool did so only once, and even on these trials, the chimpanzees began by trying to obtain the food with the bent tool. The authors concluded that the chimpanzees did not understand how the properties of the tool related to the solution.

Hang

Gaspar (1993) reported that captives Reshaped clothes by folding them in half before Hanging them from a ceiling bar to make a swing.

Contain

Bossou chimpanzees Reshaped leaves by folding them to Contain drinking water (Tonooka and Inoue-Nakamura 1993). Leaves were Detached and neatly folded in the mouth into approximately 3-centimeter pleats before dipping the leaves into water (Tonooka 2001). *Hybophrynium braunianum*, a soft, hairless leaf, was used most often as the source material (Tonooka, Inoue-Nakamura, and Matsuzawa 1994). Tonooka (2001) hypothesized that the chimpanzees may fold leaves because water may be retained between the pleats of the folded leaves.

Absorb

Wild chimpanzees Reshaped leaves to increase their absorbency as sponges. They crumpled them by briefly chewing the leaves before inserting them into water (J. Goodall 1964; K. Hunt and McGrew 2002; McGrew 1977; Reynolds 2005; Sanz and Morgan 2007; Sugiyama 1989a, 1995a; Tonooka 2001; van Lawick–Goodall 1965, 1968, 1970, 1971, 1973). Reynolds (2005) speculated that chewing also released flavor from the leaves. McGrew (1977) saw an infant do this before playfully leaf sponging. The moss used by Tongo chimpanzees to sponge was first loosely rolled into a small bundle before insertion into the tree hollow (Lanjouw 2002).

Teleki (1973b, 1973c, 1974) saw a chimpanzee using crumpled leaves to Wipe residual tissue from the skull of a prey animal. The leaves, however, had been chewed previously with bits of the meat into what has been called a wadge. The wadge itself may be a tool of sorts, serving to facilitate ingestion of soft or liquid foods or to prolong the flavor of favored foods (Beck 1977). Chimpanzees in the Mahale Mountains made wadges when eating meat, too (Nishida, Uehara, and Nyundo 1979).

Captive chimpanzees also Reshaped stems or leaves by chewing them prior to dipping for water (Bettinger and Carter 1991; Norikoshi and Kitahara 1979). Takeshita and van Hooff (1996, 2001) described captives who tore up paper, a towel, or a stuffed toy before using the remains as sponges. Two juvenile females at the Tama Zoo were provided with leafy branches and, outside their enclosure, a bottle containing juice. In the final six of twenty-one sessions, they made sponge tools by chewing and fraying the ends of the branches to separate the branch fibers, which may have facilitated juice absorption (Kitahara-Frisch and Norikoshi 1982).

Wipe

Nishida (1973) saw a chimpanzee Reshape leaves by crumpling them by hand before using them to Wipe ants.

Affix, Apply, Drape

McGrew and Marchant (1998) observed a youngster with a colobus monkey skin Draped around his neck. When they recovered the skin, they found it had been formed into a "necklace" with a knot. While they did not see the knot being tied—and it may have occurred by accident—they noted that the result was functional. They cited this as possibly the first example in which a non-human manufactured a Draped adornment. If purposive, this would be an example of manufacture through Reshaping.

ASSOCIATIVE TOOL USE: GIBBONS

No associative tool use has been documented.

ASSOCIATIVE TOOL USE: ORANGUTANS

Sequential Tool Use

Russon et al. (2010) described rehabilitants making a Reaching tool to draw in a tree branch that they then Repositioned and used to travel over water. Lethmate's (1979) subjects used sticks to Reach successively longer sticks (up to five) until they got a stick long enough to Reach the incentive. Some of the sticks had to be retrieved from platforms that were not in the same visual field as the final reward. Mulcahy, Call, and Dunbar (2005) offered orangutans an out-of-reach food, together with Reaching tools of different lengths. These apes used a tool to Reach a second tool that was then used to obtain a food incentive.

Tool Composite

A wild flanged male was observed by E. Fox and bin'Muhammad (2002) as he attempted to cross between two trees that were 3 to 4 meters apart. After swaying the vehicle tree and reaching unsuccessfully with his hand for the destination tree, he broke off a straight, 2.5-meter-long branch and used it to hook and pull in the destination tree. The vehicle tree and the branch were used simultaneously as a Tool Composite.

A young male orangutan repeatedly swayed a tree but could not reach his intended destination tree. He stopped and cracked a branch so that it was loose but still attached to the vehicle tree, and then used it successfully as a "handle" to extend his reach (Russon 2003). The handle made by this ape is a Metatool, because it improved the effectiveness of the vehicle tree.

Galdikas (1982a) provided detailed accounts of rehabilitants using dugout canoes, rafts, or floating wood to travel in or across bodies of water. Some apes moved their boats by paddling with their hands, or by reaching into the water or along the bank and pulling themselves by grasping vegetation. Others used a paddling motion with items such as sticks, boards, or dippers to propel themselves. The simultaneous use of a Repositioned floating vessel and an improvised paddle constitutes a Tool Composite.

Lethmate's (1977a, 1977b, 1977d) subjects used sticks to retrieve saturated leaf sponges from a narrow cylinder that contained juice. Lethmate (1982) also described a young orangutan who Repositioned and stood on a box while using a stick to Reach suspended food. Reuvens (1889 in Yerkes and Yerkes 1929) described a captive orangutan Throwing a sack over an orange and then using another sack to pull the first one and the orange within reach. A female orangutan made a rope from her own hair, tied a banana skin to the end as a weight, and Threw the rope over food and pulled it in (Schuster, Smits, and Ullal 2008). The rope may have worked without the weight, but it was dramatically improved by the addition of the banana skin, which was a Metatool. Walkup (2009) reported another case of a Tool Composite, involving sponging by a captive orangutan. An adult female was attempting unsuccessfully to Reach a pool of syrup outside her enclosure with a paper towel. She left the area and returned with a tree branch that she used to push the paper towel into the syrup. She then retrieved the syrup-soaked towel and licked it clean. R. Shumaker (pers. obs.) also observed a similar use of a Tool Composite by an adult female in a different facility to retrieve spilled juice outside her reach. She left the scene and came back with a branch, which she used to collect a stack of paper towels that were also out of reach (Sequential Tool Use). She returned to the spilled juice, used the stick to push the towels into the juice, drew them back, and placed the soaked towels into her mouth. In both cases, the stick was long enough to Reach the juice and could have functioned alone to retrieve the juice, albeit inefficiently. The paper towels served as Metatools, increasing the effectiveness of the stick in retrieving the juice. The behavior in both examples was spontaneous, and it did not involve tutoring or practice facilitated by humans.

Walkup (2009) replicated Visalberghi and Trinca's (1989) "short stick" variation of the tube task that had been conducted with capuchin monkeys. The orangutans were provided with a transparent tube containing a food reward, as well as sticks that were too short to Reach the incentive. The orangutans succeeded in using at least two of the short tools together as a Tool Composite.

A male orangutan at the Como Zoo removed two drinking faucets from the wall of his enclosure and used them as a hammer and chisel while attempting to break through an adjoining wall (J. Ellis 1977). Wilkie and Osborn (1912 in R. Mitchell 1999) described a captive orangutan who was allowed to smoke. The orangutan used lit cigarettes to set bags on fire, and then moved the lighted bags with a stick to set other bags on fire. This, incidentally, is one of only two cases known to us of an ape or any other animal spontaneously controlling fire (the other is a chimpanzee using one cigarette to light another [Brink 1957]).

Secondary Tool Use

A young male orangutan learned to knap stone flakes by holding and striking a hammer stone on another stone (R. Wright 1972). He subsequently used the flakes for Cutting.

Tool Crafting

Wild orangutans Crafted tools that were used for the extraction of larvae, eggs, insects, and honey from arboreal nests. After a branch was Detached (first-order modification), Subtractive modification was used to remove leaves, twigs, and sometimes bark (second, third, and perhaps more orders of modification), followed with Reshaping by fraying the ends of the branch or splitting it to create a beveled edge (final-order modification) (van Schaik, Fox, and Sitompul 1996).

An adult female zoo orangutan Reshaped the end of a stick she used to Insert and Probe in a feeder for applesauce. She shredded the distal 40 centimeters of the stick by chewing and crumpled the fibers into a "sponge" (Nakamichi 2004). She had probably Subtracted leaves and twigs from the stick before Reshaping it, and thus would have Crafted the tool.

ASSOCIATIVE TOOL USE: GORILLAS

Sequential Tool Use

Mulcahy, Call, and Dunbar (2005) reported that captive gorillas successfully used tools to obtain other tools that they then used to obtain an incentive.

Tool Composite

A 2.5-year-old female gorilla living in a sanctuary used a hammer and anvil to open nuts (Owen 2005). The author stated that she devised this on her own, without any tutelage.

In the contexts of play or displaying, gorillas Pound objects against each other to create noise (J. Gómez 1999; Nierentz 2007; R. Shumaker, pers. obs.). This behavior has been performed by infants as young as 12 months old (J. Gómez 1999). An adult male regularly Pounded two pieces of wood together rhythmically to imitate the sound of human hammering (Pettit 1997).

Gorillas build Constructions of as many as four boxes (Yerkes 1927a, 1927b, 1928–1929; Yerkes and Yerkes 1929). On one occasion, a gorilla stood on the tower, held a box in its hand, and used it to Reach up toward the food. This is a Tool Composite of a Construction and a Reaching tool (see chapter 7).

Multi-Function Tool

Breuer, Ndoundou-Hockemba, and Fishlock (2005) documented the use of a Multi-Function Tool by a single wild adult female in two separate contexts. In the first, she used a branch as she entered and moved across a pool of water. Upon entering the waist-deep water bipedally, the gorilla inserted the stick, approximately a meter long, into the pool to assess the water depth and/or the stability of the bottom. She did this a number of times while in the pool, while also relying on the stick for support while moving. Therefore, she alternated between Repositioning the stick and using it for stability, while also Reaching with the stick to assess depth and the condition of the bottom of the pool. The same authors also described the second scenario, in which she picked up and pushed a dead and detached trunk into the wet ground, and then held it with one hand to stabilize herself while she foraged with the other. After foraging, she removed the trunk and used it as a Bridge.

ASSOCIATIVE TOOL USE: BONOBOS

Secondary Tool Use

Kanzi used a hammer stone and an anvil to produce sharp-edged flakes for Cutting. He employed various techniques as he became competent at stone knapping. Kanzi initially employed a percussive movement to make flakes by holding a stone in each hand; the stone in the right hand was forcefully Hammered on the one in the left hand, which was held in a fixed position. The trajectory of the movement was either vertical or horizontal (Savage-Rumbaugh et al. 2007; Schick et al. 1999; Toth et al. 1993). Another successful technique involved holding one stone in his right hand and striking it against a stone on the ground, which Toth et al. (1993) dubbed the "anvil" technique. Kanzi also used Throwing as one mode to Detach flakes from stones. He held the stone in one hand, and then Threw it with "a very hard, rapid thrust" (Toth et al. 1993, 85) at a tile floor. This technique consistently resulted in the production of sharp flakes used as Cutting tools (Savage-Rumbaugh et al. 2007; Toth et al. 1993). Additionally, Kanzi developed a "directed-throwing technique" that was used "to produce flakes and fragments" (Toth et al. 1993, 85). In this behavior, he aimed and Threw one stone at a stationary stone that was on the ground, rather than held in the hand. According to the authors, useful flakes were produced from both the Thrown and stationary rocks.

In the most sophisticated technique, Kanzi held a stone in each hand. Moving both hands, Kanzi used the hammer stone to strike the core stone to reliably produce flakes for Cutting, which Toth and colleagues labeled "free-hand hard-hammer percussion."

Kanzi uses a stone hammer to make a stone flake that he then used for Cutting. Use of a tool to make a tool (Secondary Tool Use) is rare among animals. Photograph courtesy of Great Ape Trust.

ASSOCIATIVE TOOL USE: CHIMPANZEES

Sequential Tool Use

In some experiments on Sequential Tool Use by captives, chimpanzees had to use a tool to Reach another of a sufficient length or of a proper configuration to get an incentive. Sometimes they had to use a series of progressively longer tools, and in some cases at least some of these tools were not in the same visual field as the incentive. Hobhouse's (1926), Kats's (1972b), Köhler's (1925), and Nissen's (1949) subjects used two sticks; Jennison's (1927) used three; and those of Jackson (1942) and Jacobsen, Wolfe, and Jackson (1935) were required to use four. McGrew, Tutin, and Midgett (1975) saw a captive chimpanzee use a short stick to get a broom handle that was then used to Reach the latch on the outside of the cage door. Köhler's (1925) chimpanzees were also able to Reposition a box to reach a stick, which was then used to rake in a reward placed outside of the cage.

Tool Set

Brewer and McGrew (1990) reported the first example of a Tool Set, in which a free-ranging adolescent female in Gambia used up to four types of tools sequentially to acquire honey. She began by Detaching a stout dead branch with a sharp end, which she used as a chisel to Penetrate the beehive. She then Detached a shorter and thinner tool that she used to widen the hole. Next, she Detached a green branch and shortened it for use as a bodkin, inserting it forcefully into the nest to pierce the remaining hive barrier. Finally, she Detached a green vine and used it as a Probe to dip for the honey.

Bermejo and Illera (1999) described the Tool Set for acquiring honey used in the Lossi Forest. It consisted of three tools. The chimpanzees first used a stick as a chisel to Pound the beehive. The Pounding tool was then turned around and used as a bodkin, Penetrating the wall of the hive with the sharp end. The authors were unable to establish whether the sharp end of the bodkin was created purposefully or if it was a byproduct of Detaching the stick from the substrate. The final tool in the sequence was a dipstick, made and used as a Probing tool to extract the honey.

Sanz and Morgan (2007) documented a three-tool Set for honey extraction at Goualougo. The first was a stout branch used to Pound into the nest. Another stick was then used as a lever to widen the hole. In the final step, a dipping stick was used to extract the honey from the hole. Sanz and Morgan (2009) also noted that most honey gathering at Goualougo was accomplished by using one or two tools, and one chimpanzee used a Set of five tools in one episode. Sanz and Moran (2009, 421) argued that the sequence of honey gathering at Goualougo was "evidence of hierarchical structuring of tool use, with the opening task preceding the extraction task."

Boesch, Head, and Robbins (2009) documented Tool Sets consisting of three to five different types of tools used by chimpanzees at the Loango National Park in a hierarchical sequence to access honey from both tree and subterranean hives. Accessing honey from subterranean hives required two tools used sequentially, the "perforator" and the "collector." The perforator was used first to explore and locate the underground honey chamber, and to create an entrance tunnel for the collector. Boesch, Head, and Robbins (2009, 566) reported that the creation of the tunnel to the chamber indicated that "chimpanzees know exactly where they are aiming . . . [and have] an elaborate

A Goualougo chimpanzee uses a Tool Set. *Left*: First she used several Pounding tools to open the arboreal hive. *Right*: Then she used a slender stick Inserted into the nest to extract the honey. Photograph by Ian Nichols. Reprinted with permission of Elsevier from Sanz and Morgan 2007, p. 14.

understanding of unseen nest structure, combined with a clear appreciation that the tools permit the location of unseen resources." After use, the perforator tool has one blunt end, which is compacted with soil. The end of the collector might be frayed after use.

For extraction of honey from the arboreal hives of *Apis* or *Melipone* bees, chimpanzees used as many as four tools sequentially, including the "pounder," "enlarger," "collector," and "swabber." The chimpanzee began the sequence by using the pounder, a thick stick, to break into the hive. The Pounding resulted in blunting the end of the tool. The enlarger stick was then used to perforate and enlarge the opening, and also to break up the internal chamber within the hive. After use, this tool had signs of wear, consisting of a blunt end and wear on the sides from Levering. The honey was then retrieved from the hive with the collector and/or swabber. The collector was similar to the collector tool used to dip out honey from subterranean bee nests. The swabber was a section of bark used to dip or spoon the honey from the hive.

Boesch and Boesch (1990) reported that Taï chimpanzees used a Tool Set consisting of a stone to crack open a nut and a small stick to extract the nutmeat.

Bossou chimpanzees were documented using a Tool Set for pestle pounding. Sugiyama (1994) and Yamakoshi and Sugiyama (1995) reported that two adults and two adolescents used a leaf stalk to Pound into the tree crown to get fibrous vegetation that was then used as a sponge to Absorb the sap.

Chimpanzees from several termite-fishing populations used a two-tool Set consisting of a stick to Penetrate the subterranean termite mound and a flexible fishing probe, typically with a brushed tip. Such Tool Sets have been found at Bai Hokou (Fay and Carroll 1994), the Dja Biosphere Reserve (Deblauwe et al. 2006), the Lossi Forest (Bermejo and Illera 1999), the Ndoki Forest (Suzuki, Kuroda, and Nishihara 1995; Takenoshita 1996), and in the Goualougo Triangle population (Sanz and Morgan 2007; Sanz, Morgan, and Gulick 2004). Abwe and Morgan (2008) described a slight variation on this Set from tools found in the Ebo Forest, where they observed stout sticks used for perforation and leaf midribs that might have been used as the fishing tools.

Sanz and Morgan (2007) and Sanz, Morgan, and Gulick (2004) reported on a Tool Set used to access an epigeal termite nest. The first step involved Penetrating the nest with a perforating twig. This was followed by the Insertion of a brush-tipped fishing probe.

Sugiyama, Koman, and Sow (1988) reported one observation of a possible Tool Set for ant dipping by a Bossou female. The chimpanzee excavated an army ant nest through Digging, possibly with a stick. The Digging behavior was not observed, although stems (*Lippia* sp.) were found smeared with mud, possibly indicating Digging. Sugiyama, Koman, and Sow (1988, 58) stated, "even if the *Lippia* sticks were used in the digging, they must have only played a supplementary role." In addition to the possible Digging tools, they found slender fishing wands that had been used to dip for the ants. Sugiyama (1995b) later observed another chimpanzee at Bossou who Dug with a stick tool and was then observed using a stick as a wand to dip for the ants.

Sanz, Schöning, and Morgan (2009) reported the use of a Tool Set at Goualougo for the acquisition of army ants. The chimpanzees began by using a rigid wood sapling to perforate the ant nest, and then used a flexible herbaceous stem to dip for the ants. The authors noted that the nest-perforating tools were significantly longer than those used for dipping.

Tool Composite

The most common Tool Composite used by nonhuman animals is the hammer and repositioned anvil, used by some populations of western African chimpanzees to crack nuts. The richest examples are provided by the Bossou chimpanzees in Guinea (Sakura and Matsuzawa 1991; Sugiyama and Koman 1979). It is important to note that not all examples of nut cracking involve the use of a Tool Composite. In many locations, such as Taï (Boesch and Boesch 1990), the anvil may be a non-portable rock outcrop or a tree root, neither of which are considered to be a tool. Bossou chimpanzees, however, repositioned and transported anvils and hammers, allowing both to be considered tools (Sakura and Matsuzawa 1991). Bossou chimpanzees also modified their anvils using Metatools. Several individuals at Bossou were observed using Metatools in the "anvil-prop" form of behavior during nut cracking (Matsuzawa 1991, 1994; Sugiyama 1997). In anvil prop, a small stone is placed as a wedge under an uneven anvil, so that the anvil is horizontally level and stable. Several different chimpanzees have been observed using these Metatools. Matsuzawa (1994) considered Metatool Use involving anvil prop to be the most complex form of tool use documented in wild chimpanzees. Carvalho et al. (2008) reported the use of two wedges (both Metatools) as anvil props, resulting in a Tool Composite made up of four stones used to crack nuts (two stone wedges, one stone anvil, and one stone hammer). Chimpanzees at Bossou also repeatedly reused their Tool Composites for cracking nuts (Carvalho et al. 2009).

In another example of a Tool Composite, Matsuzawa (1991 in Sugiyama 1995a) observed a young Bossou female acquiring water from a hole by pushing in and then raking out a leaf sponge with a Detached twig. She used the two tools several times in this manner. Whiten et al. (1999) referred to this as "sponge push-pull" and noted that this behavior was also present at Gombe, Mahale M, and Taï.

In McGrew's discussion of tool composites (2004, 126), he stated that "a chimpanzee may sit on a bent sapling and use it as a vantage point for ant dipping, just as a human can use a ladder for painting a ceiling." However, McGrew did not elaborate on whether this was hypothetical or had actually been observed. In addition, since he did not state that the chimpanzee was actually responsible for bending over the sampling, we do not know if this is truly an example of a Tool Composite.

Köhler's (1925) captive chimpanzees stacked boxes and then stood on top and used a Reaching stick to obtain a suspended food reward. Although the completed tower is not a tool (but it is a Construction; see chapters 1 and 7), each stacked box in the tower is a tool. The Construction and the Reaching stick form a Tool Composite.

One of Köhler's chimpanzees stands on the top of a stack of boxes and uses a stick to Reach suspended food. This is a Tool Composite. Reprinted from Köhler 1917.

A wild chimpanzee using a Tool Composite to acquire water from a tree hole. The ape used a stick to push an absorbent leafy twig into the hole, and then used the stick to pull out the water-soaked twig. Drawing courtesy of Yukimaru Sugiyama.

Köhler (1925) may have been the first scientist to document a Tool Composite and the use of a Metatool, although he did not use this terminology. The chimpanzee Sultan was provided with two bamboo sticks that could have been Combined (see Add, Combine under Tool Manufacture) to form a longer stick, which could then have been used to Reach an out-of-reach fruit. But Sultan did not physically combine the tools, instead using them separately. Köhler described Sultan's behavior from several trials.

> He pushes one of the sticks out as far as it will go, then takes the second, and with it pokes the first one cautiously towards the objective, pushing it carefully from the nearer end and thus slowly urging it towards the fruit. This does not always succeed, but if he has got pretty close in this way, he takes

> even greater precaution; he pushes very gently, watches the movements of the stick that is lying on the ground, and actually touches the objective with its tip. Thus, all of a sudden, for the first time, the contact "animal–objective" has been established, and Sultan visibly feels (we humans can sympathize) a certain satisfaction in having even so much power over the fruit that he can touch and slightly move it by pushing the stick. (Köhler 1925, 131)

Although Sultan was not successful in obtaining the fruit in this manner, he was successful in touching the reward with the stick. For this objective, the two sticks can be viewed as a Tool Composite, and the stick with which Sultan pushed the second stick was a Metatool.

Multi-Function Tool

Sugiyama (1985, 1997) inferred that brush sticks found in association with termite mounds had been used for both Digging and Probing. He related that one end was covered in mud, indicating that it had been used to Dig into the termite mound. The Digging was confirmed by shallow holes in the mound. He noted that on fifty-two of these sticks, the other end had been frayed into a brush, which he speculated had been used for termite fishing. He argued that "the use of sticks with two functions, which appears to be habitual among chimpanzees in this area, can be considered a tool-composite because of their dual and sequential employment" (Sugiyama 1997, 24–25). However, we do not consider a single Multi-Function Tool to be Tool Composite, because both ends of the tool are not used simultaneously on the goal. Further, the classification of these sticks as Multi-Function Tools is based on inferences about their use, since there were no actual behavioral observations. Although, as noted above under Tool Manufacture, we now have evidence that chimpanzees make brush-tipped tools for Probing, as originally suggested by Sugiyama (1985), until recently we did not have evidence to support the use of any one of these tools as both probes and Penetrating/Digging tools.

Sanz, Call, and Morgan (2009) reported four observations at Goualougo that confirmed true Multi-Function Tool Use for termite fishing. The chimpanzees used a brush-tipped Probing tool to extract insects, then rotated the tool and used the blunt end to perforate the mound and clear debris from the fishing hole. Once the hole was clear, they rotated the tool again and resumed Probing with the brushed end.

In the context of nut cracking, stones may function as either hammer or anvil, thus serving as a Multi-Function Tool. "Sometimes a certain stone was used first as anvil, then as hammer [at Bossou]. This means they discriminated whether the stone should function as hammer or anvil" (Sakura and Matsuzawa 1991, 246).

Fay and Carroll (1994) observed an adult female at Bai Hokou use a blunt stick to Pound open a honey hive. After setting the tool down and retrieving some honey with her hand, she took up the Pounding stick again and made it a Multi-Function Tool by Inserting it into the hive as a Probing tool to extract the honey that adhered to it. Tutin, Ham, and Wrogemann (1995) reported three observations in which chimpanzees used a stout tool to gain access into an *Apis* beehive, and they also used the tool as a honey-extraction tool. Bermejo and Illera (1999) observed a chimpanzee at Lossi use a single tool as both pounder and bodkin to access a honey hive. After Pounding, the ape turned the tool around and used it as a bodkin, Penetrating the wall of the hive with the sharp end. We infer from Sanz and Morgan's (2009) descriptions that Goualougo chimpanzees might also have Multi-Function honey-gathering tools. Boesch, Head, and Robbins (2009) found that some of the tools used in the Tool Set for honey extraction at Loango exhibited signs of wear and use as both collectors and enlargers. These were termed Multi-Function Tools and were considered a fifth type of tool in this honey-extracting Tool Set.

Secondary Tool Use

Kitahara-Frisch, Norikoshi, and Hara (1987, 33) discussed captive chimpanzees manufacturing bone fragments as a "step towards secondary tool use." The bone fragments were the result of using a hammer to fracture a hollow bone containing a food reward. In a subsequent experiment, the apes were provided with the bone fragments and were allowed to use them to Cut through buckskin to obtain a food reward. However, as the chimpanzees did not manufacture the tools for Cutting, they were incidental byproducts of the first experiment, and we do not consider Secondary Tool Use to have occurred.

Matsuzawa (1994, 361) noted that a "chimpanzee broke a stone anvil with a hammer and the broken piece became a better hammer." He suggested that this might be the first step in stone tool manufacture by chimpanzees. If purposeful, it would be a case of Secondary Tool Use.

Tool Crafting

Much of chimpanzee tool manufacture may be considered Crafting, since after Detaching a tool from the substrate, the chimpanzees often further modify it with successive Subtractive steps and a final Reshaping. Pruetz and Bertolani (2007) were the first to classify a type of tool manufacture in chimpanzees as Crafting, citing sequenced manufacture and refinement of the tools used to hunt bushbabies. Sequenced manufacture included Detaching the branch, Subtracting leaves and side branches, trimming the end(s), and/or removing bark. The final refinement (seen in 50% of the cases) involves sharpening the stick to a point before Stabbing at the bushbabies. We agree with Pruetz and Bertolani that many cases of the manufacture of termite-fishing and ant-dipping tools should also be classified as Crafting. These sequences and final Reshaping are described above in Insert and Probe, under Tool Manufacture.

7

Seven Myths

AS WITH EVERY GOOD MYTH, THERE IS SOME TRUTH TO EACH OF THESE.

Myth #1: Tool Use is Unique From all Other Behavior

Despite our effort to define tool use and tool manufacture in a way that make them distinct behavioral categories, a number of behaviors are excluded that many have argued are tool use. We call these *borderline cases*. For example, chimpanzees, capuchin monkeys, sea otters, and white-winged choughs pound food *with* stone hammers (included) and *on* stone anvils (excluded); some birds impale insects *with* sticks (included) and others impale them *on* sticks (excluded); pig-tailed macaques use *sticks* to reach distant food (included) while long-tailed macaques use their *tails* (excluded). Reed (1977) wryly commented that a chicken raised for the pot would not be considered a tool, but it would if it were killed and used to club to death a second chicken that was then eaten. Although too numerous to review exhaustively, discussion of some of these borderline behaviors is warranted.

Constructions

The most common class of borderline cases includes nests, dens, burrows, traps, bowers, hives, webs, dams, food caches, and other feats of external "architectural" construction, described in detail by Collias and Collias (1976, 1984), Fruth and Hohmann (1996), Hansell (1984, 2000, 2007), and von Frisch (1974). Collias and Collias (1976, 2–4) provided a particularly useful and illuminating classification scheme. Recall from chapter 1 that we define a Construction as a functional thing consisting of two or more physically linked tools and / or objects. A Construction is not a tool because, once completed, it is not held or directly manipulated in its entirety. A Construction does not involve tool manufacture, because the final product is not a tool.

Constructions facilitate raising young, thermoregulation, defense against predators, and food capture and storage. They are built by a variety of invertebrates and vertebrates and are far more common than animal tool use. Constructions are made from an astonishing variety of natural and man-made materials and are dug, cemented, woven, spun, secreted, lined, decorated, camouflaged, heated, cooled, and otherwise made and altered in ways that far outnumber the modes of tool manufacture. Many variations appear to be adaptations to subtle differences in prevailing environmental conditions. Materials are sometimes brought considerable distances to the point of construction, and cooperation is often involved in construction. In some species, individuals deprived of all relevant experience can efficiently produce a species-typical Construction on the first attempt; in others, considerable experience is required. In other words, these Constructions share many of the cognitive attributes of tool use and tool manufacture. Nonetheless, they are not held or directly manipulated in their entirety by their users, and thus are not tools by our definition.

We view nests, mounds, bowers, dams, some dens, and some food caches as the cumulated products of thousands of cases of essentially identical acts of tool *use*, for example, placing a twig in a nest, dam, or bower, which changes the form of the nest, dam, or bower with each repetition. Each element is a tool, but the final Construction per se is not a tool, because it is not held or manipulated in its entirety. One could also view a Construction as the exquisite cumulated product of thousands of cases of tool *manufacture*, in which components of the putative tool are Detached, Subtracted, Added, Combined and Reshaped. The finished Construction might be a "super tool" (to borrow a term from Matsuzawa [1991], who used it in a different sense). However, since we don't regard the finished Construction as a tool, then the tool-manufacture view is not consistent.

Some will continue to be frustrated by the exclusion of Constructions as tools, even though they are made up of tools in most cases. No animal tool matches a Construction in size, complexity, or the number of elements involved. For example, one pair of bald eagles (*Haliaeetus leucocephalus*) at the Indianapolis Zoo used a nest for twelve breeding seasons, at the end of which it weighed 2.5 U.S. tons! Even more impressive are termite mounds (Emerson 1938; Noirot 1970), the nest mounds of the malleefowl (*Leipoa ocellata*) (H. Frith 1956), and beaver dams and lodges (L. Morgan 1868; P. Richard 1964, 1967; A. F. Wallace and Lathbury 1968; Wilsson in Collias and Collias 1976). Compared with the size of their Constructors, far smaller Constructions might be even more impressive: tiny weevils (*Byctiscus populi*) prepare and roll poplar leaves, excavate a hole into the roll, deposit eggs, and sometimes add outer leaf coverings. They then cut off the leaf Construction, which drops to the ground, where the eggs hatch. The larvae eat the leaf Construction from within and then pupate (Daanje 1975). These examples are chosen because they are fascinating, not because they are exceptional. Constructions are far more diverse and elaborate than tools and, though they are not held or manipulated in their entirety, they meet all of the other definitional criteria of tool use and share functional, cognitive, ontogenetic, ecological, and evolutionary attributes with tool use. Hansell and Ruxton (2008) even argue that tool behavior should be considered a form of construction behavior, rather than construction behavior being considered as some type of tool use. In this edition, we will not call these Constructions borderline cases, as did Beck (1980), but instead treat them as a category of Associative Tool Use (see table 1.2).

There are some intermediate variants that blur the distinction between tools and Constructions. Most spider webs are not included as tools, because they are not held, but *Dinopsis* spiders hold and Throw their webs. Bolas spiders (*Mastophora* spp.) carry a sticky ball made of the same secretions used by orb weavers in web construction. When prey approaches (the spider may actually Bait the prey by secreting a substance that mimics the prey's sex-attractant pheromone [Eberhard 1977]), the bolas spider swings the ball. If struck, the prey sticks to the ball (Akerman 1926; Coddington and Sobrevila 1987; Gertsch 1955; Hutchinson 1903; M. Robinson and Robinson 1971). We argue that the "webs" of these spiders are tools (see chapter 2), because they are held and manipulated in their entirety. Larvae of the hydroptilid caddisflies spin a cocoonlike tube to which they affix bits of leaf and stick, shells, pebbles, and other objects. The larva carries this casing around as it matures, so the casing is a tool. Larvae of other caddisfly families construct casings that are fixed to the substrate and are not carried (Hanna 1960; Milne and Milne 1939; H. Ross 1956, 1964). The fixed casings are not tools.

Burrows, nest cavities, and cricket "calling chambers" (Forrest 1982; Nickerson, Snyder, and Oliver 1979; Prozesky-Schulze et al. 1975) are also vexing, because objects are not Added or Combined to construct a thing, as in the construction of a nest or dam. Rather, dirt, wood, or parts of leaves are *removed*, creating a negative space. This does not literally fit our definition of Construction as a form of Associative

Tool Use. Sometimes, however, mammals and birds line a burrow with soft material. Nuthatches (*Sitta* spp.) and woodpeckers (Picidae) may deposit sticky resin around the entrances to their nest cavities or drill holes around the entrances to cause resin secretion. The resin may repel ants and predators (Dennis 1971). Thus the resin can be said to be a tool if it is actively manipulated by the bird.

Some mammals and birds cache food, usually in a cavity or hole that they construct or simply use. For example, the great spotted woodpecker (*Dendrocopos major*) and Lewis's woodpecker (*Melanerpes lewis*) excavate recesses in tree bark into which they wedge pinecones and acorns (Law 1929; Sielmann in Boswall 1977b). Some simply use existing recesses, for example, in wooden telephone poles. Kamil, Balda, and their associates have elegantly explored the positive correlation between seed caching and spatial memory in birds (e.g., Olson et al. 1995). African elephants (*Loxodonta africana*) cache water by digging and then sometimes corking small seeps with bark, leaves, or grass (Gordon 1966). The corks would be tools, but the caches are not. Caches appear to differ from most external Constructions in that they are not actually occupied, and in many cases they are not even made by the user. For the most part, caches are negative space and thus would not be considered Constructions, but the food caches of beavers (*Castor canadensis*, *C. fiber*) (P. Richard 1964, 1967; Slough 1978) and bears (*Ursus* spp.) (e.g., Mysterud 1973) are made up of Combined objects and thus would be Constructions. Caches are therefore not tools, and only some are Constructions. However, caches are sometimes camouflaged by or treated with tools.

As noted in chapter 1, there are also Constructions that arguably are directly manipulated by the user, but we do not regard them as tools because the user is not responsible for the proper and effective orientation of *all of the component parts*, that is, the user does not establish all of the mechanical linkages. These include the rifle and bullet that are used to kill antelope, as well as the reins, horse, harness, and plow, and the tractor's steering wheel, motor, drive shaft, and plow combinations that are used to till soil. These are Constructions, but not tools.

The difference between Constructions and Tool Composites, both categories of Associative Tool Use, is another borderline issue. One example, the stacking of three boxes by an ape to make a tower to reach suspended fruit, is illustrative. Beck (1980) categorized the tower as a tool (option 1 for purposes of this discussion), and he further stated that the tower was a manufactured tool, since the boxes were Combined. The ape was responsible for the proper and effective orientation of the boxes. Nonetheless, the tower cannot be a tool by definition, since it is not held or directly manipulated in its entirety. If the tower is not a tool, perhaps the highest box is a tool, and the two lower ones are not (option 2). The ape stands only on the top box, but he does climb on the other two, Repositions them, and links them by stacking. It would seem that if the third, top box is a tool, then the bottom two should also be tools. Perhaps each of the three boxes is a tool (option 3), but then what is the tower?

We were not able to resolve this without recourse to the categories we call Associative Tool Use, but the solution is still elusive. The stack of boxes fits the definition of a Construction (option 4). The boxes can be seen either as tools or as objects that are linked, in this case by gravity. Yet the tower may also fit the definition of a Tool Composite (option 5). The mode of use of each box is the same (Reposition), but this is also true of a few other Tool Composites, for example, capuchin monkeys (*Cebus* spp.) Pounding two stones together to produce a social signal. However, the ape tower differs from the archetypal Tool Composite, in that the latter comprises a tower *as well as* another tool that is not part of the tower. As described in chapter 1, the archetype includes one or two small stones (Metatools) wedged by a chimpanzee beneath a portable anvil stone (a tool) to make a multi-part anvil, which can be considered a short tower. The ape then places a nut on the anvil stone and uses another stone (another tool) to Pound the nut to open it. Thus this Tool Composite archetype (with the anvil) includes, but is *more than*, a tower. In the box tower example, if the ape climbed on the box tower with a stick, and used the stick to Reach the fruit, this behavior would then match the archetype and could comfortably be called a Tool Composite. Yet if the portable anvil stone placed on two stone wedges is not a Tool Composite per se, then the ape tower is not a Tool Composite, either. The *hammer* makes the anvil stone and wedges a Tool Composite, just as the *stick* would make the ape tower a Tool Composite. In the absence of the stick, the ape tower cannot logically be a Tool Composite. It is a Construction, and each box is a tool or Metatool!

On the other hand, consider the bowerbird's (Ptilonorhynchidae) bower, which is a Construction, comprising hundreds of sticks and other items, each a tool that is carefully linked by the bird with others. The function of this Construction is to entice a mate to approach. Some males of some species (*Sericulus*, *Ptilonorhynchus*, and *Chlamydera*) additionally "paint" the bower by Applying pigment with a piece of bark (another, different kind of tool; see chapter 3). This presumably enhances the attraction power of the bower. Is the bark like the hammer in the archetypal Tool Composite, that is, a tool applied to a Construction? If so, the painted bower is a Tool Composite, too. Yet the bark "paintbrush" really just makes the bower a better bower, whereas the stone hammer adds something functionally different from the anvil/wedge Construction. Corsican blue tits (*Parus caeruleus*) add fragrant plant parts to their nests, apparently to repel mites, bacteria, and viruses (C. Petit et al. 2002). Similarly, common waxbills (*Estrilda astrild*) add dry carnivore feces to their nests to repel predators (Schuetz 2005). Do the herbs and dung simply make the nest better nests, or are they tools that add something functionally different to the nests, which then become Tool Composites?

We authors could not agree if the painted bower or the scented nests are Tool Composites. Our definitions are simply not detailed enough to decide, and the questionable cases are too few to warrant the space and the effort required for finer distinctions. Nonetheless, our definitions allow us to determine that the ape box tower, the chimpanzee anvil/wedge stones, and the bower and nests are all Constructions. The stone hammer applied to the anvil/wedge stones results in a Tool Composite, as does the ape taking a stick to the top of the tower and using it to Reach the fruit.

Proto-Tool Use

A different class of borderline cases involves proto-tools. One type are fixed anvils: hard surfaces against which animals drop, pound, or throw foods having tough rinds or hard shells. The labrid fishes (*Cheilinus trilobatus* and *Coris angulata*) pound sea urchins on coral or stones to detach the spines and smash the exoskeletons. A particular anvil within a fish's territory may be used habitually (Fricke 1973).

Many birds use anvils (Lefebvre, Nicolakakis, and Boire 2002). One of the bowerbirds, the tooth-billed bowerbird (*Scenopoeetes dentirostris*), and the song thrush (*Turdus philomelos*) pound snails on rocks (A. Marshall 1954; D. Morris 1954; Romanes 1892). The white-winged choughs (*Corcorax melanorhamphos*) observed by Hobbs (1971) are true hammer users, but they also pound mussels against tree branches, roots, and other unopened mussels. Lewis's woodpeckers pound acorns on tree limbs to break them open (Law 1929). Boswall (1977b) noted that certain pittas, for example, the noisy pitta (*Pitta versicolor*), are anvil users, but he did not supply details. Boswall also referred to Smythies' observations of blue whistling thrushes (*Myophonus caeruleus*) breaking shells on rocks. Egyptian vultures (*Neophron percnopterus*) drop eggs of the great white pelican (*Pelecanus onocrotalus*), and red-legged seriemas (*Cariama cristata*) drop hens' eggs (L. Brown and Urban 1969; Kooij and van Zon 1964). Both predators throw the eggs down while standing on the substrate or drop them while flying. Sladen (in Boswall 1977b) observed skuas (*Stercorarius* sp.) dropping penguin eggs. New Caledonian crows (*Corvus moneduloides*) drop candlenuts (*Aleurites moluccana*) onto rocks (G. Hunt, Sakuma, and Shibata 2002). Bearded vultures (*Gypaetus barbatus*) drop bones to expose the marrow (Boswall 1977b; Hartley 1964; Huxley and Nicholson 1963; North 1948; Reiser 1936) and drop tortoises to expose the flesh (Hartley 1964, Reiser 1936). Pliny the Elder, in his *Naturalis historia* (in Leshem 1979), recounted that the Greek playwright Aeschylus died after an eagle or vulture dropped a tortoise on his head. This would be either tool use, if the bird was trying to hit Aeschylus with the tortoise, or proto-tool use, if the bird was trying to crack the shell and hit the playwright accidentally. Bald eagles likewise drop tortoises (Bindner in van Lawick-Goodall 1970). Rooks (*Corvus frugilegus*) drop mollusks and walnuts (J. Fisher and Lockley 1954). Ravens (*C. corax*) drop bones (Lorenz in van Lawick-Goodall 1970). Carrion crows (*C. corone*) drop mollusks and walnuts (Lloyd, and also Saunders, in Oldham 1930), and hooded crows (*C. cornix*) also drop mollusks (Patterson in Oldham 1930; Perry 1972). Grobecker and Pietsch (1978) and Maple (1974) saw American crows (*C. brachyrhynchos*) drop palm fruits on the street and wait for automobiles to run over and break them. Maple (1974) suggested that the crows were intentionally using the cars as nutcrackers. Cristol et al. (1997) found that crows dropping walnuts on a road were no more likely to drop them when a car was approaching than when one was not approaching. Caffrey (2001) noted that the crows at least waited until a car had passed to fly down and inspect the nut. Zach (1978, 1979) conducted a detailed study of the dropping of whelks (*Thais lamellosa*) by Northwestern crows (*C. caurinus*).

Gulls (*Larus* spp.) have frequently been noted

A herring gull drops a marine snail on rocks to break the shell. This is proto-tool use, because the rocks are not held or manipulated by the gull. Photograph courtesy of Benjamin Beck.

to drop a variety of marine shellfish to gain access to the flesh (Anthony 1906; Barash, Donovan, and Myrick 1975; Bent 1963; J. Fisher and Lockley 1954; Forbush 1925; Goethe 1958; Hartley 1964; Ingolfsson and Estrella 1978; Norton 1909; Oldham 1930; Perry 1972; Romanes 1892; Siegfried 1977; Tinbergen 1960; Townsend in Strong 1914), and there is one report of a herring gull (*L. smithsonianus*) dropping live rabbits to kill them (Young 1984). Beck (1982) conducted an observational field study of shell dropping by herring gulls. The study was specifically designed to examine the biological meaningfulness of the distinction between ape tool use and at least one type of proto-tool use. In this case, the function of the behaviors and the underlying cognitive capacities were very similar (see below).

African and Indian mongooses (*Atilax paludinosus*, *Herpestes* spp., *Helogale* spp., and *Mungos mungo*); the narrow-striped and ring-tailed "mongooses" of Madagascar (*Mungotictis lineata* and *Galidia elegans*); striped and spotted skunks (*Mephitis mephitis* and *Spilogale putorius*); a yellow-throated marten (*Martes flavigula*); and Alaskan brown bears (*Ursus arctos*) all throw or drop objects such as eggs, pill millipedes, snails, the cells of mason wasps, nuts, and bones against either hard vertical surfaces or the substrate. Ewer (1973) reviewed this literature; Acharjyo and Misra (1972), Dücker (1965), Eisner and Davis (1967) and Rasa (1973) provided additional data on mongooses; Albignac (1969) supplied observations on the ring-tailed mongoose; Wemmer and Johnson (1976) noted behaviors by the marten; and Hornaday (1922) described bone breaking by bears. Most of the observations were on captive specimens, but confirmation from the wild is common (e.g., Eisner 1968; Rasa 1973).

Wild chimpanzees (*Pan troglodytes*) at some study sites pound hard-shelled fruits against anvils (stones, tree trunks, roots, and buttresses) to open the fruits (e.g., van Lawick-Goodall 1973; Gašperšiè and Pruetz 2005; Marchant and McGrew 2005). Rehabilitant orangutans (*Pongo* spp.) even pounded hard items on conspecifics' heads (Russon 2003). Wild capuchin monkeys at some sites pound hard-shelled fruits, nuts, snails, and insect-laden branches against rocks, limbs, and tree buttresses (Freese 1977; Izawa 1979; Izawa and Mizuno 1977; Panger 1998; Struhsaker and Leland 1977; Thorington 1967). Anvil use by captive capuchins is common, and it may occur at higher rates than among wild capuchins (Panger 1998).

A variant on anvils is the use of concrete wall surfaces by members of one zoo chimpanzee colony to smear or rub fruit, producing a juicy pulp that they lick and eat (Fernández-Carriba and Loeches 2001). The first chimpanzee to show this behavior had few teeth, but others with full dentition also learned to smear fruit. Yet another variant is both wild and captive chimpanzees' slapping or drumming on tree buttresses, tree trunks, and cage surfaces to produce low-frequency sounds that accompany their vocal "pant-hoot" displays (e.g., Arcadi, Robert, and Boesch 1998; Arcadi, Robert, and Mugurusi 2004; Boesch 1991c; J. Goodall 1986).

D. Ellis and Brunson (1993) described a red-tailed hawk (*Buteo jamaicensis*) swooping at boulders while holding a snake in its talons, causing the snake's head to strike the stone. Recesses into which birds shove seeds and nuts in order to stabilize them while the bird pecks through the shell are another anvil variant (e.g., Hilton 1992; Law 1929).

Closely related to anvil use is the use of sharp projections and forks by shrikes (*Lanius* spp.) for impaling and wedging relatively large prey. The prey are thus immobilized, and their escape is prevented during dismemberment and consumption (Cade 1967; Lorenz and von Saint Paul 1968; S. Smith 1972, 1973; Wemmer 1969).

Ungulates commonly use fixed "scratching posts," which are not held or directly manipulated. Steinbacher (1965) described an analogous behavior of a young captive Indian elephant (*Elephas maximus indicus*) that had been ill for a week with what proved to be an abscess on its trunk. At the height of its illness, the elephant shoved its trunk over the end of a projecting branch and slowly inserted the branch to a depth of 50–60 centimeters. The animal then withdrew its trunk, and a clump of bloody pus remained on the tip of the branch. The animal had lanced its own abscess, and it recovered quickly.

Body Parts and Internally Employed Objects

The exclusion of tails when used to reach distant incentives, and of horns and antlers when used to scratch, have already been explained in chapter 1.

We also noted in chapter 1 that unless food is to be considered a tool to attain energy, all ingested objects must be disqualified as tools. Many crustaceans ingest sand to aid in postural orientation (Meglitsch 1972). Chia (1973) reported that juvenile sand dollars (*Dendraster excentricus*) store sand particles in an intestinal diverticulum. As the animal matures and its body size increases, proportionately less sand is stored. Chia concluded that the sand serves as a weight belt to anchor the smaller, lighter individuals in shifting sand. Iron oxide, the heaviest of all available sand-grain types, constitutes 78 percent of the diverticulum sand but only 9.8 percent of substrate sand, suggesting that sand dollars selectively store the heaviest grains. Birds ingest sand, pebbles, stones, shells, feathers, fruit pits, and other hard objects that pass directly through the crop to the muscular stomach or gizzard. The grit is assumed to grind or mechanically break down food (Farner 1960; Jenkinson and Mengel 1970; Meinertzhagen 1964; L. Miller 1962; Ziswiler and Farner 1972).

Also excluded is the ingestion of plants thought to have medicinal qualities. For example, chimpanzees and bonobos (*Pan paniscus*) suspected of having a heavy endoparasite load search for and ingest the leaves of one of a few specific plant species (Dupain et al. 2002; Huffman and Wrangham 1994). Leaves of these plants are not otherwise eaten. The leaf may be carefully folded before it is swallowed, and it may not be chewed before swallowing. The leaf may also be excreted whole. The leaves are often rough or sharp edged, suggesting that there might be a mechanical scouring, antiparasitic action (Huffman and Caton 2001; Krief, Wrangham, and Lestel 2006). Leaves of these plants may also be eaten by humans with endoparasites. The behavior arguably meets all aspects of the definition of tool use, except that the leaf is internally, not externally, employed.

Included here as tools are internally manufactured body products, provided that they are outside the body at the time of use and are held or directly manipulated by the user. Thus feces that are Thrown at an intruder are tools, and those that are simply eliminated above the intruder are not. Captive apes and elephants can become skilled feces throwers. Many captive apes learn to spit, which could be considered a variant of aimed Throwing. Captive and wild hippopotamuses (*Hippopotamus amphibius*) spread their feces, presumably as a social signal, by vigorously twirling their tails as they defecate (B. Beck, pers. obs.). The behavior and its effect are barely visible underwater but quite evident on dry land, especially if there is a vertical surface (a veritable canvas) behind the hippopotamus. If the hippopotamus is directly manipulating and applying its feces, this might be tool use.

Many mammals scent mark, or deposit bodily secretions on themselves and on environmental surfaces (e.g., Eisenberg 1981; Ewer 1973). Scent marks can be considered tools only when the substances are actually held and applied by the animal, for example, some scent marking by lemurs (see chapter 5).

Soil and Water

Many birds dust bathe or sand bathe, which is thought to aid in plumage maintenance (see reviews by Goodwin 1956 and Levine, Hunter, and Borchelt 1974). The birds stir up the soil with their wings and feet, and they manage to get the soil on their plumage. Nonetheless, we do not see these behaviors as meeting either the "hold or directly manipulate" criterion or the "proper and effective orientation" criterion for true tool use. Berthold (1967) documented that the rusty coloration of many birds, and a few mammals, derives not from internal pigmentation but from adhering iron oxide. It is not clear if these animals actually apply the tint or simply get it passively from the environment, and its purpose is also not evident.

Discussions of tool use frequently allude to the washing of dirty food or the soaking of hard food. The most celebrated examples are sweet potato washing and dip-seasoning, as well as wheat rinsing, by the Japanese macaques (*Macaca fuscata*) of Koshima Islet (Itani and Nishimura 1973; Kawai 1965; Kawamura 1959; Watanabe 1994). Raccoons (*Procyon lotor*) commonly submerge food in water, but the behavior seems to be an artifact of captivity rather than functional washing or soaking (Ewer 1973; Lyall-Watson 1963). Ewer (1973) also relays Duplaix-Hall's observations of similar behavior by clawless otters (*Aonyx*

A Japanese macaque from Koshima Island rinses a sweet potato in the ocean. Rinsing cleans dirt from the potato and adds flavoring. Photograph courtesy of Satoshi Hirata. Reprinted with kind permission of Springer Science+Business Media from Hirata, Watanabe, and Kawai in Matsuzawa 2001, p. 491.

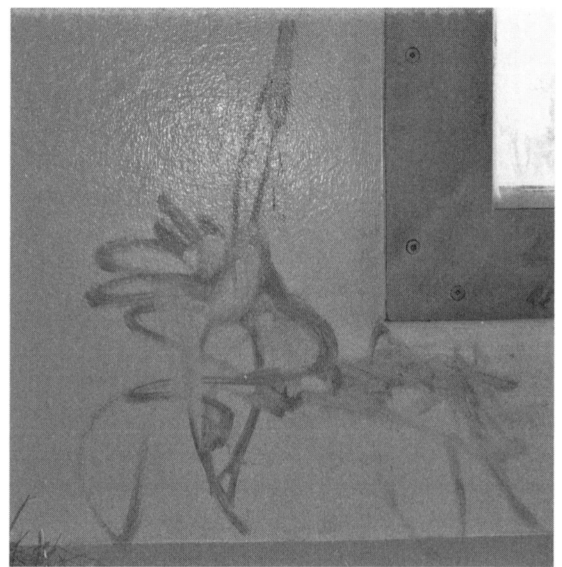

A "painting" produced spontaneously by an orangutan at Great Ape Trust. The work is 1 meter high. Photograph courtesy of Robert W. Shumaker.

capensis). Captive monkeys and apes commonly soak hard biscuits, and captive polar bears (*Thalarctos maritimus*) soak hard crusty bread (B. Beck, pers. obs.). Wild grackles (*Quiscalus quiscula*) soak stale bread, and wild herring gulls dip noxious, sand-covered, or very large gastropods (Beck 1982). Marabou storks (*Leptoptilos crumeniferus*) wash captured dung beetles before eating them (Seibt and Wickler 1978). W. McGrew (pers. comm.) saw a goshawk (*Micronisus gabar*) capture a bat and drown it in a stream. None of these behaviors can be included as tool use as presently defined, since the water is not held or carried.

Painting

Some captive apes, monkeys, dolphins, elephants, and birds apply pigments of various types to different types of flat surfaces. The result is often called a "painting" or "art" (Boysen, Berntson, and Prentice 1987; R. Fouts 1997; Gucwa and Ehmann 1985; C. Hayes 1951; Kohts 1935; Levy 1992; D. Morris 1962; Russon 2000; Schiller 1951; Temerlin 1975; Vancatova 2008). Animals are reported to have used brushes, sticks, pencils, crayons, pens, charcoal, and a smoldering mosquito coil to apply the pigment, which some observers have called tools. Some apes make or modify the "tool." Most animals have to be trained to perform these behaviors with a modicum of control, and sometimes they are manually guided by humans. The animals seem to perform these behaviors willingly, sometimes even eagerly, and the products are greatly esteemed by people. Some people infer that the animals copy exemplars, and some people see elemental aesthetic capabilities in the products. The animals clearly hold the "tool" and apply the pigment with it, but it is unclear whether the animal has a purpose, that is, whether the animal intends to alter the condition of the surface. Like Boysen, Berntson, and Prentice (1987), in the absence of unambiguous evidence of purpose, we regard this set of behaviors as object play or object manipulation, not tool use. However, we do regard the application of pigment to the inside of bowers with bits of bark by some wild bowerbirds as tool use (see chapter 3).

Object Play

Sometimes animals manipulate objects in ways that resemble tool use, but seem purposeless or merely playful. Seemingly functionless "stone-play" or "stone handling" by Japanese macaques is a well-documented example (e.g., Huffman 1984; Huffman and Quiatt 1986; Leca, Gunst, and Huffman 2008). Wild and captive chimpanzees carry twigs, sticks, leaves, and even a sapling in ways that J. Goodall (1963a) and McGrew

"Stone handling" by a Japanese macaque. Some individuals rub stones together habitually, but since the behavior has no apparent purpose, it is not considered tool use. Photograph courtesy of Jean-Baptiste Leca.

(1977, 1992) called playful. McGrew (1993b) and Nishida (1980a) called the objects "toys." Young chimpanzees may steal these objects from each other in play (Boesch 1995; McGrew 1977, 1993b). Chimpanzees have been documented on several occasions to capture a prey animal and hold or manipulate it rather than kill and eat it (Boesch and Boesch 1989; Hirata et al. 2001; van Lawick-Goodall 1968). Herring gulls drop shells and other objects in play (Beck 1982; Gamble and Cristol 2002). Captive bottlenose dolphins (*Tursiops truncatus*) voluntarily create rings of bubbles which they then manipulate as the bubbles rise to the surface. The dolphins tend to produce these rings at specific sites in their pools. The behavior seems to be playful but otherwise purposeless (McCowan et al. 2000). A captive soft-shelled turtle (*Trionyx triunguis*) named Pigface played extensively with a basketball, a ring, and sticks in a rare case of object play by a reptile (Burghardt, Ward, and Rosscoe 1996). All of these behaviors fit our definition of object manipulation or object play: "holding or directly manipulating unattached or manipulable attached environmental objects with no evident purpose." Because of their purposelessness, they are not tool use. However, when authors stipulate that objects are used to trigger or enhance social play, we do include the events as tool use.

Captive bottlenose dolphins produce and manipulate rings of bubbles. The rings have no apparent purpose and thus are not tools. Photograph courtesy of Brenda McCowan.

Pre-Positioned Strings and Rakes

Recall from chapter 1 that we do not regard animals pulling in strings with food attached as tool use, because they are not responsible to the proper and effective orientation of the tool to the incentive (e.g., Beck 1967; Dücker and Rensch 1977; Fischer and Kitchener 1965; Klüver 1933). Boswall (1977b) called string pulling by birds borderline, W. Thorpe (1963) reviewed the literature and called pulling an elementary form of tool use, and Millikan and Bowman (1967) called it true tool use. We do not. Likewise, we exclude experimental results with primates and birds where reaching tools were pre-positioned behind or in front of the incentive, and the animal simply had to pull or push (respectively) the tool to get the incentive. C. Carpenter (1963) raised the possibility that bridging behavior by spider and howler monkeys is tool use. Analogous behavior by ants has been excluded, because the bridge of linked conspecifics is pre-positioned and not held or manipulated by those who cross. Likewise, a young monkey simply walks over the solicitous adult that spans an arboreal gap. Bard (1994) and Rijksen (1978) documented social bridging by orangutans, but the bridge animal is again not held or manipulated by the recipient.

There undoubtedly are other behaviors that some would at least wish to consider borderline. Some might prefer to move behaviors that we have called tool use to the borderline category, and vice versa. Yet in the end we are forced to conclude that tool use, in terms of topography, function, or causal dynamics, dovetails imperceptibly with other categories of behavior. In contrast, Marchant and McGrew (2005) characterize as a "cognitive leap" the progression of chimpanzees (and human ancestors) from pounding hand-held nuts *on* anvils (borderline) to laying them on anvils and pounding them *with* hammers (tool use). Simply put, they would claim that in this case tool use evidences more intelligence (although some chimpanzees do both forms of pounding). Lefebvre, Nicolakakis, and Boire (2002) found that the relative brain sizes of bird taxa reported to use true tools tended to be larger than those that used only proto-tools or showed only borderline behaviors. However, even here there was overlap: some of the borderline or proto-tool users had larger brains than some of the true tool users.

Myth #2: Tool Behavior is Intelligent

The myth is that tool use is cognitively more complex, or more intelligent, than behavior that does not involve tools (e.g., Bird and Emery 2009a; Brosnan 2009; van Schaik, Deaner, and Merrill 1999). Further, tool manufacture is said to be more intelligent than tool use (e.g., G. Hunt and Gray 2004a; Matsuzawa 1996; Weir, Chappell, and Kacelnik 2002). Metatools, Tool Composites, and Tool Sets are said to be equally if not more intelligent than simple tool manufacture (e.g., Clayton 2007; Matsuzawa 1996; McGrew 1993b; Taylor et al. 2007), and Secondary Tool Use, the use of a tool to make a tool, is said to be more intelligent yet (R. Wright 1972).

Beck (1980, 237) argued against these conclusions: "There is no consistent association between tool behavior and intelligence in living animals." Most scientists have abandoned the myth, since Beck showed that many animals, not just intelligent ones, use and make tools. Further, following Hall (1963), Beck argued that much animal tool behavior appears to be genetically programmed or based on simple association learning. Hansell (2007) has been a strong and convincing advocate of the "tools don't necessarily imply intelligence" position. McGrew (1993b, 167) wrote that "there is no obvious connection between degree of intelligence and complexity of tool use." But the myth survives in media and popular accounts, and it is periodically reinvented by scientists wanting to show that the species with which they work are intelligent ("Since tool use is intelligent, and X species uses tools, then X species is intelligent"). Both the premise and the reasoning are fallacious. A similar and equally fallacious argument is sometimes implied: "Since chimpanzees use a (specified) type of tool, and X species uses the same type of tool, then X species is as intelligent as chimpanzees."

An exhaustive treatment of "intelligence" is beyond our scope, but subsumed under intelligence are learning sets (rapid, hypothesis-based or concept-based discrimination learning); identification of an object with each of two or more senses (implying some central cognitive representation of the object); self-recognition and self-awareness; cognitive spatial maps; mental representation of objects and others and of relationships among them (e.g., kin, dominance); understanding of functional relationships and causality; ability to solve problems insightfully; ability to innovate; ability to comprehend the perceptions and intentions of others; ability to learn by observing others; persistent, population-wide behavioral traditions ("cultures"); teaching; deception; understanding and use of symbols in a rule-based fashion; numerical competence; counting and adding; and use of tools (e.g., Bitterman 1960, 1965; Boysen 1997; R. Byrne 1997; R. Byrne and Whiten 1988; Cheney and Seyfarth 2007; Essock-Vitale and Seyfarth 1987; Hodos 1970; Hodos and Campbell 1969; Panger 2007; Rumbaugh 1970, 1971; Rumbaugh and Washburn 2003; Russon 1998; Savage-Rumbaugh and Lewin 1994; Shillito et al. 2005; Shumaker et al. 2001; Visalberghi, Fragaszy, and Savage-Rumbaugh 1995; Warren 1973; Woodruff and Premack 1979). If the myth is correct, we should find considerable evidence of "intelligence" in tool behavior, and less in non-tool behavior. We should find no tool behavior with little evidence of intelligence, and no impressively intelligent non-tool behavior.

Some try to settle the issue through definitional caveat. Jalles-Filho, Teixeira da Cunha, and Salm (2001, 366) included as tool use only those cases that "demonstrate great cognitive flexibility and complexity." Urbani and Garber (2002) included as "true" tool use only those cases that evidence causal understanding. Aside from the logical fallacy of absence of evidence of cognitive complexity being evidence of absence of cognitive complexity, this definitional requirement is circular and would, for the most part, make relatively large-brained mammals and birds the only tool users. Matsuzawa (2001, 11) stated it plainly: "tool use is limited to primates." Van Schaik, Deaner, and Merrill (1999, 719) were more nuanced; they focused on feeding tools, because "intelligence explains why in the wild only apes are known to make and use feeding tools routinely." Thus there is little surprise when they concluded that "intelligent species" such as chimpanzees and orangutans are the only ones to use and manufacture tools on a population-wide basis in the wild, while capuchin monkeys "show virtually no sustained tool use in the wild" (Van Schaik, Deaner, and Merrill 1999, 722). Definition is crucial here. To avoid circularity in evolutionary, taxonomic, ontogenetic, and behavioral-ecological analyses of tool behavior, the definition of tool use and tool manufacture must be free of any reference to or dependence upon intelligence.

It seems parsimonious to use an intelligence-free definition of tool use and tool manufacture, and then to recognize that some tool use and manufacture is intelligent, while some is less intelligent or not intelligent at all. The tool behavior of apes, for the most part, can be confidently inferred to be highly intelligent. Viewing our catalog, the crows (*Corvus moneduloides*) that made Probing tools, the badgers (*Taxidea taxus*) that used tools to Block the burrow entrances of ground squirrels (*Urocitellus richardsonii*), the California sea otters (*Enhydra lutris*) that Pounded shellfish on portable anvils, the elephant that plugged its cache of rare water also all seemed to have been using tools intelligently. Crabs wearing shells and Waving actinians (a Tool Composite), and assassin bugs (*Salyavata variegata*) using dead termites as Bait, are probably following genetically programmed routines, albeit ones that are intricate, highly adaptive, and essential for survival. The baboons (*Papio* spp.) that Beck (1973a, 1973b) observed using Reaching tools so dexterously learned to do so by simple trial-and-error learning. The New Caledonian crows tested by Wimpenny et al. (2009) learned complex Sequential Tool Use tasks by association rather than through higher-level cognitive processes. Clearly, not all tool use is intelligent!

Further, much non-tool behavior is remarkably intelligent. The chimpanzees and orangutans that are using and making tools so intelligently are also following cognitive maps, finding and processing widely distributed and seasonally available foods, building complex nests, and conducting complex and nuanced behavioral interactions with conspecifics, all without tools. Baboons, which do not often use tools, keep large mental inventories of other individuals and of the kinship and dominance relationships among them. The baboons make predictions about the outcomes of interactions between and among those individuals, and they strategize about how to optimize outcomes for themselves in interacting with them (Cheney and Seyfarth 2007). Wild gorillas (*Gorilla* spp.) rarely use tools in feeding (one of the "inconvenient truths" for those wishing to champion the tool-using abilities of great apes), but R. Byrne (1997) has demonstrated the cognitive complexity of their feeding behavior (at least of mountain gorillas), and Russon (1998) did likewise for orangutan feeding and locomotor behavior.

Observable attributes of predatory shell dropping by herring gulls (Beck 1982) strongly supported inferences of the involvement of such advanced cognitive processes as extended concentration, purposefulness, selectivity, premeditation, imagery, plasticity with regard to subtle variations in environmental conditions, and strategies. Social learning and play appear to be involved in ontogeny. Beck (1980, 1982) showed direct comparability of these inferred cognitive processes with those still thought to underlie the use of feeding tools by great apes.

We can only reiterate Beck's plea (1980, 210) that we "finally put to rest the notion [myth] that tool behavior necessitates cognitive sophistication [intelligence] that is unique and more advanced than that subserving non-tool behavior." Let us cease claiming that tool use is intelligent by defining tool use in a way that includes only intelligent tool use, while excluding cases in which association learning and genetic programming play dominant ontogenetic roles. Let us cease claiming that apes, monkeys, and corvids are the only tool users, because an intelligence-free definition of tool use clearly refutes the claim. And let us carefully qualify even the claim that apes, monkeys, and corvids are the only intelligent tool users, and impartially investigate whether other animals also sometimes use tools intelligently.

Myth #3: Only Primates Use and Make Tools

Jouffroy (1993, 6) asserted that "the concept of 'tool' is essentially and inseparably associated with the concept of hand." This might lead some to assume that only primates, or often only apes, use tools. For example, as noted above, Matsuzawa (2001, 11) stated that "the taxonomic distribution of tool use is limited to primates." However, our catalog confirms and expands on Beck's (1980) demonstration that many animals other than primates, including many animals without hands, use and make tools (tables 7.1 and 7.2). While the presence of hands certainly allows primates to manipulate objects and use tools in more human-like ways, our catalog shows that we cannot discount beaks, mouths, trunks, pseudopods, claws, paws, antlers, horns, or other appendages in tool use. Gavin Hunt and his colleagues' multiple accounts of the remarkable tool making and tool using of wild New Caledonian crows is a recent example. Without hands, tool use may be limited in its complexity (Fragaszy 1998). For example, few animals other than primates use Tool Composites (use of two tools simultaneously on a single task), in part because few are physically capable of handling two objects simultaneously (table 7.3). But crabs use discarded shells, sponges, and anemones as Tool Composites, placing and holding their tools with chelipeds (a large pincerlike claw or leg). Many species, such as assassin bugs and digger wasps (*Sphex* spp.) use more than one kind of tool and thus are said to have Tool Kits, although there is no confirmation yet that all of these tools are used by a single population of these animals.

Some would readily acknowledge that non-primates *use* tools, but then assert that only primates *make* them. McKenna (1982, 77) stated that "the ability to modify a natural item from the environment, that is, to change its natural form to an unnatural, modified one, rarely occurs except among higher primates." Our catalog shows that tool manufacture is not so rare, and that some non-primates are tool makers. Given the sheer number of invertebrates, there are probably millions more individual invertebrates making tools than there are primate tool makers.

A variant of this myth is that while other animals make and use tools, only humans depend on tools for survival. But we have shown that tool use is customary or habitual among many types of animals, suggesting that tools are important, if not critical, for survival. Could California sea otters, Bossou chimpanzees, and Tietê capuchin monkeys survive without the calorie- and protein-rich hard-shelled foods that they open with stone hammers? Hermit crabs (*Diogenes edwardsi*) most certainly could not survive without inhabiting gastropod shells. Many types of invertebrate larvae, and probably decorator crabs (Majidae), could not survive without using debris for camouflage or protection.

Tables 7.1, 7.2, and 7.3 list major taxonomic categories for which tool use, manufacture, and Associative Tool Use (respectively) have been documented. We must point out several cautions to readers. First, the categories for which tool behavior has not been reported are not listed, due to space limitations. This confounds efforts to extract overarching trends in the evolutionary mosaic of animal tool behavior. And, of course, tool behavior might occur and be unreported in the unlisted taxa.

A second caution is that the taxonomic level is not uniform. Invertebrates and non-primate mammals are broken down at the level of order. Fish, amphibians, reptiles, and birds are presented by class. Prosimians, Old World monkeys, and New World monkeys are presented as separate categories. Ape tool users and makers are presented by genus, except in the case of *Pan*, where we used species to distinguish bonobos and chimpanzees.

A third caution is that sampling effort and methodologies have not been equivalent across taxonomic categories.

The tables are intended to document the discussion of this myth, summarize our catalog as a whole, and assist readers in thinking about ontogenetic, ecological, evolutionary, and cognitive aspects of tool behavior. We have refrained from making such analyses ourselves because of the sampling issues, but Bentley-Condit and Smith (2010), though citing McGrew's (1992) caution about sampling, attempted an analysis of the frequency of tool use across taxa. However, they included borderline tool use; did not adjust absolute frequencies of occurrence for the number of individuals, species, genera, and orders within the classes (more of any of these would lead to higher expected

Table 7.1
Phyletic differences in animal tool use. Cases of tool use in captivity and in the wild are combined.

Mode of use	*Insects*	*Echinoids*	*Crustaceans*	*Arachnids*	*Cephalopods*	*Gastropods*	*Fish*	*Amphibians*	*Reptiles*	*Birds*	*Rodents*	*Carnivores*	*Ungulates*	*Elephants*	*Cetaceans*	*Prosimians*	*New World monkeys*	*Old World monkeys*	*Gibbons*	*Orangutans*	*Gorillas*	*Bonobos*	*Chimpanzees*
Drop	X									X				X			X	X	X	X	X	X	X
Throw	X			X	X		X			X	X	X		X	X		X	X	X	X	X	X	X
Drag, Roll, Kick, Slap, Push Over										X	X	X		X			X	X		X	X	X	X
Brandish, Wave, Shake			X								X		X	X	X	X	X	X	X	X	X	X	X
Bait, Entice	X			X					?*	X		X					X	X		X	X	X	X
Club, Beat										X				X			X	X		X	X	X	X
Pound, Hammer	X									X		X					X	X		X	X	X	X
Pry, Apply Leverage						X				X		X					X	X		X	X	X	X
Dig					X					X	X		X		X		X	X		X	X	X	X
Jab, Stab, Penetrate										X							X	X		X	X	X	X
Reach										X	X	X		X			X	X	X	X	X	X	X
Insert and Probe	X									X				X			X	X	X	X	X	X	X
Scratch, Rub										X		X	X	X			X	X	X	X	X	X	X
Cut																	X			X		X	X
Block			X		X					X	X	X		X			X			X	X		X
Prop and Climb, Balance and Climb, Bridge, Reposition										X	X			X	X	X	X	X		X	X	X	X
Hang																	X	X	X	X		X	X
Contain	X						X			X		X			X		X	X		X	X	X	X
Absorb										X							X	X	X	X	X	X	X
Wipe										X				X	X		X	X		X	X	X	X
Affix, Apply, Drape	X	X	X	X	X	X		X		X	X	X	X	X		X	X	X		X	X	X	X
Symbolize																	X			X	X	X	X

*Davis and Zickefoose (1998) reported Mooney's observation that crocodiles Bait birds with fish. It is unknown if the crocodiles actively place the fish on the water as a tool to lure in the bird.

frequencies); and omitted a number of reports of tool use (see Preface). They do, however, confirm our conclusion that many animals other than primates use and make tools.

Myth #4: New Caledonian Crows Use and Make Tools as Well as, or Even Better Than, Chimpanzees

G. Hunt and Gray (2006, 622) identified numerous aspects of tool making by New Caledonian crows that they claimed were "more similar to human toolmaking than that of any other non-human species, including chimpanzees." They argued that New Caledonian crows were more humanlike in their tool behavior regarding (1) distinct tool types, (2) standardized tool form, (3) hook tools, (4) the manufacture of tools following a rule system, (5) cumulative evolution in tool design, (6) species-wide manufacture, and (7) targeting raw materials (G. Hunt 1996; G. Hunt and Gray 2003, 2006). While we agree that these birds are sophisticated toolmakers, G. Hunt and Gray's (2006) argument is

Table 7.2
Phyletic differences in animal tool manufacture. Cases of tool manufacture in captivity and in the wild are combined.

Mode of manufacture	*Insects*	*Echinoids*	*Crustaceans*	*Arachnids*	*Cephalopods*	*Gastropods*	*Fish*	*Amphibians*	*Reptiles*	*Birds*	*Rodents*	*Carnivores*	*Ungulates*	*Elephants*	*Cetaceans*	*Prosimians*	*New World monkeys*	*Old World monkeys*	*Gibbons*	*Orangutans*	*Gorillas*	*Bonobos*	*Chimpanzees*
Detach	X		X							X	X	X		X			X	X	X	X	X	X	X
Subtract										X	?*			X			X	X		X	X	X	X
Add, Combine	X									X							X	X		X	X	X	X
Reshape			X							X		X					X	X		X	X	X	X

*North American beavers may use subtractive modification to alter Blocking tools. Current reports are not definitive on this matter.

Table 7.3
Phyletic differences in Associative Tool Use. Cases of Associative Tool Use in captivity and in the wild are combined.

	Insects	*Echinoids*	*Crustaceans*	*Arachnids*	*Cephalopods*	*Gastropods*	*Fish*	*Amphibians*	*Reptiles*	*Birds*	*Rodents*	*Carnivores*	*Ungulates*	*Elephants*	*Cetaceans*	*Prosimians*	*New World monkeys*	*Old World monkeys*	*Gibbons*	*Orangutans*	*Gorillas*	*Bonobos*	*Chimpanzees*
Sequential Tool										X							X	X		X	X		X
Tool Set																	X						X
Tool Composite			X		X							X					X			X	X		X
Metatool																	X			X			X
Multi-Function Tool			X														X	X			X		X
Secondary Tool																	X			X		X	X
Tool Crafting										X										X			X

not supported by our knowledge of the tool behavior of other animals, particularly chimpanzees. To argue that "crow tool manufacture had three features new to tool use in free-living non-humans, and that only first appeared in early human tool-using cultures after the Lower Paleolithic" (G. Hunt 1996, 251) mischaracterizes tool manufacture documented in chimpanzees and orangutans.

Manufacture of Distinct Types of Tools

Despite G. Hunt and Gray's (2006) assertion, New Caledonian crows are not the only tool users to manufacture distinct types of tools and to make tools for different purposes. Chimpanzees, orangutans, and capuchin monkeys also have Tool Kits that include different tools manufactured for different purposes (E. Fox, Sitompul, and van Schaik 1999; McGrew 1992; Ottoni and Izar 2008). These primate tools may be for altogether different tasks (ant-dipping wand and termite-fishing rod), or may constitute a Tool Set (two or more different tools used sequentially to accomplish one goal) or a Tool Composite (two or more tools used simultaneously to accomplish one goal). For example, as reviewed in chapter 6, the chimpanzees living in Loango National Park in Gabon use a Tool Set comprising up to five different tools for accessing honey (Boesch, Head, and Robbins 2009). Tool Sets and Tool Composites have not yet been documented in birds.

In addition, New World monkeys (all species combined), chimpanzees, and orangutans, each use and make tools for a much wider variety of tasks than crows, and even all birds combined. An examination of the modes of tool use and manufacture (tables 7.1 and 7.2) makes this point. While chimpanzees, orangutans, and New World monkeys have been documented to use tools within all of our twenty-two categorical modes of tool use, the entire class of birds has been documented to use tools in only eighteen (Brandish/Wave/Shake, Cut, Hang, and Symbolize are not represented). The most diverse bird tool users, the corvids (which include New Caledonian crows) have demonstrated only thirteen of the twenty-two modes of tool use (see table 3.1). New Caledonian crows use tools in only two modes: Drop and Insert and Probe. One could make the argument that sampling bias is partially responsible for this difference, but there are no comparative data on sampling effort.

The point is made clearer when we examine the modes of tool manufacture and Associative Tool Use. Both corvids and chimpanzees are documented to use all four modes of tool manufacture, and both New Caledonian crows and chimpanzees use a sophisticated form of manufacture, Tool Crafting. However, while all of the forms of Associative Tool Use (see table 7.3) have been documented in chimpanzees, only Sequential Tool Use has been documented in corvids. As noted above, use of a Tool Set or a Tool Composite has not yet been documented for any bird. The use of Metatools has been claimed for captive New Caledonian crows (Taylor et al. 2007), but in our definitional scheme the observations are only of Sequential Tool Use, which is common in many apes and monkeys. It is inaccurate to state that "New Caledonian crows in the wild manufacture and use tools in ways more sophisticated than do other non-humans" (G. Hunt, Rutledge, and Gray 2006, 307).

Standardized Tool Form

G. Hunt (1996) argued that only New Caledonian crows and humans manufacture tools to a standardized form. Yet different tool types manufactured and used by chimpanzees (McGrew 1992) and orangutans (e.g., E. Fox, Sitompul, and van Schaik 1999) have statistically significant distinctions in terms of their manufactured form and shape. These physically distinctive tool types indicate tool standardization.

Hook Tools

We agree with G. Hunt and Gray (2006) that at present only humans and New Caledonian crows sculpt a tool into the shape of a hook. This is a sophisticated form of tool manufacture that distinguishes crows from other non-human toolmakers. However, in reviewing our catalog, we find other examples of either tool use or multi-step toolmaking that happen to be shared by humans and one other animal taxon. For example, only humans and chimpanzees make tools to Stab at other animals (Pruetz and Bertolani 2007), and only humans and chimpanzees create brush-tipped tools for termite collection (Sanz, Call, and Morgan 2009). Only humans and some invertebrates Apply or Affix objects and pigments to their bodies for camouflage (see chapter 2).

The *Pandanus* tools of New Caledonian crows are also unique, as there is not a similar type of tool made by chimpanzees or by any other animal. To make these tools, the crows are described as controlling length and width shape variables simultaneously (G. Hunt 2000a). They snip out a tool of the appropriate width and length through a series of alternating cuts and tears with their beaks. The finished tool is said to be predetermined before the process begins, since the final Detached tool is not refinished. This contrasts with chimpanzee tool manufacture, in which the length and width of manufactured tools are produced sequentially. However, as in hook tool manufacture, the presence of this unique sort of tool manufacture in New Caledonian crows should be compared with all other forms of manufacture that might be shared uniquely by one animal taxon and humans. Only chimpanzees and humans create stabbing tools, brush-tipped fishing wands, or complex anvils that include Metatools. Only bowerbirds and humans make and use tools to apply pigment. In addition, while there is no evidence that wild chimpanzees are controlling multiple shape variables simultaneously in tool manufacture, they do, as noted above, use two or more tools simultaneously as Tool Composites, something not seen in New Caledonian crows. Further, in addition to New Caledonian crows, there is evidence suggesting predetermined templates and goals in the manufacture and use of tools by chimpanzees (Boesch and Boesch 1984a; Mulcahy and Call 2006a; Osvath 2009; Osvath and Osvath 2008).

Manufacture of Tools Following a Rule System

G. Hunt and Gray (2006) argued that the use of rule systems (evidenced by a fixed hierarchical sequence of successive steps) in tool manufacture is limited to New Caledonian crows. However, chimpanzees also follow a fixed hierarchical sequence of successive steps when crafting Stabbing tools (Pruetz and Bertolani 2007). Furthermore, Pruetz and Bertolani argued that the manufacture of tools for termite fishing and ant dipping should additionally be considered as Tool Crafting, as these tools also require a number of sequential steps and final Reshaping to create the final tool. We agree, noting, in particular, the crafting of brush-tipped fishing tools by chimpanzees.

We are not even sure that rule-based systems of tool manufacture and highly standardized tools are significant or necessarily humanlike. Is it more cognitively adaptive to manufacture a tool by choosing dimensions to optimally meet the demands of a particular instance (not strictly rule based), or to make a standardized, strictly rule-based tool that must be made to work in every instance?

Cumulative Evolution in Tool Design

G. Hunt and Gray (2006) and Tomasello (1999) noted that chimpanzees do not show cumulative change in their tool forms, which is a distinguishing feature of human tool use. *Pandanus* tool manufacture, Detaching a section of a leaf from an attached tree and leaving behind a negative cutout leaf portion on the living tree, has led researchers to infer cumulative change in New Caledonian crows (G. Hunt and Gray 2003). By analyzing the cutouts, G. Hunt and Gray found that the *Pandanus* tools occurred in three different forms: stepped tools, narrow tools, and wide tools. The stepped tools occur in either a one-step or a multi-step form. They opined that the manufacture of multi-stepped tools is "more complex" (G. Hunt and Gray 2006, 624), but they admitted that there are no data on the functional advantages of each type. They also concluded that the three types have a "common historical origin" and have been subject to a "population-level diversification" (G. Hunt and Gray 2006, 623). They argued that the multi-stepped tools are the most advanced and result from cumulative evolution over time. However, their conclusion is speculative, because they have no data on change over time. The differences may represent interpopulation or interindividual differences, or even ontogenetic stages (intraindividual differences). New Caledonia is small, and there is only one New Caledonian crow *population*, although there are many study sites. Why would the more primitive forms of *Pandanus* tools continue to exist if the advanced form is so advantageous? In brief, the cumulative evolution in tool form can only be documented by physical evidence of change over time, not from the synchronous existence of behavioral variants. As noted in chapters 1 and 7, some members of some contemporary chimpanzee populations pound fruits *on* hard objects to open them (proto-tool use), and some pound fruit *with* hard objects (true tool use), but one cannot infer cumulative evolution from proto-tool to true tool use from such variation.

Species-Wide Manufacture

G. Hunt and Gray (2006, 624) noted that "species-wide manufacture" was only documented for New Caledonian crows and chimpanzees. However, it appears that Detaching probes for extracting insects is present throughout the range of at least one other species, the Galápagos woodpecker finch (*Camarhynchus pallidus*) (see chapter 3). Detaching branches to Drop or Throw onto pursuers appears to be present throughout the range of orangutans (see chapter 6). In addition, examination of the conclusions about nuthatches, woodpecker finches, and orangutans in G. Hunt and Gray (2006, table 1) suggests that these authors might be more mindful of a point that we have already made—absence of evidence is not evidence of absence—before concluding that some populations of a given species do not perform a particular form of tool manufacture. Even if certain populations do not, we ask why "species-wide" distribution is of such interest? Primatologists and anthropologists are far more interested in interpopulational *differences* in tool behavior, since they are indications of possible cultural differences. We humans do not Craft chopsticks as eating utensils throughout our geographical range. The ontogeny of many tool behaviors in nonhuman primates (especially great apes) and in humans is related to opportunities for socially mediated learning (van Schaik, Deaner, and Merrill 1999; van Schaik, Laland, and Galef 2009). It is therefore expected that a particular form of manufacture may not be found in every population of a species, especially given the large and fragmented geographic ranges of many animals in comparison with that of New Caledonian crows.

Targeting Raw Materials

G. Hunt and Gray (2006) argued that the selection of specific materials from which to make tools was present in crows, but only possibly present in chimpanzees, yet they overlooked evidence that chimpanzees purposefully select raw materials for tools. For example, 98 percent of puncturing sticks manufactured and used by Goualougo Triangle chimpanzees are of the species *Thomandersia hensii*, whereas 96 percent of fishing probes at this site are manufactured from *Sarcophrynium* spp. (Sanz, Morgan, and Gulick 2004).

New Caledonian crows make and use some forms of tools that great apes do not use, but the reverse is also true. Only in the great apes do we see all of the known modes of animal tool use and Associative Tool Use. In addition, crows are not unique in sharing a specific form of tool behavior only with humans; other types of animals also use tools in ways we do not see in great apes, but which we do see in humans. We agree with Emery and Clayton (2004a, 2004b) that New Caledonian crows and other birds are very useful for studying convergent evolution of cognition and ontogeny of tool behavior between birds and primates. Yet we see no reason to search selectively for grounds to elevate one or diminish another species based on the similarity of their tool behavior to that of modern humans.

Myth #5: Apes That Spend More Time Around People Are More Proficient Tool Users

Beck (1980, 76) stated that "except for draping and dropping or throwing down, tool use by wild orangutans does not approach in frequency or diversity that exhibited by captives or rehabilitants." McGrew (1989, 463; 1991) characterized long-term field studies as revealing "no evidence of tool use" among wild orangutans. In contrast, he described rehabilitant orangutans as demonstrating a "rich array of tool use, some of it remarkably inventive" (McGrew 1989, 463), and he concluded that orangutans in captivity are "unparalleled tool users" (1989, 464). Beck's and McGrew's assessments of the tool behavior of wild orangutans were premature. Years later, E. Fox, Sitompul, and van Schaik (1999, 99) noted that "until recently, there was limited evidence of tool use (*sensu* Beck 1980) by wild orangutans (*Pongo pygmaeus*) and no evidence of tool manufacture." Soon thereafter, van Schaik et al. (2003) summarized a wealth of new observations documenting rich tool repertoires among wild orangutans.

We have examined the tool behavior of wild, rehabilitant, and captive orangutans to assess the effects of human contact and influence. Both rehabilitant orangutans (those living in and around sanctuaries, see below) and captives (those living in zoos, laboratories, and human homes) have indeed demonstrated *more diversity of tool use* than wild orangutans. Rehabilitant and captive orangutans have been documented to engage in all of our twenty-two modes of tool use, while wild orangutans have been documented to engage in only sixteen. Further, because our modes tend to lump some different behaviors, some variation may be masked; the descriptions by Russon et al. (2009) suggest that the diversity of tool behavior at the level of mode underestimates the amount of diversity in the tool use of rehabilitants. Nonetheless, there are no examples of tool use by wild orangutans in the modes of Bait/Entice, Dig, Cut, Block, Hang, and Contain. Wild, rehabilitant, and captive orangutans have each been documented to engage in all four of our modes of tool *manufacture*. There are *more examples* of use and manufacture by rehabilitants and captives, but frequency data are confounded by greater sampling effort and better observational conditions with rehabilitants and captives. With regard to Associative Tool Use, wild, rehabilitant, and captive orangutans all used Tool Composites, only wild orangutans and possibly one captive (Nakamichi 2004) engaged in Tool Crafting, and only rehabilitants and captives showed Sequential Tool Use.

Appreciation of these comparisons requires some elaboration about the rehabilitant and captive categories. Rehabilitants are those living in and around sanctuaries. Most were born in the wild, and then poached for the pet trade or left homeless by habitat conversion. They were subsequently confiscated or rescued, and taken to sanctuaries for protection. Most sanctuaries are in or near forests. Rehabilitants being prepared for reintroduction are often allowed to live freely in the forests, but they still depend on humans for food and protection. Captive orangutans are those living in zoos, laboratories, and human homes. Before 1950, most of these had been born in the wild, captured, and exported. With successful captive breeding programs and legal prohibition of capture and exportation, an ever-growing proportion of captives have been born in captivity.

These distinctions correlate with at least four factors that influence the ontogeny and functional value of tool behavior (Russon et al. 2009; van Schaik and Pradhan 2003). The factors are degree of contact with older, experienced orangutans and/or humans as models for socially mediated learning; exposure to objects that can serve as tools; degree of terrestriality and proximity to standing water; and the types of challenges the orangutans face to survive. For example, a wild orangutan infant stays in very close contact with its mother for seven to nine years and has very little interaction with other orangutans, people, and human-made objects. Wild orangutans infrequently come to the ground, and they have little contact with standing water. They live in an environment with challenges to subsistence, such as extracting honey from beehives, that have been met by orangutans for millennia. Rehabilitants generally have some experience with their mothers in the wild, but they then were separated and brought into an environment dominated by humans and human-made objects. In sanctuaries they also have a significant amount of contact with other orangutans and, during the process of reintroduction, they can access the full array of objects afforded by both a natural and a human-influenced environment. They spend considerable time on the ground or on flat substrates, and around water. When in the forest, they face challenges faced by generations of wild orangutans, but because they are still in the company of people, they face other challenges, such as locks and keys. From an orangutan's perspective, rehabilitants live in the most challenging of both worlds. Captives may be reared by their mothers or by nurturing humans. As they gain in size and strength, they have progressively less contact with humans and human-made objects, but they continue to be totally dependent on people for survival. They spend very much of their lives on the ground in outdoor enclosures or on flat substrates in cages, and they have considerable exposure to water. Many captives have the opportunity to interact with only two or three other orangutans in a lifetime. Captive orangutans face challenges related only to getting what they need and want in an environment that offers few choices, little change, limited mobility, and dependency on humans.

The lesser variety in tool use by wild orangutans may be explained by the environmental challenges they face. When reported for great apes, the mode Bait, Entice is frequently associated with subtle signals to attract the attention of a conspecific, such as chim-

Supinah, a rehabilitant orangutan, tried to make a fire. She carefully poured kerosene into a cup. Photograph courtesy of Anne E. Russon.

After dipping a burning stick in the kerosene, Supinah fanned the kerosene with a metal lid. The technique was imperfect, because the stick was extinguished when dipped into the kerosene. Photograph courtesy of Anne E. Russon.

Then Supinah tried to blow on the stick to reignite it. She imitated steps used by humans to make a fire. Photograph courtesy of Anne E. Russon.

panzees clipping leaves to invite copulation in a competitive social setting. However, wild orangutans live in very low densities and probably do not use subtle invitations. Bait, Entice also includes using objects to attract the attention of a human and/or to trade with humans. Wild orangutans have no need or opportunity for either.

Digging is ecologically irrelevant for highly arboreal wild orangutans. Hanging is also ecologically irrelevant for wild orangutans, since the forest provides limitless natural opportunities to climb and hang. When rehabilitants and captives use tools in the Contain mode, they do so almost exclusively to carry food or fluids, often in socially competitive contexts. Wild orangutans seem not to carry unattached foods or water, and they rarely have to carry food away to escape competition. Most cases of Cutting in apes or other animals have involved close human tutelage; wild orangutans have no opportunity for this. Blocking is functional primarily in human-influenced settings with mechanical devices, such as doors or automated water dispensers, although a juvenile rehabilitant did Block the approach of a conspecific by bending and holding on to a tree. Therefore, the six modes of tool use not documented for wild orangutans are ecologically irrelevant for them, characteristic of human-influenced environments, and/or functionally advantageous only in unnatural, socially competitive situations.

This, plus our knowledge that wild orangutans do use an assortment of tools, and that they make and combine tools in nearly the same variety of ways as rehabilitants and captives, strongly indicate that past conclusions about the tool abilities of wild orangutans should be reconsidered.

It does appear that rehabilitant orangutans use tools more frequently than wild orangutans. Indeed, as noted above, more people have observed rehabilitants, and this observation takes place under conditions that make detection of tool behavior more likely. Moreover, rehabilitants have access to a much greater variety of objects, have more time to manipulate objects (because they are provisioned), spend more time on the ground and around standing water, and face challenges in a human environment where tools are more functionally advantageous. These factors all increase the probability of tool behavior, but differences in the opportunity for socially mediated learning may also be involved.

Van Schaik and Pradhan (2003, 645) suggested that the "probability of socially biased learning" and the "number of potential experts at close proximity" are critical in the acquisition of tool behavior and in the emergence and preservation of tool traditions. The tool behaviors of wild orangutans are canalized toward long-established, effective, and traditional patterns by means of socially mediated learning from their mothers. Russon et al. (2009, 294), discussing innovation in general (including tool use) among orangutans, hypothesized that because rehabilitants lack maternal guidance about traditional methods of tool use, they "have little choice but to invent their own solutions." Additionally, they have the opportunity to learn from many different humans, each of whom has unique ways of using tools (van Schaik 2006).

In conclusion, rehabilitant, and probably captive, orangutans do use tools in a greater variety of ways than wild orangutans, and they probably do so more frequently, although there is now documentation of an impressive diversity and frequency of tool use by wild orangutans. The differences that exist can be explained by different opportunities and functional advantages offered by the environment, and by differences in the opportunities and models for socially mediated learning.

There are more orangutans in sanctuaries in Malaysia and Indonesia than there are chimpanzees, gorillas, and bonobos combined in African sanctuaries (Beck 2010). There is much less published research about the tool behavior of sanctuary-living African apes than about orangutans, but this might simply be a matter of opportunity. There are as many or more captive chimpanzees and gorillas as there are captive orangutans, although there are fewer captive bonobos. Nonetheless, it is difficult to compare efforts to describe the tool behavior of captive African apes with those for captive orangutans. To examine the effects of human influence on the tool behavior of the other apes, we will therefore combine data from our catalog for rehabilitants and captives. We also include capuchin monkeys in this analysis, since tool use by these monkeys is also frequently reported in captivity and the wild.

As we have already shown for orangutans, in all cases rehabilitants and captives use tools in more diverse ways than their wild counterparts (see table 7.4). We suspect that many of the same ecological and ontogenetic dynamics that we used to explain the difference for orangutans are also applicable to these other primate taxa. However, the difference in diversity (number of modes) is minimal for chimpanzees, and the frequen-

Table 7.4
Comparison of the diversity of tool use by wild and captive great apes, gibbons, and capuchin monkeys

	Number of modes of tool use in the wild	*Number of modes of tool use in captivity (including rehabilitants)*
Capuchin monkeys	13	18
Gibbons	2	7
Gorillas	9	19
Bonobos	13	20
Orangutans	16	22
Chimpanzees	20	21

cy of tool behavior by wild chimpanzees is probably greater than that of their rehabilitant and captive counterparts. This may result from wild chimpanzees being the most frequent and diverse tool users and tool makers of all of the great apes (see myth #6, below).

Myth #6: Chimpanzees Are The Most Proficient Animal Tool Users

As noted in chapter 6, Beck (1980, 79) described chimpanzees as showing "the greatest frequency and diversity of tool behavior of any non-human animal." McGrew (1989, 468) considered the then existing tool-using literature for all great apes and speculated that chimpanzees may be "the only true tool user." In a subsequent analysis that included capuchin monkeys, McGrew (1993b, 152) concluded, "captive capuchin monkeys perform better than most apes but fail in nature, as do all others but chimpanzees." Are these conclusions still valid? Here we present an updated analysis that addresses the question of chimpanzee tool-using proficiency as compared with the other apes and capuchin monkeys, the most prolific monkey tool users. In the process, we offer a response to McGrew's (1991, 177) question: "Are there real differences in tool use across great apes?"

As described on pages 74–78, capuchin monkeys have been documented to use tools in twenty of our twenty-two modes. As shown in table 7.1, gibbons have been shown to use tools in eight of our twenty-two modes, gorillas in twenty, bonobos in twenty-one, and orangutans and chimpanzees in all twenty-two. Thus, in terms of the diversity of tool use that has been documented in all settings, chimpanzees and orangutans are equivalent and slightly more proficient than bonobos, gorillas, and capuchin monkeys. Gibbons demonstrate little tool-using proficiency. The rankings remain about the same when examining tool-use diversity in the wild only (table 7.4), but the absolute difference between chimpanzees and the others is greater.

There are apparent discrepancies between table 7.4 and table 7.1. For example, wild and captive gibbons combined are listed in table 7.1 as using tools in eight modes. In table 7.4, they are listed as using tools in two modes in the wild and seven in captivity. But only one mode (Drop) seen in the wild has not been reported in captivity, and six modes (Throw, Reach, Insert and Probe, Scratch/Rub, Hang, Absorb) seen in captivity have not been reported in the wild. One mode (Brandish/Wave/Shake) has been seen in both the wild and captivity. Thus wild gibbons have been seen to use tools in two modes and captive gibbons in seven (table 7.4), but all gibbons combined have been seen to use tools in only eight modes (table 7.1). Also, capuchin monkeys (twenty modes) are included within New World monkeys (twenty-two modes) in table 7.1 but are listed separately in table 7.4 since they are by far the most diverse and frequent New World monkey tool users.

Capuchin monkeys, orangutans, gorillas, bonobos, and chimpanzees manufacture tools in all four of our manufacture modes, while gibbons are represented in only the simplest, Detach.

Dramatic differences emerge when forms of Associative Tool Use are considered (see table 7.3). Gibbons are absent from all categories, and bonobos are poorly represented. Gorillas use Sequential Tools, Tool Composites, and Multi-Function Tools. Orangutans are represented in five of the seven categories, not having demonstrated use of Tool Sets or Multi-Function Tools. Capuchin monkeys have not been documented to Craft tools, but they show all other forms of As-

A capuchin monkey uses a stone hammer to Pound a small fruit on a stone anvil. Capuchin monkeys use tools in many of the same ways as great apes. Photograph courtesy of Tiago Falótico.

sociative Tool Use. Chimpanzees are represented in all forms. Orangutans, capuchin monkeys, and chimpanzees have been seen to use Metatools.

Overall, our catalog indicates that chimpanzees are one of the two most diverse, and are the most frequent, tool users of the apes and capuchin monkeys. The differences are more pronounced when considering only cases from the wild. All of the great apes and capuchin monkeys are equivalent with regard to diversity of tool manufacture, although our modes may mask some diversity by lumping similar but varying behaviors. Chimpanzees, orangutans, and capuchin monkeys seem to stand out with regard to Associative Tool Use, but chimpanzees again are at the top. This is especially clear when we consider the frequency of some forms of chimpanzee Associative Tool Use, namely their frequent use of Tool Sets.

The "myth" that chimpanzees are the most proficient tool users may not be a myth at all, at least for wild chimpanzees, but the differences between chimpanzees, the other great apes, and capuchin monkeys appear to be smaller than is generally thought. In making these sorts of comparisons, we remind readers of all of the methodological cautions that we offered in the discussion of myth #5.

We do not have the space to consider the many evolutionary, ecological, and cognitive explanations that have been offered to account for differences in the frequency and diversity of tool behavior among apes and monkeys. Variances in observational methodology; dietary protein content; extractive foraging on embedded foods; time and activity budgets; time spent on the ground; abilities, opportunities, and models for socially mediated learning; manipulative abilities; curiosity; the capacity to understand causality; evolutionary parallelism and convergence; and the capacity for mirror self-recognition are among the explanatory concepts that have been offered. These are all fruitful areas for further research and synthesis.

Myth #7: Human Tool Behavior Is Unique

We have shown here that tool behavior is not unique to humans, but we have not yet examined the uniqueness of human tool behavior. "Biologists who study the behavior of non-human animals are fond of presenting us with apparently irrefutable evidence of the manufacture and use of tools by a great variety of creatures, whereas anthropologists—more committed to the premise of absolute human uniqueness—are inclined to discover reasons why such evidence can be discounted" (Ingold 1997, 109). We will not argue for absolute human uniqueness in terms of tool behavior, nor for minimizing the significance of animal tool behavior. Animal tool behavior should not be discounted as tool behavior, nor should it be considered less significant by lack of evidence for "intelligent use" in some cases (see myth # 2). But there are aspects of tool behavior that remain unique to the human lineage, going back to the earliest stone tool users, such as *Homo habilis*. These unique forms of tool behavior are described below. We recognize that just as some of the conclusions from the first edition of this book have had to be amended by subsequent discoveries and research, some of these conclusions may have to be altered as we learn more.

Distinction #1: Non-human animals do not typically use tools to make tools

Beck stated (1980, 218), "unquestionably man is the only animal that to date has been observed to use a tool to make a tool. However, we have seen in the last few decades a steady erosion of putative hallmarks of man: it would not be surprising to find that some animals can use a tool to make a tool." Indeed, there are now a few cases of Secondary Tool Use by primates. An orangutan (R. Wright 1972) and a bonobo (Schick et al. 1999; Toth et al. 1993) learned to use a hammer stone to Detach a flake from a core stone, which they then used as a Cutting tool. Matsuzawa (1994, 361) documented one clear example of stone tool manufacture in the wild, noting that "the chimpanzee broke a stone anvil with a hammer and the broken piece became a better hammer." Bortolini and Bicca-Marques (2007) described Secondary Tool Use in which a wild capuchin monkey modified a twig Probing tool with a stone. These examples demonstrate that the capacity for Secondary Tool Use is present in great apes and capuchin monkeys. However, it is rare in the wild, and tutelage was required in captivity. Secondary Tool Use is not customary or habitual in any population of animals other than humans. The customary *use* of a wide variety of tools to *make* a wide variety of tools appears to be a form of tool behavior that distinguishes humans from non-humans, and it was critical in human evolution.

Distinction #2: Non-human animals other than great apes rarely use tools in combinations for a single task

Humans, chimpanzees, and orangutans use tools in combinations known as Tool Composites, Metatools, and Tool Sets (for definitions, see table 1.2). While documentation of the use of Metatools and Tool Composites is relatively rare, the use of Tool Sets (two or more tools used sequentially for a single task) is well documented and habitual in some chimpanzee populations, such as those in the Goualougo Triangle (Sanz and Morgan 2007, 2009). Of course, many non-humans make and use Constructions such as nests, but we do not consider these to be tools (see myth #1 in this chapter; also see chapter 1).

Distinction #3: Non-human animals rarely cooperate in using or making tools

Cooperative tool behavior may have been an important factor in the development of communication in hominin evolution (Schick and Toth 1993). Moreover, despite the advantages that may result from sharing tools and cooperating in their use, cooperative tool use under natural conditions seems to be limited to humans. For example, although wild chimpanzees regularly use stone hammers and anvils, they do not cooperate in the transport and use of the stones (Boesch and Boesch 1990; Sakura and Matsuzawa 1991). Cooperation by wild chimpanzees in tool use has been observed only rarely and is limited to mothers and their offspring (Boesch 1991c).

Captive primates have used tools cooperatively to gain a food reward in experimental settings. Beck (1973a) found spontaneous cooperation in a pair of captive hamadryas baboons (*Papio hamadryas*), with a female baboon learning to bring a male a tool he could not access, which he then used to retrieve food that they shared. Savage-Rumbaugh, Rumbaugh, and Boysen (1978) found that two subadult male chimpanzees could use abstract symbols to request tools from one another in order to solve a problem. Westergaard and Suomi (1997c) reported that members of one captive capuchin monkey (*Cebus apella*) group transferred stones to individuals of another group, which then used the stones to acquire a food reward that they then shared with the first group.

As with Secondary Tool Use, non-human primates have the capacity to use tools cooperatively, but only humans do so frequently and in a variety of ways.

Distinction #4: Non-humans power their tools only with gravity and their own energy

Only humans harness other forms of mechanical energy—such as that found in fossil fuels, wind, the sun, water, or nuclear reactions—to power their tools. Related to this, only humans build and use tools or Constructions to increase the mechanical efficiency of their tools, for example, guns and bows to propel bullets and arrows, respectively. Non-human animals use only their own metabolic energy and gravity, and they use only simple amplifications of mechanical force.

As this book serves to both catalog and celebrate the diverse forms of tool behavior in non-human animals, we are not diminishing the significance of non-human animal tool behavior by acknowledging the differences between human and non-human tool use. It is true that non-human tools are relatively simple and made of either naturally occurring (or human-made) objects or lightly manufactured objects, such as digging sticks, leaf sponges, fishing probes, and hammers and anvils. In contrast, human technological tools, such as a portable drill or laptop computer, may include multiple interconnecting parts and may contain materials that have been highly processed and complexly manufactured. Their workings are understood by their makers, but not by all of their users. They are the products of cumulative cultural knowledge, Tomasello's (1999; Tennie, Call, and Tomasello 2009) "ratchet effect." The hypothetical human isolated from the cultural matrix that transmits the knowledge of how to create and use these tools would not use or make them. Cultural processes are operative in some tool and proto-tool behaviors of some non-human animals, mainly great apes and monkeys (Itani and Nishimura 1973; Kawamura 1959; van Schaik et al. 2003, 2009; Whiten et al. 1999, 2001), and there is evidence for the ratchet effect in the water-related behavior of Japanese macaques on Koshima Island (Kawai 1965; Laland and Hoppitt 2006; Watanabe 1994). However, cumulative cultural processes are far more prevalent and influential in the tool behavior of humans.

Acknowledging both the similarities and the differences in human and non-human tool behavior is a critical step in unraveling its evolutionary and adaptive significance. The differences noted above are pivotal to the evolutionary history and the future of our species. We humans use tools to make better tools, make and use tools to work in synergistic combinations, make and use them cooperatively to produce effects beyond the ability of a single individual, and elaborate on and improve them over generations of innovators and imitators. We eventually began to power our tools with more than gravity and our own metabolic energy, thereby inventing *technology*. The following few sentences, which served as the concluding remarks of the first edition of this text, are presciently apt for our twenty-first-century world, in which we face a future that could be destroyed or saved by our technological abilities.

> The unprecedented impact of technology on the biological and physical environment and the uncertainty of continued energy sources to support technological endeavor have led many to predict the extinction of the human life style, the species, or even the planet itself. In a final irony, tool behavior would not only be stripped of its role as a primary shaper of our species, but would also be recast as the progenitor of the engine of our destruction. (Beck 1980, 247)

REFERENCES

Abordo, E.J. 1976. The learning skills of gibbons. *Gibbon and Siamang* 4:106–134.

Abwe, E.E., and B.J. Morgan. 2008. The Ebo forest: Four years of preliminary research and conservation of the Nigeria-Cameroon chimpanzee (*Pan troglodytes vellerosus*). *Pan Africa News* 15 (2):26–29.

Acharjyo, L., and R. Misra. 1972. On the feeding habits of crab-eating mongoose (*Herpestes urva*) in captivity. *Journal of the Bombay Natural History Society* 69:411–412.

Acuña, F.H., A.C. Excoffon, and M.A. Scelzo. 2003. Mutualism between the sea anemone *Antholoba achates* (Drayton, 1846) (Cnidaria: Actiniaria: Actinostolidae) and the spider crab *Libinia spinosa* (Milne-Edwards, 1834) (Crustacea: Decapoda, Majidae). *Belgian Journal of Zoology* 133 (1):85–87.

Addessi, E., L. Crescimbene, and E. Visalberghi. 2008. Food and token quantity discrimination in capuchin monkeys (*Cebus apella*). *Animal Cognition* 11 (2):275–282.

Agbogba, C. 1985. Observations sur la récolte de substances liquides et de sucs animaux chez deux espèces d'*Aphaenogaster*: *A. senilis* et *A. subterranea* (Hym.: Formicidae). *Insect Society* 32 (4):427–434.

Akerman, C. 1926. On the spider *Menneus camelus* Pocock, which constructs a moth-catching, expanding snare. *Annals of the Natal Museum* 5:411–422.

Albignac, R. 1969. Notes éthologiques sur quelques carnivores malgaches: Le *Galidia elegans* I. Geoffroy. *Terre et la Vie* 23:202–215.

Albrecht, H., and S.C. Dunnett. 1971. *Chimpanzees in western Africa*. Munich: R. Piper.

Alcock, J. 1970. Origin of tool-using by Egyptian vultures *Neophron percnopterus*. *Ibis* 112 (4):542.

———. 1972. The evolution of the use of tools by feeding animals. *Evolution* 26:464–473.

———. 1981. Book review: *Animal Tool Behavior*. *Quarterly Review of Biology* 56 (2):231.

Alexander, J.E. 1838. *Expedition of discovery into the interior of Africa*, vol. 1. London: H. Colburn.

Allsop, K. 1949. Anting of green woodpecker. *British Birds* 42:390.

Alp, R. 1993. Meat eating and ant dipping by wild chimpanzees in Sierra Leone. *Primates* 34 (4):463–468.

———. 1997. "Stepping-sticks" and "seat-sticks": New types of tools used by wild chimpanzees (*Pan troglodytes*) in Sierra Leone. *American Journal of Primatology* 41 (1):45–52.

Ambrose, D.P. 1986. Impact of nymphal camouflaging on predation and cannibalism in the bug *Acanthaspis siva*. *Environment and Ecology* 4:197–200.

———. 1999. *Assassin bugs*. Enfield: Science Publishers.

Andersen, A., G. Cook, and R. Williams. 2003. *Fire in tropical savannas: The Kapalga experiment*. New York: Springer.

Anderson, J.D., IV, M.J. Owren, and S. Boinski. 2007. Impact-amplitude of controlled percussive strikes provides evidence of site-selection for enhancement of acoustic display behavior in brown capuchins (*Cebus apella*) in Suriname. *American Journal of Physical Anthropology* 132 (S44):64.

Anderson, J.R. 1985. Development of tool-use to obtain food in a captive group of *Macaca tonkeana*. *Journal of Human Evolution* 14 (7):637–645.

———. 1990. Use of objects as hammers to open nuts by capuchin monkeys (*Cebus apella*). *Folia Primatologica* 54 (3–4):138–145.

Anderson, J.R., and M.C. Henneman. 1994. Solutions to a tool-use problem in a pair of *Cebus apella*. *Mammalia* 58 (3):351–361.

Anderson, J.R., C. Lamarque, and J. Fagot. 1996. A multitask assessment of hand lateralization in capuchin monkeys (*Cebus apella*). *Primates* 37 (1):97–103.

Anderson, J.R., E.A. Williamson, and J. Carter. 1983. Chimpanzees of Sapo Forest, Liberia: Density, nests, tools, and meat-eating. *Primates* 24 (4):594–601.

Anderson, K.L., J.E. Seymour, and R. Rowe. 2003. Influence of a dorsal trash-package on interactions between larvae of *Mallada signata* (Schneider) (Neuroptera: Chrysopidae). *Australian Journal of Entomology* 42:363–366.

Anderson, R.C., J.A. Mather, and C.W. Steele. 2004. Burying and associated behaviors of *Rossia pacifica* (Cephalopoda: Sepiolidae). *Vie et Milieu* 54 (1):13–19.

Andersson, S. 1989. Tool use by the fan-tailed raven *Corvus rhipidurus*. *Condor* 91 (4):999.

Anonymous. 1971. Siamang also a tool user. *Yerkes Newsletter* 8:12.

Antevs, A. 1948. Behaviour of gila woodpecker, ruby-crowned kinglet, and broad-tailed hummingbird. *Condor* 50:91–92.

Anthony, A. 1906. Random notes on Pacific coast gulls. *Auk* 23:129–137.

Antinucci, F., G. Spinozzi, and F. Natale. 1986. Stage 5 cognition in an infant gorilla. In *Current perspectives in primate social dynamics*, edited by D.M. Taub and F.A. King. New York: Van Nostrand Reinhold, 403–415.

Antinucci, F., and E. Visalberghi. 1986. Tool use in *Cebus apella*: A case study. *International Journal of Primatology* 7 (4):351–363.

Arcadi, A.C., D. Robert, and C. Boesch. 1998. Buttress drumming by wild chimpanzees: Temporal patterning, phrase integration into loud calls, and preliminary evidence for individual distinctiveness. *Primates* 39 (4):505–518.

Arcadi, A.C., D. Robert, and F. Mugurusi. 2004. A comparison of buttress drumming by male chimpanzees from two populations. *Primates* 45:135–139.

Armbrust, W. 1958. Haltung und zucht von *Hoplosternum thoracatum*. *Die Aquarien- und Terrarienzeitschrift* 11:97–100.

Armbruster, L. 1921. Über Werkzeuggebrauch bei Tieren. *Naturwissenschaften* 9:303–305.

Armstrong, E. 1965. *Bird display and behaviour*. New York: Dover.

Artaud, Y., and M. Bertrand. 1984. Unusual manipulatory activity and tool-use in a captive crab-eating macaque (*Macaca fascicularis*). In *Current primate researches*, edited by M.L. Roonwal, S.M. Mohnot, and N.S. Rathore. Jodhpur, India: University of Jodhpur, 423–428.

Asano, T. 1994. Tool using behavior and language in primates. In *Behavior analysis of language and cognition*, edited by S. Hayes, L. Hayes, and M. Sato. Reno, NV: Context Press, 145–148.

Assersohn, C., A. Whiten, Z.T. Kiwede, J. Tinka, and J. Karamagi. 2004. Use of leaves to inspect ectoparasites in wild chimpanzees: A third cultural variant? *Primates* 45 (4):255–258.

Auffenberg, W. 1981. Behavior of *Lissemys punctata* (Reptilia: Testudinata: Trionychidae) in a drying lake in Rajasthan, India. *Journal of the Bombay Natural History Society* 78 (3):487–493.

Aumann, T. 1990. Use of stones by the black-breasted buzzard *Hamirostra melanosternon* to gain access to egg contents for food. *Emu* 90 (3):141–144.

Babitz, M.A. 2000. Object manipulation and tool use in Sulawesi crested black macaques. *American Journal of Primatology* 51 (S1):38.

Bacher, S., and S. Luder. 2005. Picky predators and the function of the faecal shield of a cassidine larva. *Functional Ecology* 19 (2):263–272.

Badrian, N., A. Badrian, and R. Susman. 1981. Preliminary observations of the feeding behavior of *Pan paniscus* in the Lomako Forest of Central Zaïre. *Primates* 22 (2):173–181.

Baker, M. 1996. Fur rubbing: Use of medicinal plants by capuchin monkeys (*Cebus capucinus*). *American Journal of Primatology* 38:263–270.

———. 1997. Identification and selection of fur rubbing materials by white-faced capuchin monkeys (*Cebus capucinus*). *American Journal of Primatology* 42 (2):93.

———. 2000. Cognitive components of plant selection for fur rubbing in white-faced capuchin monkeys, *Cebus capucinus*. *American Journal of Primatology* 51:39–40.

Balda, R.P. 2007. Corvids in combat: With a weapon? *Wilson Journal of Ornithology* 119 (1):100–102.

Baldwin, J.D., and J.I. Baldwin. 1977. Observations on *Cebus capucinus* in southwestern Panama. *Primates* 18:937–941.

Baldwin, L.A., and G. Teleki. 1976. Patterns of gibbon behavior on Hall's Island, Bermuda: A preliminary ethogram for *Hylobates lar*. *Gibbon and Siamang* 4: 21–105.

Baldwin, P.J. 1979. The natural history of the chimpanzee (*Pan troglodytes verus*) at Mt. Assirik, Senegal. Ph.D. diss., University of Stirling.

Bandres, J., J.J. Campos, and R. Llavona. 1989. Behavioral observation in America: The Spanish pioneers in the 16th and 17th centuries. *Bulletin of the Psychonomic Society* 27 (2):184–187.

Bania, A.E., S. Harris, H.R. Kinsley, and S.T. Boysen. 2009. Constructive and deconstructive tool modification by chimpanzees (*Pan troglodytes*). *Animal Cognition* 12 (1):85–95.

Banschbach, V.S., A. Brunelle, K.M. Bartlett, J.Y. Grivetti, and R.L. Yeamans. 2006. Tool use by the forest ant *Aphaenogaster rudis*: Ecology and task allocation. *Insectes Sociaux* 53:463–471.

Barash, D.P., P. Donovan, and R. Myrick. 1975. Clam dropping behavior of glaucous-winged gull (*Larus glaucescens*). *Wilson Bulletin* 87 (1):60–64.

Barber, J., E. Ellgaard, L. Thien, and A. Stack. 1989. The use of tools for food transportation by the imported fire ant, *Solenopsis invicta*. *Animal Behaviour* 38 (3): 550–552.

Bard, K.A. 1993. Cognitive competence underlying tool use in free-ranging orang-utans. In *The use of tools by human and non-human primates*, edited by A. Berthelet and J. Chavaillon. Oxford: Clarendon Press, 103–113.

———. 1994. The use of tools by free-ranging orangutans: Cognitive competence. In *Proceedings of the International Conference on "Orangutans: The Neglected Ape,"* edited by J. Ogden, L. Perkins, and L. Sheeran. San Diego: Zoological Society of San Diego, 177–183.

———. 1995. Sensorimotor cognition in young feral orangutans (*Pongo pygmaeus*). *Primates* 36 (3):297–321.

Bard, K.A., D. Fragaszy, and E. Visalberghi. 1995. Acquisition and comprehension of a tool-using behavior by young chimpanzees (*Pan troglodytes*): Effects of age and modeling. *International Journal of Comparative Psychology* 8 (2):47–68.

Bard, K.A., and J. Vauclair. 1989. What's the tool and where's the goal? *Behavioral and Brain Sciences* 12 (3):590–591.

Barnes, D.M. 2005. Possible tool use by beavers, *Castor canadensis*, in a northern Ontario watershed. *Canadian Field Naturalist* 119 (3):441–443.

Barnes, J.L., L.A. Edwards, L.M. Hallett, and L.R. Santos. 2006. Non-human primates? Sensitivity to the causal features of tool-use. *International Journal of Primatology* 27 (Suppl. 1):Abstract #344.

Bartlett, D., and J. Bartlett. 1973. Beyond the north wind with the snow goose. *National Geographic* 144:822–847.

Bauer, H. 2001. Use of tools by lions in Waza National Park, Cameroon. *African Journal of Ecology* 39 (3):317.

Bauer, H.R. 1977. Chimpanzee bipedal locomotion in the Gombe National Park, East Africa. *Primates* 18:913–921.

Baxter, R.H., S.K. Urban, and L.H. Brown. 1969. A nineteenth-century reference to the use of tools by the Egyptian vulture. *Journal of the East African Natural History Society and National Museum* 27:231.

Bayart, F. 1982. Un cas d'utilisation d'outil chez un macaque (*Macaca tonkeana*) énlevé en semi-liberté. *Mammalia* 46 (4):541–544.

Bayart, F., and J.R. Anderson. 1985. Mirror-image reactions in a tool-using, adult male *Macaca tonkeana*. *Behavioural Processes* 10 (3):219–227.

Beatty, H. 1951. A note on the behavior of the chimpanzee. *Journal of Mammalogy* 32:118.

Beck, B.B. 1967. A study of problem solving by gibbons. *Behaviour* 28:95–109.

———. 1972. Tool use in captive hamadryas baboons. *Primates* 13:277–295.

———. 1973a. Cooperative tool use by captive hamadryas baboons. *Science* 182:594–597.

———. 1973b. Observation learning of tool use by captive Guinea baboons (*Papio papio*). *American Journal of Physical Anthropology* 38 (2):579–582.

———. 1974. Baboons, chimpanzees, and tools. *Journal of Human Evolution* 3:509–516.

———. 1975. Primate tool behavior. In *Socioecology and psychology of primates*, edited by R.H. Tuttle. The Hague, Netherlands: Mouton, 352–360.

———. 1976. Tool use by captive pigtailed macaques. *Primates* 17:301–310.

———. 1977. Köhler's chimpanzees: How did they really perform? *Zoologische Garten* 47:352–360.

———. 1980. *Animal tool behavior*. New York: Garland.

———. 1982. Chimpocentrism: Bias in cognitive ethology. *Journal of Human Evolution* 11 (1):3–17.

———. 1986. *L'abilità tecnica degli animali: Uso e costruzione di arnesi*. Turin: Bollati Boringhiere.

———. In press. Chimpanzee orphans: Sanctuaries, reintroduction and cognition. In *The Mind of the Chimpanzee: Ecological and Experimental Perspectives*, edited by E. Lonsdorf, S. Ross, and T. Matsuzawa. Chicago: University of Chicago Press.

Beck, B.B., B. Leja, and K. Zemanek. 1978. Horn and antler adornment. Paper presented at the Annual Meeting of the Animal Behavior Society, June, Seattle, Washington.

Beck, B.B., K. Walkup, M. Rodrigues, S. Unwin, D. Travis, and T. Stoinski. 2007. *Best practice guidelines for the re-introduction of great apes*. Gland, Switzerland: SSC Primate Specialist Group of the World Conservation Union.

Becker, C. 1984. *Orang-utans und Bonobos im Spiel*. Munich: Profil.

Becker, P.R. 1993. *Werkzeuggebrauch im Tierreich*. Stuttgart, Germany: S. Hirzel.

Bekoff, M. 2009. Animal emotions, wild justice, and why they matter: Grieving magpies, a pissy baboon, and empathic elephants. *Emotion, Space and Society* 2 (2):82–85.

Bekoff, M., and R. Dorr. 1976. Predation by "shooting" in archer fish, *Toxotes jaculatrix*: Accuracy and sequences. *Bulletin of the Psychonomic Society* 7:167–168.

Belt, T. 1874. *The naturalist in Nicaragua*. London: John Murray.

Benchley, B.J. 1940. *My life in a man-made jungle*. Boston: Little, Brown.

Benhar, E.E., and D. Samuel. 1978. A case of tool use in captive olive baboons (*Papio anubis*). *Primates* 19:385–389.

Bent, A. 1963. *Life histories of North American gulls and terns*. New York: Dover.

Bentley-Condit, V.K., and E.O. Smith. 2010. Animal tool use: Current definitions and an updated comprehensive catalog. *Behaviour* 147 (2):185–221.

Berg, C.J. 1975. Behavior and ecology of conch (superfamily Strombacea) on a deep subtidal algal plain. *Bulletin of Marine Science* 25 (3):307–317.

Berger, J. 2008. *The better to eat you with: Fear in the animal world*. Chicago: University of Chicago Press.

Bermejo, M., and G. Illera. 1999. Tool-set for termite-fishing and honey extraction by wild chimpanzees in the Lossi Forest, Congo. *Primates* 40 (4):619–627.

Bermejo, M., G. Illera, and J. Sabater-Pi. 1989. New observations on the tool-behavior of chimpanzees from Mt. Assirik (Senegal, West Africa). *Primates* 30 (1):65–73.

Berney, F. 1905. Field notes on birds of the Richmond District, North Queensland. *Emu* 5:15–20.

Bernstein, I.S. 1962. Response to nesting materials of wild born and captive born chimpanzees. *Animal Behaviour* 10:1–6.

Bertagnolio, P. 1994. Tool-using by parrots: The palm cockatoo and the hyacinthine macaw. *Avicultural Magazine* 100:68–73.

Berthold, P. 1967. Über Haftfarben bei Vögeln: Rostfärbung durch Eisenoxid beim Bartgeier (*Gypaetus barbatus*) und bei anderen Arten. *Zoologische Jahrbucher* 93:507–595.

Bertolani, P., C. Scholes, J.D. Pruetz, and W.C. McGrew. 2007. Laterality in termite-fishing by Fongoli chimpanzees: Preliminary report. *Pan Africa News* 14 (1): 1–3.

Bertrand, M. 1976. Acquisition by a pigtail macaque of behavior patterns beyond the natural repertoire of the species. *Zeitschrift für Tierpsychologie* 42:139–169.

Besch, D. 1981. Wild behavior at the Tulsa Zoo: Chimp Island residents tap tool-using talents. *Animal Kingdom* 84 (3):30–33.

Bettinger, T., and T. Carter. 1991. Behavior exhibited by captive adult chimpanzees. *Chimpanzoo Conference Proceedings* 1991:33–44.

Bhat, H.R. 1990. Additional information on tool use by lion-tailed macaques. *Lion-Tailed Macaque Newsletter* 7 (1):6.

Bierens de Haan, J.A. 1926. Versuch über den Farbensinn und das psychische Leben von *Octopus vulgaris*. *Zeitschrift für Vergleichende Physiologie* 4:766–796.

———. 1931. Werkzeuggebrauch und Werkzeugherstellung bei einem niederen Affen (*Cebus hypoleucus* Humb.). *Zeitschrift für Vergleichende Physiologie* 13:639–695.

Bila-Isa, I. 2003. Bonobos dig termite mounds: A field example of tool use by wild bonobos of the Etate, northern sector of the Salonga National Park. In *2003 Bonobo Workshop: Behaviour, Ecology and Conservation of Wild Bonobos*, edited by J. Thompson, G. Hohmann, and T. Furuichi. Inuyama, Japan: Kyoto University Primate Research Institute, 6.

Bindner, C.M., Jr. 1968. Bald eagles use tools. *Florida Naturalist* 41 (4):169.

Bingham, H. 1929. *Chimpanzee translocation by means of boxes*. Comparative Psychology Monographs, vol. 5, no. 3, serial no. 25. Baltimore: Johns Hopkins Press.

Birch, H.G. 1945. The relation of previous experience to insightful problem-solving. *Journal of Comparative Psychology* 38:367–383.

Bird, C.D., and N.J. Emery. 2009a. Insightful problem solving and creative tool modification by captive non-tool-using rooks. *Proceedings of the National Academy of Sciences of the United States of America* 106 (25): 10370–10375.

———. 2009b. Rooks use stones to raise the water level to reach a floating worm. *Current Biology* 19 (16):1410–1414.

Birkinshaw, C.R. 1999. Use of millipedes by black lemurs to anoint their bodies. *Folia Primatologica* 70 (3): 170–171.

Biro, D., C. Sousa, and T. Matsuzawa. 2006. Ontogeny and cultural propagation of tool use by wild chimpanzees at Bossou, Guinea: Case studies in nut cracking and leaf folding. In *Cognitive development in chimpanzees*, edited by T. Matsuzawa, M. Tomonaga, and M. Tanaka. New York: Springer, 476–508.

Biro, D., N. Inoue-Nakamura, R. Tonooka, G. Yamakoshi, C. Sousa, and T. Matsuzawa. 2003. Cultural innovation and transmission of tool use in wild chimpanzees: Evidence from field experiments. *Animal Cognition* 6 (4):213–223.

Bitterman, M.E. 1960. Toward a comparative psychology of learning. *American Psychologist* 15 (11):704–712.

———. 1965. Phyletic differences in learning. *American Psychologist* 20 (6):396–410.

Blair, C.L. 1981. Ferruginous hawk using rock in nest defense. *Raptor Research* 15 (4):120.

Bodamer, M.D. 1990. The use of raisin tubes as an enrichment device. *American Journal of Primatology* 20 (3):175.

Boehm, C. 1992. Two methods of dipping for army ants. *Bulletin of the Chicago Academy of Sciences* 15 (1):10.

Böer, M. 1990. Verhaltensuntersuchungen an Flachland-gorillas. *Zoologische Garten* 60 (3–4):137–189.

Boesch, C. 1978. Nouvelles observations sur les chimpanzés de la forêt de Taï (Côte d'Ivoire). *Terre et la Vie* 32: 195–201.

———. 1991a. The effect of leopard predation on grouping patterns in forest chimpanzees. *Behaviour* 117 (3–4): 220–242.

———. 1991b. Handedness in wild chimpanzees. *International Journal of Primatology* 12 (6):541–558.

———. 1991c. Teaching among wild chimpanzees. *Animal Behaviour* 41 (3):530–532.

———. 1995. Innovation in wild chimpanzees. *International Journal of Primatology* 16 (1):1–16.

———. 1996. The emergence of cultures among wild chimpanzees. *Proceedings of the British Academy* 88: 251–268.

Boesch, C., and H. Boesch. 1981. Sex differences in the use of natural hammers by wild chimpanzees: A preliminary report. *Journal of Human Evolution* 10:585–593.

———. 1983. Optimisation of nut-cracking with natural hammers by wild chimpanzees. *Behaviour* 83 (3–4): 265–286.

———. 1984a. Mental map in wild chimpanzees: An analysis of hammer transports for nut cracking. *Primates* 25 (2):160–170.

———. 1984b. Possible causes of sex differences in the use of natural hammers by wild chimpanzees. *Journal of Human Evolution* 13 (5):415–440.

———. 1989. Hunting behavior of wild chimpanzees in the Taï National Park. *American Journal of Physical Anthropology* 78 (4):547–573.

———. 1990. Tool use and tool making in wild chimpanzees. *Folia Primatologica* 54 (1–2):86–99.

———. 1993a. Different hand postures for pounding nuts with natural hammers by wild chimpanzees. In *Hands of primates*, edited by H. Preuschoft and D.J. Chivers. New York: Springer, 31–43.

———. 1993b. Diversity of tool use and tool-making in wild chimpanzees. In *The use of tools by human and non-human primates*, edited by A. Berthelet and J. Chavaillon. Oxford: Clarendon Press, 158–168.

———. 1994. L'utilisation des outils par les chimpanzés de la forêt de Taï. *Sempervira* 2:1–23.

Boesch, C., and H. Boesch-Achermann. 1991a. Dim forest, bright chimps. *Natural History* 100 (9):50–57.

———. 1991b. Les chimpanzés et l'outil. *La Recherche* 22 (233):724–731.

———. 1994. Technique et culture chez les chimpanzés sauvages. *Techniques and Culture* 23–24:1–27.

———. 2000. *The chimpanzees of the Taï Forest: Behavioral ecology and evolution*. New York: Oxford University Press.

Boesch, C., J. Head, and M.M. Robbins. 2009. Complex tool sets for honey extraction among chimpanzees in Loango National Park, Gabon. *Journal of Human Evolution* 56 (6):560–569.

Boesch, C., P. Marchesi, N. Marchesi, B. Fruth, and F. Joulian. 1994. Is nut cracking in wild chimpanzees a cultural behaviour? *Journal of Human Evolution* 26 (4):325–338.

Boesch-Achermann, H., and C. Boesch. 1993. Tool use in wild chimpanzees: New light from dark forests. *Current Directions in Psychological Science* 2 (1):18–21.

Bogart, S.L. 2009. Behavioral ecology of savanna chimpanzees (*Pan troglodytes verus*) with respect to insectivory at Fongoli, Senegal. Ph.D. diss., Iowa State University.

Bogart, S.L., and J.D. Pruetz. 2008. Ecological context of savanna chimpanzee (*Pan troglodytes verus*) termite fishing at Fongoli, Senegal. *American Journal of Primatology* 70 (6):605–612.

Bogart, S.L., J.D. Pruetz, and W.C. McGrew. 2005. Termite de jour: Termite fishing by West African chimpanzees (*Pan troglodytes verus*) at Fongoli, Senegal. *American Journal of Physical Anthropology* 126 (S40):75.

Boggs, C.L. 1995. Male nuptial gifts: Phenotypic consequences and evolutionary implications. In *Insect reproduction*, edited by S.R. Leather and J. Hardie. New York: CRC Press, 215–242.

Boinski, S. 1988. Use of a club by a wild white-faced capuchin (*Cebus capucinus*) to attack a venomous snake (*Bothrops asper*). *American Journal of Primatology* 14 (2):177–179.

Boinski, S., R.P. Quatrone, and H. Swartz. 2000. Substrate and tool use by brown capuchins in Suriname: Ecological contexts and cognitive bases. *American Anthropologist* 102 (4):741–761.

Bolwig, N. 1961. An intelligent tool-using baboon. *South African Journal of Science* 57 (6):147–152.

———. 1963. Observations on the mental and manipulative abilities of a captive baboon (*Papio doguera*). *Behaviour* 22:24–40.

Borchelt, P.L. 1975. Organization of dustbathing components in bobwhite quail (*Colinus virginianus*). *Behaviour* 53:217–237.

Borsari, A., and E. Ottoni. 2005. Preliminary observations of tool use in captive hyacinth macaws (*Anodorhynchus hyacinthinus*). *Animal Cognition* 8:48–52.

Bortolini, T.S., and J.C. Bicca-Marques. 2007. A case of spontaneous tool-making by a captive capuchin monkey. *Neotropical Primates* 14 (2):74–76.

Boswall, J. 1977a. Notes on tool-using by Egyptian vultures. *Bulletin of the British Ornithologists' Club* 97:77–78.

———. 1977b. Tool using by birds and related behaviour. *Avicultural Magazine* 83:88–97, 146–159, 220–228.

———. 1978. Further notes on tool using by birds and related behaviour. *Avicultural Magazine* 84 (3):162–166.

———. 1983a. Tool-using and related behaviour in birds: More notes. *Avicultural Magazine* 89 (2):94–108.

———. 1983b. Tool-using and related behaviour in birds: Yet more notes. *Avicultural Magazine* 89 (3):170–181.

Boulenger, E.G. 1936. *Apes and monkeys*. London: G.G. Harrap.

Bourne, G.H. 1971. *The ape people*. New York: G.P. Putnam's Sons.

Bowman, R. 1961. *Morphological differentiation and adaptation in the Galápagos finches*. University of California Publications in Zoology, vol. 58. Berkeley: University of California Press.

Boycott, B.B. 1954. Learning in *Octopus vulgaris* and other Cephalopods. *Pubblicazione della Stazione Zoologica di Napoli* 25:67–93.

Boysen, S. 1997. Representation of quantity by apes. *Advances in the Study of Behavior* 26:435–462.

Boysen, S., G.G. Berntson, and J. Prentice. 1987. Simian scribbles: A reappraisal of drawing in the chimpanzee (*Pan troglodytes*). *Journal of Comparative Psychology* 101 (1):82–89.

Boysen, S., and D. Frisch. 1987. Extractive tool use in captive lowland gorillas. *American Journal of Primatology* 12 (3):332.

Boysen, S., V.A. Kuhlmeier, P.O. Halliday, and Y.M. Halliday. 1999. Tool use in captive gorillas. In *The mentalities of gorillas and orangutans: Comparative perspectives*, edited by S.T. Parker, R.W. Mitchell, and H.L. Miles. Cambridge: Cambridge University Press, 179–187.

Bradford, P.V., and H. Blume. 1992. *Ota Benga: The pygmy in the zoo*. New York: St. Martin's Press.

Brandibas, P., R. Chalmeau, and A. Gallo. 1995. Tool use in a captive group of chimpanzees. *Folia Primatologica* 64 (1–2):76.

Brandt, M., and D. Mahsberg. 2002. Bugs with a backpack: The function of nymphal camouflage in the West African assassin bugs *Paredocla* and *Acanthaspis* spp. *Animal Behaviour* 63:277–284.

Bräuer, J., J. Call, and M. Tomasello. 2009. Are apes inequity averse? New data on the token-exchange paradigm. *American Journal of Primatology* 71:175–181.

Brent, L., M.A. Bloomsmith, and S.D. Fisher. 1995. Factors determining tool-using ability in two captive chimpanzee (*Pan troglodytes*) colonies. *Primates* 36 (2):265–274.

Bresard, B. 1993. Some one- and two-handed functions and processes in tool use by pongids. In *The use of tools by human and non-human primates*, edited by A. Berthelet and J. Chavaillon. Oxford: Clarendon Press, 78–94.

Breuer, T., M. Ndoundou-Hockemba, and V. Fishlock. 2005. First observation of tool use in wild gorillas. *PLoS Biology* 3 (11):2041–2043.

Brewer, S.M. 1976. Chimpanzee rehabilitation. *International*

Primate Protection League Special Report, December: 1–10.
———. 1978. *The chimps of Mt. Assirik*. New York: Alfred A. Knopf.
———. 1982. Essai de réhabilitation de chimpanzés au Niokola-Koba de chimpanzés auparavant en captivité. *Mémoires de l'Institut Fondamental d'Afrique Noire* 92:341–362.
Brewer, S.M., and W.C. McGrew. 1990. Chimpanzee use of a tool-set to get honey. *Folia Primatologica* 54 (1–2): 100–104.
Bril, B., G. Dietrich, J. Foucart, K. Fuwa, and S. Hirata. 2009. Tool use as a way to assess cognition: How do captive chimpanzees handle the weight of the hammer when cracking a nut? *Animal Cognition* 12 (2): 217–235.
Brink, A.S. 1957. The spontaneous fire-controlling reactions of two chimpanzee smoking addicts. *South African Journal of Science* 53:241–247.
Bristowe, W.S. 1929. The mating habits of spiders, with special reference to the problems surrounding sex dimorphism. *Proceedings of the Zoological Society of London* 1929:309–358.
———. 1971. *The world of spiders*, rev. ed. New Naturalist, vol. 38. London: Collins.
Brockie, R. 1976. Self-anointing by wild hedgehogs, *Erinaceus europaeus*, in New Zealand. *Animal Behaviour* 24:68–71.
Brockmann, H.J. 1985. Tool use in digger wasps: Hymenoptera: Sphecinae. *Psyche* 92 (2–3):309–330.
Brodie, E.D. 1977. Hedgehogs use toad venom in their own defense. *Nature* 268 (5621):627–628.
Brooke, R.K. 1979a. Predation on ostrich eggs by tool-using crows and Egyptian vultures. *Ostrich* 48 (4):257–258.
———. 1979b. Tool using by the Egyptian vulture to the detriment of the ostrich. *Ostrich* 50 (2):119–120.
Brooks, R. 1990. The hermit crab's special forces. *Sea Frontiers* 36 (4):50–55.
Brosnan, S.F. 2009. Animal behavior: The right tool for the job. *Current Biology* 19 (3):R124–R125.
Brosnan, S.F., and M.J. Beran. 2009. Trading behavior between conspecifics in chimpanzees, *Pan troglodytes*. *Journal of Comparative Psychology* 123:181–194.
Brosnan, S.F., and F.B.M. de Waal. 2004. A concept of value during experimental exchange in brown capuchin monkeys. *Folia Primatologica* 75:317–330.
———. 2005. Responses to a simple barter task in chimpanzees, *Pan troglodytes*. *Primates* 46 (3):173–182.
Brown, L., and E. Urban. 1969. The breeding biology of the great white pelican *Pelecanus onocrotalus roseus* at Lake Shala, Ethiopia. *Ibis* 111:199–237.
Brown, N.R., and R.C. Clark. 1962. Studies of predators of the balsam wooly aphid, *Adelges piceae* (Ratz.) (Homoptera: Adelgidae): X. Field Identification of *Laricobius erichsonii* Rosen. (Coleoptera: Derodontidae). *Canadian Entomologist* 94:191–193.
Brown, S.G., W.P. Dunlap, and T.L. Maple. 1982. Notes on water-contact by a captive male lowland gorilla. *Zoo Biology* 1 (3):243–249.
Bshary, R., W. Wickler, and H. Fricke. 2002. Fish cognition: A primate's eye view. *Animal Cognition* 5:1–13.
Buckley, C. 2003. Captive orang-utans (*Pongo pygmaeus*) and environmental enrichment. *Ratel* 30 (1):11–22.
Buckley, J.S. 1983. The feeding behavior, social behavior, and ecology of the white-faced monkey, *Cebus capucinus*, at Trujillo, northern Honduras, Central America. Ph.D. diss., University of Texas at Austin.
Bürger, O. 1903. Ueber das Zusammenleben von *Antholoba reticulata* Couth. und *Hepatus chilensis* M.E. *Biologishe Centralblatt* 23:678–679.
Burghardt, G.M., B. Ward, and R. Rosscoe. 1996. Problem of reptile play: Environmental enrichment and play behavior in a captive Nile soft-shelled turtle (*Trionyx triunguis*). *Zoo Biology* 15:223–238.
Burton, M. 1957. Hedgehog self-anointing. *Proceedings of the Zoological Society of London* 129:452–453.
———. 1959. *Phoenix re-born*. London: Hutchinson.
Butler, E.A. 1923. *A biology of the British Hemiptera-Heteroptera*. London: H.F. & G. Witherby.
Bygott, J.D. 1979. Agonistic behavior, dominance, and social structure in wild chimpanzees of the Gombe National Park. In *The great apes: Perspectives on human evolution*, edited by D.A. Hamburg and E.R. McCown. Menlo Park, CA: Benjamin/Cummings, 405–407.
Byrne, G., and S.J. Suomi. 2004. Influences of manipulative style, age, gender, and cortisol reactivity on tufted capuchins' (*Cebus apella*) performance in a tool-using task. *Advances in Psychology Research* 31:79–93.
Byrne, R.W. 1997. The technical intelligence hypothesis: An additional evolutionary stimulus to intelligence? In *Machiavellian intelligence II: Extensions and evaluations*, edited by A. Whiten and R.W. Byrne. Cambridge: Cambridge University Press, 289–311.
Byrne, R.W., and A. Whiten. 1988. *Machiavellian intelligence: Social expertise and the evolution of intellect in monkeys, apes, and humans*. Oxford: Clarendon Press.
Cade, T. 1967. Ecological and behavioral aspects of predation by the northern shrike. *Living Bird* 6:43–86.
Caffrey, C. 2000. Tool modification and use by an American crow. *Wilson Bulletin* 112 (2):283–284.
———. 2001. Goal-directed use of objects by American crows. *Wilson Bulletin* 113 (1):114–115.
Call, J., and M. Tomasello. 1994. The social learning of tool use by orangutans (*Pongo pygmaeus*). *Human Evolution* 9 (4):297–313.
Camacho, L. 1907. Do animals reason? *Scientific American* 96:515.
Campbell, A., and H. Barnard. 1917. Birds of the Rockingham Bay District, North Queensland. *Emu* 17:2–38.
Campbell, C.J. 2000. Fur rubbing behavior in free-ranging black-handed spider monkeys (*Ateles geoffroyi*) in Panama. *American Journal of Primatology* 51 (3):205–208.
Canale, G.R., C.E. Guidorizzi, M.C.M. Kierulff, and C. Gatto. 2009. First record of tool use by wild popu-

lations of the yellow-breasted capuchin monkey (*Cebus xanthosternos*) and new records for the bearded capuchin (*Cebus libidinosus*). *American Journal of Primatology* 71 (5):366–372.
Candland, D.K. 1987. Tool use. In *Behavior, cognition, and motivation*, edited by G. Mitchell, J. Erwin, and A.R. Liss. Vol. 2, part B of *Comparative primate biology*. New York: Alan R. Liss, 85–103.
Candland, D.K., J.A. French, and C.N. Johnson. 1978. Object-play: Test of a categorized model by the genesis of object-play in *Macaca fuscata*. In *Social play in primates*, edited by E.O. Smith. New York: Academic Press, 259–296.
Candland, D.K., P. Weldon, G. Lorinc, and K. O'Connor. 1978. Creative manipulation of the environment among *Macaca fuscata*. In *Behaviour*, edited by J. Chivers and J. Herbert. Vol. 1 of *Recent advances in primatology*. New York: Academic Press, 965–966.
Cantalupo, C., H. Freeman, W. Rodes, and W. Hopkins. 2008. Handedness for tool use correlates with cerebellar asymmetries in chimpanzees (*Pan troglodytes*). *Behavioral Neuroscience* 122 (1):191–198.
Carpenter, A. 1887. Monkeys opening oysters. *Nature* 36:53.
Carpenter, C.R. 1934. *A field study of the behavior and social relations of howling monkeys*. Comparative Psychology Monographs, vol. 10, no. 2, serial no. 48. Baltimore: Johns Hopkins Press.
———. 1935. Behavior of red spider monkeys in Panama. *Journal of Mammalogy* 16:171–180.
———. 1937. An observational study of two captive gorillas. *Human Biology* 9:175–196.
———. 1940. *A field study in Siam of the behavior and social relations of the gibbon (*Hylobates lar*)*. Comparative Psychology Monographs, vol. 16, no. 5, serial no. 84. Baltimore: Johns Hopkins Press.
———. 1963. Comments on "Tool-using performances as indicators of behavioral adaptability." *Current Anthropology* 4:488.
Carpenter, C.R., and N.M. Locke. 1937. Notes on symbolic behavior in a cebus monkey (*Capucinus apella*). *Journal of General Psychology* 51:267–278.
Carvalho, S., C. Sousa, and T. Matsuzawa. 2007. New nut-cracking sites in Diécké Forest, Guinea: An overview of the surveys. *Pan Africa News* 14 (1):11–13.
Carvalho, S., E. Cunha, C. Sousa, and T. Matsuzawa. 2008. Chaînes operatoires and resource-exploitation strategies in chimpanzee (*Pan troglodytes*) nut cracking. *Journal of Human Evolution* 55 (1):148–163.
Carvalho, S., D. Biro, W.C. McGrew, and T. Matsuzawa. 2009. Tool-composite reuse in wild chimpanzees (*Pan troglodytes*): Archaeologically invisible steps in the technological evolution of early hominins? *Animal Cognition* 12 (Suppl. 1):103–114.
Casanova, C., R. Mondragon-Ceballos, and P.C. Lee. 2008. Innovative social behavior in chimpanzees (*Pan troglodytes*). *American Journal of Primatology* 70:54–61.
Celli, M., S. Hirata, and M. Tomonaga. 1999. [Tool use in captive chimpanzees in a honey dipping and a juice drinking task]. *Reichorui Kenkyu* 15 (3):446.
———. 2004. Socioecological influences on tool use in captive chimpanzees. *International Journal of Primatology* 25 (6):1267–1281.
Celli, M., M. Tomonaga, T. Udono, M. Teramoto, and K. Nagano. 2001. Learning processes in the acquisition of a tool using task by captive chimpanzees. *Psychologia* 44 (1):70–81.
Chaffer, N. 1931. The satin birds of National Park, N.S.W. *Emu* 30:277–285.
Chalmeau, R., and P. Peignot. 1998. Exchange of objects between humans and captive western lowland gorillas. *Primates* 39 (4):389–398.
Chamove, A.S. 1989. Enrichment in chimpanzees: Unpredictable ropes and tools. *Ratel* 16 (5):139–141.
Chapman, C.A. 1986. Boa constrictor predation and group response in white-faced *Cebus* monkeys. *Biotropica* 18 (2):171–172.
Chappell, J., and A. Kacelnik. 2002. Tool selectivity in a non-primate, the New Caledonian crow (*Corvus moneduloides*). *Animal Cognition* 5 (2):71–78.
———. 2004. Selection of tool diameter by New Caledonian crows *Corvus moneduloides*. *Animal Cognition* 7 (2):121–127.
Charpentier, M.R.E., M. Boulet, and C.M. Drea. 2008. Smelling right: The scent of male lemurs advertises genetic quality and relatedness. *Molecular Ecology* 17:3225–3233.
Chauvin, A., and B. Muckensturm-Chauvin. 1980. *Behavioral complexities*. New York: International Universities Press.
Cheney, D.L., and R.M. Seyfarth. 2007. *Baboon metaphysics: The evolution of a social mind*. Chicago: University of Chicago Press.
Chevalier-Skolnikoff, S. 1977. A Piagetian model for describing and comparing socialization in monkey, ape, and human infants. In *Primate bio-social development: Biological, social, and ecological determinants*, edited by S. Chevalier-Skolnikoff and F.E. Poirier. New York: Garland, 159–187.
———. 1989. Spontaneous tool use and sensorimotor intelligence in *Cebus* compared with other monkeys and apes. *Behavioral and Brain Sciences* 12 (3):561–588.
———. 1990. Tool use by wild cebus monkeys at Santa Rosa National Park, Costa Rica. *Primates* 31 (2): 375–383.
Chevalier-Skolnikoff, S., B.M.F. Galdikas, and A.Z. Skolnikoff. 1982. The adaptive significance of higher intelligence in wild orang-utans: A preliminary report. *Journal of Human Evolution* 11 (7):639–652.
Chevalier-Skolnikoff, S., and J. Liska. 1993. Tool use by wild and captive elephants. *Animal Behaviour* 46 (2):209–219.
Chia, F. 1973. Sand dollar: A weight belt for the juvenile. *Science* 181:73–74.
Chiang, M. 1967. Use of tools by wild macaque monkeys in Singapore. *Nature* 214:1258–1259.

Chintiroglou, C.C., D. Doumenc, and D. Guinot. 1996. Anemone-carrying behaviour in a deep-water homolid crab (Brachyura: Podotremata). *Crustaceana* 69:19–25.

Chisholm, A.H. 1954. The use by birds of "tools" or "instruments." *Ibis* 96:380–383.

———. 1959. The history of anting. *Emu* 59:101–130.

———. 1971. The use by birds of tools and playthings. *Victorian Naturalist* 88:180–188.

Chivers, D. 1974. *The siamang in Malaya: A field study of a primate in tropical rain forest.* Contributions to Primatology, vol. 4. Basel, Switzerland: S. Karger.

Choe, D., and M.K. Rust. 2007. Use of plant resin by a bee assassin bug, *Apiomerus flaviventris* (Hemiptera: Reduviidae). *Annals of the Entomological Society of America* 100 (2):320–326.

Clapham, P. 2000. The humpback whale: Seasonal feeding and breeding in a baleen whale. In *Cetacean societies*, edited by J. Mann, R. Connor, P.L. Tyack, and H. Whitehead. Chicago: University of Chicago Press, 173–196.

Clark, C.C., and L. Clark. 1990. "Anting" behavior by common grackles and European starlings. *Wilson Bulletin* 102 (1):167–169.

Clayton, D.H., and J.G. Vernon. 1993. Common grackle anting with lime fruit and its effect on ectoparasites. *Auk* 110 (4):951–952.

Clayton, N. 2007. Animal cognition: Crows spontaneously solve a metatool task. *Current Biology* 17:R894–R895.

Clayton, N., and A. Jolliffe. 1996. Marsh tits (*Parus palustris*) use tools to store food. *Ibis* 138 (3):554.

Cleveland, A., A. Rocca, E.L. Wendt, and G.C. Westergaard. 2003. Throwing behavior and mass distribution of stone selection in tufted capuchin monkeys (*Cebus apella*). *American Journal of Primatology* 61 (4):159–172.

———. 2004. Transport of tools to food sites in tufted capuchin monkeys (*Cebus apella*). *Animal Cognition* 7 (3):193–198.

Clucas, B., D.H. Owings, and M.P. Rowe. 2008. Donning your enemy's cloak: Ground squirrels exploit rattlesnake scent to reduce predation risk. *Proceedings of the Royal Society B-Biological Sciences* 275 (1636):847–852.

Clucas, B., M.P. Rowe, D.H. Owings, and P.C. Arrowood. 2008. Snake scent application in ground squirrels, *Spermophilus* spp.: A novel form of antipredator behaviour? *Animal Behaviour* 75:299–307.

Clunie, F. 1976. Jungle mynah "anting" with a millipede. *Notornis* 23:77.

Coddington, J.A., and C. Sobrevila. 1987. Web manipulation and two stereotyped attack behaviors in the ogre-faced spider *Deinopis spinosus* Marx (Araneae: Deinopidae). *Journal of Arachnology* 15:213–225.

Colahan, B.D. 1981. Anting in the cape white-eye. *Ostrich* 52 (3):186.

Cole, M. 1987. How we keep our gorillas occupied. *Animal Keepers' Forum* 14 (12):401–403.

———. 1989. Examples of tool use in a captive group of western lowland gorillas (*Gorilla gorilla gorilla*). *American Journal of Primatology* 18 (2):140–141.

Cole, P.D. 2004. The ontogenesis of innovative tool use in an American crow (*Corvus brachyrhynchos*). Ph.D. diss., Dalhousie University.

Collias, N., and E. Collias. 1976. *External constructions by animals.* Stroudsburg, PA: Dowden, Hutchinson, and Ross.

———. 1984. *Nest building and bird behavior.* Princeton, NJ: Princeton University Press.

Collins, D.A., and W.C. McGrew. 1985. Chimpanzees' (*Pan troglodytes*) choice of prey among termites (Macrotermitinae) in western Tanzania. *Primates* 26 (4):375–389.

———. 1987. Termite fauna related to differences in tool-use between groups of chimpanzees (*Pan troglodytes*). *Primates* 28 (4):457–471.

Cooper, L.R., and H.F. Harlow. 1961. Note on a cebus monkey's use of a stick as a weapon. *Psychological Reports* 8:418.

Corp, N., H. Hayaki, T. Matsusaka, S. Fujita, K. Hosaka, N. Kutsukake, M. Nakamura, M. Nakamura, H. Nishie, M. Shimada, K. Zamma, W. Wallauer, and T. Nishida. 2009. Prevalence of muzzle-rubbing and hand-rubbing behavior in wild chimpanzees in Mahale Mountains National Park, Tanzania. *Primates* 50 (2):184–189.

Coss, R.G., and J.E. Biardi. 1997. Individual variation in the antisnake behavior of California ground squirrels (*Spermophilus beecheyi*). *Journal of Mammalogy* 78 (2):294–310.

Coss, R.G., and D. Owings. 1978. Snake-directed behavior by snake naive and experienced California ground squirrels in a simulated burrow. *Zeitschrift für Tierpsychologie* 48:421–435.

Costa, G., A. Petralia, E. Conti, C. Hänel, and M.K. Seely. 1993. Seven-stone spiders on the gravel plains of the Namib Desert. *Bollettino della Accademia Gioenia di Scienze Naturali* 26:77–83.

Costello, M. 1987. Tool use and manufacture in manipulanda-deprived capuchins (*Cebus apella*): Ontogenetic and phylogenetic issues. *American Journal of Primatology* 12 (3):337.

Cowles, J.T. 1937. *Food-tokens as incentives for learning by chimpanzees.* Comparative Psychology Monographs, vol. 14, no. 5, serial no. 71. Baltimore: Johns Hopkins Press.

Cowper, A. 1971. Observations on chimpanzees in Chester Zoo. *Revue de Comportement Animal* 5:39–44.

Cox, K. 1962. *California abalone, family Haliotidae.* Fish Bulletin 118. Sacramento: California Department of Fish and Game.

Craig, A. 1999. Anting in Afrotropical birds: A review. *Ostrich* 70 (3–4):203–207.

Cristol, D.A., P.V. Switzer, K.L. Johnson, and L.S. Walke. 1997. Crows do not use automobiles as nutcrackers: Putting an anecdote to the test. *Auk* 114 (2):296–298.

Cummins-Sebree, S., and D.M. Fragaszy. 2001. The right

stuff: Capuchin monkeys perceive affordances of tools. In *Studies in perception and action VI*, edited by G.A. Burton and R.C. Schmidt. Mahwah, NJ: Lawrence Erlbaum, 89–92.

———. 2005. Choosing and using tools: Capuchins (*Cebus apella*) use a different metric than tamarins (*Saguinus oedipus*). *Journal of Comparative Psychology* 119 (2):210–219.

Cunningham, A. 1921. John Daniel, civilized gorilla. *Literary Digest* 71:44–48.

Cunningham, C. 2006. Cognitive flexibility in gibbons (Hylobatidae): Object manipulation and tool-use. Ph.D. diss., University of Stirling.

Cunningham, C., and J. Anderson. 2004a. Tool manipulation to gain a reward in gibbons: Insight, learning, and understanding. *Folia Primatologica* 75 (Suppl. 1):252.

———. 2004b. Tool manipulation to gain a reward in gibbons: Insight, learning, and understanding. *Primate Eye* 83:18–19.

Cunningham, C., J. Anderson, and A. Mootnick. 2006. Object manipulation to obtain a food reward in hoolock gibbons, *Bunopithecus hoolock*. *Animal Behaviour* 71 (3):621–629.

Curio, E., and P. Kramer. 1964. Vom Mangrovefinken (*Cactospiza heliobates* Snodgrass und Heller). *Zeitschrift für Tierpsychologie* 21:223–234.

Cushing, J. 1939. Sea otters and abalones. *Journal of Mammalogy* 20:371.

Daanje, A. 1975. Some special features of the leaf-rolling technique of *Bysticus populi* L. (Coleoptera: Rhynchitini). *Behaviour* 53 (3–4):285–316.

Dare, R. 1974. The social behaviour and ecology of spider monkeys, *Ateles geoffroyi*, on Barro Colorado Island. Ph.D. diss., University of Oregon.

Darrow, M. 1979. Odds and ends on woollies. *Simian*, February:3–6.

Darwin, C. 1871. *The descent of man and selection in relation to sex*. London: Murray.

Darwin, E. 1794. *Zoonomia, or Laws of organic life*. London: J. Johnson.

Dater, E.E. 1953. Anting of a blue-winged warbler (*Vermivora pinus*) at Ramsey, New Jersey. *Auk* 70:89.

Dathe, H. 1961. Werkzeuggebrauch durch Braunbären. *Wissenschaftliche Zeitschrift der Karl-Marx-Universität Leipzig* 10:243–244.

Davenport, R.K., Jr. 1967. The orang-utan in Sabah. *Folia Primatologica* 5:247–263.

Davis, W.E., Jr., and J. Zickefoose. 1998. Bait-fishing by birds: A fascinating example of tool use. *Bird Observer* 26 (3):139–143.

Dayton, P.K., G.A. Robillard, and R.T. Paine. 1970. Benthic faunal zonation as a result of anchor ice at McMurdo Sound, Antarctica. In *Antarctic ecology*, edited by M.W. Holdgate. London: Academic Press, 244–258.

de Beer, B. 1948. Secrets of the ant-lion. *African Wildlife* 2:57–61.

de Resende, B.D., and E.B. Ottoni. 2002. Ontogeny of nut-cracking behavior in a semi-free-ranging group of tufted capuchin monkeys. In *Caring for primates: Abstracts of the XIXth Congress of the International Primatological Society*. Beijing: Mammalogical Society of China, 319–320.

de Silva, G.S. 1970. Training orang-utans for the wild. *Oryx* 10:389–393.

de Waal, F.B.M. 1989. Food sharing and reciprocal obligations among chimpanzees. *Journal of Human Evolution* 18:433–459.

———. 1996. *Good natured: The origins of right and wrong in humans and other animals*. Cambridge, MA: Harvard University Press.

———. 1997. *Bonobo: The forgotten ape*. Berkeley: University of California Press.

———. 2007. *Chimpanzee politics*. Baltimore: Johns Hopkins University Press.

de Waal, F.B.M., and J.A. Hoekstra. 1980. Contexts and predictability of aggression in chimpanzees. *Animal Behaviour* 28 (3):929–937.

Deag, J., and J. Crook. 1971. Social behavior and "agonistic buffering" in the wild Barbary macaque. *Folia Primatologica* 15:183–200.

Deblauwe, I., P. Guislain, J. Dupain, and L. van Elsacker. 2006. Use of a tool-set by *Pan troglodytes troglodytes* to obtain termites (*Macrotermes*) in the periphery of the Dja Biosphere Reserve, southeast Cameroon. *American Journal of Primatology* 68 (12):1191–1196.

Debus, S.J.S. 1991. Further observations on the black-breasted buzzard *Hamirostra melanosternon* using stones to break eggs. *Australian Bird Watcher* 14: 138–143.

Defler, T.R. 1979a. Notes on the interactions between the tayra (*Eira barbara*) and the white-fronted capuchin (*Cebus albifrons*). *Journal of Mammalogy* 61 (1):156.

———. 1979b. On the ecology and behavior of *Cebus albifrons* in eastern Columbia: II. Behavior. *Primates* 20 (4):491–502.

DeJoseph, M., R.S. Taylor, M. Baker, and M. Aregulin. 2002. Fur-rubbing behavior of capuchin monkeys. *Journal of the American Academy of Dermatology* 46:924–925.

Dennett, D.C. 1988. Précis of the intentional stance. *Behavioral and Brain Sciences* 11:495–546.

Dennis, J. 1971. Species using red-cockaded woodpecker holes in northeastern South Carolina. *Bird Banding* 42:79–87.

———. 1985. Commentary on grackles anting with marigold blossoms. *Blue Jay* 43 (3):175–177.

Derr, S., and D. Quiatt. 1992. Japanese macaque stone handlers in the New World. *American Journal of Physical Anthropology* 35 (S14):70.

Devos, C., S. Gatti, and F. Levréro. 2002. New record of algae feeding and scooping by *Pan t. troglodytes* at Lokoué Bai in Odzala National Park, Republic of Congo. *Pan Africa News* 9 (2):19–21.

Dewsbury, D.A. 2009. Samuel Fernberger's rejected doctoral dissertation: A neglected resource for the history of ape research in America. *History of Psychology* 12 (1):1–6.

D'Have, H., J. Scheirs, R. Verhagen, and W. De Coen. 2005. Gender, age, and seasonal dependent self-anointing in the European hedgehog *Erinaceus europaeus*. *Acta Theriologica* 50 (2):167–173.

Dick, J.A., and M.B. Fenton. 1979. Tool-using by a black eagle? *Bokmakierie* 31:17.

Dill, L.M. 1977. Refraction and the spitting behavior of the archerfish (*Toxotes chatareus*). *Behavioral Ecology and Sociobiology* 2:169–184.

Dittrich, L. 1983. Menschenaffen als Erfinder und Handwerker. *Der Zoofreund* 50:2–7.

Dix, T.G. 1970. Covering response of echinoid *Evechinus chloroticus*. *Pacific Science* 24 (2):187–193.

Dixon, G.Y., and A.F. Dixon. 1891. Report on the marine invertebrate fauna near Dublin. *Proceedings of the Royal Irish Academy*, Series 3, 2:19–33.

Döhl, J. 1966. Manipulierfähigkeit und "einsichtiges" Verhalten eines Schimpansen bei komplizierten Handlungsketten. *Zeitschrift für Tierpsychologie* 23:77–113.

———. 1968. Über die Fähigkeit einer Schimpansin, Umwege mit selbständigen Zwischenzielen zu überblicken. *Zeitschrift für Tierpsychologie* 25:89–103.

———. 1969. [Research with a chimpanzee on shortening of detours with self-selected intermediate goals]. *Zeitschrift für Tierpsychologie* 26:200–207.

Döhl, J., and J. Lethmate. 1986. Werkzeuggebrauch und einsichtiges Handeln eines Orang-utans. In *Publikationen zu Wissenschaftlichen Filmen*. Göttingen, Germany: Institut für den Wissenschaftlichen Film, 3–14.

Döhl, J., and D. Podolczak. 1973. Versuche zur Manipulierfreudigkeit von zwei jungen Orang-utans (*Pongo pygmaeus*) im Frankfurter Zoo. *Zoologische Garten* 43: 81–94.

Domínguez, K., and M.L. Jiménez. 2005. Mating and self-burying behavior of *Homalonychus theologus* Chamberlin (Araneae: Homalonychidae) in Baja California Sur. *Journal of Arachnology* 33 (1):167–174.

Douglas, C.A. 1976. Availability of drift materials and covering response of sea-urchin *Strongylocentrotus purpuratus* (Stimpson). *Pacific Science* 30 (1):83–89.

Douglas-Hamilton, I., and O. Douglas-Hamilton. 1975. *Among the elephants*. New York: Viking Press.

Dow, D.D. 1980. Primitive weaponry in birds: The Australian brush-turkey's defense. *Emu* 80:91–92.

Downman, M. 2000. Introducing gorillas to a naturalistic environment. *Gorilla Gazette* 14 (1):20–24.

Drescher, K., and W. Trendelenburg. 1927. Weiterer Beitrag zur Intelligenz-prüfung an Affen (einschließlich Anthropoiden). *Zeitschrift für Vergleichende Physiologie* 5:613–642.

Dubois, C.A. 1969. Grackle anting with a mothball. *Auk* 86 (1):131.

Dubois, M. 2001. Spatial facilitation in a probing task in *Cebus olivaceus*. In *XVIIIth Congress of the International Primatological Society: Primates in the New Millennium*. Adelaide, Australia: International Primatological Society, 439.

Dubois, M., J.F. Gerard, E. Sampaio, O.F. Galvuo, and C. Guilhem. 2001. Spatial facilitation in a probing task in wedge-capped capuchins (*Cebus olivaceus*). *International Journal of Primatology* 22 (6):993–1006.

Dücker, G. 1965. Das Verhalten der Viverriden. *Handbuch der Zoologie*. 8:1–48.

Dücker, G., and B. Rensch. 1977. The solution of patterned string problems by birds. *Behaviour* 62 (1–2):162–173.

Ducoing, A.M., and B. Thierry. 2005. Tool-use learning in Tonkean macaques (*Macaca tonkeana*). *Animal Cognition* 8 (2):103–113.

Duerden, J. 1905. On the habits and reactions of crabs bearing actinians in their claws. *Proceedings of the Zoological Society of London* 2:494–511.

Dufour, V., and E.H.M. Sterck. 2008. Chimpanzees fail to plan in an exchange task but succeed in a tool-using procedure. *Behavioural Processes* 79 (1):19–27.

Dufour, V., M. Pelé, M. Neumann, B. Thierry, and J. Call. 2009. Calculated reciprocity after all: Computation behind token transfers in orang-utans. *Biology Letters* 5 (2):172–175.

Dumont, C.P., D. Drolet, I. Deschênes, and J.H. Himmelman. 2007. Multiple factors explain the covering behaviour in the green sea urchin, *Strongylocentrotus droebachiensis*. *Animal Behaviour* 73:979–986.

Dupain, J., L. van Elsacker, C. Nell, P. Garcia, F. Ponce, and M. Huffman. 2002. New evidence for leaf swallowing and *Oesophagostomum* infection in bonobos (*Pan paniscus*). *International Journal of Primatology* 23 (5):1053–1062.

Duyck, I., and J. Duyck. 1984. Great tit using instrument to pick up food. *Wielewaal* 50 (11):416.

Eaton, G. 1972. Snowball construction by a feral troop of Japanese macaques (*Macaca fuscata*) living under semi-natural conditions. *Primates* 13:411–414.

Eaton, R.L. 1978. Do chimpanzees use weapons or mimic hominid weaponry? *Carnivore* 1 (2):82–89.

Eberhard, W.G. 1977. Aggressive chemical mimicry by a bolas spider. *Science* 198 (4322):1173–1175.

———. 1980. The natural history and behavior of the bolas spider. *Psyche* 87:143–170.

Ebert, E. 1968. A food habits study of the southern sea otter, *Enhydra lutris nereis*. *California Fish and Game* 54:33–42.

Ehrlich, P.R., D.S. Dobkin, and W. White. 1986. The adaptive significance of anting. *Auk* 103:835.

Eibl-Eibesfeldt, I. 1961. Über den Werkzeuggebrauch des Spechtfinken *Camarhynchus pallidus*. *Zeitschrift für Tierpsychologie* 18:343–346.

———. 1970. *Ethology: The biology of behavior*. New York: Holt, Rinehart and Winston.

Eibl-Eibesfeldt, I., and H. Sielmann. 1962. Beobachtungen am Spechtfinken *Camarhynchus pallidus* (Sclater und Salvin). *Journal für Ornithologie* 103:92–101.

Eisenberg, J. 1981. *The mammalian radiation: An analysis of trends in evolution, adaptation, and behavior*. Chicago: University of Chicago Press.

Eisenberg, J., and E. Gould. 1966. The behavior of *Solenodon paradoxus* in captivity with comments on the behavior of the Insectivora. *Zoologica* 51:49–58.

Eisenberg, J., and D. Kleiman. 1977. The usefulness of behavior studies in developing captive breeding programmes for mammals. In *International Zoo Yearbook*, edited by P. Olney. London: Zoological Society of London, 81–89.

Eisentraut, M. 1933. Biologische Studien im Bolivianischen Chaco. *Zeitschrift für Säugetierekunde* 8:47–69.

———. 1953. Sichbespucken bei Igeln. *Zeitschrift für Tierpsychologie* 10:50–55.

Eisner, T. 1968. Mongooses and millipedes. *Science* 150:1367.

Eisner, T., and D. Aneshansley. 2008. "Anting" in blue jays: Evidence in support of a food-preparatory function. *Chemoecology* 18 (4):197–203.

Eisner, T., and J.A. Davis. 1967. Mongoose throwing and smashing millipedes. *Science* 155:577–579.

Eisner, T., and M. Eisner. 2000. Defensive use of a fecal thatch by a beetle larva (*Hemisphaerota cyanea*). *Proceedings of the National Academy of Sciences of the United States of America* 97 (6):2632–2636.

———. 2002. Coiling into a sphere: Defensive behavior of a trash-carrying chrysopid larva *Leucochrysa* (*Nodita*) *pavida* (Neuroptera: Chrysopidae). *Entomological News* 113 (1):6–10.

Eisner, T., M. Eisner, and M. Siegler. 2005. *Secret weapons: Defenses of insects, spiders, scorpions, and other many-legged creatures*. Cambridge, MA: Belknap Press.

Eisner, T., and R.E. Silberglied. 1988. A chrysopid larva that cloaks itself in mealybug wax. *Psyche* 9:15–19.

Eisner, T., E. van Tassell, and J.E. Carrel. 1967. Defensive use of a "fecal shield" by a beetle larva. *Science* 158:1471–1473.

Eisner, T., K. Hicks, M. Eisner, and D.S. Robson. 1978. "Wolf-in-sheep's-clothing" strategy of a predaceous insect larva. *Science* 199 (4330):790–794.

Eisner, T., J.E. Carrel, E. Van Tassell, E.R. Hoebeke, and M. Eisner. 2002. Construction of a defensive trash packet from sycamore leaf trichomes by a chrysopid larva (Neuroptera: Chrysopidae). *Proceedings of the Entomological Society of Washington* 104 (2):437–446.

Ellis, D.H., and S. Brunson. 1993. "Tool" use by the red-tailed hawk (*Buteo jamaicensis*). *Journal of Raptor Research* 27 (2):128.

Ellis, J. 1975. Orangutan tool use at Oklahoma City Zoo. *Keeper* 1:5–6.

———. 1977. Further notes on orangutan tool use. *Ratel* 4 (2–3):21.

Emerson, A. 1938. Termite nests: A study in the phylogeny of behavior. *Ecological Monographs* 8:248–284.

Emerton, J.H. 1912. Four burrowing *Lycosa* (*Geolycosa* Montg., *Scaptocosa* Banks) including one new species. *Pysche* 19:25–36.

Emery, N.J., and N.S. Clayton. 2004a. Comparing the complex cognition of birds and primates. In *Comparative vertebrate cognition: Are primates superior to non-primates?*, edited by L.J. Rogers and G. Kaplan. New York: Kluwer Academic/Plenum, 3–55.

———. 2004b. The mentality of crows: Convergent evolution of intelligence in corvids and apes. *Science* 306 (5703):1903–1907.

Emlen, J.T. 1962. The display of the gorilla. *Proceedings of the American Philosophical Society* (106):516–519.

English, M. 1987. More on fishing green-backed herons. *Bokmakierie* 39:124–125.

Epstein, R., and S.D. Medalie. 1983. The spontaneous use of a tool by a pigeon. *Behaviour Analysis Letters* 3 (4):241–247.

Epstein, R., C.E. Kirshnit, R.P. Lanza, and L.C. Rubin. 1984. Insight in the pigeon: Antecedents and determinants of an intelligent performance. *Nature* 308 (5954):61–62.

Escola, L., I. Intskirveli, M.A. Umiltà, A. Nasoyan, G. Rizzolatti, and V. Gallese. 2004a. Goal-relatedness in area F5 of the macaque monkey during tool use. *Society for Neuroscience Abstracts* 2004:Abstract #191.198.

———. 2004b. Tool use and goal-relatedness in areas F5 and F1 of the macaque monkey. *FENS Forum Abstracts* 2:A025.025.

Essock-Vitale, S., and R.M. Seyfarth. 1987. Intelligence and social cognition. In *Primate societies*, edited by B. Smuts, D. Cheney, R. Seyfarth, R. Wrangham, and T. Struhsaker. Chicago: University of Chicago Press, 452–461.

Evans, H.E. 1959. Observations of the nesting behavior of digger wasps of the genus *Ammophila*. *American Midland Naturalist* 62:449–473.

———. 1965. Simultaneous care of more than one nest by *Ammophila azteca* Cameron (Hym.: Sphecidae). *Psyche* 72:8–23.

Evans, H.E., and M.J. West-Eberhard. 1970. *The wasps*. Ann Arbor: University of Michigan Press.

Evans, S., P. Weldon, J. Gioannetti, C. Moody, and E. Vicaria. 2003. Anointing in owl monkeys. *American Journal of Primatology* 60 (S1):135.

Evans, T.A., and G.C. Westergaard. 2004. Discrimination of functionally appropriate and inappropriate throwing tools by captive tufted capuchins (*Cebus apella*). *Animal Cognition* 7 (4):255–262.

———. 2006. Self-control and tool use in tufted capuchin monkeys (*Cebus apella*). *Journal of Comparative Psychology* 120 (2):163–166.

Ewer, R.F. 1973. *The carnivores*. Ithaca, NY: Cornell University Press.

Fagen, R.M. 1981. *Animal play behavior*. New York: Oxford University Press.

Falótico, T. 2006. Estudo experimental do uso de ferramentas para quebra de frutos encapsulados por macacos-prego (*Cebus apella*) em semi-liberdade. Master's thesis, University of São Paulo.

Falótico, T., and E.B. Ottoni. 2005. [Choice experiment of

tools for breaking coconuts by semi-free capuchin monkeys (*Cebus apella*)]. In *Programa e livro de resumos do XI Congresso Brasileiro de Primatologia*. Porto Alegre, Brazil: Sociedade Brasileira de Primatologia, 98.

Falótico, T., M.P. Verderane, B.D. de Resende, M.B. Labruna, P. Izar, and E.B. Ottoni. 2004. Anting in a semi-free-ranging group of brown capuchin monkeys (*Cebus apella*). *Folia Primatologica* 75 (Suppl. 1):372.

Farner, D. 1960. Digestion and the digestive system. In *Biology and comparative physiology of birds*, edited by A. Marshall. New York: Academic Press, 411–467.

Fay, J.M., and R.W. Carroll. 1994. Chimpanzee tool use for honey and termite extraction in Central Africa. *American Journal of Primatology* 34 (4):309–317.

Felce, W. 1948. *Apes: An account of personal experiences in a zoological garden*. London: Chapman and Hall.

Fellers, J.H., and G.M. Fellers. 1976. Tool use in a social insect and its implications for competitive interactions. *Science* 192 (4234):70–72.

Fernandes, M.E.B. 1991. Tool use and predation of oysters (*Crassostrea rhizophorae*) by the tufted capuchin, *Cebus apella apella*, in brackish water mangrove swamp. *Primates* 32 (4):529–531.

Fernández, L., J. Parapar, E. González-Gurriarán, and R. Muiño. 1998. Epibiosis and ornamental cover patterns of the spider crab *Maja squinado* on the Galician coast, northwestern Spain: Influence of behavioral and ecological characteristics of the host. *Journal of Crustacean Biology* 18 (4):728–737.

Fernández-Carriba, S., and A. Loeches. 2001. Fruit smearing by captive chimpanzees: A newly observed food-processing behavior. *Current Anthropology* 42 (1): 143–147.

Ferrari, P., S. Rozzi, G. Rizzolatti, and L. Fogassi. 2004. Neurophysiological and behavioral aspects of tool-use understanding in macaques. *Folia Primatologica* 75: 385–414.

Fiedler, W. 1957. Beobachtungen zum Markerungs-Verhalten Einiger Säugetiere. *Zeitschrift für Säugetierekunde* 22:57–76.

Field, M.Y. 2007. If you give a monkey an onion: An introduction to fur rubbing in human-commensal white-fronted capuchin monkeys (*Cebus albifrons*). *American Journal of Physical Anthropology* 132 (S44):108.

Finch, B.W. 1982. Sulphur-crested cockatoo *Cacatua galerita* using small branches for intimidation. *PNG Bird Society Newsletter* 193 / 194:29.

Finn, J.K., T. Tregenza, and M.D. Norman. 2009. Defensive tool use in a coconut-carrying octopus. *Current Biology* 19 (23): R1069–R1070.

Firsov, L.A. 1972. *Pamiat' u antropoidov: Fiziologicheskii analiz*. Leningrad: Nauka.

———. 1982. *I.P. Pavlov and experimental primatology*. Leningrad: Nauka.

Fischer, G.J., and S.L. Kitchener. 1965. Comparative learning in young gorillas and orangutans. *Journal of Genetic Psychology* 107:337–348.

Fisher, E. 1939. Habits of the southern sea otter. *Journal of Mammalogy* 20:21–36.

Fisher, J., and R. Lockley. 1954. *Sea-birds: An introduction to the natural history of the sea-birds of the North Atlantic*. Boston: Houghton, Mifflin.

Fitch-Snyder, H., and J. Carter. 1993. Tool use to acquire drinking water by free-ranging lion-tailed macaques (*Macaca silenus*). *Laboratory Primate Newsletter* 32 (1): 1–2.

Flemming, T.M., M.J. Rattermann, and R.K.R. Thompson. 2006. Differential individual access to and use of reaching tools in social groups of capuchin monkeys (*Cebus apella*) and human infants (*Homo sapiens*). *Aquatic Mammals* 32 (4):491–499.

Fletcher, A.W., and J.A. Weghorst. 2005. Laterality of hand function in naturalistically housed chimpanzees (*Pan troglodytes*). *Laterality* 10 (3):219–242.

Fontaine, B., P.Y. Moisson, and E.J. Wickings. 1995. Observations of spontaneous tool making and tool use in a captive group of western lowland gorillas (*Gorilla gorilla gorilla*). *Folia Primatologica* 65 (4):219–223.

Fontenot, M.B., S.L. Watson, K.A. Roberts, and R.W. Miller. 2007. Effects of food preferences on token exchange and behavioral responses to inequality in tufted capuchin monkeys, *Cebus apella*. *Animal Behaviour* 74 (3):487–496.

Fooden, J. 1995. Systematic review of Southeast Asian longtail macaques, *Macaca fasicularis*, (Raffles, 1821). *Fieldiana Zoology* 81:1–206.

Forbush, E. 1925. *Birds of Massachusetts and other New England states*. Norwood, MA: Berwick and Smith.

Formanowicz, D.R., Jr., and P.K. Ducey. 1991. Burrowing behavior and soil manipulation by a tarantula, *Rhechostica hentzi* (Girard, 1853) (Araneida: Theraphosidae). *Canadian Journal of Zoology* 69:2916–2918.

Fornalé, F., C. Spiezio, and D. Grassi. 2008. Learning new tool-use behaviour in captive pig-tailed macaques. *Folia Primatologica* 79 (3):143.

Forrest, T.G. 1982. Acoustic communication and baffling behaviors of crickets. *Florida Entomologist* 65 (1):33–44.

Fossey, D. 1970. Making friends with mountain gorillas. *National Geographic* 137:48–67.

———. 1981. A grim struggle for survival: The imperiled mountain gorilla. *National Geographic* 159:500–523.

Foucart, J., B. Bril, S. Hirata, N. Morimura, C. Houki, Y. Ueno, and T. Matsuzawa. 2005. A preliminary analysis of nut-cracking movements in a captive chimpanzee: Adaptation to the properties of tools and nuts. In *Stone knapping: The necessary conditions for a uniquely hominin behaviour*, edited by V. Roux and B. Bril. McDonald Institute Monographs. Cambridge: McDonald Institute for Archaeological Research, University of Cambridge, 147–157.

Fouts, D.H. 1983. Loulis tries his hand at surgery. *Friends of Washoe* 3 (1):4.

Fouts, R.S. 1997. *Next of kin: My conversations with chimpanzees*. New York: Harper Paperbacks.

Fouts, R.S., and D.H. Fouts. 1989. Loulis in conversation with the cross-fostered chimpanzees. In *Teaching sign language to chimpanzees*, edited by A. Gardner, B.T. Gardner, and T.E. van Cantfort. Albany: State University of New York, 293–307.

Fowler, A., and V. Sommer. 2007. Subsistence technology of Nigerian chimpanzees. *International Journal of Primatology* (28):997–1023.

Fowler, H.G. 1982. Tool use by *Aphaenogaster* ants: A reconsideration of its role in competitive interactions. *Bulletin of the New Jersey Academy of Science* 27:81–82.

Fox, E.A., and I. bin'Muhammad. 2002. New tool use by wild Sumatran orangutans (*Pongo pygmaeus abelii*). *American Journal of Physical Anthropology* 119 (2): 186–188.

Fox, E.A., A.F. Sitompul, and C.P. van Schaik. 1999. Intelligent tool use in wild Sumatran orangutans. In *The mentalities of gorillas and orangutans: Comparative perspectives*, edited by S.T. Parker, R.W. Mitchell, and H.L. Miles. Cambridge: Cambridge University Press, 99–116.

Fox, M.W. 1971. Possible examples of high-order behavior in wolves. *Journal of Mammalogy* 52 (3):640–641.

Foxall, C.D., and D. Drury. 1987. Green backed heron "bait-fishing" in Nairobi National Park. *East African Natural Society Bulletin* 17:11.

Fragaszy, D.M. 1998. How non-human primates use their hands. *Clinic in Developmental Medicine* 147:77–96.

Fragaszy, D.M., and L.E. Adams-Curtis. 1991. Environmental challenges in groups of capuchins. In *Primate responses to environmental change*, edited by H.O. Box. New York: Chapman and Hall, 239–264.

Fragaszy, D.M., and E. Visalberghi. 1989. Social influences on the acquisition of tool-using behaviors in tufted capuchin monkeys (*Cebus apella*). *Journal of Comparative Psychology* 103 (2):159–170.

Fragaszy, D.M., A.F. Vitale, and B. Ritchie. 1994. Variation among juvenile capuchins in social influences on exploration. *American Journal of Primatology* 32 (4): 249–260.

Fragaszy, D.M., P. Izar, E. Visalberghi, E.B. Ottoni, and M. Gomes de Oliveira. 2004. Wild capuchin monkeys (*Cebus libidinosus*) use anvils and stone pounding tools. *American Journal of Primatology* 64 (4):359–366.

Fredman, T., and A. Whiten. 2008. Observational learning from tool using models by human-reared and mother-reared capuchin monkeys (*Cebus apella*). *Animal Cognition* 11 (2):295–309.

Freeman, H., and J. Alcock. 1973. Play behavior of a mixed group of juvenile gorillas and orang-utans. *International Zoo Yearbook* 13:189–194.

Freese, C.H. 1977. Food habits of white-faced capuchins *Cebus capucinus* L. (Primates: Cebidae) in Santa Rosa National Park, Costa Rica. *Brenesia* 10:43–56.

Fricke, H. 1973. Behaviour as part of ecological adaptation. *Helgolander Wissenschaftliche Meeresunters* 24:120–144.

Frisch, J.A. 1940. Did the Peckhams witness the invention of a tool by *Ammophila urnaria*? *American Midland Naturalist* 24:345–350.

Frith, C.B., and D.W. Frith. 2004. *Bowerbirds*. Oxford: Oxford University Press.

Frith, H. 1956. Temperature regulation in the nesting mounds of the mallee-fowl, *Leipoa ocellata* Gould. *Commonwealth Scientific and Industrial Research Organization [CSIRO] Wildlife Research* 1:79–95.

Frolov, I.U.P. 1937. *Fish who answer the telephone, and other studies in experimental biology*. London: Kegan Paul, Trench, Trubner.

Fruth, B., and G. Hohmann. 1996. Nest building behavior in the great apes: The great leap forward? In *Great ape societies*, edited by W.C. McGrew, L. Marchant, and T. Nishida. Cambridge: Cambridge University Press, 225–240.

Fuentes, A., K.G. Suaryana, I.G.A. Artaputra, I.D.K. Haryaputra, and A.L.T. Rompis. 2001. The behavioral ecology and distribution of long-tailed macaques (*Macaca fascicularis*) in Bali, Indonesia. *American Journal of Physical Anthropology* 114 (S32):66.

Fujita, K., H. Kuroshima, and S. Asai. 2003. How do tufted capuchin monkeys (*Cebus apella*) understand causality involved in tool use? *Journal of Experimental Psychology: Animal Behaviour Processes* 29 (3):233–242.

Funk, M.S. 2002. Problem solving skills in young yellow-crowned parakeets (*Cyanoramphus auriceps*). *Animal Cognition* 5 (3):167–176.

Furlong, E.E., K.J. Boose, and S.T. Boysen. 2008. Raking it in: The impact of enculturation on chimpanzee tool use. *Animal Cognition* 11 (1):83–97.

Furness, W. 1916. Observations of the mentality of chimpanzees and orang-utans. *Proceedings of the American Philosophical Society* 55 (3):281–290.

Furniss, J. 1879a. Intellect in brutes. *Nature* 19:385.

———. 1879b. Intellect in brutes. *Nature* 20:21.

Fushimi, T., O. Sakura, T. Matsuzawa, H. Ono, and Y. Sugiyama. 1991. Nut-cracking behavior of wild chimpanzees (*Pan troglodytes*) in Bossou, Guinea (West Africa). In *Primatology today*, edited by A. Ehara, T. Kimuar, O. Takenaka, and M. Iwamoto. Amsterdam: Elsevier Science, 695–696.

Gaddis, P. 1981. Tool use by a mountain chickadee. *Continental Birdlife* 2:19–20.

Gadow, H. 1901. *Amphibia and reptiles*. Vol. 8 of *Cambridge natural history*. London: Macmillan.

Galat-Luong, A. 1984. L'utilisation spontanée d'outils pour le toilettage chez des *Cercopithecidae* africains captifs. *Revue d'Écologie: La Terre et la Vie* 39 (2):231–236.

Galat-Luong, A., and G. Galat. 2000. Chimpanzees and baboons drink filtrated water. *Folia Primatologica* 71 (4):258.

Galdikas, B.M.F. 1978a. Orang-utan adaptation at Tanjung Puting Reserve, Central Borneo. Ph.D. diss., University of California, Los Angeles.

———. 1978b. Orang-utans and hominid evolution. In *Spectrum: Essays presented to Sutan Takdir Alisjahbana on his seventieth birthday*, edited by S. Udin. Jakarta: Dian Rakyat, 287–309.

———. 1982a. Orang-utan tool-use at Tanjung Puting Reserve, central Indonesian Borneo (Kalimantan Tengah). *Journal of Human Evolution* 11 (1):19–33.

———. 1982b. An unusual instance of tool-use among wild orang-utans in Tanjung Puting Reserve, Indonesian Borneo. *Primates* 23:138–139.

———. 1983. The orangutan long call and snag crashing at Tanjung Puting Reserve. *Primates* 24 (3):371–384.

———. 1989. Orangutan tool use. *Science* 243 (4888):152.

Galdikas-Brindamour, B. 1975. Orangutans, Indonesia's "people of the forest." *National Geographic* 148:444–473.

Gamble, J.R., and D.A. Cristol. 2002. Drop-catch behaviour is play in herring gulls, *Larus argentatus*. *Animal Behaviour* 63:339–345.

Gannon, R. 1930. Observations on the satin bower bird with regard to the material used by it in painting its bower. *Emu* 30:39–41.

Garber, P.A., and E. Brown. 2004. Wild capuchins (*Cebus capucinus*) fail to use tools in an experimental field study. *American Journal of Primatology* 62 (3):165–170.

Garner, R. 1892. *The speech of monkeys*. New York: Charles L. Webster.

Garstang, W. 1890. Foreign substances attached to crabs. *Nature* 41:417–418, 490, 538.

Gaspar, A.D. 1993. First steps of enrichment at Lisboa Zoo: Testing the effects of two materials on chimpanzee behavior. *Chimpanzoo Conference Proceedings* 1993: 26–35.

Gaspar, A.D., and E.T. Reis. 1993. Tool use at Panzoo (Lisbon): Experiments and opportunistic observations. *Chimpanzoo Conference Proceedings* 1993:36–40.

Gašperšiè, M., and J.D. Pruetz. 2005. Chimpanzee (*Pan troglodytes verus*) and baobab (*Adansonia digitata*) at Fongoli, southeastern Senegal: Preliminary results on percussive technology. *Primate Report* 72:38–39.

Gatinot, B.L. 1974. Note sur l'observation d'une utilization spontanée d'outil chez *Erythrocebus patas* en captivité. *Mammalia* 38:557–558.

Gayou, D. 1982. Tool uses by green jays. *Wilson Bulletin* 94 (4):593–594.

Geissmann, T. 2000. Gibbon songs and human music from an evolutionary perspective. In *The origins of music*, edited by N. Wallin, B. Merker, and S. Brown. Cambridge MA: MIT Press, 103–123.

———. 2009. Door slamming: Tool-use by a captive white-handed gibbon (*Hylobates lar*). *Gibbon Journal* 5:53–60.

Geist, V. 1966. The evolution of horn-like organs. *Behaviour* 27:175–214.

———. 1971. *Mountain sheep: A study in behavior and evolution*. Chicago: University of Chicago Press.

Gertsch, W. 1949. *American spiders*. London: D. Van Nostrand.

———. 1955. The North American bolas spiders of the genera *Mastophora* and *Agatostichus*. *Bulletin of the American Museum of Natural History* 106:221–254.

Getty, T., and B.A. Hazlett. 1978. Decoration behavior in *Microphrys bicornutus* (Latreille, 1825) (Decapoda: Brachyura). *Crustaceana* 34:105–108.

Gewalt, W. 1975. Orang-Utans (*Pongo pygmaeus*) als "Seiler." *Zeitschrift für Säugetierekunde* 40:320–321.

Gibson, K.R. 1990. Tool use, imitation, and deception in a captive cebus monkey. In *"Language" and intelligence in monkeys and apes: Comparative developmental perspectives*, edited by S.T. Parker and K.R. Gibson. New York: Cambridge University Press, 205–218.

Gifford, E. 1919. Field notes on the land birds of the Galápagos Islands and of Cocos Islands, Costa Rica. *Proceedings of the California Academy of Sciences* 2: 189–258.

Gilbert, T.M., D.A. Brown, and S.T. Boysen. 1998. Social effects on anointing behavior in capuchins (*Cebus apella*). *American Journal of Primatology* 45:182.

Gill, F., M. Wright, and D. Donsker. 2009. *IOC world bird names* (version 2.2). www.worldbirdnames.org [accessed 18 Nov. 2009].

Gill, T. 1909. The archer-fish and its feats. *Smithsonian* 52:277–286.

Girndt, A., T. Meier, and J. Call. 2008. Task constraints mask great apes' ability to solve the trap-table task. *Journal of Experimental Psychology: Animal Behaviour Processes* 34 (1):54–62.

Giudice, A.M., and R. Pavé. 2007. *Cebus paraguayanus* in zoos: The spontaneous expression of species-specific behaviors. *Neotropical Primates* 14 (2):65–71.

Glickman, S.E., and R.W. Sroges. 1966. Curiosity in zoo animals. *Behaviour* 26:151–188.

Goethe, F. 1958. Anhäufungen unversehrter muscheln durch silbermöwen. *Natur und Volk* 88:181–187.

Gold, K.C. 2002. Ladder use and clubbing by a bonobo (*Pan paniscus*) in Apenheul Primate Park. *Zoo Biology* 21 (6):607–611.

Goldenberg, G., and A. Iriki. 2007. From sticks to coffeemaker: Mastery of tools and technology by human and non-human primates. *Cortex* 43 (3):285–288.

Gómez, J.C. 1988. Tool-use and communication as alternative strategies of problem-solving in the gorilla. *Primate Report* 19:25–28.

———. 1990. The emergence of intentional communication as a problem-solving strategy in the gorilla. In *"Language" and intelligence in monkeys and apes: Comparative developmental perspectives*, edited by S.T. Parker and K.R. Gibson. New York: Cambridge University Press, 333–355.

———. 1999. Development of sensorimotor intelligence in infant gorillas: The manipulation of objects in problem-solving and exploration. In *The mentalities of gorillas and orangutans: Comparative perspectives*, edited by S.T. Parker, R.W. Mitchell, and H.L. Miles. Cambridge: Cambridge University Press, 160–178.

Gómez, J.C., and B. Martin-Andrade. 2002. Possible precursors of pretend play in nonpretend actions of captive gorillas. In *Pretending and imagination in animals and children*, edited by R.W. Mitchell. Cambridge: Cambridge University Press, 255–268.

Gómez, N.E., L. Witte, and T. Hartmann. 1999. Chemical defense in larval tortoise beetles: Essential oil composition of fecal shields of *Eurypedus nigrosignata* and foliage of its host plant, *Cordia curassavica*. *Journal of Chemical Ecology* 25:1007–1027.

Gompper, M.E., and A.M. Hoylman. 1993. Grooming with *Trattinnickia* resin: Possible pharmaceutical plant use by coatis in Panama. *Journal of Tropical Ecology* 9 (4):533–540.

Gonzalez-Kirchner, J.P., and M. Sainz de la Maza. 1992. Sticks used by wild chimpanzees: A new locality in Rio Muni. *Folia Primatologica* 58 (2):99–102.

Goodall, A. 1979. *The wandering gorillas*. London: Collins.

Goodall, J. 1963a. Feeding behaviour of wild chimpanzees: A preliminary report. *Symposia of the Zoological Society of London* 10:39–47.

———. 1963b. My life among wild chimpanzees. *National Geographic* 124:272–308.

———. 1964. Tool-using and aimed throwing in a community of free-living chimpanzees. *Nature* 201:1264–1266.

———. 1965. Chimpanzees of the Gombe Stream Reserve. In *Primate behavior: Field studies of monkeys and apes*, edited by I. Devore. New York: Holt, Rinehart and Winston, 425–473.

———. 1979. Life and death at Gombe. *National Geographic* 155:592–621.

———. 1986. *The chimpanzees of Gombe: Patterns of behavior*. Cambridge, MA: Belknap Press.

Goodman, A., and E. Fisk. 1973. Breeding behaviour of captive striped owls (*Rhinoptynx clamator*). *Avicultural Magazine* 79:158–162.

Goodman, J.M. 1960. Aves incendiaria. *Wilson Bulletin* 72 (4):400–401.

Goodwin, D. 1956. Care of the body surface—preening, bathing, dusting, and anting. In *Ornithologists' Guide*, edited by H.P.W. Hutson. London: British Ornithological Union, 101–104.

Gordon, J. 1966. Elephants do think. *African Wildlife* 20: 75–79.

Gould, J. 1865. *Handbook to the birds of Australia*. London: J. Gould.

Gould, J.E., and T. Snyder. 1990. Tool use, tool manufacturing, and tests of mirror-image self-recognition in two captive gorillas. *American Journal of Primatology* 20 (3):195.

Goustard, M. 1986. L'utilisation d'instruments et la capacité d'adaptabilitié chez le chimpanzé (*Pan troglodytes schweinfurthii*) observé en semi-liberté. *Journal de Psychologie* 81 (3–4):395–412.

Green, C. 1972. Use of tools by orange-winged sittella. *Emu* 72:185–186.

Greenfield, P.M., and E.S. Savage-Rumbaugh. 1984. Perceived variability and symbol use: A common language-cognition interface in children and chimpanzees (*Pan troglodytes*). *Journal of Comparative Psychology* 98 (2):201–218.

Greenfield, P.M., A.E. Maynard, C. Boehm, and E.Y. Schmidtling. 2000. Cultural apprenticeship and cultural change: Tool learning and imitation in chimpanzees and humans. In *Biology, brains, and behavior: The evolution of human development*, edited by S.T. Parker, J. Langer, and M. McKinney. Santa Fe, NM: SAR Press, 237–277.

Greenhood, W., and R.L. Norton. 1999. Novel feeding technique of the woodpecker finch. *Journal of Field Ornithology* 70 (1):104–106.

Gressitt, J.L. 1952. The tortoise beetles of China (Chrysomelidae: Cassidinae). *Proceedings of the California Academy of Sciences* 27 (17):433–592.

Grobecker, D., and T. Pietsch. 1978. Crows use automobiles as nutcrackers. *Auk* 95:760–761.

Groves, C.P. 1970. *Gorillas*. New York: Arco.

———. 2001. *Primate taxonomy*. Washington, DC: Smithsonian Institution Press.

Gruber, A. 1969. A functional definition of primate toolmaking. *Man* 4:573–579.

Grummt, W. 1963. Werkzeuggebrauch beim Wasserbüffel (*Bubalus arnee* f. *bubalis* L.). *Zoologische Garten* 27: 262–263.

Grzimek, B. 1941. Beobachtungen an einem kleinen Schimpansenmädchen. *Zeitschrift für Tierpsychologie* 4:295–306.

———. 1956. Ein merkwürdiges Verhalten von afrikanischen Elefanten. *Zeitschrift für Tierpsychologie* 13: 151–152.

———. 1970. *Grzimek unter Afrikas Tieren*. New York: Stein and Day.

Gucwa, D., and J. Ehmann. 1985. *To whom it may concern: An investigation of the art of elephants*. New York: W.W. Norton.

Guillaume, P., and I. Meyerson. 1930. Recherches sur l'usage de l'instrument chez les singes: 1. Le problème du détour. *Journal de Psychologie* 27:177–236.

———. 1934. Recherches sur l'usage de l'instrument chez les singes: 3. L'intermédiaire indépendant de l'objet. *Journal de Psychologie* 31:497–554.

Guinot, D., D. Doumenc, and C.C. Chintiroglou. 1995. A review of the carrying behavior in brachyuran crabs, with additional information on the symbioses with sea anemones. *Raffles Bulletin of Zoology* 43 (2):377–416.

Gumert, M.D., M. Kluck, and S. Malaivijitnond. 2009. The physical characteristics and usage patterns of stone axe and pounding hammers used by long-tailed macaques in the Andaman Sea region of Thailand. *American Journal of Primatology* 71 (7):594–608.

Günther, M.M., and C. Boesch. 1993. Energetic cost of nutcracking behaviour in wild chimpanzees. In *Hands of primates*, edited by H. Preuschoft and D.J. Chivers. Vienna: Springer, 109–129.

Haeseler, V. 1985. Werkzeuggebrauch bei der europäischen Grabwespe *Ammophila hungarica* Mocsary, 1883 (Hymenoptera: Sphecidae). *Zoologischer Anzeiger* 215 (5–6):279–286.

Haggerty, M. 1910. Preliminary experiments as indicators of behavioral adaptability. *Current Anthropology* 4: 479–494.

———. 1913. Plumbing the minds of apes. *McClure's Magazine* 41:151–154.

Hailman, J.P. 1960. Anting of a captive slate-colored junco. *Wilson Bulletin* 72:398–399.

Hall, K.R.L. 1961. Feeding habits of the chacma baboon. *Advancement of Science* 17:559–567.

———. 1963. Tool-using performances as indicators of behavioral adaptability. *Current Anthropology* 4:479–494.

Hall, K.R.L., and G.B. Schaller. 1964. Tool-using behavior of the California sea otter. *Journal of Mammalogy* 45 (2):287–298.

Hamilton, W., R. Buskirk, and W. Buskirk. 1975. Defensive stoning by baboons. *Nature* 256:488–489.

———. 1978. Environmental developmental determinants of object manipulation by chacma baboons (*Papio ursinus*) in two southern African environments. *Journal of Human Evolution* 7:205–216.

Hancocks, D. 1983. Gorillas go natural. *Animal Kingdom* 86 (3):10–16.

Hanna, H. 1960. Methods of case-building and repair by larvae of cadis flies. *Proceedings of the Royal Entomological Society of London, Series A* 35:97–106.

Hannah, A.C., and W.C. McGrew. 1987. Chimpanzees using stones to crack open oil palm nuts in Liberia. *Primates* 28 (1):31–46.

Hansell, M.H. 1984. *Animal architecture and building behavior*. London: Longman.

———. 2000. *Bird nests and construction behaviour*. Cambridge: Cambridge University Press.

———. 2007. *Built by animals*. Oxford: Oxford University Press.

Hansell, M.H., and G.D. Ruxton. 2008. Setting tool use within the context of animal construction behaviour. *Trends in Ecology and Evolution* 23 (2):73–78.

Harcourt, C. 1981. An examination of the function of urine washing in *Galago senegalensis*. *Zeitschrift für Tierpsychologie* 55:119–128.

Harington, C. 1962. A bear fable? *Beaver* 1962:4–7.

Harlow, H.F. 1951. Primate learning. In *Comparative psychology*, edited by C.P. Stone. New York: Prentice-Hall, 183–238.

Harper, P. 1992. Behavioural enrichment of Adelaide Zoo's sole male orangutan. *Australian Primatology* 6 (4):2–10.

———. 2001. Eight years of environmental enrichment for Adelaide Zoo's adult male orangutan. *Australian Primatology* 15 (1):15–23.

Harris, L.J. 1993. Handedness in apes and monkeys: Some views from the past. In *Primate laterality: Current behavioral evidence of primate asymmetries*, edited by J.P. Ward and W.D. Hopkins. New York: Springer, 1–42.

Harris, R., and K. Duff. 1970. *Wild deer in Britain*. New York: Taplinger.

Harris, W.C. 1838. *Narrative of an expedition into southern Africa*. Bombay: American Mission Press.

Harrison, R.M., and P. Nystrom. 2008. Handedness in captive bonobos (*Pan paniscus*). *Folia Primatologica* 79 (5):253–268.

Harrisson, B. 1962a. *Orang-utan*. New York: Doubleday.

———. 1962b. A study of orang-utan behaviour in the semi-wild state. In *International Zoo Yearbook*, edited by C. Jarvis. London: Zoological Society of London, 57–68.

———. 1963. Education to wild living of young orangutans at Bako National Park, Sarawak. *Sarawak Museum Journal* 11:220–258.

Hart, B.L., and L.A. Hart. 1994. Fly switching by Asian elephants: Tool use to control parasites. *Animal Behaviour* 48 (1):35–45.

Hart, B.L., L.A. Hart, and N. Pinter-Wollman. 2008. Large brains and cognition: Where do elephants fit in? *Neuroscience and Biobehavioral Reviews* 32 (1):86–98.

Hart, B.L., L.A. Hart, M. McCoy, and C.R. Sarath. 2001. Cognitive behaviour in Asian elephants: Use and modification of branches for fly switching. *Animal Behaviour* 62 (5):839–847.

Hartley, P. 1964. Feeding habits. In *A new dictionary of birds*, edited by A. Thompson. New York: McGraw-Hill, 286–289.

Hartman, C. 1905. Observations of the habits of some solitary wasps of Texas. *Bulletin of the University of Texas* 65:1–73.

Harvey, M.J. 1998. Secondary tool modification by captive chimpanzees. Master's thesis, Central Washington University.

Harvey, R. 1999. Tool use by green heron. *Connecticut Warbler* 19:172.

Hashimoto, C. 1998. Chimpanzees of the Kalinzu Forest, Uganda. *Pan Africa News* 5 (1):6–8.

Hashimoto, C., T. Furuichi, and Y. Tashiro. 2000. Ant dipping and meat eating by wild chimpanzees in the Kalinzu Forest, Uganda. *Primates* 41 (1):103–108.

Hauser, M.D. 1988. Invention and social transmission: New data from wild vervet monkeys. In *Machiavellian intelligence: Social expertise and the evolution of intellect in monkeys, apes, and humans*, edited by R.W. Byrne and A. Whiten. Oxford: Oxford University Press, 327–343.

Hauser, M.D., J. Kralik, and C. Botto-Mahan. 1999. Problem solving and functional design features: Experiments on cotton-top tamarins, *Saguinus oedipus oedipus*. *Animal Behaviour* 57 (3):565–582.

Hauser, M.D., H. Pearson, and D. Seelig. 2002. Ontogeny of tool use in cottontop tamarins, *Saguinus oedipus*: Innate recognition of functionally relevant features. *Animal Behaviour* 64 (2):299–311.

Hauser, M.D., L.R. Santos, G.M. Spaepen, and H.E. Pearson. 2002. Problem solving, inhibition, and domain-specific experience: Experiments on cottontop tamarins, *Saguinus oedipus*. *Animal Behaviour* 64 (3):387–396.

Hayaki, H. 1985. Social play of juvenile and adolescent

chimpanzees in the Mahale Mountains National Park, Tanzania. *Primates* 26 (4):343–360.
Hayashi, M., and T. Matsuzawa. 2003. Cognitive development in object manipulation by infant chimpanzees. *Animal Cognition* 6 (4):225–233.
Hayashi, M., Y. Mizuno, and T. Matsuzawa. 2005. How does stone-tool use emerge? Introduction of stones and nuts to naive chimpanzees in captivity. *Primates* 46 (2):91–102.
Hayes, C. 1951. *The ape in our house*. New York: Harper.
Hayes, K.J., and C. Hayes. 1952. Imitation in a home-raised chimpanzee. *Journal of Comparative and Physiological Psychology* 45:450–459.
———. 1954. The cultural capacity of chimpanzees. *Human Biology* 26:288–303.
Hayes, K.J., and C.H. Nissen. 1971. Higher mental functions of a home-raised chimpanzee. In *Behavior of non-human primates: Modern research trends*, vol. 4, edited by A.M. Schrier and F. Stollnitz. New York: Academic Press, 59–115.
Hays, H., and G. Donaldson. 1970. Sand-kicking camouflages young black skimmers. *Wilson Bulletin* 82 (1):100.
Hazlett, B.A. 1981. The behavioral ecology of hermit crabs. *Annual Review of Ecology and Systematics* 12:1–22.
Heaton, J.L., and T. Rayne Pickering. 2006. Archaeological analysis does not support intentionality in the production of brushed ends on chimpanzee termiting tools. *International Journal of Primatology* 27 (6):1619–1633.
Hebel, N. 2002. The keeping and breeding of hyacinth macaws. In *Vth International Parrot Convention, Tenerife, Spain*. Puerto de la Cruz, Tenerife: Loro Parque Fundacíon, 94–100.
Heck, H. 1970. Der milu. *Milu* 3:1–15.
Heidegger, M. 1927/1962. *Being and time*. Translated by J. Macquarrie and E. Robinson. New York: Harper and Row.
Heinrich, B. 1988. Raven tool use? *Condor* 90 (1):270–271.
———. 1999. *Mind of the raven: Investigations and adventures with wolf-birds*. New York: Harper Perennial.
Heinrich, B., and T. Bugnyar. 2005. Testing problem solving in ravens: String-pulling to reach food. *Ethology* 111 (10):962–976.
Hendrichs, H., and U. Hendrichs. 1971. *Dikdik und elefanten*. Munich: Piper.
Hennessy, D., and D. Owings. 1978. Snake species discrimination and the role of olfactory cues in the snake-directed behavior of the California ground squirrel. *Behaviour* 65:115–124.
Henry, P., and J. Aznar. 2006. Tool-use in Charadrii: Active bait-fishing by a herring gull. *Waterbirds* 29 (2):233–234.
Henschel, J.R. 1995. Tool use by spiders: Stone selection and placement by corolla spiders *Ariadna* (Segestriidae) of the Namib Desert. *Ethology* 101 (3):187–199.
———. 1998. Dune spiders of the Negev Desert with notes on *Cerbalus psammodes* (Heteropodidae). *Israel Journal of Zoology* 44 (2):243–251.
Herald, E. 1956. How accurate is the archer fish? *Pacific Discovery* 9:12–13.
Hernandez-Aguilar, R.A., J. Moore, and T.R. Pickerings. 2007. Savanna chimpanzees use tools to harvest the underground storage organs of plants. *Proceedings of the National Academy of Sciences of the United States of America* 104 (49):19210–19213.
Hernandez-Camacho, J., and R.W. Cooper. 1976. The non-human primates of Colombia. In *Neotropical primates: Field studies and conservation*, edited by R. Thorington. Washington, DC: National Academy of Sciences, 35–69.
Herrmann, E., J. Call, M. Hernandez-Lloreda, B. Hare, and M. Tomasello. 2007. Humans have evolved specialized skills of social cognition: The cultural intelligence hypothesis. *Science* 317 (5843):1360–1366.
Herter, K. 1972. The insectivores. In *Grzimek's animal life encyclopedia*, edited by B. Grzimek. New York: Van Nostrand and Reinhold, 176–257.
Herzfeld, C., and D. Lestel. 2005. Knot tying in great apes: Etho-ethnology of an unusual tool behavior. *Social Science Information* [alternate title: *Information sur les Sciences Sociales*] 44 (4):621–653.
Heymann, E.W. 1995. Urine washing and related behaviour in wild moustached tamarins, *Saguinus mystax* (Callitrichidae). *Primates* 36 (2):259–264.
Hicks, C. 1932a. Nesting habits of *Sphex xanthopterus* (Cam.) (Hymen.). *Canadian Entomologist* 64:193–198.
———. 1932b. Notes on *Sphex aberti* (Hald.). *Canadian Entomologist* 64:145–151.
Hicks, T.C., R.S. Fouts, and D.H. Fouts. 2005. Chimpanzee (*Pan troglodytes troglodytes*) tool use in the Ngotto Forest, Central African Republic. *American Journal of Primatology* 65 (3):221–237.
Higuchi, H. 1986. Bait-fishing by the green-backed heron *Ardeola striata* in Japan. *Ibis* 128:285–290.
———. 1988a. Bait-fishing by green-backed herons in south Florida. *Florida Field Naturalist* 16:8–9.
———. 1988b. Individual differences in bait-fishing by the green-backed heron *Ardeola striata* associated with territory quality. *Ibis* 130 (1):39–44.
Hihara, S., S. Obayashi, M. Tanaka, and A. Iriki. 2003a. Rapid learning of sequential tool use by macaque monkeys. *Physiology & Behavior* 78 (3):427–434.
Hihara, S., H. Yamada, A. Iriki, and K. Okanoya. 2003b. Spontaneous vocal differentiation of coo-calls for tools and food in Japanese monkeys. *Neuroscience Research* 45 (4):383–389.
Hihara, S., T. Notoya, M. Tanaka, S. Ichinose, and A. Iriki. 2003c. Sprouting of terminal arborizations of the temporoparietal afferents by tool-use learning in adult monkeys. *Society for Neuroscience Abstracts* 2003: Abstract #939.915.
Hihara, S., T. Notoya, M. Tanaka, S. Ichinose, H. Ojima, S. Obayashi, N. Fujii, and A. Iriki. 2006. Extension of

corticocortical afferents into the anterior bank of the intraparietal sulcus by tool-use training in adult monkeys. *Neuropsychologia* 44 (13):2636–2646.

Hill, W. 1960. *Primates*, vol. 4, part A. Edinburgh: Edinburgh University Press.

———. 1967. Self-anointing in primates with special reference to capuchins (*Cebus*). *Laboratory Primate Newsletter* 6:20–21.

Hillix, W.A., and D. Rumbaugh. 2004. *Animal bodies, human minds: Ape, dolphin, and parrot language skills*. New York: Kluwer.

Hilton, B., Jr. 1992. Tool-making and tool-using by a brown thrasher (*Toxostoma rufum*). *Chat* (Raleigh) 56 (1):4–5.

Hingston, R.W.G. 1932. *A naturalist in the Guiana forest*. New York: Longmans, Green.

Hirata, S. 2006. Chimpanzee learning and transmission of tool use to fish for honey. In *Cognitive development in chimpanzees*, edited by T. Matsuzawa, M. Tomonaga, and M. Tanaka. New York: Springer, 201–213.

Hirata, S., and M.L. Celli. 2003. Role of mothers in the acquisition of tool-use behaviors by captive infant chimpanzees. *Animal Cognition* 6 (4):235–244.

Hirata, S., and N. Morimura. 2000. Naive chimpanzees' (*Pan troglodytes*) observation of experienced conspecifics in a tool-using task. *Journal of Comparative Psychology* 114 (3):291–296.

Hirata, S., N. Morimura, and C. Houki. 2009. How to crack nuts: Acquisition process in captive chimpanzees (*Pan troglodytes*) observing a model. *Animal Cognition* 12:S87–S101.

Hirata, S., M. Myowa, and T. Matsuzawa. 1998. Use of leaves as cushions to sit on wet ground by wild chimpanzees. *American Journal of Primatology* 44 (3):215–220.

Hirata, S., and G. Ohashi. 2003. [An experimental study of tool use in orangutans]. *Reichorui Kenkyu* 19 (1):87–95.

Hirata, S., K. Watanabe, and M. Kawai. 2001. "Sweet-potato washing" revisited. In *Primate origins of human cognition and behavior*, edited by T. Matsuzawa. Tokyo: Springer, 487–508.

Hirata, S., G. Yamakoshi, S. Fujita, F. Ohashi, and T. Matsuzawa. 2001. Capturing and toying with hyraxes (*Dendrohyrax dorsalis*) by wild chimpanzees (*Pan troglodytes*) at Bossou, Guinea. *American Journal of Primatology* 53 (2):93–97.

Hladik, C.M. 1973. Alimentation et activité d'un groupe de chimpanzés réintroduits en forêt gabonaise. *Terre et la Vie* 27:343–413.

———. 1975. Ecology, diet, and social patterning in Old and New World primates. In *Socioecology and psychology of primates*, edited by R. Tuttle. The Hague, Netherlands: Mouton, 3–35.

———. 1977. Chimpanzees of Gabon and chimpanzees of Gombe: Some comparative data on the diet. In *Primate ecology: Studies of feeding and ranging behaviour in lemurs, monkeys, and apes*, edited by T.H. Clutton-Brock. New York: Academic Press, 481–501.

Hladik, C.M., and G. Viroben. 1974. L'alimentation protéique du chimpanzé dans son environnement forestier naturel. *Comptes rendus de l'Académie des sciences, Paris* 279:1475–1478.

Hobbie, R. 1993. Tool-use by aquatic snails. *Creation Research Society Quarterly* 29 (4):193.

Hobbs, J. 1971. Use of tools by the white-winged chough. *Emu* 71:84–85.

Hobhouse, L.T. 1926. *Mind in evolution*, 3rd ed. London: Macmillan.

Hockings, K.J., T. Humle, J.R. Anderson, D. Biro, C. Sousa, G. Ohashi, and T. Matsuzawa. 2007. Chimpanzees share forbidden fruit. *PLoS ONE* 2 (9):e886.

Hodos, W. 1970. Evolutionary interpretation of neural and behavioral studies of living vertebrates. In *Neurosciences: Second study program*, edited by F. Schmidt. New York: Rockefeller University Press, 26–39.

Hodos, W., and C. Campbell. 1969. *Scala naturae*: Why there is no theory in comparative psychology. *Psychological Review* 76:337–350.

Hohmann, G. 1988. A case of simple tool use in wild lion-tailed macaques (*Macaca silenus*). *Primates* 29 (4): 565–567.

Hohmann, G., and B. Fruth. 2003. Culture in bonobos? Between-species and within-species variation in behavior. *Current Anthropology* 44 (4):563–571.

Holdrege, C. 2001. Elephantine intelligence. *Context* 5: 10–13.

Hölldobler, B., and E.O. Wilson. 1977. Weaver ants. *Scientific American* 237:146–154.

———. 1990. *The ants*. Cambridge, MA: Harvard University Press.

Holmes, S.J. 1911. *The evolution of animal intelligence*. New York: Henry Holt.

Holzhaider, J., G. Hunt, V. Campbell, and R. Gray. 2008. Do wild New Caledonian crows (*Corvus moneduloides*) attend to the functional properties of their tools? *Animal Cognition* 11 (2):243–254.

Hooton, E. 1942. *Man's poor relations*. Garden City, NY: Doubleday, Doran.

Hopkins, W.D., and D.M. Rabinowitz. 1997. Manual specialization and tool use in captive chimpanzees (*Pan troglodytes*): The effect of unimanual and bimanual strategies on hand preference. *Laterality* 2 (3/4): 267–277.

Hopkins, W.D., J.L. Russell, and C. Cantalupo. 2007. Neuroanatomical correlates of handedness for tool use in chimpanzees (*Pan troglodytes*): Implication for theories on the evolution of language. *Psychological Science* 18 (11):971–977.

Hopkins, W.D., K.A. Bard, A. Jones, and S.L. Bales. 1993. Chimpanzee hand preference in throwing and infant cradling: Implications for the origin of human handedness. *Current Anthropology* 34 (5):786–790.

Hopkins, W.D., S. Fernandez-Carriba, M.J. Wesley, A. Hostetter, D. Pilcher, and S. Poss. 2001. The use of bouts and frequencies in the evaluation of hand preferences for a coordinated bimanual task in chimpanzees (*Pan troglodytes*): An empirical study comparing two different indices of laterality. *Journal of Comparative Psychology* 115 (3):294–299.

Hopkins, W.D., J.L. Russell, C. Cantalupo, H. Freeman, and S.J. Schapiro. 2005. Factors influencing the prevalence and handedness for throwing in captive chimpanzees (*Pan troglodytes*). *Journal of Comparative Psychology* 119 (4):363–370.

Hopper, L.M., A. Spiteri, S.P. Lambeth, S.J. Schapiro, V. Horner, and A. Whiten. 2007. Experimental studies of traditions and underlying transmission processes in chimpanzees. *Animal Behaviour* 73 (6):1021–1032.

Hornaday, W. 1922. *The minds and manners of wild animals*. New York: Scribner.

Horner, V., and A. Whiten. 2005. Causal knowledge and imitation/emulation switching in chimpanzees (*Pan troglodytes*) and children (*Homo sapiens*). *Animal Cognition* 8 (3):164–181.

———. 2007. Learning from others' mistakes? Limits on understanding a trap-tube task by young chimpanzees (*Pan troglodytes*) and children (*Homo sapiens*). *Journal of Comparative Psychology* 121 (1):12–21.

Horr, D.A. 1975. The Borneo orangutan: Population structure and dynamics in relationship to ecology and reproductive strategy. In *Primate behaviour: Developments in field and laboratory research*, edited by L. Rosenblum. New York: Academic Press, 307–323.

———. 1977. Orang-utan maturation: Growing up in a female world. In *Primate bio-social development: Biological, social, and ecological determinants*, edited by S. Chevalier-Skolnikoff and F.E. Poirier. New York: Garland, 289–321.

Hostetter, A.B., M. Cantero, and W.D. Hopkins. 2001. Differential use of vocal and gestural communication by chimpanzees (*Pan troglodytes*) in response to the attentional status of a human (*Homo sapiens*). *Journal of Comparative Psychology* 115 (4):337–343.

Houk, J.L., and J.J. Geibel. 1974. Observation of underwater tool use by the sea otter, *Enhydra lutris* Linnaeus. *California Fish and Game* 60 (4):207–208.

Hrubesch, C., S. Preuschoft, and C. van Schaik. 2009. Skill mastery inhibits adoption of observed alternative solutions among chimpanzees (*Pan troglodytes*). *Animal Cognition* 12 (2):209–216.

Hueter, R.E., D.A. Mann, K.P. Maruska, J.A. Sisneros, and L.S. Demski. 2004. Sensory biology of elasmobranchs. In *Biology of sharks and their relatives*, edited by J.C. Carrier, J.A. Musick, and M.R. Heithaus. Boca Raton, FL: CRC Press, 448–510.

Huffman, M.A. 1984. Stone-play of *Macaca fuscata* in Arashiyama B troop: Transmission of a non-adaptive behavior. *Journal of Human Evolution* 13:725–735.

Huffman, M.A., and J.M. Caton. 2001. Self-induced increase of gut motility and the control of parasitic infections in wild chimpanzees. *International Journal of Primatology* 22:329–346.

Huffman, M.A., and M.S. Kalunde. 1993. Tool-assisted predation on a squirrel by a female chimpanzee in the Mahale Mountains, Tanzania. *Primates* 34 (1):93–98.

Huffman, M.A., and D. Quiatt. 1986. Stone handling by Japanese macaques (*Macaca fuscata*): Implications for tool use of stone. *Primates* 27 (4):413–423.

Huffman, M.A., and R.W. Wrangham. 1994. Diversity of medicinal plant use by chimpanzees in the wild. In *Chimpanzee Cultures*, edited by R.W. Wrangham, W.C. McGrew, F.B.M. de Waal, and P.G. Heltne. Cambridge, MA: Harvard University Press, 129–148.

Huffman, M.A., C.J. Campbell, A. Fuentes, K.C. MacKinnon, M. Panger, and S.K. Bearder. 2007. Primate self-medication. In *Primates in perspective*, edited by C.J. Campbell, A. Fuentes, K.C. MacKinnon, M. Panger, and S.K. Bearder. New York: Oxford University Press, 677–690.

Hughes, J., and M. Redshaw. 1974. Cognitive, manipulative, and social skills in gorillas: I. The first year. *Annual Report of the Jersey Wildlife Preservation Trust* 11:53–60.

Humle, T. 1999. New record of fishing for termites (*Macrotermes*) by the chimpanzees of Bossou (*Pan troglodytes verus*), Guinea. *Pan Africa News* 6 (1):3–4.

———. 2003. Culture and variation in wild chimpanzee behaviour: A study of three communities in West Africa. Ph.D. diss., University of Stirling.

———. 2006. Ant dipping in chimpanzees: An example of how microecological variables, tool use, and culture reflect the cognitive abilities of chimpanzees. In *Cognitive development in chimpanzees*, edited by T. Matsuzawa, M. Tomonaga, and M. Tanaka. New York: Springer, 452–475.

Humle, T., and T. Matsuzawa. 2001. Behavioural diversity among the wild chimpanzee populations of Bossou and neighboring areas, Guinea and Côte d'Ivoire, West Africa. A preliminary report. *Folia Primatologica* 72 (2):57–68.

———. 2002. Ant-dipping among the chimpanzees of Bossou, Guinea, and some comparisons with other sites. *American Journal of Primatology* 58 (3):133–148.

———. 2004. Oil palm use by adjacent communities of chimpanzees at Bossou and Nimba Mountains, West Africa. *International Journal of Primatology* 25 (3):551–581.

———. 2009. Laterality in hand use across four tool-use behaviors among the wild chimpanzees of Bossou, Guinea, West Africa. *American Journal of Primatology* 71 (1):40–48.

Hundley, M. 1963. Notes on methods of feeding and the use of tools in the Geospizinae. *Auk* 80:372–373.

Hungerford, H.B., and F.X. Williams. 1912. *Ammophila*, sp. undetermined. *Entomological News Philadelphia* 23:245.

Hunt, G.R. 1996. Manufacture and use of hook-tools by New Caledonian crows. *Nature* 379 (18):249–251.

———. 2000a. Human-like, population-level specialization in the manufacture of *Pandanus* tools by New Caledonian crows *Corvus moneduloides*. *Proceedings of the Royal Society B-Biological Sciences* 267 (1441):403–413.

———. 2000b. Tool use by the New Caledonian crow *Corvus moneduloides* to obtain Cerambycidae from dead wood. *Emu* 100 (2):109–114.

———. 2008. Introduced *Lantana camara* used as tools by New Caledonian crows (*Corvus moneduloides*). *New Zealand Journal of Zoology* 35 (2):115–118.

Hunt, G.R., M.C. Corballis, and R.D. Gray. 2001. Laterality in tool manufacture by crows. *Nature* 414:707.

———. 2006. Design complexity and strength of laterality are correlated in New Caledonian crows' *Pandanus*-tool manufacture. *Proceedings of the Royal Society B-Biological Sciences* 273 (1590):1127–1133.

Hunt, G.R., and R.D. Gray. 2002. Species-wide manufacture of stick-type tools by New Caledonian crows. *Emu* 102 (4):349–353.

———. 2003. Diversification and cumulative evolution in New Caledonian crow tool manufacture. *Proceedings of the Royal Society B-Biological Sciences* 270 (1517): 867–874.

———. 2004a. The crafting of hook tools by wild New Caledonian crows. *Proceedings of the Royal Society B-Biological Sciences* 271 (Suppl. 3):S88–S90.

———. 2004b. Direct observations of *Pandanus*-tool manufacture and use by a New Caledonian crow (*Corvus moneduloides*). *Animal Cognition* 7 (2):114–120.

———. 2006. Tool manufacture by New Caledonian crows: Chipping away at human uniqueness. *Acta Zoologica Sinica* 52 (S):622–625.

———. 2007. Parallel tool industries in New Caledonian crows. *Biology Letters* 3 (2):173–175.

Hunt, G.R., C. Lambert, and R.D. Gray. 2007. Cognitive requirements for tool use by New Caledonian crows (*Corvus moneduloides*). *New Zealand Journal of Zoology* 34 (1):1–7.

Hunt, G.R., R.B. Rutledge, and R.D. Gray. 2006. The right tool for the job: What strategies do wild New Caledonian crows use? *Animal Cognition* 9 (4):307–316.

Hunt, G.R., F. Sakuma, and Y. Shibata. 2002. New Caledonian crows drop candle-nuts onto rock from communally used forks on branches. *Emu* 102 (3):283–290.

Hunt, G.R., J. Abdelkrim, M.G. Anderson, J.C. Holzhaider, A.J. Marshall, N.J. Gemmell, and R.D. Gray. 2007. Innovative *Pandanus*-tool folding by New Caledonian crows. *Australian Journal of Zoology* 55 (5):291–298.

Hunt, K.D., and W.C. McGrew. 2002. Chimpanzees in the dry habitats of Assirik, Senegal, and Semliki Wildlife Reserve, Uganda. In *Behavioural diversity in chimpanzees and bonobos*, edited by C. Boesch, G. Hohmann, and L. Marchant. New York: Cambridge University Press, 35–51.

Hunter, M.L., Jr., A. Calhoun, and D.S. Wilcove. 2004. Goliath heron fishing with an artificial bait? *Waterbirds* 27 (3):312–313.

Hutchinson, C.E. 1903. A bolas-throwing spider. *Scientific American* 89 (10):172.

Huxley, J., and E. Nicholson. 1963. Lammergeier *Gypaetus barbatus* breaking bones. *Ibis* 105:106–107.

Hyatt, C.W., and W.D. Hopkins. 1998. Interspecies object exchange: Bartering in apes? *Behavioural Processes* 42 (2–3):177–187.

Iankov, P. 1983. Un percnoptère d'Égypte (*Neophron percnopterus*) en Bulgarie se sert d'instruments. *Alauda* 51:228.

Ikeo, M. 1993. *Kanzi: Ape of genius*. NHK [Japanese Broadcasting Corporation], documentary film.

Ilany, G. 1982. Egyptian vultures and dabb lizards. *Israel—Land and Nature* 8:37.

Ingmanson, E.J. 1988. The context of object manipulation by captive pygmy chimpanzees (*Pan paniscus*). *American Journal of Physical Anthropology* 75 (2):224.

———. 1989. Branch dragging by pygmy chimpanzees at Wamba, Zaïre: The use of objects to facilitate social communications in the wild. *American Journal of Physical Anthropology* 78 (2):244.

———. 1992. Tool use among bonobos in the wild and the social dimensions of intelligence. *Bulletin of the Chicago Academy of Sciences* 15 (1):26.

———. 1994. Learning and the use of rain hats by *Pan paniscus* at Wamba, Zaïre. *American Journal of Physical Anthropology* 37 (S18):111–112.

———. 1996. Tool-using behavior in wild *Pan paniscus*: Social and ecological consideration. In *Reaching into thought: The minds of the great apes*, edited by A.R. Russon, K.A. Bard, and S.T. Parker. New York: Cambridge University Press, 190–210.

———. 1997. Chimpanzee ant-dipping tools from West Cameroon. *American Journal of Primatology* 42 (2):117.

Ingmanson, E.J., and T. Kano. 1993. Waging peace: Among the humanlike bonobos of Zaïre's central rain forest, cooperation, not conflict, orders society. *International Wildlife* 23 (6):30–37.

Ingold, T. 1997. Eight themes in the anthropology of technology. *Social Analysis* 41:106–138.

Ingolfsson, A., and B. Estrella. 1978. The development of shell-cracking behavior in herring gulls. *Auk* 95: 577–579.

Inoue, Y., and E. Inoue. 2002. The trap-table problem with a young white-handed gibbon (*Hylobates lar*). Paper presented at the Joint Symposium of COE2/SAGA5, "Evolution of apes and the origins of human beings," November 14–17, Inuyama, Japan.

Inoue-Nakamura, N., and T. Matsuzawa. 1997. Development of stone tool use by wild chimpanzees (*Pan troglodytes*). *Journal of Comparative Psychology* 111 (2):159–173.

Inoue-Nakamura, N., R. Tonooka, and T. Matsuzawa. 1996. Developmental processes of nut-cracking skill among infant chimpanzees in the wild. *Japanese Journal of Developmental Psychology* 7 (2):148–158.

Iriki, A., M. Tanaka, and Y. Iwamura. 1996a. Assimilation of a tool to the hand: Its neuronal correlates in monkeys. *Society for Neuroscience Abstracts* 22 (3):1855.

———. 1996b. Coding of modified body schema during tool use by macaque postcentral neurones. *Neuroreport* 7 (14):2325–2330.

Ishibashi, H., S. Hihara, and A. Iriki. 2000. Acquisition and development of monkey tool-use: Behavioral and kinematic analyses. *Canadian Journal of Physiology and Pharmacology* 78 (11):958–966.

Ishibashi, H., S. Hihara, M. Takahashi, and A. Iriki. 1999. Immediate-early-gene expression by the training of tool-use in the monkey intraparietal cortex. *Society for Neuroscience Abstracts* 25 (1):889.

———. 2001. Induction of neurotrophic factor expression in the monkey parietal cortex during tool use learning. *Society for Neuroscience Abstracts* 27 (1):502.

Ishibashi, H., S. Hihara, M. Takahashi, T. Heike, T. Yokota, and A. Iriki. 2002. Tool-use learning selectively induces expression of brain-derived neurotrophic factor, its receptor *trk*B, and neurotrophin 3 in the intraparietal multisensory cortex of monkeys. *Cognitive Brain Research* 14 (1):3–9.

Isra, D. 2006. New evidence of honey-stick use by chimpanzees in southeast Cameroon. *Pan Africa News* 13 (1): 2–4.

Itani, J. 1959. Paternal care in the wild Japanese monkey, *Macaca fuscata fuscata*. *Primates* 2:61–93.

Itani, J., and A. Nishimura. 1973. The study of infrahuman culture in Japan: A review. In *Precultural Primate Behavior*, edited by E.W. Menzel. Basel, Switzerland: S. Karger, 26–50.

Itani, J., and A. Suzuki. 1967. The social unit of chimpanzees. *Primates* 8:355–381.

Iwata, K. 1976. *Evolution of instinct: Comparative ethology of Hymenoptera*. New Delhi: Amerind.

Izawa, K. 1979. Foods and feeding behavior of wild black-capped capuchin (*Cebus apella*). *Primates* 20:57–76.

Izawa, K., and J. Itani. 1966. Chimpanzee in Kasakati Basin, Tanganyika: I. Ecological study in the rainy season 1963–1964. *Kyoto University African Studies* 1:73–156.

Izawa, K., and A. Mizuno. 1977. Palm-fruit cracking behaviour of wild black-capped capuchin (*Cebus apella*). *Primates* (18):773–792.

Jackson, T.A. 1942. Use of the stick as a tool by young chimpanzees. *Journal of Comparative Psychology* 34: 223–235.

Jacobsen, C., J. Wolfe, and T. Jackson. 1935. An experimental analysis of the functions of frontal association areas in primates. *Journal of Nervous and Mental Disease* 82:1–14.

Jacobson, E., and E. Wasmann. 1904–1905. Beobachtungen ueber *Polyrhachis dives* auf Java, die ihre Larven zum Spinnen der Nester benutzt. *Notes from the Leyden Museum* 25 (3):133–140.

Jalles-Filho, E. 1995. Manipulative propensity and tool use in capuchin monkeys. *Current Anthropology* 36 (4): 664–667.

Jalles-Filho, E., R. Teixeira da Cunha, and R.A. Salm. 2001. Transport of tools and mental representation: Is capuchin monkey tool behaviour a useful model of Plio-Pleistocene hominid technology? *Journal of Human Evolution* 40 (5):365–377.

Janes, S. 1976. The apparent use of rocks by a raven in nest defense. *Condor* 78:409.

Jantschke, F. 1972. *Orang-utans in zoologischen gaerten*. Munich: R. Piper.

Janzen, M.J., D.H. Janzen, and C.M. Pond. 1976. Tool-using by the African grey parrot (*Psittacus erithacus*). *Biotropica* 8:70.

Jarvis, K.A., and J.E. Gould. 2007. Effects of a complex enrichment device on stereotypic behaviors, tool use, tool manufacturing, and activity budgets in captive western lowland gorillas (*Gorilla gorilla gorilla*). *American Journal of Primatology* 69 (S1):51.

Jay, P.C. 1968. Primate field studies and human evolution. In *Primates studies in adaptation and variability*, edited by P. Jay. New York: Holt, Rinehart and Winston, 487–519.

Jenkinson, M., and R. Mengel. 1970. Ingestion of stones by goatsuckers (Caprimulgidae). *Condor* 72:236–237.

Jennison, G. 1927. *Natural history: Animals*. London: A. & C. Black.

Jensvold, M.L.A., and R.S. Fouts. 1993. Imaginary play in chimpanzees (*Pan troglodytes*) *Human Evolution* 8 (3):217–227.

Jewett, S.G. 1924. An intelligent crow. *Condor* 26:72.

Johnson, E.C. 1996. Flexibility of tool-using skills in capuchin monkeys (*Cebus apella*). *IPS/ASP Congress Abstracts* 1996:Abstract #316.

Johnston, A.F. 1985. Anting-like behavior of dipper with fish. *British Birds* 78 (5):242.

Joines, S. 1976. The gorilla conservation program at the San Diego Wild Animal Park. *Zoonooz* 49 (10):5–12.

Jolly, A. 1966. Breeding synchrony in wild *Lemur catta*. In *Social communication among primates*, edited by S.A. Altmann. Chicago: University of Chicago Press, 3–14.

Jones, A. 1986. Chimpanzoo at the Cheyenne Mountain Zoo. *Chimpanzoo Conference Proceedings* 1986:35–49.

Jones, C., and J. Sabater-Pi. 1969. Sticks used by chimpanzees in Rio Muni, West Africa. *Nature* 233:100–101.

Jones, C.B. 2003. Urine-washing behaviors as condition-dependent signals of quality by adult mantled howler monkeys (*Alouatta palliata*). *Laboratory Animal Newsletter* 42 (1):12–14.

Jones, R. 1951. Present status of the sea otter in Alaska. In *Transactions of the Sixteenth North American Wildlife Conference*, edited by E. Quee. Washington, DC: Wildlife Management Institute, 376–383.

Jones, T.B., and A.C. Kamil. 1973. Tool-making and tool-using in the northern blue jay. *Science* 180 (4090):1076–1078.

Jordan, C. 1982. Object manipulation and tool-use in captive pygmy chimpanzees (*Pan paniscus*). *Journal of Human Evolution* 11 (1):35–39.

Jouffroy, F.K. 1993. Primate hands and the human hand: The tool of tools. In *The use of tools by human and non-human primates*, edited by A. Berthelet and J. Chavaillon. Oxford: Clarendon Press, 6–33.

Joulian, F. 1995. Mise en évidence de différences traditionnelles dans le cassage des noix chez les chimpanzés de la Côte d'Ivoire: Implications paléoanthropologiques. *Journal des Africanistes* 65 (2):57–77.

———. 1996. Comparing chimpanzee and early hominid techniques: Some contributions to cultural and cognitive questions. In *Modeling the early human mind*, edited by P. Mellars and K. Gibson. Cambridge: University of Cambridge Press, 173–189.

———. 2000. Techniques du corps et traditions chimpanzières. *Terrain* 34:37–54.

Judd, W.W. 1975. A blue jay in captivity for 18 years. *Bird Banding* 46 (3):250.

Judson, O.P., and A.T.D. Bennett. 1992. "Anting" as food preparation: Formic acid is worse on an empty stomach. *Behavioral Ecology and Sociobiology* 31 (6):437–439.

Kano, T. 1979. A pilot study on the ecology of pygmy chimpanzees, *Pan paniscus*. In *The great apes: Perspectives on human evolution*, edited by D.A. Hamburg and E. McCown. Menlo Park, CA: Benjamin/Cummings, 123–135.

———. 1982. The use of leafy twigs for rain cover by the pygmy chimpanzees of Wamba. *Primates* 23 (3): 453–457.

———. 1992. *The last ape: Pygmy chimpanzee behavior and ecology*. Stanford, CA: Stanford University Press.

———. 1997. Leaf-dropping sexual display exhibited by a male bonobo at Wamba. *Pan Africa News* 4 (1):3–4.

Kaplan, G., and L. Rogers. 1994. *Orang-utans in Borneo*. Armidale, Australia: University of New England Press.

Kats, A. 1972a. An experimental investigation of aimed throwing in some primates (chimpanzees and hamadryas baboons) and its significance in the anthropogenesis. *Transactions of the Moscow Society of Naturalists* 18:58–76.

———. 1972b. Intellectual behavior of primates as a biological prerequisite of instrumental activity. Paper presented at the Twentieth International Congress of Psychology, August 13–19, Tokyo, Japan.

Kattmann, U. 2001. Piecing together the history of our knowledge of chimpanzee tool use. *Nature* 411 (6836):413.

Katz, D., and R. Katz. 1936. Some problems concerning the feeding behaviour of monkeys. *Proceedings of the Zoological Society of London* 1936:579–582.

Katz, S. 1975. Tool-using behavior of the pocket gopher (*Thomomys bottae*). Paper presented at the Fifty-fifth Annual Meeting of the American Society of Mammalogists, June 16–19, Missoula, Montana.

———. 1980. Tool-using behavior of the pocket gopher, *Thomomys bottae* (Geomyidae). *Southwestern Naturalist* 25 (2):270–271.

Kaufmann, J. 1962. Ecology and social behavior of the coati, *Nasua narica*, on Barro Colorado Island, Panama. *University of California Publications in Zoology* 60:95–222.

Kawai, M. 1965. Newly acquired pre-cultural behavior of the natural troop of Japanese monkeys on Koshima Islet. *Primates* 6 (1):1–30.

Kawamura, S. 1954. A new type of behaviour of the wild Japanese monkeys: An analysis of an animal culture. *Seibutsu Shinka* 2 (1):10–13.

———. 1959. The process of sub-culture propagation among Japanese macaques. *Primates* (2):43–60.

Kawanaka, K. 1990. Age differences in ant-eating by adult and adolescent males. In *The chimpanzees of the Mahale Mountains*, edited by T. Nishida. Tokyo: University of Tokyo Press, 207–222.

Keenan, W.J. 1981. Green heron fishing with mayflies. *Chat* (Raleigh) 45:41.

Keenleyside, M.H.A., and C.E. Prince. 1976. Spawning-site selection in relation to parental care of eggs in *Aequidens paraguayensis* (Pisces: Cichlidae). *Canadian Journal of Zoology* 54:2135–2139.

Kellogg, W.N., and L.A. Kellogg. 1933. *The ape and the child*. New York: McGraw-Hill.

Kelso, L., and M.M. Nice. 1963. A Russian contribution to anting and feather mites. *Wilson Bulletin* 75 (1):23–26.

Kenward, B., A.A.S. Weir, C. Rutz, and A. Kacelnik. 2005. Tool manufacture by naive juvenile crows. *Nature* 433 (7022):121.

Kenward, B., C. Rutz, A.A.S. Weir, and A. Kacelnik. 2006. Development of tool use in New Caledonian crows: Inherited action patterns and social influences. *Animal Behaviour* 72 (6):1329–1343.

Kenyon, K. 1969. *The sea otter in the eastern Pacific Ocean*. Washington, DC: U.S. Bureau of Sport Fisheries and Wildlife.

Kessel, E.L. 1955. Mating activities of balloon flies. *Systematic Zoology* 4 (3):97–104.

Khroustov, G.F. 1964. Formation and highest frontier of the implemental activity of anthropoids. Vol. 3 of *VIIth International Congress of Anthropological and Ethnological Sciences*. Moscow: Nauka, 503–509.

Kilham, L. 1968. Reproductive behavior of white-breasted nuthatches: I. Distraction display, bill-sweeping, and nest hole defense. *Auk* 85:477–492.

———. 1971. Use of blister beetle in bill-sweeping by white-breasted nuthatches. *Auk* 88:175–176.

———. 1974. Covering of stores by white-breasted and red-breasted nuthatches. *Condor* 76:108–109.

Kiliaan, H. 1974. The possible use of tools by polar bears to obtain their food. *Arbok [Norsk Polarinstitutt]* 1972:177–178.

Killington, F.J. 1936–1937. *A monograph of the British Neuroptera*, vols. 1 and 2. London: Ray Society.

King, B.J. 1986. Individual differences in tool-using by two captive orangutans (*Pongo pygmaeus*). In *Current perspectives in primate social dynamics*, edited by D.M. Taub and F.A. King. New York: Van Nostrand Reinhold, 469–475.

King, N.E., V.J. Stevens, and J.D. Mellen. 1980. Social behavior in a captive chimpanzee (*Pan troglodytes*) group. *Primates* 21 (2):198–210.

King, W.B., and C.B. Kepler. 1970. Active anting in the Puerto Rican tanager. *Auk* 87 (2):376–378.

Kingery, H.E., and C.K. Ghalambor. 2001. *Pygmy nuthatch (*Sitta pygmaea*)*. Vol. 15, no. 567 of *The Birds of North America*, edited by A. Poole, P. Stettenheim, and F. Gill. Philadelphia: Birds of North America.

Kitabayashi, N., Y. Kusunoki, and Y.P. Gunji. 1999. The emergence of the concept of a tool in food-retrieving behavior of the ants *Formica japonica* Motschulsky. *Biosystems* 50 (2):143–156.

Kitahara-Frisch, J. 1993. The origin of secondary tools. In *The use of tools by human and non-human primates*, edited by A. Berthelet and J. Chavaillon. Oxford: Claredon Press, 239–246.

Kitahara-Frisch, J., and K. Norikoshi. 1982. Spontaneous sponge-making in captive chimpanzees. *Journal of Human Evolution* 11:41–47.

Kitahara-Frisch, J., K. Norikoshi, and K. Hara. 1987. Use of a bone fragment as a step towards secondary tool use in captive chimpanzee. *Primate Report* 18:33–37.

Klein, L.L. 1974. Agonistic behavior in neotropical primates. In *Primate aggression, territoriality, and xenophobia*, edited by R.L. Holloway. New York: Academic Press, 77–122.

Klüver, H. 1933. *Behavior mechanisms in monkeys*. Editor's foreword by Ernest W. Burgess. Chicago: University of Chicago Press.

———. 1937. Re-examination of implement-using behavior in a cebus monkey after an interval of three years. *Acta Psychologica* 2:347–397.

Knott, C. 1999. Orangutan behavior and ecology. In *The nonhuman primates*, edited by P. Dolhinow and A. Fuentes. Mountain View, CA: Mayfield, 50–57.

Kobayashi, T., and M. Watanabe. 1986. An analysis of snake-scent application behavior in Siberian chipmunks (*Eutamias sibiricus asiaticus*). *Ethology* 72 (1): 40–52.

Koehler, W. 1993. The mentality of orangs. *International Journal of Comparative Psychology* 6 (4):189–229.

Koenig, W.D. 1985. Dunking of prey by Brewer's blackbirds: A novel source of water for nestlings. *Condor* 87 (3):444–445.

Köhler, W. 1917. *Intelligenzprüfungen an Anthropoiden*. Abhandlungen der Königlich Preussischen Akademie der Wissenschafen, Jg. 1917, Nr. 1. Berlin: G. Reimer.

———. 1925. *The mentality of apes*. Translated by E. Winter. London: K. Paul, Trench, Trubner.

Kohts, N. 1923. *Untersuchungen über die Erkenntnisfähigkeiten des Schimpansen*. Moscow: Museum of Darwinianum.

———. 1935. *Infant ape and human child: Instincts, emotions, play, habits*. Moscow: Scientific Memoirs of the Museum Darwinianum.

Kollar, E.J. 1972. Object relations and the origin of tools. *Archives of General Psychiatry* 26:23–27.

Kooij, M., and J. van Zon. 1964. Gooiende seriema's. *Artis* 9:197–201.

Koops, K., W.C. McGrew, and T. Matsuzawa. 2010. Do chimpanzees (*Pan troglodytes*) use cleavers and anvils to fracture *Treculia africana* fruits? Preliminary data on a new form of percussive technology. *Primates* 51 (2):175–178.

Kortlandt, A. 1962. Chimpanzees in the wild. *Scientific American* 206 (5):128–134, 137–138.

———. 1965. How do chimpanzees use weapons when fighting leopards? *Yearbook of the American Philosophical Society* 1964:327–332.

———. 1966. On tool-use among primates. *Current Anthropology* 7:215–216.

———. 1967a. Experimentation with chimpanzees in the wild. In *Neue ergebnisse der primatologie*, edited by D. Starck, R. Schneider, and H.J. Kuhn. Stuttgart, Germany: Gustav Fischer, 208–224.

———. 1967b. Handgebrauch bei freilebenden schimpansen. In *Handgebrauch und verständigung bei affen und frühmenschen*, edited by B. Rensch. Bern, Switzerland: Huber, 59–102.

———. 1972. *New perspectives on ape and human evolution*. Amsterdam: Stichting voor Psychobiologie.

———. 1986. The use of stone tools by wild-living chimpanzees and earliest hominids. *Journal of Human Evolution* 15 (2):77–132.

Kortlandt, A., and E. Holzhaus. 1987. New data on the use of stone tools by chimpanzees in Guinea and Liberia. *Primates* 28 (4):473–496.

Kortlandt, A., and M. Kooij. 1963. Protohominid behavior in primates (preliminary communication). *Symposia of the Zoological Society of London* 10:61–88.

Krief, S., R.W. Wrangham, and D. Lestel. 2006. Diversity of items of low nutritional value ingested by chimpanzees from Kanyawara, Kibale National Park, Uganda: An example of the etho-ethnology of chimpanzees. *Social Science Information* 45 (2):227–263.

Krieg, H. 1930. Biologische reisestudien in Sudamerika: VXI. Die affen des Gran Chaco und seiner Grenzgebiete. *Zeitschrift für Morphologie und Ökologie der Tiere* 18:760–785.

Krützen, M., J. Mann, M.R. Heithaus, R.C. Connor, L. Bejder, and W.B. Sherwin. 2005. Cultural transmission of tool use in bottlenose dolphins. *Proceedings of the National Academy of Sciences of the United States of America* 102 (25):8939–8943.

Kuba, M.J., R.A. Byrne, and G.M. Burghardt. 2010. A new method for studying problem solving and tool use in stingrays (*Potamotrygon castexi*). *Animal Cognition* 13(3):507–513.

Kühme, W. 1962. Ethology of the African elephant (*Loxodonta africana* Blumenbach 1791) in captivity. In *International Zoo Yearbook*, edited by C. Jarvis and D. Morris. London: Hutchinson, 113–121.

———. 1963. Erganzende beobachtungen an afrikanischen Elephanten (*Loxodonta africana* Blumenbach 1791) im Freigehege. *Zeitschrift für Tierpsychologie* 20:66–79.

Kumar, S., H.N. Kumara, and M. Singh. 2008. Observations on tool use in captive lion-tailed macaque (*Macaca silenus*). *Current Science* 94 (7):925–928.

Kummer, H. 1967. Tripartite relations in hamadryas

baboons. In *Social communication among primates*, edited by S. Altmann. Chicago: University of Chicago Press, 63–71.

———. 1971. *Primate societies: Group techniques of ecological adaptation*. Chicago: Aldine-Atherton.

Kummer, H., and J. Goodall. 1985. Conditions of innovative behaviour in primates. *Philosophical Transactions of the Royal Society B, Biological Sciences* 308 (1135): 203–214.

Kummer, H., and F. Kurt. 1965. A comparison of social behavior in captive and wild hamadryas baboons. In *The baboon in medical research: Proceedings of the first international symposium on the baboon and its use as an experimental animal*. Austin: University of Texas Press, 65–80.

Kuroda, S. 1980. Social behavior of the pygmy chimpanzees. *Primates* 21 (2):181–197.

———. 1982. [*The unknown ape: The pygmy chimpanzees*]. Tokyo: Chikuma-Shobo.

———. 1991. Preliminary report on the socio-ecology of the Tschego chimpanzee (*Pan troglodytes troglodytes*) in the Ndoki Nouabale Reserve, northern Congo. Paper presented at the Great Apes Conference, December 15–22, Jakarta, Indonesia.

———. 1992. Ecological interspecies relationships between gorillas and chimpanzees in the Ndoki-Nouabale Reserve, northern Congo. In *Behavior, ecology and conservation*, edited by N. Itoigawa, Y. Sugiyama, G. Sackett, and R. Thompson. Vol. 2 of *Topics in primatology.* Tokyo: University of Tokyo Press, 385–394.

Kuroda, S., S. Suzuki, and T. Nishihara. 1996. Preliminary report on predatory behavior and meat sharing in Tschego chimpanzees (*Pan troglodytes troglodytes*) in the Ndoki Forest, northern Congo. *Primates* 37 (3):253–259.

Kurt, F., and G.B. Hartl. 1995. Asian elephants (*Elephas maximus*) in captivity: A challenge for zoo biological research. In *Research and captive propagation*, edited by U. Ganslosser, J.K. Hodges, and W. Kaumanns. Fürth, Germany: Filander, 310–326.

Kushlan, J.A. 1978. Feeding ecology of wading birds. In *Wading Birds*, edited by A. Sprunt, J.C. Ogden, and S. Winckler. New York: National Audubon Society, 249–296.

Kyes, R.C. 1988. Grooming with a stone in sooty mangabeys (*Cercocebus atys*). *American Journal of Primatology* 16 (2):171–175.

Lack, D. 1940. Courtship feeding in birds. *Auk* 57:169–178.

———. 1947. *Darwin's finches*. Cambridge: Cambridge University Press.

———. 1953. Darwin's finches. *Scientific American* 188: 66–72.

Ladygina-Kohts, N. 1959. [*Constructive and instrumental activities of higher primates (chimpanzees)*]. Moscow: Scientific Publishing House.

———. 2002. *Infant chimpanzee and human child: A classic 1935 comparative study of ape emotions and intelligence*. New York: Oxford University Press.

Ladygina-Kohts, N., and Y.N. Dembovskii. 1969. The psychology of primates. In *A handbook of contemporary Soviet psychology*, edited by M. Cole and I. Maltzman. London: Basic Books, 41–70.

Laidre, M. 2008. Spontaneous performance of wild baboons on three novel food-access puzzles. *Animal Cognition* 11 (2):223–230.

Laland, K.N., and W. Hoppitt. 2003. Do animals have culture? *Evolutionary Anthropology* 12 (3):150–159.

Lane, F. 1957. *Kingdom of the octopus*. London: Jarrold.

Lang, E. 1974. Werkzeuggebrauch beim Brillenbären (*Ursus ornatus*). *Zoologische Garten* 44:324–328.

Langguth, A., and C. Alonso. 1997. Capuchin monkeys in the Caatinga: Tool use and food habits during drought. *Neotropical Primates* 5 (3):77–78.

Lanjouw, A. 2002. Behavioural adaptations to water scarcity in Tongo chimpanzees. In *Behavioural diversity in chimpanzees and bonobos*, edited by C. Boesch, G. Hohmann, and L. Marchant. New York: Cambridge University Press, 52–60.

Laska, M., V. Bauer, and L.T. Hernandez Salazar. 2007. Self-anointing behavior in free-ranging spider monkeys (*Ateles geoffroyi*) in Mexico. *Primates* 48 (2):160–163.

Lau, D. 1965. Werkzeuggebrauch beim Kerabau, *Bubalus arnee* f. *bubalis* L. *Zoologische Garten* 31:269–271.

Laursen, L. 1975. *The ethology of the African elephant*. Boulder: University of Colorado.

Lavallee, A.C. 1999. Capuchin (*Cebus apella*) tool use in a captive naturalistic environment. *International Journal of Primatology* 20 (3):399–414.

Law, E. 1929. Another Lewis' woodpecker stores acorns. *Condor* 31:233–238.

Leca, J.B., N. Gunst, and M.A. Huffman. 2008. Of stones and monkeys: Testing ecological constraints on stone handling, a behavioral tradition in Japanese macaques. *American Journal of Physical Anthropology* 135 (2):233–244.

———. 2009. The first case of dental flossing by a Japanese macaque (*Macaca fuscata*): Implications for the determinants of behavioral innovation and the constraints on social transmission. *Primates* 51 (1):13–22.

Leca, J.B., N. Gunst, and O. Petit. 2007. Social aspects of fur rubbing in *Cebus capucinus* and *C. apella*. *International Journal of Primatology* 28:801–817.

Leca, J.B., C. Nahallage, N. Gunst, and M. Huffman. 2008. Stone-throwing by Japanese macaques: Form and functional aspects of a group-specific behavioral tradition. *Journal of Human Evolution* 55 (6):989–998.

Lefebvre, L. 1982. Food exchange strategies in an infant chimpanzee. *Journal for Human Evolution* 11:195–204.

Lefebvre, L., and T.A. Hewitt. 1986. Food exchange in captive chimpanzees. In *Current perspectives in primate social dynamics*, edited by D.M. Taub and F.A. King. New York: Van Nostrand Reinhold, 476–486.

Lefebvre, L., N. Nicolakakis, and D. Boire. 2002. Tools and brains in birds. *Behaviour* 139:939–973.

Legge, W.V. 1917. A new raptor (*Gypoictinia melanosterna*) for Tasmania. *Emu* 17:103–106.

Leighton, T., P. White, and D. Finfer. 2008. Hypotheses regarding exploitation of bubble acoustics by cetaceans. Paper presented at the 9th European Conference on Underwater Acoustics, June 29–July 4, Paris, France.

Leitch, G.F. 1953. Buzzards destroying emu eggs. *North Queensland Naturalist* 21:4.

Leschen, R.A.B., and C.E. Carlton. 1993. Debris cloaking in Endomychidae: New species from Peru (Coleoptera). *Zoological Journal of the Linnean Society* 109 (1):35–51.

Leshem, Y. 1979. Golden eagles in our backyard. *Israel—Land and Nature* 5:70–75.

Lestel, D., and E. Grundmann. 1999. Tools, techniques, and animals: The role of mediations of actions in the dynamics of social behaviours. *Social Science Information* [alternate title: *Information sur les Sciences Sociales*] 38 (3):367–407.

Lethmate, J. 1976a. Gebrauch und Herstellung von Trinkwerkzeugen bei Orang-Utans. *Zoologischer Anzeiger* 197:251–263.

———. 1976b. Versuche zur Doppelstockhandlung mit einem jungen Orang-Utan. *Zoologischer Anzeiger* 197:264–271.

———. 1976c. Werkzeugintelligenz von Orang-Utans. *Umschau in Wissenschaft und Technik* 76 (24):782–784.

———. 1976d. Werkzeugverhalten von Orang-Utans. *Biologie in Unserer Zeit* 6:33–40.

———. 1977a. Instrumentelles Verhalten zoolebender Orang-Utans. *Zeitschrift für Morphologie und Anthropologie* 68:57–87.

———. 1977b. Problemlöseverhalten von Orang-Utans (*Pongo pygmaeus*). *Fortschritte der Verhaltensforschung* 19:1–70.

———. 1977c. Versuche zum Schlagstockverfahren mit zwei jungen Orang-Utans. *Zoologischer Anzeiger* 199:209–226.

———. 1977d. Weitere Versuche zum Manipulier- und Werkzeugverhalten junger Orang-Utans. *Primates* 18:531–543.

———. 1977e. Werkzeugherstellung eines jungen Orang-Utan. *Behaviour* 62:174–189.

———. 1978. Versuche zum "vorbedingten" Handeln mit einem jungen Orang-Utan. *Primates* 19:727–736.

———. 1979. Instrumental behaviour of zoo orang-utans. *Journal of Human Evolution* 8:741–744.

———. 1982. Tool-using skills of orang-utans. *Journal of Human Evolution* 11:49–64.

Levey, D.J., R.S. Duncan, and C.F. Levins. 2004. Use of dung as a tool by burrowing owls. *Nature* 431 (7004):39–39.

Levin, L.A., A.J. Gooday, and D.W. James. 2001. Dressing up for the deep: Agglutinated protists adorn an irregular urchin. *Journal of the Marine Biological Association of the United Kingdom* 81 (5):881–882.

Levine, R.L., J.E. Hunter, and P.L. Borchelt. 1974. Dustbathing as a regulatory mechanism. *Bulletin of Mathematical Biology* 36 (5–6):545–553.

Levy, B.A. 1992. Psychoaesthetics dolphin project. *Art Therapy* 9:193–197.

Limbaugh, C. 1961. Observations on the California sea otter. *Journal of Mammalogy* 42 (2):271–273.

Limongelli, L., S.T. Boysen, and E. Visalberghi. 1995. Comprehension of cause-effect relations in a tool-using task by chimpanzees (*Pan troglodytes*). *Journal of Comparative Psychology* 109 (1):18–26.

Lin, N. 1964–1965. The use of sand grains by the pavement ant, *Tetramorium caespitum*, while attacking halictine bees. *Bulletin of the Brooklyn Entomological Society* 59–60:30–34.

Linden, E. 1999. *The parrot's lament*. New York: Dutton.

———. 2002. *The octopus and the orangutan*. New York: Dutton.

Lindshield, S.M., and M.A. Rodrigues. 2009. Tool use in wild spider monkeys (*Ateles geoffroyi*). *Primates* 50 (3):269–272.

Liu, Q. D. Fragaszy, and K. Simpson. 2006. Kinematics of nut-cracking in wild capuchin monkeys in Piauí, Brazil. *American Journal of Primatology* 68 (S1):111.

Liu, Q., K. Simpson, P. Izar, E. Ottoni, E. Visalberghi, and D. Fragaszy. 2009. Kinematics and energetics of nut-cracking in wild capuchin monkeys (*Cebus libidinosus*) in Piauí, Brazil. *American Journal of Physical Anthropology* 138 (2):210–220.

Livingstone, D., and D.P. Ambrose. 1978. Feeding behaviour and predatory efficiency of some reduviids from the Palghat Gap. *Journal of Madras University (B)* 41: 1–25.

Lockwood, D. 1963. *We, the aborigines*. Melbourne, Australia: Cassell.

Longino, J.T. 1984. True anting by the capuchin, *Cebus capucinus*. *Primates* 25:243–245.

Longman, H. 1922. The magnificent spider, *Dicrostichus magnificus*: Notes on cocoon spinning and method of catching prey. *Proceedings of the Royal Society of Queensland* 33:91–98.

Lonsdorf, E.V. 2005. Sex differences in the development of termite-fishing skills in the wild chimpanzees, *Pan troglodytes schweinfurthii*, of Gombe National Park, Tanzania. *Animal Behaviour* 70 (3):673–683.

———. 2006. What is the role of mothers in the acquisition of termite-fishing behaviors in wild chimpanzees (*Pan troglodytes schweinfurthii*)? *Animal Cognition* 9 (1):36–46.

Lonsdorf, E.V., L.E. Eberly, and A.E. Pusey. 2004. Sex differences in learning in chimpanzees. *Nature* 428 (6984):715–716.

Lonsdorf, E.V., and W.D. Hopkins. 2005. Wild chimpanzees show population-level handedness for tool use. *Proceedings of the National Academy of Sciences of the United States of America* 102 (35):12634–12638.

Lonsdorf, E.V., S.R. Ross, S.A. Linick, M.S. Milstein, and T.N. Melber. 2009. An experimental, comparative investigation of tool use in chimpanzees and gorillas. *Animal Behaviour* 77 (5):1119–1126.

Lorenz, K., and U. von Saint Paul. 1968. Die Entwicklung des Spiessens und Klemmens bei den drei Würgerarten, *Lanius collurio, L. senator,* und *L. excubitor. Journal für Ornithologie* 109:137–156.

Louis, D. 1974. Biology of Reduviidae of cocoa farms in Ghana. *American Midland Naturalist* 91:68–89.

Lovell, H. 1958. Baiting of fish by a green heron. *Wilson Bulletin* 70 (3):280–281.

Lozano, G.A. 1998. Parasitic stress and self-medication in wild animals. *Advances in the Study of Behavior* 27: 291–317.

Ludes, E., and J.R. Anderson. 1995. Peat-bathing by captive white-faced capuchin monkeys *Cebus capucinus. Folia Primatologica* 65:38–42.

Lüling, K. 1958. Morphologische-anatomische und histologische untersuchungen am auge des schützenfisches *Toxotes jaculatrix* nebst bemerkungen zum spuckgehaben. *Zeitschrift für Morphologie und Ökologie der Tiere* 47:529–610.

———. 1963. The archer fish. *Scientific American* 209: 100–108.

Lunt, N., P.E. Hulley, and A. Craig. 2004. Active anting in captive Cape white-eyes *Zosterops pallidus. Ibis* 146 (2):360–362.

Lyall-Watson, M. 1963. A critical re-examination of food "washing" behaviour in the raccoon. *Proceedings of the Zoological Society of London* 141:371–393.

Lydekker, R. 1910. *Mammals.* Vol. 1 of *Library of natural history.* New York: Saalfield.

MacDonell, J.R. 2000. The learning capabilities of *Corvus brachyrhynchos* (the common crow) as observed through the use of tools. *AAAS Annual Meeting and Science Innovation Exposition* 166:A96.

Machida, S. 1990. Standing and climbing a pole by members of a captive group of Japanese monkeys. *Primates* 31 (2):291–298.

MacKinnon, J. 1971. The orang-utan in Sabah today: A study of a wild population in the Ulu Segama reserve. *Oryx* 11:141–191.

———. 1974a. The behaviour and ecology of wild orang-utans (*Pongo pygmaeus*). *Animal Behaviour* 22:3–74.

———. 1974b. *In search of the red ape.* New York: Holt, Rinehart and Winston.

———. 1978. *The ape within us.* London: Collins.

Maki, S., P.L. Alford, M.A. Bloomsmith, and J. Franklin. 1989. Food puzzle device simulating termite fishing for captive chimpanzees (*Pan troglodytes*). *American Journal of Primatology* S1:71–78.

Malaivijitnond, S., C. Lekprayoon, N. Tandavanittj, S. Panha, C. Cheewatham, and Y. Hamada. 2007. Stone-tool usage by Thai long-tailed macaques (*Macaca fascicularis*). *American Journal of Primatology* 69 (2):227–233.

Mannu, M., and E.B. Ottoni. 1999. [Posture and laterality in spontaneous cracking of palm fruits by a group of tufted capuchins (*Cebus apella*) in semi-captive conditions]. In *Livro de resumos do IX Congresso Brasileiro de Primatologia.* Santa Teresa, Brazil: Congresso Brasileiro de Primatologia, 31.

———. 2005. [Tool use by two groups of capuchins (*Cebus apella*) in the Caatinga: Preliminary data]. In *Programa e livro de resumos XI do Congresso Brasileiro de Primatologia.* Porto Alegre, Brazil: Sociedade Brasileira de Primatologia, 45.

———. 2009. The enhanced tool-kit of two groups of wild bearded capuchin monkeys in the Caatinga: Tool making, associative use, and secondary tools. *American Journal of Primatology* 71 (3):242–251.

Maple, T. 1974. Do crows use automobiles as nutcrackers? *Western Birds* 5:97–98.

———. 1975. Aggressive object displays of captive baboons. *Journal of Mammalogy* 56:949–950.

Marais, E. 1969. *The soul of the ape.* New York: Atheneum.

Maravita, A., and A. Iriki. 2004. Tools for the body (schema). *Trends in Cognitive Sciences* 8 (2):79–86.

Marchant, L.F., and W.C. McGrew. 1999. Innovative behavior at Mahale: New data on nasal probe and nipple press. *Pan Africa News* 6 (2):16–18.

———. 2005. Percussive technology: Chimpanzee baobab smashing and the evolutionary modeling of hominid knapping. In *Stone knapping: The necessary conditions for a uniquely hominin behaviour,* edited by V. Roux and B. Bril. McDonald Institute Monographs. Cambridge: McDonald Institute for Archaeological Research, University of Cambridge, 341–350.

———. 2007. Ant fishing by wild chimpanzees is not lateralized. *Primates* 48 (1):22–26.

Marchant, L.F., W.C. McGrew, and K.D. Hunt. 2007. Ethoarchaeology of manual laterality: Well-digging by wild chimpanzees. *American Journal of Physical Anthropology* 132 (S44):163.

Mariyama, T., S. Hihara, H. Ishibashi, A. Iriki, and K. Nakamura. 2002. A reinforcement learning model with the role of parietal association area for macaque monkey tool use. *Neuroscience Research* 45 (Suppl. 1):S89.

Marks, J.S., and C.S. Hall. 1992. Tool use by bristle-thighed curlews feeding on albatross eggs. *Condor* 94 (4): 1032–1034.

Marshall, A. 1954. *Bower-birds: Their displays and breeding cycles; A preliminary statement.* Oxford: Clarendon Press.

Marshall, B.E. 1982. A possible example of tool usage by the Marabou stork. *Ostrich* 53 (3):181–182.

Marshall-Pescini, S., and A. Whiten. 2004. Social learning of nut-cracking behaviour in infant and juvenile chimpanzees (*Pan troglodytes*). *Folia Primatologica* 75 (1):49.

———. 2008. Chimpanzees (*Pan troglodytes*) and the question of cumulative culture: An experimental approach. *Animal Cognition* 11:449–456.

Martin, A. 1890. *Home life on an ostrich farm.* London: George Philip.

Martin, A.R., V.M.F. da Silva, and P. Rothery. 2008. Object carrying as socio-sexual display in an aquatic mammal. *Biology Letters* 4:243–245.

Martinelli, M., B. Calcinai, and G. Bavestrello. 2006. Use of sponges in the decoration of *Inachus phalangium* (Decapoda: Majidae) from the Adriatic Sea. *Italian Journal of Zoology* 73 (4):347–353.

Martin-Ordas, G., J. Call, and F. Colmenares. 2008. Tubes, tables, and traps: Great apes solve two functionally equivalent trap tasks but show no evidence of transfer across tasks. *Animal Cognition* 11:423–430.

Masataka, N., H. Koda, N. Urasopon, and K. Watanabe. 2009. Free-ranging macaque mothers exaggerate tool-using behaviour when observed by offspring. *PLoS One* 4 (3):e4768.

Mason, R.T., H.M. Fales, M. Eisner, and T. Eisner. 1991. Wax of a whitefly and its utilization by a chrysopid larva. *Naturwissenschaften* 78 (1):28–30.

Mastro, E. 1981. Algal preferences for decoration by the Californian kelp crab, *Pugettia producta* (Randall) (Decapoda: Majidae). *Crustaceana* 41:64–70.

Matevia, M., G.P. Patterson, and W.A. Hillix. 2002. Pretend play in a signing gorilla. In *Pretending and imagination in animals and children*, edited by R.W. Mitchell. Cambridge: Cambridge University Press, 285–305.

Mather, J.A. 1992. Interactions of juvenile *Octopus vulgaris* with scavenging and territorial fishes. *Marine Behavior and Physiology* 19:175–182.

———. 1994. "Home" choice and modification by juvenile *Octopus vulgaris* (Mollusca: Cephalopoda): Specialized intelligence and tool use? *Journal of Zoology* 233:359–368.

———. 1995. Cognition in cephalopods. *Advances in the Study of Behavior* 24:317–353.

Matheson, M.D., H.M. Mack, L.K. Sheeran, T. Yenter, and P. Schulz. 2008. Captive ex-pet *Macaca fascicularis* use hair and dental floss to floss their teeth. *American Journal of Primatology* 70 (S1):56.

Mathieu, M., and G. Bergeron. 1977. Direct observation of cognitive development in nursery-reared chimpanzees. Paper presented at the Annual Meeting of the Animal Behavior Society, June 6–7, University Park, Pennsylvania.

Mathieu, M., N. Daudelin, Y. Dagenais, and T.G. Decarie. 1980. Piagetian causality in two house-reared chimpanzees (*Pan troglodytes*). *Canadian Journal of Psychology* 34:179–186.

Matschie, P. 1904. Bemerkungen über die gattung *Gorilla*. In *Sitzungsberichte der Gesellschaft Naturforschender Freunde zu Berlin* 1904 (3):45–53.

Matsumoto-Oda, A. 2000. Chimpanzees in the Rubondo Island National Park, Tanzania. *Pan Africa News* 7 (2):16–17.

Matsumoto-Oda, A., and M. Tomonaga. 2005. "Intentional" control of sound production found in leaf-clipping display of Mahale chimpanzees. *Journal of Ethology* 23 (2):109–112.

Matsusaka, T. 2007. Exploratory-threat behaviors in wild chimpanzees encountering a porcupine. *Pan African News* 14 (2):29–31.

Matsusaka, T., and N. Kutsukake. 2002. Use of leaf-sponge and leaf-spoon by juvenile chimpanzees at Mahale. *Pan Africa News* 9 (1):6–9.

Matsusaka, T., H. Nishie, M. Shimada, N. Kutsukake, K. Zamma, M. Nakamura, and T. Nishida. 2006. Tool-use for drinking water by immature chimpanzees of Mahale: Prevalence of an unessential behavior. *Primates* 47 (2):113–122.

Matsuzawa, T. 1991. Nesting cups and metatools in chimpanzees. *Behavioral and Brain Sciences* 14 (4):570–571.

———. 1994. Field experiments on use of stone tools by chimpanzees in the wild. In *Chimpanzee cultures*, edited by R.W. Wrangham, W.C. McGrew, F.B.M. de Waal, and P.G. Heltne. Cambridge, MA: Harvard University Press, 351–370.

———. 1996. Chimpanzee intelligence in nature and in captivity: Isomorphism of symbol use and tool use. In *Great ape societies*, edited by W.C. McGrew, L. Marchant, and T. Nishida. Cambridge: Cambridge University Press, 196–209.

———. 1997. The death of an infant chimpanzee at Bossou, Guinea. *Pan Africa News* 4 (1):4–6.

———. 1999. Communication and tool use in chimpanzees: Cultural and social contexts. In *Design of animal communication*, edited by M. Hauser and M. Konishi. Cambridge, MA: MIT Press, 645–671.

———. 2001. Primate foundations of human intelligence: A view of tool use in nonhuman primates and fossil hominids. In *Primate origins of human cognition and behavior*, edited by T. Matsuzawa. Tokyo: Springer, 3–25.

———. 2008. *Essay on evolutionary neighbors: 5. Doll play in wild chimpanzees*. www.pri.kyoto-u.ac.jp/ai/essay/5_hyrax.htm [accessed 11 May 2008].

Matsuzawa, T., and G. Yamakoshi. 1996. Comparison of chimpanzee material culture between Bossou and Nimba, West Africa. In *Reaching into thought: The minds of the great apes*, edited by A.E. Russon, K.A. Bard, and S.T. Parker. New York: Cambridge University Press, 211–232.

Matsuzawa, T., G. Yamakoshi, and T. Humle. 1996. [A newly found tool-use by wild chimpanzees: Algae scooping]. *Reichorui Kenkyu* 12 (3):283.

Matsuzawa, T., H. Takemoto, S. Hayakawa, and M. Shimada. 1999. Diécké forest in Guinea. *Pan Africa News* 6 (1):10–11.

Matsuzawa, T., D. Biro, T. Humle, N. Inoue-Nakamura, R. Tonooka, and G. Yamakoshi. 2001. Emergence of culture in wild chimpanzees: Education by master-apprenticeship. In *Primate origins of human cognition and behavior*, edited by T. Matsuzawa. Tokyo: Springer, 557–574.

Matsuzawa, T., T. Humle, K. Koops, D. Biro, M. Hayashi, C. Sousa, Y. Mizuno, A. Kato, G. Yamakoshi, G. Ohashi, A. Sugiyama, and M. Kourouma. 2004. Wild chimpanzees at Bossou-Nimba: Deaths through a flu-like epidemic in 2003 and the green-corridor project. *Primate Research* 20 (1):45–55.

Matthews, R., and J. Matthews. 1978. *Insect behavior*. New York: John Wiley & Sons.

Maulik, S. 1919. *The fauna of British India, Chrysomelidae*

(Hispinae and Cassidinae), vol. 1. London: Taylor and Francis.

Mayaud, N. 1983. Réflexions sur le comportement de jet de pierre du percnoptère. *Alauda* 51:229–231.

McAtee, W.L. 1938. "Anting" by birds. *Auk* (55):98–105.

McBeath, N.M., and W.C. McGrew. 1982. Tools used by wild chimpanzees to obtain termites at Mt. Assirik, Senegal: The influence of habitat. *Journal of Human Evolution* 11:65–72.

McCowan, B., L. Marino, E. Vance, L. Walke, and D. Reiss. 2000. Bubble ring play of bottlenose dolphins (*Tursiops truncatus*): Implications for cognition. *Journal of Comparative Psychology* 114 (1):98–106.

McCullough, D.D., and R.J. Beasley. 1996. Bait-fishing herons. *Bird Watcher's Digest* 18 (3):48–51.

McCullough, D.R. 1971. *The tule elk: Its history, behavior, and ecology*, California Library reprint series ed. University of California Publications in Zoology, vol. 88. Berkeley: University of California Press.

McDonald, P. 1984. Tool use by the ant, *Novomessor albisetosus* (Mayr). *Journal of the New York Entomological Society* 92 (2):156–161.

McGrew, W.C. 1974. Tool use by wild chimpanzees in feeding upon driver ants. *Journal of Human Evolution* 3: 501–508.

———. 1977. Socialization and object manipulation of wild chimpanzees. In *Primate bio-social development: Biological, social, and ecological determinants*, edited by S. Chevalier-Skolnikoff and F.E. Poirier. New York: Garland, 261–288.

———. 1979. Evolutionary implications of sex differences in chimpanzee predation and tool use. In *The great apes: Perspectives on human evolution*, edited by D.A. Hamburg and E. McCown. Menlo Park, CA: Benjamin/Cummings, 441–463.

———. 1983. Animal foods in the diets of wild chimpanzees (*Pan troglodytes*): Why cross-cultural variation? *Journal of Ethology* 1:46–61.

———. 1989. Why is ape tool use so confusing? In *Comparative socioecology: The behavioural ecology of humans and other mammals*, edited by V. Standen and R.A. Foley. Oxford: Blackwell Scientific, 457–472.

———. 1991. Are there real differences in tool-use across great apes? In *The great apes conference proceedings, December 15–22, 1991, Bohorok, Jakarta, Pangkalan Bun and Tanjung Puting, Central Kalimantan, Indonesia*. Indonesia: Ministry of Forestry, 177–178.

———. 1992. *Chimpanzee material culture: Implications for human evolution*. Cambridge: Cambridge University Press.

———. 1993a. Brains, hands, and minds: Puzzling incongruities in ape tool use. In *The use of tools by human and non-human primates*, edited by A. Berthelet and J. Chavaillon. Oxford: Clarendon Press, 143–153.

———. 1993b. The intelligent use of tools: Twenty propositions. In *Tools, language, and cognition in human evolution*, edited by K.R. Gibson and T. Ingold. Cambridge: Cambridge University Press, 151–170.

———. 1994. Tools compared: The material of culture. In *Chimpanzee cultures*, edited by R.W. Wrangham, W.C. McGrew, F.B.M. de Waal, and P.G. Heltne. Cambridge, MA: Harvard University Press, 25–40.

———. 2004. *The cultured chimpanzee: Reflections on cultural primatology*. New York: Cambridge University Press.

McGrew, W.C., and D.A. Collins. 1985. Tool use by wild chimpanzees (*Pan troglodytes*) to obtain termites (*Macrotermes herus*) in the Mahale Mountains, Tanzania. *American Journal of Primatology* 9 (1):47–62.

McGrew, W.C., and L.F. Marchant. 1992. Chimpanzees, tools, and termites: Hand preference or handedness? *Current Anthropology* 33 (1):114–119.

———. 1998. Chimpanzee wears a knotted skin "necklace." *Pan Africa News* 5 (1):8–9.

———. 1999. Laterality of hand use pays off in foraging success for wild chimpanzees. *Primates* 40 (3):509–513.

McGrew, W.C., J.D. Pruetz, and S.J. Fulton. 2005. Chimpanzees use tools to harvest social insects at Fongoli, Senegal. *Folia Primatologica* 76 (4):222–226.

McGrew, W.C., and M.E. Rogers. 1983. Chimpanzees, tools, and termites: New record from Gabon. *American Journal of Primatology* 5 (2):171–174.

McGrew, W.C., and C.E.G. Tutin. 1972. Chimpanzee dentistry. *Journal of the American Dental Association* 85:1198–1204.

———. 1973. Chimpanzee tool use in dental grooming. *Nature* 241:477–478.

McGrew, W.C., C.E.G. Tutin, and P.J. Baldwin. 1979. Chimpanzees, tools, and termites: Cross-cultural comparisons of Senegal, Tanzania, and Rio Muni. *Man* 14 (2):185–214.

McGrew, W.C., C.E.G. Tutin, and P.S. Midgett, Jr. 1975. Tool use in a group of captive chimpanzees: I. Escape. *Zeitschrift für Tierpsychologie* 37:145–162.

McGrew, W.C., P.J. Baldwin, L.F. Marchant, J.D. Pruetz, S.E. Scott, and C.E.G. Tutin. 2003. Ethoarchaeology and elementary technology of unhabituated wild chimpanzees at Assirik, Senegal, West Africa. *Paleoanthropology* 1:1–20.

McHugh, J.V., and T.G. Kiselyova. 2003. First descriptions for larval stages of *Eurysphindus* (Coleoptera: Cucujoidea: Sphindidae). *Coleopterists Bulletin* 57 (1):17–25.

McKay, G. 1973. Behavior and ecology of the Asiatic elephant in southeastern Ceylon. *Smithsonian Contributions to Zoology* 125:1–113.

McKenna, J.J. 1982. Primate field studies: The evolution of behavior and its socioecology. In *Primate behavior*, edited by J.L. Fobes and J.E. King. New York: Academic Press, 53–83.

McLay, C.L. 1983. Dispersal and use of sponges and ascidians as camouflage by *Cryptodromia hilgendorfi* (Brachyura: Dromiacea). *Marine Biology* 76:17–32.

McLean, I.G. 1978. Plugging of nest burrows by female *Spermophilus columbianus*. *Journal of Mammalogy* 59 (2):437–439.

McLean, R. 1974. Direct shell acquisition by hermit crabs from gastropods. *Experientia* 30:206–208.

McMahan, E.A. 1982. Bait-and-capture strategy of a termite-eating assassin bug. *Insectes Sociaux* 29 (2):346–351.

———. 1983a. Adaptations, feeding preferences, and biometrics of a termite-baiting assassin bug (Hemiptera: Reduviidae). *Annals of the Entomological Society of America* 76 (3):483–486.

———. 1983b. Bugs angle for termites. *Natural History* 92 (5):40–47.

Meglitsch, P. 1972. *Invertebrate zoology*. New York: Oxford University Press.

Mehlman, P.T. 1996. Branch shaking and related displays in wild Barbary macaques. In *Evolution and ecology of macaque societies*, edited by J.E. Fa and D.G. Lindburg. Cambridge: Cambridge University Press, 503–526.

Meinertzhagen, R. 1964. Grit. In *A new dictionary of birds*, edited by A. Thomson. New York: McGraw-Hill, 341–342.

Mendes, F.D.C., L.B.R. Martins, J.A. Pereira, and R.F. Marquezan. 2000. Fishing with a bait: A note on behavioural flexibility in *Cebus apella*. *Folia Primatologica* 71 (5):350–352.

Mendes, N., D. Hanus, and J. Call. 2007. Raising the level: Orangutans use water as a tool. *Biology Letters* 3: 453–455.

Mentz, I., and K. Perret. 1999. Environmental enrichment bei Flachlandgorillas (*Gorilla g. gorilla*): Beobachtungen zur Nahrungsaufnahme und zum Manipulationsverhalten. *Zoologische Garten* 69 (1):49–63.

Menzel, E.W., Jr. 1971. Communication about the environment in a group of young chimpanzees. *Folia Primatologica* 15:220–232.

———. 1972. Spontaneous invention of ladders in a group of young chimpanzees. *Folia Primatologica* 17:87–106.

———. 1973a. Further observations on the use of ladders in a group of young chimpanzees. *Folia Primatologica* 19:450–457.

———. 1973b. Leadership and communication in young chimpanzees. In *Precultural primate behavior*, edited by E. Menzel. Vol. 1 of *Symposia of the Fourth International Congress of Primatology*. Basel, Switzerland: S. Karger, 192–225.

Menzel, E.W., Jr., R.K. Davenport, and C.M. Rogers. 1970. The development of tool using in wild-born and restriction-reared chimpanzees. *Folia Primatologica* 12:273–283.

Mercader, J., M. Panger, and C. Boesch. 2002. Excavation of a chimpanzee stone tool site in the African rainforest. *Science* 296 (5572):1452–1455.

Mercader, J., H. Barton, J. Gillespie, J. Harris, S. Kuhn, R. Tyler, and C. Boesch. 2007. 4,300-year-old chimpanzee sites and the origins of percussive stone technology. *Proceedings of the National Academy of Sciences of the United States of America* 104 (9):3043–3048.

Merfield, F.G. 1956. *Gorilla hunter*. New York: Farrar, Straus, and Cudahy.

Merlen, G., and G. Davis-Merlen. 2000. Whish: More than a tool-using finch. *Noticias de Galápagos* (61):2–9.

Merz, E. 1978. Male-male interactions with dead infants in *Macaca sylvana*. *Primates* 19:749–754.

Messeri, P. 1978. Some observations on a littoral troop of yellow baboons. *Monitore Zoologico Italiano* 12:69.

Messinger, D. 2007. *Grains of golden sand*. Honolulu: Fine Print Press.

Metevier, C., K.M. Stonemetz, and M.A. Novak. 2006. The use of two different tools in series in captive rhesus macaques (*Macaca mulatta*). *American Journal of Primatology* 68 (S1):52.

Meunier, H., O. Petit, and J.L. Deneubourg. 2007. Social facilitation of fur rubbing behavior in white-faced capuchins. *American Journal of Primatology* 69:1–14.

———. 2008. Resource influence on the form of fur rubbing behaviour in white-faced capuchins. *Behavioural Processes* 77 (3):320–326.

Meyerriecks, A.J. 1972. Tool-using by a double-crested cormorant. *Wilson Bulletin* 84 (4):482–483.

Michener, G.R. 2004. Hunting techniques and tool use by North American badgers preying on Richardson's ground squirrels. *Journal of Mammalogy* 85 (5):1019–1027.

Mignault, C. 1985. Transition between sensorimotor and symbolic activities in nursery-reared chimpanzees (*Pan troglodytes*). *Journal of Human Evolution* 14:747–758.

Milburn, O., and R. Alexander. 1976. The performance of the muscles involved in spitting by the archerfish *Toxotes*. *Journal of Zoology* 180:243–251.

Miles, H.L.W. 1993. Language and the orang-utan: The old "person" of the forest. In *The great ape project: Equality beyond humanity*, edited by P. Cavalieri and P. Singer. New York: St. Martin's Press, 42–57.

———. 1994. Me Chantek: The development of self-awareness in a signing orangutan. In *Self-awareness in animals and humans: Developmental perspectives*, edited by S.T. Parker, R.W. Mitchell, and M. Boccia. Cambridge: Cambridge University Press, 254–272.

Miller, K.E., K. Laszlo, and S.J. Suomi. 2008. Why do captive tufted capuchins (*Cebus apella*) urine wash? *American Journal of Primatology* 70:119–126.

Miller, L. 1962. Stomach stones. *Zoonooz* 35:10–13.

Miller, L., and D. Quiatt. 1983. Tool use by a captive orangutan. *Laboratory Primate Newsletter* 22 (1):10.

Miller, N.C.E. 1956. *The biology of the Heteroptera*. London: Leonard Hill.

Millikan, G., and R. Bowman. 1967. Observations on Galápagos tool-using finches in captivity. *Living Bird* 6:23–41.

Millott, N. 1956. The covering reaction of sea-urchins: 1. A preliminary account of covering in the tropical echinoid *Lytechinus variegatus* (Lamarck) and its relation to light. *Journal of Experimental Biology* 33 (3):508–523.

Milne, M., and L. Milne. 1939. Evolutionary trends in caddis worm case construction. *Annals of the Entomological Society of America* 32:533–542.

Mitani, J.C., and D.P. Watts. 2001. Why do chimpanzees hunt and share meat? *Animal Behaviour* 61:915–924.

Mitchell, H. 1972. Further recording of a tool-using bird. *Australian Bird Watcher* 4:237.

Mitchell, R.W. 1999. Scientific and popular conceptions of the psychology of great apes from the 1700s to the 1970s: Deja vu all over again. *Primate Report* 53:3–118.

Mitchell, T.L. 1993. Tool use by a white-breasted nuthatch. *Bulletin of the Oklahoma Ornithological Society* 26 (1):6–7.

Möbius, Y., C. Boesch, K. Koops, T. Matsuzawa, and T. Humle. 2008. Cultural differences in army ant predation by West African chimpanzees? A comparative study of microecological variables. *Animal Behaviour* 76:37–45.

Modahl, K.B., and G.G. Eaton. 1977. Display behavior in a confined troop of Japanese macaques (*Macaca fuscata*). *Animal Behaviour* 25:525–535.

Moglich, M.H.T., and G.D. Alpert. 1979. Stone dropping by *Conomyrma bicolor* (Hymenoptera: Formicidae): A new technique of interference competition. *Behavioral Ecology and Sociobiology* 6:105–113.

Molitor, A. 1931. Neue Beobachtungen und Versuche mit Grabwespen. *Biologisches Zentralblatt* 51:412–424.

Montagu, A. 1970. A remarkable case of tool-using in a bird. *American Anthropologist* 72 (3):610.

Montevecchi, W.A. 1978. Corvids using objects to displace gulls from nests. *Condor* 80 (3):349.

Morand-Ferron, J., M. Veillette, and L. Lefebvre. 2006. Stealing of dunked food in Carib grackles (*Quiscalus lugubris*). *Behavioural Processes* 73 (3):342–347.

Morand-Ferron, J., L. Lefebvre, S.M. Reader, D. Sol, and S. Elvin. 2004. Dunking behaviour in Carib grackles. *Animal Behaviour* 68:1267–1274.

Moreno, J., M. Soler, A.P. Møller, and M. Linden. 1994. The function of stone carrying in the black wheatear, *Oenanthe leucura*. *Animal Behaviour* 47 (6):1297–1309.

Morgan, B.J., and E.E. Abwe. 2006. Chimpanzees use stone hammers in Cameroon. *Current Biology* 16 (16):R632–R633.

Morgan, L.H. 1868. *The American beaver and his works*. Philadelphia: J.B. Lippincott.

Morimura, N. 2003. A note on enrichment for spontaneous tool use by chimpanzees (*Pan troglodytes*). *Applied Animal Behaviour Science* 82 (3):241–247.

Morimura, N., S. Sekine, K. Fuwa, and G. Idani. 2002. Selectability on tool-using behavior by chimpanzees. *Anthropological Science* 110 (1):137.

Morrill, W. 1972. Tool-using behavior of *Pogonomyrmex badius* (Hymenoptera: Formicidae). *Florida Entomologist* 55:59–60.

Morris, D. 1954. The snail-eating behaviour of thrushes and blackbirds. *British Birds* 47:33–49.

———. 1962. *The biology of art*. New York: Alfred A. Knopf.

Morris, K., and J. Goodall. 1977. Competition for meat between chimpanzees and baboons of the Gombe National Park. *Folia Primatologica* 28:109–121.

Morrogh-Bernard, H.C. 2008. Fur-rubbing as a form of self-medication in *Pongo pygmaeus*. *International Journal of Primatology* 29 (4):1059–1064.

Morse, D.H. 1968. Use of tools by brown-headed nuthatches. *Wilson Bulletin* 80 (2):220–224.

Morton, T.C., and F.V. Vencl. 1998. Larval beetles form a defense from recycled host plant chemicals discharged as fecal wastes. *Journal of Chemical Ecology* 24:765–785.

Moss, C. 1988. *Elephant memories: Thirteen years in the life of an elephant family*. New York: William Morrow.

Mottershead, G. 1959. Experiments with a chimpanzee colony at Chester Zoo. *International Zoo Yearbook* 1:18–20.

———. 1963. Experiences with chimpanzees at liberty on islands. *Zoologische Garten* 28:31–33.

Moura, A.C. 2002. Tool use by wild groups of *Cebus apella libidinosus* living in the dry Caatinga forest of northeastern Brazil. In *Caring for primates: Abstracts of the XIXth Congress of the International Primatological Society*. Beijing: Mammalogical Society of China, 244–245.

———. 2003. Ecological pressures driving tool use in capuchin monkeys. *Folia Primatologica* 74 (4):209.

———. 2007. Stone banging by wild capuchin monkeys: An unusual auditory display. *Folia Primatologica* 78 (1):36–45.

Moura, A.C., and P.C. Lee. 2004a. Capuchin stone tool use in Caatinga dry forest. *Science* 306:1909.

———. 2004b. The cognitive abilities of *Cebus apella libidinosus*: Tool use and survival in a harsh environment. *Folia Primatologica* 75 (1):49.

———. 2004c. Tool-use as an ecological strategy for survival in a harsh environment among *Cebus apella libidinosus*. *Folia Primatologica* 75:437.

Moynihan, M. 1976. *The New World primates: Adaptive radiation and the evolution of social behavior, languages, and intelligence*. Princeton, NJ: Princeton University Press.

Mulcahy, N., and J. Call. 2006a. Apes save tools for future use. *Science* 312:1038–1040.

———. 2006b. How great apes perform on a modified trap-tube task. *Animal Cognition* 9 (3):193–199.

Mulcahy, N., J. Call, and R. Dunbar. 2005. Gorillas (*Gorilla gorilla*) and orangutans (*Pongo pygmaeus*) encode relevant problem features in a tool-using task. *Journal of Comparative Psychology* 119 (1):23–32.

Müller, C., and M. Hilker. 1999. Unexpected reactions of a generalist predator towards defensive devices of cassidine larvae (Coleoptera: Chrysomelidae). *Oecologia* 118:166–172.

———. 2003. Advantages and disadvantages of larval abdominal shields of Chrysomelidae: Mini-review. In *Special topics in leaf beetle biology: Proceedings of the Fifth International Symposium on the Chrysomelidae, August 25–27, 2000, Iguassu Falls, Brazil*, edited by D.G. Furth. Sofia, Bulgaria: Pensoft, 243–259.

———. 2004. Ecological relevance of fecal matter in Chrysomelidae. In *Developments in the biology of Chrysomelidae*, edited by H. Jolivet, J. Santiago-Blay, and M. Schmitt. The Hague, Netherlands: SPB Academic, 693–705.

Murie, O. 1940. Notes on the sea otter. *Journal of Mammalogy* 21:119–131.

Muroyama, Y. 1991. Chimpanzees' choices of prey between two sympatric species of *Macrotermes* in the Campo Animal Reserve, Cameroon. *Human Evolution* 6 (2):143–151.

Murphy, D.E. 1976. Enrichment and occupational devices for orangutans and chimpanzees. *International Zoo News* 23 (5):24–26.

Mwanza, N., J. Yamagiwa, T. Maruhashi, and T. Yumoto. 1992. Animal eating and tool-use by chimpanzees in the Kahuzi-Biega National Park, Zaïre. In *XIVth Congress of the International Primatological Society*. Strasbourg, France: International Primatological Society, 153.

Mysterud, I. 1973. Behaviour of the brown bear (*Ursus actos*) at moose kills. *Norwegian Journal of Zoology* 21:267–272.

Nagell, K., R. Olguin, and M. Tomasello. 1993. Processes of social learning in the tool use of chimpanzees (*Pan troglodytes*) and human children (*Homo sapiens*). *Journal of Comparative Psychology* 107 (2):174–186.

Nakahira, K., and R. Arakawa. 2006. Defensive functions of the trash-package of a green lacewing, *Mallada desjardinsi* (Neuroptera: Chrysopidae), against a ladybird, *Harmonia axyridis* (Coleoptera: Coccinellidae). *Applied Entomology and Zoology* 41 (1):111–115.

Nakamichi, M. 1998. Stick throwing by gorillas (*Gorilla gorilla gorilla*) at the San Diego Wild Animal Park. *Folia Primatologica* 69 (5):291–295.

———. 1999. Spontaneous use of sticks as tools by captive gorillas (*Gorilla gorilla gorilla*). *Primates* 40 (3):487–498.

———. 2004. Tool-use and tool-making by captive, group-living orangutans (*Pongo pygmaeus abelii*) at an artificial termite mound. *Behavioural Processes* 65 (1):87–93.

Nakamichi, M., E. Kato, Y. Kojima, and N. Itoigawa. 1998. Carrying and washing of grass roots by free-ranging Japanese macaques at Katsuyama. *Folia Primatologica* 69 (1):35–40.

Nakamura, H., and T. Tabata. 1988. Why does the broad-billed roller *Eurystomus orientalis* bring strange objects to the nest? *Japanese Journal of Ornithology* 36 (4): 137–152.

Nakamura, M., and N. Itoh. 2008. Hunting with tools by Mahale chimpanzees. *Pan Africa News* 15 (1):3–6.

Nash, V.J. 1981. Fishin' chimps. *Annual Report, The Royal Zoological Society of Scotland* 69:41–44.

———. 1982. Tool use by captive chimpanzees at an artificial termite mound. *Zoo Biology* 1 (3):211–221.

Natale, F. 1989. Causality II: The stick problem. In *Cognitive structure and development in nonhuman primates*, edited by F. Antinucci. Hillsdale, NJ: Lawrence Erlbaum Associates, 121–133.

Natale, F., P. Potì, and G. Spinozzi. 1988. Development of tool use in a macaque and a gorilla. *Primates* 29 (3):413–416.

Neary, P. 1997. Tool use, object manipulation, and nest building by captive bonobos (*Pan paniscus*). *American Journal of Primatology* 42 (2):135–136.

Nellman, H., and W. Trendelenburg. 1926. Ein beitrag zur intelligenzprüfung niederer affen. *Zeitschrift für Vergleichende Physiologie* 4:142–200.

New, T.R. 1969. Note on the debris-carrying habit in larvae of British Chrysopidae (Neuroptera). *Entomologist's Gazette* 20:119–124.

Nicholson, B. 1955. The African elephant (*Loxodonta africana*). *African Wildlife* 9:31–40.

Nickelson, S.A., and J.S. Lockard. 1978. Ethogram of Celebes monkeys (*Macaca nigra*) in two captive habitats. *Primates* 19:437–447.

Nickerson, J.C., D.E. Snyder, and C.C. Oliver. 1979. Acoustical burrows constructed by mole crickets. *Annals of the Entomological Society of America* 72:438–440.

Niemeyer, M., and H. Kingery. 2003. Tool use by European starling. *Colorado Birds* 37 (2):71.

Nierentz, D. 2007. Gorillas at Zurich Zoo. *Gorilla Gazette* 20:50–52.

Niio, G. 2000. *Kanzi II*. NHK [Japanese Broadcasting Corporation], documentary film.

Nishida, T. 1968. The social group of wild chimpanzees in the Mahale mountains. *Primates* 9:167–224.

———. 1970. Social behavior and relationships among wild chimpanzees of the Mahale Mountains. *Primates* 11:47–87.

———. 1973. The ant-gathering behaviour by the use of tools among wild chimpanzees of the Mahale Mountains. *Journal of Human Evolution* 2:357–370.

———. 1977. [Chimpanzee social structure]. In *Reichorui*. Jinruigaku Koza 2. Tokyo: Yuzankaku Shuppan, 277–314.

———. 1980a. The leaf-clipping display: A newly-discovered expressive gesture in wild chimpanzees. *Journal of Human Evolution* 9:117–128.

———. 1980b. Local differences in responses to water among wild chimpanzees. *Folia Primatologica* 33:189–209.

———. 1981. Tool use—fundamentals and philosophy. *Nature* 289:616.

———. 1987. Local traditions and cultural transmissions. In *Primate societies*, edited by B. Smuts, D. Cheney, R. Seyfarth, R. Wrangham, and T. Struhsaker. Chicago: University of Chicago Press, 462–474.

———. 1990. A quarter century of research in the Mahale Mountains: An overview. In *The chimpanzees of the Mahale Mountains*, edited by T. Nishida. Tokyo: University of Tokyo Press, 3–35.

———. 1997. Sexual behaviour of adult male chimpanzees of the Mahale Mountains National Park, Tanzania. *Primates* 38:379–398.

———. 2002. A self-medicating attempt to remove the sand flea from a toe by a young chimpanzee. *Pan Africa News* 9 (1):5–6.

———. 2003a. Harassment of mature female chimpanzees by young males in the Mahale Mountains. *International Journal of Primatology* 24 (3):503–514.

———. 2003b. Individuality and flexibility of cultural behavior patterns in chimpanzees. In *Animal social com-*

plexity: Intelligence, culture, and individualized societies, edited by F. de Waal and P. Tyack. Cambridge, MA: Harvard University Press, 392–413.

Nishida, T., and M. Hiraiwa. 1982. Natural history of a tool-using behavior by wild chimpanzees in feeding upon wood-boring ants. *Journal of Human Evolution* 11 (1):73–99.

Nishida, T., and K. Hosaka. 1996. Coalition strategies among adult male chimpanzees of the Mahale Mountains, Tanzania. In *Great ape societies*, edited by W.C. McGrew, L. Marchant, and T. Nishida. Cambridge: Cambridge University Press, 114–134.

Nishida, T., T. Matsusaka, and W.C. McGrew. 2009. Emergence, propagation, or disappearance of novel behavioral patterns in the habituated chimpanzees of Mahale: A review. *Primates* 50 (1):23–36.

Nishida, T., and M. Nakamura. 1993. Chimpanzee tool use to clear a blocked nasal passage. *Folia Primatologica* 61 (4):218–220.

Nishida, T., and S. Uehara. 1980. Chimpanzees, tools, and termites: Another example from Tanzania. *Current Anthropology* 21 (5):671–672.

Nishida, T., S. Uehara, and R. Nyundo. 1979. Predatory behavior among wild chimpanzees of the Mahale Mountains. *Primates* 20 (1):1–20.

Nishida, T., and W. Wallauer. 2003. Leaf-pile pulling: An unusual play pattern in wild chimpanzees. *American Journal of Primatology* 60:167–173.

Nishida, T., T. Hasegawa, H. Hayaki, Y. Takahata, and S. Uehara. 1992. Meat-sharing as a coalition strategy by an alpha male chimpanzee? In *Human origins*, edited by T. Nishida, W.C. McGrew, P. Marler, M. Pickford, and F.B.M. de Waal. Vol. 2 of *Topics in primatology*. Tokyo: University of Tokyo Press, 159–174.

Nishida, T., T. Kano, J. Goodall, W.C. McGrew, and M. Nakamura. 1999. Ethogram and ethnography of Mahale chimpanzees. *Anthropological Science* 107 (2):141–188.

Nishie, H., N. Itoh, and T. Nishida. 2006. Natural history of *Camponotus* ant fishing behavior by M-group chimpanzees at Mahale. *Reichorui Kenkyu* 22 (S):14.

Nishimura, T., N. Okayasu, Y. Hamada, and J. Yamagiwa. 2003. A case report of a novel type of stick use by wild chimpanzees. *Primates* 44 (2):199–201.

Nissani, M. 2004. Theory of mind and insight in chimpanzees, elephants, and other animals? In *Comparative vertebrate cognition: Are primates superior to non-primates?* edited by L.J. Rogers and G. Kaplan. New York: Kluwer Academic/Plenum, 227–258.

Nissen, H.W. 1949. The psychology of apes. In *Vertebrates*, edited by M. Burton. Vol. 2 of *The story of animal life*. London: Elsevier, 355–366.

———. 1956. Individuality in the behavior of chimpanzees. *American Anthropologist* 58 (3):407–413.

Nogge, G. 1984. Jahresbericht 1983 der Aktiengesellschaft Zoologischer Garten Köln. *Zeitschrift des Kölner Zoo* 27 (1):3–27.

———. 1989. Jahresbericht 1988 der Aktiengesellschaft Zoologischer Garten Köln. *Zeitschrift des Kölner Zoo* 32 (1):3–26.

Nogueira-de-Sá, F., and J.R. Trigo. 2002. Do fecal shields provide physical protection to larvae of the tortoise beetles *Plagiometriona flavescens* and *Stolas chalybea* against natural enemies? *Entomologia Experimentalis et Applicata* 104 (1):203–206.

———. 2005. Faecal shield of the tortoise beetle *Plagiometriona* aff. *flavescens* (Chrysomelidae: Cassidinae) as chemically mediated defense against predators. *Journal of Tropical Ecology* 21:189–194.

Noirot, C. 1970. The nests of termites. In *Biology of termites*, edited by K. Krishna and F. Weesner. New York: Academic Press, 73–125.

Nolte, A. 1958. Beobachtungen über das Instinktverhalten von Kapuzineraffen (*Cebus apella* L.) in der Gefangenschaft. *Behaviour* 12:183–207.

Nonaka, K. 2000. Lopburi no hamigamizaru. *Ecosophia* 5:70–73.

Norikoshi, K. 1994. Hand preference of wild chimpanzees observed in their tool behavior. *Reichorui Kenkyu* 10 (3):315–319.

Norikoshi, K., and T. Kitahara. 1979. [An experiment in spontaneous tool making behavior by captive chimpanzees]. *Annual of Animal Psychology* 29:85–94.

Norris, D. 1975. Green heron goes fishing. *Florida Wildlife* 29:16–17.

North, M. 1948. The lammergeyer in Kenya colony. *Ibis* 90:138–141.

Norton, A. 1909. The food of several Maine water-birds. *Auk* 26:438–440.

Noske, R.A. 1985. Left-footedness and tool-using in the varied sittella *Daphoenositta chrysoptera* and crested shrike-tit *Falcunculus frontatus*. *Corella* 9:63–64.

O'Hara, S.J., and P.C. Lee. 2006. High frequency of postcoital penis cleaning in Budongo chimpanzees. *Folia Primatologica* 77 (5):353–358.

O'Malley, R.C., and W.C. McGrew. 2000. Oral tool use by captive orangutans (*Pongo pygmaeus*). *Folia Primatologica* 71 (5):334–341.

———. 2006. Hand preferences in captive orangutans (*Pongo pygmaeus*). *Primates* 47 (3):279–283.

Oake, K. 1992. More bait-fishing by green backed heron in Botswana. *Babbler* 23:49.

Oakley, K. 1961. On man's use of fire, with comments on tool-making and hunting. In *Social life of early man*, edited by S. Washburn. Chicago: Aldine, 176–193.

———. 1967. *Man the tool-maker*, 5th ed. London: British Museum of Natural History.

Obayashi, S., T. Suhara, K. Kawabe, T. Okauchi, J. Maeda, H. Onoe, Y. Iwamura, and A. Iriki. 2000. PET imaging of the monkey brain during tool-use. *Society for Neuroscience Abstracts* 26 (2):1500.

Obayashi, S., T. Suhara, K. Kawabe, T. Okauchi, J. Maeda, and A. Iriki. 2001a. Cerebellar contribution to bilateral transfer of tool use learning for monkey revealed by positron emission tomography (PET). *Society for Neuroscience Abstracts* 27 (1):502.

Obayashi, S., T. Suhara, K. Kawabe, T. Okauchi, J. Maeda, Y. Akine, H. Onoe, and A. Iriki. 2001b. Functional brain mapping of monkey tool use. *Neuroimage* 14 (4):853–861.

Obayashi, S., T. Suhara, Y. Nagai, J. Maeda, S. Hihara, and A. Iriki. 2002a. Contribution of monkey prefrontal cortex in the master of the versatile tools. *Society for Neuroscience Abstracts* 2002:Abstract #285.212.

Obayashi, S., Y. Nagai, T. Suhara, J. Maeda, S. Hihara, and A. Iriki. 2002b. Functional specialization of monkey prefrontal cortex involved in cognitive set-shifting of tool function in the context. *Neuroscience Research* 45 (S1):S126.

Obayashi, S., T. Suhara, K. Kawabe, T. Okauchi, J. Maeda, Y. Nagai, and A. Iriki. 2002c. Macaque prefrontal activity associated with extensive tool use. *Neuroreport* 13 (17):2349–2354.

———. 2003. Fronto-parieto-cerebellar interaction associated with intermanual transfer of monkey tool-use learning. *Neuroscience Letters* 339 (2):123–126.

Ochiai, T., and T. Matsuzawa. 1998. Planting trees in an outdoor compound of chimpanzees for an enriched environment. In *Proceedings of the Third International Conference on Environmental Enrichment*, edited by V.J. Hare and K.E. Worley. San Diego: Shape of Enrichment, 355–364.

Odell, G. 2004. *Lithic analysis*. New York: Kluwer Academic/Plenum.

Odhiambo, T.R. 1958. Some observations on the natural history of *Acanthaspis petax* Stål (Hemiptera: Reduviidae) living in termite mounds in Uganda. *Proceedings of the Royal Entomological Society of London, Series A* 33:167–175.

Ohashi, G. 2006. Behavioral repertoire of tool use in the wild chimpanzees at Bossou. In *Cognitive development in chimpanzees*, edited by T. Matsuzawa, M. Tomonaga, and M. Tanaka. New York: Springer, 439–451.

Okano, T., C. Asami, Y. Haruki, M. Sasaki, N. Itoigawa, S. Shinohara, and T. Tsuzuki. 1973. Social relations in a chimpanzee colony. In *Behavioral regulators of behavior in primates*, edited by C. Carpenter. Lewisburg, PA: Bucknell University Press, 85–105.

Okanoya, K., N. Tokimoto, N. Kumazawa, S. Hihara, and A. Iriki. 2008. Tool-use training in a species of rodent: The emergence of an optimal motor strategy and functional understanding. *PLoS One* 3 (3):e1860.

Oldham, C. 1930. The shell-smashing habit of gulls. *Ibis* 6:239–243.

Oliver, L. 1996. New enrichment toy. *Gorilla: Journal of the Gorilla Foundation* 19 (2):10.

Olmstead, K.L. 1991. Effectiveness and cost of larval defense in tortoise beetles (Coleoptera: Chrysomelidae: Cassidinae), Ph.D. diss., University of Maryland.

———. 1994. Waste products as chrysomelid defenses. In *Novel aspects of the biology of Chrysomelidae*, edited by P.H. Jolivet, M.L. Cox, and E. Petitpierre. Dordrecht, Netherlands: Kluwer Academic, 311–318.

———. 1996. Cassidine defenses and natural enemies. In *Chrysomelidae biology*, edited by M.L.C.P.H. Jolivet. Vol. 2 of *Ecological studies*. Amsterdam: SPB Academic, 3–21.

Olmstead, K.L., and R.F Denno. 1992. Cost of shield defense for tortoise beetles (Coleoptera: Chrysomelidae). *Ecological Entomology* 17:237–243.

———. 1993. Effectiveness of tortoise beetle larval shields against different predator species. *Ecology* 74:1394–1405.

Olson, D.J., A.C. Kamil, R.P. Balda, and P.J. Nims. 1995. Performance of four seed-caching corvid species in operant tests of nonspatial and spatial memory. *Journal of Comparative Psychology* 109:173–181.

Oppenheimer, J. 1968. Behavior and ecology of the white faced monkey, *Cebus capucinus*, on Barro Colorado Island. Ph.D. diss., University of Illinois at Urbana–Champaign.

———. 1973. Social and communicatory behavior in the cebus monkey. In *Behavioral regulators of behavior in primates*, edited by C. Carpenter. Lewisburg, PA: Bucknell University Press, 251–271.

———. 1977. Forest structure and its relation to activity of the capuchin monkey (*Cebus*). In *Use of non-human primates in biomedical research*, edited by T.C.A. Kumar and M.R.N. Prasad. New Delhi: Indian National Science Academy, 74–84.

Orenstein, R.I. 1972. Tool-use by the New Caledonian crow (*Corvus moneduloides*). *Auk* 89 (3):674–676.

Osborn, S.A.H. 1998. Anting by an American dipper (*Cinclus mexicanus*). *Wilson Bulletin* 110 (3):423–425.

Osten-Sacken, C.R. 1877. A singular habit of *Hilara*. *Entomology Monthly Magazine* 14:126–127.

Osvath, M. 2009. Spontaneous planning for future stone throwing by a male chimpanzee. *Current Biology* 19 (5):R190–R191.

Osvath, M., and H. Osvath. 2008. Chimpanzee (*Pan troglodytes*) and orangutan (*Pongo abelii*) forethought: Self-control and pre-experience in the face of future tool use. *Animal Cognition* 11 (4):661–674.

Oswalt, W.H. 1973. *Habitat and technology: The evolution of hunting*. New York: John Wiley & Sons.

———. 1976. *An anthropological analysis of food-getting technology*. New York: John Wiley & Sons.

Ottoni, E.B., B.D. de Resende, and P. Izar. 2005. Watching the best nutcrackers: What capuchin monkeys (*Cebus apella*) know about others' tool-using skills. *Animal Cognition* 8 (4):215–219.

Ottoni, E.B., and P. Izar. 2008. Capuchin monkey tool use: Overview and implications. *Evolutionary Anthropology* 17 (4):171–178.

Ottoni, E.B., and M. Mannu. 2001. Semi-free-ranging tufted capuchins (*Cebus apella*) spontaneously use tools to crack open nuts. *International Journal of Primatology* 22 (3):347–358.

———. 2003. Spontaneous use of tools by semi-free-ranging capuchin monkeys. In *Animal social complexity: Intelligence, culture, and individualized societies*, edited by F. de Waal and P. Tyack. Cambridge, MA: Harvard University Press, 440–443.

Ottoni, E.B., M. Mannu, and B.D. de Resende. 2002. Developmental aspects of the spontaneous use of tools by semi-free-ranging brown capuchin monkeys. In *39th Annual Meeting of the Animal Behaviour Society, July 13–17, Bloomington, Indiana*. www.animalbehavior.org/ABS/Media/AbstractStatus.html.

Ottoni, E.B., B.D. de Resende, M. Mannu, C. Aquino, A.E. Sestini, and P. Izar. 2001. Tool use, social structure, and information transfer in capuchin monkeys. *Advances in Ethology* 36:234.

Owen, J. 2005. Nut-cracking ape may boost gorillas' IQ rep. *National Geographic News*, October 25. http://news.nationalgeographic.com/news/.

Owings, D., M. Borchert, and R. Virginia. 1977. The behavior of California ground squirrels. *Animal Behaviour* 25:221–230.

Owings, D., and R. Coss. 1977. Snake mobbing by California ground squirrels: Adaptive variation and ontogeny. *Behaviour* 62:50–69.

Owings, D., R.G. Coss, D. McKernon, M.P. Rowe, and P.C. Arrowood. 2001. Snake-directed antipredator behavior of rock squirrels (*Spermophilus variegatus*): Population differences and snake-species discrimination. *Behaviour* 138 (5):575–595.

Oyen, O.J. 1978. Stone-eating and tool-use among olive baboons. *Texas Journal of Science* 30:295.

———. 1979. Tool-use in free-ranging baboons of Nairobi National Park. *Primates* 20:595–597.

Pandolfi, S.S., C.P. van Schaik, and A.E. Pusey. 2003. Sex differences in termite fishing among Gombe chimpanzees. In *Animal social complexity: Intelligence, culture, and individualized societies*, edited by F. de Waal and P. Tyack. Cambridge, MA: Harvard University Press, 414–418.

Panger, M.A. 1998. Object-use in free-ranging white-faced capuchins (*Cebus capucinus*) in Costa Rica. *American Journal of Physical Anthropology* 106 (3):311–321.

———. 1999. Capuchin object manipulation. In *The nonhuman primates*, edited by P. Dolhinow and A. Fuentes. Mountain View, CA: Mayfield, 115–120.

———. 2007. Tool use and cognition in primates. In *Primates in perspective*, edited by C.J. Campbell, A. Fuentes, K.C. MacKinnon, M. Panger, and S. K. Bearder. New York: Oxford University Press, 665–677.

Panger, M.A., A.S. Brooks, B.G. Richmond, and B. Wood. 2002. Older than the Oldowan? Rethinking the emergence of hominin tool use. *Evolutionary Anthropology* 11 (6):235–245.

Paquette, D. 1992. Discovering and learning tool-use for fishing honey by captive chimpanzees. *Human Evolution* 7 (3):17–30.

———. 1994. Can chimpanzees use tools by observational learning? In *The ethological roots of culture*, edited by R.A. Gardner, A.B. Chiarelli, B.T. Gardner, and F.X. Plooij. Dordrecht, Netherlands: Kluwer Academic, 155–172.

Parish, A.R. 1994. Sex and food control in the "uncommon chimpanzee": How bonobo females overcome a phylogenetic legacy of male dominance. *Ethology and Sociobiology* 15 (3):157–179.

Parker, C.E. 1968. The use of tools by apes. *Zoonooz* 41 (7):10–13.

———. 1969a. Bob orangutan—from a Sarawak jungle to a college education. *Zoonooz* 42 (3):12–18.

———. 1969b. Responsiveness, manipulation, and implementation behavior in chimpanzees, gorillas, and orang-utans. In *Behavior*, edited by C.R. Carpenter. Vol. 1 of *Proceedings of the Second International Congress of Primatology*. Basel, Switzerland: S. Karger, 160–166.

———. 1973. Manipulatory behavior and responsiveness. *Gibbon and Siamang* 2:185–207.

———. 1974. Behavioral diversity in ten species of nonhuman primates. *Journal of Comparative and Physiological Psychology* 87:930–937.

Parker, S.T., and K.R. Gibson. 1977. Object manipulation, tool use, and sensorimotor intelligence as feeding adaptations in cebus monkeys and great apes. *Journal of Human Evolution* 6:623–641.

Parker, S.T., M. Kerr, H. Markowitz, and J. Gould. 1999. A survey of tool use in zoo gorillas. In *The mentalities of gorillas and orangutans: Comparative perspectives*, edited by S.T. Parker, R.W. Mitchell, and H.L. Miles. Cambridge: Cambridge University Press, 188–193.

Parks, K., and M. Novak. 1993. Observations of increased activity and tool use in captive rhesus monkeys exposed to troughs of water. *American Journal of Primatology* 29 (1):13–25.

Parnell, R., and H. Buchanan-Smith. 2001. An unusual social display by gorillas. *Nature* 412 (6844):294.

Parra, G.J. 2007. Observations of an Indo-Pacific humpback dolphin carrying a sponge: Object play or tool use? *Mammalia* 71 (3):147–149.

Patterson, F. 1985. Nut logs: Safe, natural diversion stimulates tool use in captive great apes. *Gorilla: Journal of the Gorilla Foundation* 8 (2):8–9.

Patterson, F., and E. Linden. 1981. *The education of Koko*. New York: Holt, Rinehart and Winston.

Pavlov, I.P. 1949. [*Pavlovian Wednesdays*], vols. 1–3. Moscow: Izdatel'stvo Akademii Nauk SSSR.

———. 1955. *Selected works*. Edited by J. Gibbons under the supervision of Kh. S. Koshtoyants, translated by S. Belsky. Moscow: Foreign Languages Publishing House.

———. 1957. *Experimental psychology and other essays*. New York: Philosophical Library.

Payne, S., and R. Jameson. 1984. Early behavioral development of the sea otter, *Enhydra lutris*. *Journal of Mammalogy* 65 (3):527–531.

Peal, S. 1879. Intellect in brutes. *Nature* 21:34.

Pechstein, L.A., and D.F.D. Brown. 1939. An experimental analysis of the alleged criteria of insight learning. *Journal of Educational Psychology* 30:38–52.

Peckham, G., and E. Peckham. 1898. *On the instincts and habits of the solitary wasps*. Bulletin, Wisconsin Geological Natural History Survey 2. Madison: Wisconsin Geological Natural History Survey.

Pepperberg, I.M., and H.R. Shive. 2001. Simultaneous development of vocal and physical object combinations by a grey parrot (*Psittacus erithacus*): Bottle caps, lids, and labels. *Journal of Comparative Psychology* 115 (4):376–384.

Pepper-Edwards, D.L., and E. Notley. 1991. Observations of a captive black-breasted buzzard *Hamirostra melanosternon* using stones to break open eggs. *Australian Bird Watcher* 14:103–106.

Pergande, T. 1912. *The life history of the alder blight aphis*. United States Department of Agriculture, Bureau of Entomology, Technical Series No. 24. Washington, DC: U.S. Government Printing Office.

Perret, K., S. Buechner, and H.J. Adler. 1998. Beschäftigungsprogramme für Schimpansen (*Pan troglodytes*) im Zoo. *Zoologische Garten* 68 (2):95–111.

Perry, R. 1966. *The world of the polar bear*. Seattle: University of Washington Press.

———. 1972. *At the turn of the tide: A book of wild birds*. New York: Taplinger.

Pescetta, M., C. Spiezio, D. Grassi, and E.P. Previde. 2008. Spread of new tool-use behaviour in captive chimpanzees: Observational learning. *Folia Primatologica* 79 (3):145–146.

Peters, H.H. 2001. Tool use to modify calls by wild orangutans. *Folia Primatologica* 72 (4):242–244.

Petit, C., M. Hossaert-McKey, P. Perret, J. Blondel, and M.M. Lambrechts. 2002. Blue tits use selected plants and olfaction to maintain an aromatic environment for nestlings. *Ecological Letters* 5:585–589.

Petit, D.R., L.J. Petit, and K.E. Petit. 1989. Winter catching behavior of deciduous woodland birds and adaptations for protection of store food. *Condor* 91:766–776.

Petit, O., and B. Thierry. 1993. Use of stones in a captive group of Guinea baboons (*Papio papio*). *Folia Primatologica* 61 (3):160–164.

Pettet, A. 1975. Matters arising: Defensive stoning by baboons. *Nature* 258:549–550.

Pettit, D. 1997. Creative use of objects as tools and playthings. *Gorilla: Journal of the Gorilla Foundation* 20 (1):3–4.

Pfeiffer, A.J., and L.J. Koebner. 1978. The resocialization of single-caged chimpanzees and the establishment of an island colony. *Journal of Medical Primatology* (7):70–81.

Phillips, K.A. 1998. Tool use in wild capuchin monkeys (*Cebus albifrons trinitatis*). *American Journal of Primatology* 46 (3):259–261.

Phillips, T. 1950. Man's relation to the apes. *Man* 272:168.

Pickford, M. 1975. Matters arising: Defensive stoning by baboons. *Nature* 258:549.

Pierce, J.D., Jr. 1986. A review of tool use in insects. *Florida Entomologist* 69 (1):95–104.

Pietsch, R. 1989. Lion-tailed macaque group at Lawry Park Zoo. *Lion-Tailed Macaque Newsletter* 6 (1):3.

Pika, S., and K. Zuberbühler. 2008. Social games between bonobos and humans: Evidence for shared intentionality? *American Journal of Primatology* 70 (3):207–210.

Pilleri, G. 1983. Ingenious tool use by the Canadian beaver (*Castor canadensis*) in captivity. *Investigations of Beavers* 1:99–102.

Plooij, F.X. 1978. Tool-use during chimpanzees' bushpig hunt. *Carnivore* 1 (May):103–106.

Pluta, G., and B.B. Beck. 1979. Object play by captive polar bears. Paper presented at the annual meeting of the Animal Behavior Society, June, New Orleans, Louisiana.

Poirier, F.E., and D.M. Davidson. 1979. A preliminary study of the Taiwan macaque (*Macaca cyclopis*). *Quarterly Journal of the Taiwan Museum* 32:123–191.

Pollack, D. 1998. Spontaneous tool use in a vervet monkey (*Cercopithecus aethiops sabaeus*). *American Journal of Primatology* 45 (2):201.

Portmann, A. 1959. *Animal camouflage*. Ann Arbor: University of Michigan Press.

Porter, S. 1936. The kea. *Avicultural Magazine* 5 (1):186–189.

Post, W., and M. Browe. 1982. Active anting by the yellow-shouldered blackbird. *Wilson Bulletin* 94 (1):89–90.

Potì, P., F. Natale, and S.T. Parker. 1990. The development of tool use in *Cebus apella* and a comparison with other primate species. *International Journal of Anthropology* 5 (2):148.

———. 1991. The development of tool use in *Cebus apella* and a comparison with other primate species. In *Prospettive nello studio del primati*, vol. 2, edited by A. Baca. Trieste, Italy: Università degli Studi di Trieste, 103–108.

Potter, E.F. 1970. Anting in wild birds, its frequency and probable purpose. *Auk* 87:692–713.

Poulsen, H. 1974–1975. Keeping chimpanzees occupied in captivity. *International Zoo News* (21):19–20.

Pouydebat, E., C. Berge, P. Gorce, and Y. Coppens. 2005. Use and manufacture of tools to extract food by captive *Gorilla gorilla gorilla*: Experimental approach. *Folia Primatologica* 76 (3):180–183.

Povinelli, D.J., J.E. Reaux, L.A. Theall, and S. Giambrone. 2000. *Folk physics for apes: The chimpanzee's theory of how the world works*. New York: Oxford University Press.

Powell, R.W., and W. Kelly. 1975. Method for objective study of tool-using behavior. *Journal of the Experimental Analysis of Behavior* 24 (2):249–253.

Pranty, B. 1995. Tool use by brown-headed nuthatches in two Florida slash pine forests. *Florida Field Naturalist* 23 (2):33–34.

Pravosudov, V.V., and T.C. Grubb, Jr. 1993. *White-breasted nuthatch (*Sitta carolinensis*)*. Vol. 2, no. 54 of *The Birds of North America*, edited by A. Poole, P. Stettenheim, and F. Gill. Philadelphia: Birds of North America.

Premack, D. 1976. Language and intelligence in ape and man. *American Scientist* 64:674–683.

Preston, B. 1998. Cognition and tool use. *Mind and Language* 13 (4):513–547.

Preston, C.R., H. Moseley, and C. Moseley. 1986. Green-backed heron baits fish with insects. *Wilson Bulletin* 98 (4):613–614.

Priddey, M.W. 1977. Blackbird using tool. *British Birds* 70:262–263.

Prince, E., S. Lambeth, S. Schapiro, and A. Whiten. 2009. A potent effect of observational learning on chimpanzee tool construction. *Proceedings of the Royal Society B-Biological Sciences* 276 (1671):3377–3383.

Prince-Hughes, D. 2001. *Gorillas among us: A primate ethnographer's book of days*. Tucson: University of Arizona Press.

Prozesky-Schulze, L., O.P.M. Prozesky, F. Anderson, and G.J.J. Van Der Merwe. 1975. Use of a self-made sound baffle by a tree cricket. *Nature* 255:142–143.

Pruetz, J.D. 2006. Feeding ecology of savanna chimpanzees (*Pan troglodytes verus*) at Fongoli, Senegal. In *Feeding ecology of apes and other primates: Ecological, physical, and behavioral aspects*. New York: Cambridge University Press, 161–182.

Pruetz, J.D., and P. Bertolani. 2007. Savanna chimpanzees, *Pan troglodytes verus*, hunt with tools. *Current Biology* 17 (5):412–417.

Pruitt, W. 1954. Rutting behavior of the whitetail deer (*Odocoileus virginianus*). *Journal of Mammalogy* 35: 129–130.

Prytherch, R. 1980. Squacco heron possibly using insects as bait. *British Birds* 73 (4):183–184.

Quiatt, D. 2001. Leaf-sponging by chimpanzees of the Sonso community. *Primate Eye* 75:22–23.

———. 2006. Instrumental leaf use by chimpanzees of the Budongo Forest (Sonso community). In *Primates of western Uganda*, edited by J.J. Vea, J. Serrallonga, D. Turbon, J.M. Fullola, and D. Serrat. New York: Springer, 313–325.

Quick, R. 1976. Gorilla habitat display: A follow-up report on the indoor moated exhibit for gorillas at Houston. *International Zoo News* 23 (6):13–16.

Quinn, J.P. 2004. Fur rubbing behavior in a captive group of *Cebus apella apella* monkeys. Master's thesis, California State University–Fullerton.

Rahm, U. 1971. L'emploi d'outils par les chimpanzés de l'ouest de la Cote-d'Ivoire. *Terre et la Vie* 25:506–509.

Rajan, S.A., and P. Balasubramanian. 1989. Tool-using behaviour in Indian house crow (*Corvus splendens*). *Journal of the Bombay Natural History Society* 86 (3):450.

Rajchard, J. 2008. Exogenous chemical substances in bird perception: A review. *Veterinární Medicina* 53 (8): 412–419.

Ransom, T.W., and B.S. Ransom. 1971. Adult male-infant relations among baboons (*Papio anubis*). *Folia Primatologica* 16 (3–4):179–195.

Rasa, O. 1973. Prey capture, feeding techniques, and their ontogeny in the African dwarf mongoose, *Helogale undulata rufula*. *Zeitschrift für Tierpsychologie* 32:449–488.

Rau, P. 1937. A note on the nesting habits of the roach-hunting wasp, *Podium (Parapodium) carolina* Rohwer (Hym.). *Entomology News* 48:91–94.

Rau, P., and N. Rau. 1918. *Wasp studies afield*. Princeton, NJ: Princeton University Press.

Razran, G. 1961. Raphael's "idealess" behavior. *Journal of Comparative and Physiological Psychology* 54 (4):366–367.

Redshaw, M. 1975. Cognitive, manipulative, and social skills in gorillas: II. The second year. *Annual Report of the Jersey Wildlife Preservation Trust* 12:56–60.

———. 1978. Cognitive development in human and gorilla infants. *Journal of Human Evolution* 7:133–141.

Reed, C.A. 1977. Prologue to *The origins of agriculture*, edited by C. Reed. Chicago: Aldine, 7–19.

———. 1985. Energy-traps and tools. In *Hominid evolution: Past, present, and future*, edited by P.V. Tobias. New York: Alan R. Liss, 89–97.

Reese, E.S. 1962. Shell selection behaviour of hermit crabs. *Animal Behaviour* 10 (3–4):347–360.

Reid, J.B. 1982. Tool use by a rook (*Corvus frugilegus*) and its causation. *Animal Behaviour* 30 (4):1212–1216.

———. 1985. Tool-use by elephants. *Journal of the Bombay Natural History Society* 82 (2):402–403.

Reinhardt, V. 1991. Uncommon tool usage by captive primates. *International Zoo News* 38 (5):13–14.

Reiser, O. 1936. Erlebnisse mit dem Bartgeier (*Gypaetus barbatus*) in Bosnien und der Herzegowina. *Journal für Ornithologie* 84 (1):159–172.

Rengger, J. 1830. *Naturgeschichte der Säugetiere von Paraguay*. Basel, Switzerland: Schweighauser.

Rensch, B. 1973. Play and art in apes and monkeys. In *Precultural primate behavior*, edited by E. Menzel. Vol. 1 of *Symposia of the Fourth International Congress of Primatology*. Basel, Switzerland: S. Karger, 102–123.

Rensch, B., and R. Altevogt. 1954. Zähmung und Dressurleistungen indischer Arbeitselefanten. *Zeitschrift für Tierpsychologie* 11:497–510.

Rensch, B., and J. Döhl. 1967. Spontanes Öffnen verschiedener Kistenverschlüsse durch einen Schimpansen. *Zeitschrift für Tierpsychologie* 24:476–489.

Rensch, B., and G. Dücker. 1966. Manipulierfähigkeit eines jungen Orang-utans und eines jungen Gorillas. *Zeitschrift für Tierpsychologie* 23:874–887.

Rensenbrink, H.P. 1960. [Sarina and the chair]. *Artis* 5: 206–210.

Reynolds, V. 2005. *The chimpanzees of the Budongo Forest: Ecology, behaviour, and conservation*. New York: Oxford University Press.

Reynolds, V., and F. Reynolds. 1965. Chimpanzees of the Budongo Forest. In *Primate behavior: Field studies of monkeys and apes*, edited by I. Devore. New York: Holt, Rinehart and Winston, 368–424.

Richard, A. 1970. A comparative study of the activity patterns and behavior of *Alouatta villosa* and *Ateles geoffroyi*. *Folia Primatologica* 12:241–263.

Richard, P. 1964. Les matériaux de construction du castor (*Castor fiber*), leur signification pour ce rongeur. *Zeitschrift für Tierpsychologie* 21:592–601.

———. 1967. Le déterminisme de la construction des barrages chez le castor du Rhône. *Terre et la Vie* 21:339–470.

Richard-Hansen, C., N. Bello, and J.C. Vie. 1998. Tool use by a red howler monkey (*Alouatta seniculus*) towards a two-toed sloth (*Choloepus didactylus*). *Primates* 39 (4):545–548.

Richards, B. 1971. Strike-tit using twig. *Australian Bird Watcher* 4:97–98.

Riedman, M.L., and J.A. Estes. 1990. The sea otter (*Enhydra lutris*): Behavior, ecology, and natural history. *U.S. Fish and Wildlife Service Biological Report* 90 (14):126.

Riehl, C. 2001. Black-crowned night heron fishes with bait. *Waterbirds* 24 (2):285–286.

Rijksen, H.D. 1974. Orang-utan conservation and rehabilitation in Sumatra. *Biological Conservation* 6:20–25.

———. 1978. *A field study on Sumatran orang-utans (*Pongo pygmaeus abelii *Lesson, 1827): Ecology, behaviour, and conservation*. Wageningen, Netherlands: H. Veenman and Zonen.

Riss, D., and J. Goodall. 1977. Recent rise to alpha-rank in a population of free-living chimpanzees. *Folia Primatologica* 27 (2):134–151.

Ritchie, B.G., and D.M. Fragaszy. 1988. Capuchin monkey (*Cebus apella*) grooms her infant's wound with tools. *American Journal of Primatology* 16 (4):345–348.

Roberts, G.J. 1982. Apparent baiting behavior by a black kite. *Emu* 82:53–54.

Robinson, M.H., and B. Robinson. 1971. Predatory behavior of the ogre-faced spider *Dinopis longipes* F. Cambridge (Araneae: Dinopidae). *American Midland Naturalist* 85 (1):85–96.

Robinson, S.K. 1994. Use of bait and lures by green-backed herons in Amazonian Peru. *Wilson Bulletin* 106 (3):567–569.

Rocha, V.J., N.R. dos Reis, and M.L. Sekiama. 1998. [Tool use in *Cebus apella* Linnaeus (Primate: Cebidae) to get *Coleoptera* larvae that parasite seeds of *Syagrus romanzoffianum* (Cham.) Glassm. (Arecaceae)]. *Revista Brasileira de Zoologia* 15 (4):945–950.

Roche, J.P. 1996. The use of a rock by an osprey in an agonistic encounter. *Journal of Raptor Research* 30 (1): 42–43.

Rodrigues, M.R., and S.L. Lindshield. 2007. Scratching the surface: Observations of tool use in wild spider monkeys. *American Journal of Physical Anthropology* 132 (S34):201–202.

Rogers, L.J., and G. Kaplan. 1993. Köhler and tool use in orang-utans. *International Journal of Comparative Psychology* 6 (4):234–241.

———. 1994. A new form of tool use by orang-utans in Sabah, East Malaysia. *Folia Primatologica* 63 (1):50–52.

Rolando, A., and M. Zunino. 1992. Observations of tool use in corvids. *Ornis Scandinavica* 23 (2):201–202.

Romanes, G. 1892. *Animal intelligence*. New York: Appleton.

Ross, D. 1971. Protection of hermit crabs (*Dardanus* spp.) from octopus by commensal sea anemones (*Calliactis* spp.). *Nature* 230:401–402.

———. 1983. Symbiotic relations. In *The biology of Crustacea*, edited by F.J. Vernberg and W.B. Vernberg. New York: Academic Press, 163–200.

Ross, H. 1956. *Evolution and classification of the mountain caddisflies*. Urbana: University of Illinois Press.

———. 1964. Evolution of the caddisworm cases and nets. *American Zoologist* 4:209–220.

Rowe, M., and D. Owings. 1978. The meaning of the sound of rattling by rattlesnakes to California ground squirrels. *Behaviour* 66:252–267.

Ruiz, J.C. 2005. Relative humidity, ambient temperature, and urine washing behavior in Bolivian squirrel monkeys, *Saimiri boliviensis boliviensis*. *Primate Report* 71:57–61.

Rumbaugh, D.M. 1970. Learning skills of anthropoids. In *Primate behavior: Developments in field and laboratory research*, vol. 1, edited by L.A. Rosenblum. New York: Academic Press, 1–70.

———. 1971. Evidence of qualitative differences in learning processes among primates. *Journal of Comparative and Physiological Psychology* 76 (2):250–255.

———. 2006. Survey of the primates: A gorilla's first year, tool use by a gibbon. *American Journal of Primatology* 68 (S1):57.

Rumbaugh, D.M., and D. Washburn. 2003. *Intelligence of apes and other rational beings*. New Haven, CT: Yale University Press.

Russon, A.E. 1998. The nature and evolution of intelligence in orangutans (*Pongo pygmaeus*). *Primates* 39 (4): 485–503.

———. 1999a. Naturalistic approaches to orangutan intelligence and the question of enculturation. *International Journal of Comparative Psychology* 12 (4):181–202.

———. 1999b. Orangutans' imitation of tool use: A cognitive interpretation. In *The mentalities of gorillas and orangutans: Comparative perspectives*, edited by S.T. Parker, R.W. Mitchell, and H.L. Miles. Cambridge: Cambridge University Press, 117–146.

———. 2000. *Orangutans: Wizards of the rain forest*. Toronto: Firefly Books.

———. 2002. Comparative developmental perspectives on culture: The great apes. In *Between culture and biology: Perspectives on ontogenetic development*, edited by H. Keller, Y.H. Poortinga, and A. Schölmerich. New York: Cambridge University Press, 30–56.

———. 2003. Innovation and creativity in forest-living rehabilitant orang-utans. In *Animal innovation*, edited by S.M. Reader and K.N. Laland. Oxford: Oxford University Press, 279–306.

———. 2006. Acquisition of complex foraging skills in juvenile and adolescent orangutans (*Pongo pygmaeus*): Developmental influences. *Aquatic Mammals* 32 (4):500–510.

———. 2009. Orangutan rehabilitation and reintroduction: Successes, failures, and role in conservation. In *Orangutans: Geographic variation in behavioral ecology and conservation*, edited by S.A. Wich, S.S. Utami Atmoko, T.M. Setia, and C.P. van Schaik. New York: Oxford University Press, 327–350.

Russon, A.E., and B.M.F. Galdikas. 1993. Imitation in free-ranging rehabilitant orangutans (*Pongo pygmaeus*). *Journal of Comparative Psychology* 107 (2):147–161.

———. 1995. Imitation and tool use in rehabilitant orangutans. In *The neglected ape*, edited by R. Nadler, B. Galdikas, L. Sheeran, and N. Rosen. New York: Plenum Press, 191–197.

Russon, A.E., P. Vasey, and C. Gauthier. 2002. Seeing with the mind's eye: Eye-covering play in orangutans and Japanese macaques. In *Pretending and imagination in animals and children*, edited by R.W. Mitchell. Cambridge: Cambridge University Press, 241–254.

Russon, A.E., D.P. Handayani, P. Kuncoro, and A. Ferisa. 2007. Orangutan leaf-carrying for nest-building: Toward unraveling cultural processes. *Animal Cognition* 10:189–202.

Russon, A.E., C.P. van Schaik, P. Kuncoro, A. Ferisa, D.P. Handayani, and M.A. van Noordwijk. 2009. Innovation and intelligence in orangutans. In *Orangutans: Geographic variation in behavioral ecology and conservation*, edited by S.A. Wich, S.S. Utami Atmoko, T.M. Setia, and C.P. van Schaik. New York: Oxford University Press, 279–298.

Russon, A.E., P. Kuncoro, A. Ferisa, and D.P. Handayani. 2010. How orangutans (*Pongo pygmaeus*) innovate for water. *Journal of Comparative Psychology* 124 (1):14–28.

Rutledge, R., and G.R. Hunt. 2004. Lateralized tool use in wild New Caledonian crows. *Animal Behaviour* 67 (2):327–332.

Rutz, C., L.A. Bluff, A.A.S. Weir, and A. Kacelnik. 2007. Video cameras on wild birds. *Science* 318 (5851):765.

Rye, E.C. 1866. *British beetles: An introduction to the study of our indigenous Coleoptera*. London: Lovell Reeve.

Sabater-Pi, J. 1972. Bastones fabricados y usados por los chimpancés de las montañas de Okorobikó. *Ethnica* 4:189–199.

———. 1974. An elementary industry of the chimpanzees in the Okorobikó Mountains, Rio Muni (Republic of Equatorial Guinea), West Africa. *Primates* 15: 351–364.

Sakura, O., and T. Matsuzawa. 1991. Flexibility of wild chimpanzee nut-cracking behavior using stone hammers and anvils: An experimental analysis. *Ethology* 87 (3–4):237–248.

Salmons, S., and G. Gough. 2007. *Tool use by chestnut-backed chickadee*. Migratory Birds Center Spotlight on Birds Archive, April 2007. http://nationalzoo.si.edu/SCBI/MigratoryBirds/Science_Article/archives.cfm [accessed 1 Nov. 2009].

Santos, L.R., N. Mahajan, and J.L. Barnes. 2005. How prosimian primates represent tools: Experiments with two lemur species (*Eulemur fulvus* and *Lemur catta*). *Journal of Comparative Psychology* 119 (4):394–403.

Santos, L.R., A. Rosati, C. Sproul, B. Spaulding, and M.D. Hauser. 2005. Means-means-end tool choice in cotton-top tamarins (*Saguinus oedipus*): Finding the limits on primates' knowledge of tools. *Animal Cognition* 8 (4):236–246.

Santos, L.R., H.M. Pearson, G.M. Spaepen, F. Tsao, and M.D. Hauser. 2006. Probing the limits of tool competence: Experiments with two non-tool-using species (*Cercopithecus aethiops* and *Saguinus oedipus*). *Animal Cognition* 9 (2):94–109.

Sanyal, R.K.B. 1902. Notes on animals kept in the Alipore Zoological Garden. *Journal of the Asiatic Society of Bengal* 71:92–93.

Sanz, C.M. 2004. Behavioral ecology of chimpanzees in a Central African forest: *Pan troglodytes troglodytes* in the Goualougo Triangle, Republic of Congo. Ph.D. diss., Washington University.

Sanz, C.M., J. Call, and D.B. Morgan. 2009. Design complexity in termite-fishing tools of chimpanzees (*Pan troglodytes*). *Biology Letters* 5 (3):293–296.

Sanz, C.M., and D.B. Morgan. 2007. Chimpanzee tool technology in the Goualougo Triangle, Republic of Congo. *Journal of Human Evolution* 52 (4):420–433.

———. 2009. Flexible and persistent tool-using strategies in honey-gathering by wild chimpanzees. *International Journal of Primatology* 30 (3):411–427.

Sanz, C.M., D.B. Morgan, and S. Gulick. 2004. New insights into chimpanzees, tools, and termites from the Congo basin. *American Naturalist* 164 (5):567–581.

Sanz, C.M., C. Schöning, and D.B. Morgan. 2009. Chimpanzees prey on army ants with specialized tool set. *American Journal of Primatology* 71:1–8.

Savage, C. 1995. *Bird brains*. San Francisco: Sierra Club Books.

Savage, S. 1976. Pygmy chimpanzees: Update. *Yerkes News* 13:25–26.

Savage, T., and J. Wyman. 1843–1844. Observations on the external characters and habits of the *Troglodytes niger* Geoff. and on its organization. *Boston Journal of Natural History* 4:362–386.

Savage-Rumbaugh, E.S. 1986. *Ape language: From conditioned response to symbol*. New York: Columbia University Press.

———. 1993. *Bonobo people*. NHK [Japanese Broadcasting Corporation], documentary film.

Savage-Rumbaugh, E.S., and R. Lewin. 1994. *Kanzi: The ape at the brink of the human mind*. New York: John Wiley & Sons.

Savage-Rumbaugh, E.S., D.M. Rumbaugh, and S. Boysen. 1978. Linguistically mediated tool use and exchange by chimpanzees (*Pan troglodytes*). *Behavioral and Brain Sciences* 1:539–554.

Savage-Rumbaugh, E.S., J. Murphy, R.A. Sevcik, K.E. Brakke, S.L. Williams, and D.M. Rumbaugh. 1993. *Language comprehension in ape and child*. With commentary by E. Bates. Monographs of the Society for Research in Child Development vol. 58, nos. 3–4, serial no. 233. Chicago: University of Chicago Press.

Savage-Rumbaugh, E.S., S.L. Williams, T. Furuichi, and T. Kano. 1996. Language perceived: *Paniscus* branches out. In *Great ape societies*, edited by W.C. McGrew, L. Marchant, and T. Nishida. Cambridge: Cambridge University Press, 173–184.

Savage-Rumbaugh, E.S., N. Toth, K. Schick, and D.A. Washburn. 2007. Kanzi learns to knap stone tools. In *Primate perspectives on behavior and cognition*, edited by D.A. Washburn. Washington, DC: American Psychological Association, 279–291.

Schaffner, U., and C. Müller. 2001. Exploitation of the faecal shield of lily leaf beetle, *Lilioceris lilii* (Coleoptera: Chrysomelidae), by the specialist parasitoid *Lemophagus pulcher* (Hymenoptera: Ichneumonidae). *Journal of Insect Behavior* 14:739–757.

Schaller, G.B. 1961. The orang-utan in Sarawak. *Zoologica* 46:73–82.

———. 1963. *The mountain gorilla: Ecology and behavior*. Chicago: University of Chicago Press.

———. 1967. *The deer and the tiger: A study of wildlife in India*. Chicago: University of Chicago Press.

———. 1973. *Golden shadows, flying hooves*. New York: Knopf.

Schaller, G.B., and A. Hamer. 1978. Rutting behavior of Père David's deer, *Elaphurus davidianus*. *Zoologische Garten* 48:1–15.

Schastnyi, A.I. 1963. Rate of differentiation of inhibition during the development of complex forms of behavior in the chimpanzee. *Doklady Akademii Nauk SSSR* 150 (6):1404–1407.

———. 1972. *Complex behavior of the apes*. Leningrad: Nauka.

Schick, K., and N. Toth. 1993. *Making silent stone speak: Human evolution and the dawn of technologies*. New York: Simon & Schuster.

Schick, K., N. Toth, G. Garufi, E.S. Savage-Rumbaugh, D. Rumbaugh, and R. Sevcik. 1999. Continuing investigations into the stone tool-making and tool-using capabilities of a bonobo (*Pan paniscus*). *Journal of Archaeological Science* 26 (7):821–832.

Schiller, P.H. 1951. Figural preferences in the drawings of a chimpanzee. *Journal of Comparative and Physiological Psychology* 44:101–111.

———. 1952. Innate constituents of complex responses in primates. *Psychological Review* 59 (3):177–191.

———. 1957. Innate motor action as a basis of learning: Manipulative patterns in the chimpanzee. In *Instinctive behavior*, edited by C.H. Schiller. New York: International Universities Press, 264–287.

Schneider, L., A.L. Serbena, and N.M.R. Guedes. 2002. Manipulaçao de frutos de acuri e bocaiúva por araras-azuis no Pantanal Sul. In *Resumos, XX Encontro Anual de Etologia, Natal, R[io Grande do Norte, Brazil]*, 378.

Schöning, C., D. Ellis, A. Fowler, and V. Sommer. 2007. Army ant prey availability and consumption by chimpanzees (*Pan troglodytes vellerosus*) at Gashaka (Nigeria). *Journal of Zoology* 271 (2):125–133.

Schöning, C., T. Humle, Y. Möbius, and W.C. McGrew. 2008. The nature of culture: Technological variation in chimpanzee predation on army ants revisited. *Journal of Human Evolution* 55:48–59.

Schrauf, C., L. Huber, and E. Visalberghi. 2008. Do capuchin monkeys use weight to select hammer tools? *Animal Cognition* 11 (3):413–422.

Schuck-Paim, C., A. Borsari, and E.B. Ottoni. 2009. Means to an end: Neotropical parrots manage to pull strings to meet their goals. *Animal Cognition* 12 (2):287–301.

Schuetz, J.G. 2005. Common waxbills use carnivore scat to reduce the risk of nest predation. *Behavioral Ecology* 16 (1):133–137.

Schulman, S. 1973. Impaired tool-using behavior in monkeys from bilateral destruction of the dorsomedial nuclei of the thalamus. *Transactions of the American Neurological Association* 98:138–141.

Schultz, A. 1961. Some factors influencing the social life of primates in general and of early man in particular. In *Social life of early man*, edited by S. Washburn. Chicago: Aldine, 58–90.

Schultz, G.W. 1982. Soil-dropping behavior of the pavement ant, *Tetramorium caespitum* (L.) (Hymenoptera: Halictidae). *Journal of the Kansas Entomological Society* 55 (2):277–282.

Schuster, G., W. Smits, and J. Ullal. 2008. *Thinkers of the jungle*. Translated by Marrie Powell and Michael Scuffil. Cologne: H.F. Ullmann/Tandem.

Schuster, S., S. Wohl, M. Griebsch, and I. Klostermeier. 2006. Animal cognition: How archer fish learn to down rapidly moving targets. *Current Biology* 16 (4):378–383.

Seed, A.M., J. Call, N.J. Emery, and N.S. Clayton. 2009. Chimpanzees solve the trap problem when the confound of tool-use is removed. *Journal of Experimental Psychology: Animal Behavior Processes* 35 (1):23–34.

Segerdahl, P., W. Fields, and E.S. Savage-Rumbaugh. 2005. *Kanzi's primal language: The cultural initiation of primates into language*. New York: Palgrave Macmillan.

Seibt, U., and W. Wickler. 1978. Marabou storks wash dung beetles. *Zeitschrift für Tierpsychologie* 46 (3):324–327.

Sept, J.M., and G.E. Brooks. 1994. Reports of chimpanzee natural history, including tool use, in 16th- and 17th-century Sierra Leone. *International Journal of Primatology* 15 (6):867–878.

Shafer, D.D. 1988. Handedness in gorillas. *Gorilla: Journal of the Gorilla Foundation* 11 (2):2–5.

Sharpe, F.A., and L.M. Dill. 1997. The behavior of Pacific herring schools in response to artificial humpback whale bubbles. *Canadian Journal of Zoology* 75:725–730.

Sheak, W. 1922. Disposition and intelligence of the orangutan. *Journal of Mammalogy* 3:47–51.

———. 1924. Some further observations on the chimpanzee. *Journal of Mammalogy* 5:122–129.

Shelford, R.W.C. 1917. *A naturalist in Borneo*. Edited by E.B. Poulton. New York: E.P. Dutton.

Shepherd, W. 1910. *Some mental processes of the rhesus monkeys*. Psychological Monographs vol. 12, no. 5. Lancaster, PA: Psychological Review.

———. 1923. Some observations and experiments of the intelligence of the chimpanzee and orang. *American Journal of Psychology* 34:590–591.

Sherrow, H.M. 2005. Tool use in insect foraging by the chimpanzees of Ngogo, Kibale National Park, Uganda. *American Journal of Primatology* 65 (4):377–383.

Shillito, D.J., R.M. Shumaker, G.G. Gallup, and B.B. Beck. 2005. Understanding visual barriers: evidence for Level 1 perspective taking in an orangutan, *Pongo pygmaeus*. *Animal Behaviour* 69:679–687.

Shimada, M.K. 2000. A survey of the Nimba Mountains, West Africa, from three routes: Confirmed new habitat and ant-catching wand use of chimpanzees. *Pan Africa News* 7 (1):7–10.

Shoshani, J., and J.E. Eisenberg. 2000. Intelligence and survival. In *Elephants: Majestic creatures of the wild*, edited by J. Shoshani. New York: Checkmark Books, 134–137.

Shumaker, R.W., and B.B. Beck. 2003. *Primates in question*. Washington, DC: Smithsonian Books.

Shumaker, R.W., A.M. Palkovich, B.B. Beck, G.A. Guagnano, and H. Morovitz. 2001. Spontaneous use of magnitude discrimination and ordination by the orangutan (*Pongo pygmaeus*). *Journal of Comparative Psychology* 115 (4):385–391.

Shurcliff, A., D. Brown, and F. Stollnitz. 1971. Specificity of training required for solution of a stick problem by rhesus monkeys (*Macaca mulatta*). *Learning and Motivation* 2:255–270.

Shuster, G., and P. Sherman. 1998. Tool use by naked mole-rats. *Animal Cognition* 1:71–74.

Sick, H. 1957. Anting by two tanagers in Brazil. *Wilson Bulletin* 69:187–188.

Siegfried, W. 1977. Mussel-dropping behaviour of kelp gulls. *South African Journal of Science* 73:337–341.

Sikes, S. 1971. *The natural history of the African elephant*. New York: American Elsevier.

Simmons, K. 1957. A review of the anting behaviour of passerine birds. *British Birds* 50:401–424.

———. 1966. Anting and the problem of self-stimulation. *Journal of Zoology* 149:145–162.

Sinclair, J.C. 1984. Baiting behaviour in a captive lesser black-backed gull, *Larus fuscus*. *Cormorant* 12: 105–106.

Sinha, A. 1997. Complex tool manufacture by a wild bonnet macaque, *Macaca radiata*. *Folia Primatologica* 68 (1):23–25.

Sisson, R. 1974. Aha! It really works! *National Geographic* 145:142–147.

Sitompul, A.F., E.A. Fox, and C.P. van Schaik. 1998. Intelligent tool use in wild Sumatran orangutans (*Pongo pygmaeus abelii*). *American Journal of Physical Anthropology* 105 (S26):202–203.

Skaife, S. 1957. Demons of the dust. *African Wildlife* 11: 191–197.

Skead, C.J. 1971. Use of tools by the Egyptian vulture. *Ostrich* 42:226.

Skorepa, A.C., and A.J. Sharp. 1971. Lichens in "packets" of lacewing larvae (Chrysopidae). *Bryologist* 74:363–364.

Skutch, A.F. 1948. Anting by some Costa Rican birds. *Wilson Bulletin* 60:115–116.

———. 1996. *The minds of birds*. College Station: Texas A&M University Press.

Slough, B. 1978. Beaver food cache structure and utilization. *Journal of Wildlife Management* 42:644–646.

Smith, G. 1970. Tool-using by birds. *Avicultural Magazine* 76:171.

———. 1971. Tool-using by birds. *Avicultural Magazine* 77:47–48.

Smith, H. 1936. The archer fish. *Natural History* 38:2–11.

Smith, M., and C. Conway. 2007. Use of mammal manure by nesting burrowing owls: A test of four functional hypotheses. *Animal Behaviour* 73:65–73.

Smith, R.C. 1926. The trash-carrying habit of certain lacewing larvae. *Scientific Monthly* 23 (3):265–267.

Smith, S. 1972. The ontogeny of impaling behaviour in the loggerhead shrike, *Lanius ludovicianus* L. *Behaviour* 42:232–247.

———. 1973. A study of prey-attack behaviour in young loggerhead shrikes, *Lanius ludovicianus* L. *Behaviour* 44:113–141.

Smolker, R., A. Richards, R. Connor, J. Mann, and P. Berggren. 1997. Sponge carrying by dolphins (Delphinidae: *Tursiops* spp.): A foraging specialization involving tool use? *Ethology* 103 (6):454–465.

Someren, G.R.C. 1996. Active anting by weavers and sunbirds at Karen, Kenya. *Ostrich* 67 (3–4):165.

Sousa, C., and T. Matsuzawa. 2001. The use of tokens as rewards and tools by chimpanzees (*Pan troglodytes*). *Animal Cognition* 4:213–221.

———. 2004. Development of leaf-using skills to drink water in Bossou chimpanzees (*Pan troglodytes verus*). *Folia Primatologica* 75 (S1):336–337.

———. 2006. Token use by chimpanzees (*Pan troglodytes*): Choice, metatool, cost. In *Cognitive development in chimpanzees*, edited by T. Matsuzawa, M. Tomonaga, and M. Tanaka. New York: Springer, 411–438.

Southern, W. 1963. Three species observed anting on a wet lawn. *Wilson Bulletin* 75:275–276.

Souza, C.A., F.D.C. Mendes, and N. Jorge da Silva, Jr. 2002. [Tool use by free-ranging *Cebus apella libidinosus*]. In *Livro de resumos do X Congresso Brasileiro de Primatologia*. Belém, Brazil: Congresso Brasileiro de Primatologia, 69.

Spaulding, B., and M. Hauser. 2005. What experience is required for acquiring tool competence? Experiments with two callitrichids. *Animal Behaviour* 70 (3):517–526.

Spence, K. 1937. Experimental studies of learning and the higher mental processes in infra-human primates. *Psychological Bulletin* 34:806–850.

St Amant, R., and T.E. Horton. 2008. Revisiting the definition of animal tool use. *Animal Behaviour* 75 (4): 1199–1208.

St Amant, R., and A. Wood. 2005. Tool use for autonomous agents. Paper presented at the Twentieth National Conference on Artificial Intelligence, July 9–13, Pittsburg, Pennsylvania.

Stachowicz, J.J., and M.E. Hay. 2000. Geographic variation in camouflage specialization by a decorator crab. *American Naturalist* 156 (1):59–71.

Stanford, C.B. 1996. The hunting ecology of wild chimpanzees: Implications for the evolutionary ecology of Pliocene hominids. *American Anthropologist* 98 (1): 96–113.

———. 1998. *Chimpanzee and red colobus: The ecology of predator and prey*. Cambridge, MA: Harvard University Press.

Stanford, C.B., J. Wallis, E. Mpongo, and J. Goodall. 1994. Hunting decisions in wild chimpanzees. *Behaviour* 131 (1–2):1–18.

Stanford, C.B., C. Gambaneza, J.B. Nkurunungi, and M.L. Goldsmith. 2000. Chimpanzees in Bwindi-Impenetrable National Park, Uganda, use different tools to obtain different types of honey. *Primates* 41 (3):337–341.

Stanford, F. 1995. Sponge/shell switching by hermit crabs, *Pagurus impressus*. *Invertebrate Biology* 114 (1):73–78.

Stanford, F.G. 1949. "Anting" of green woodpecker. *British Birds* 42:390.

Starin, E.D. 1990. Object manipulation by wild red colobus monkeys living in the Abuko National Reserve, Gambia. *Primates* 31 (3):385–391.

Steinbacher, G. 1965. Bemerkenswerter werkzeuggebrauch einer erkrankten indischen elefanten. *Zoologische Garten* 31:271–272.

Sterling, E.J., and D.J. Povinelli. 1999. Tool use, aye-ayes, and sensorimotor intelligence. *Folia Primatologica* 70 (1):8–16.

Stevens, L. 1996. Lipids and their metabolism. In *Avian biochemistry and molecular biology*, edited by L. Stevens. Cambridge: Cambridge University Press, 46–64.

Stirling, I. 1974. Midsummer observations on the behavior of wild polar bears. *Canadian Journal of Zoology* 52:1191–1198.

Stirling, I., and P. Latour. 1978. Comparative hunting abilities of polar bear cubs of different ages. *Canadian Journal of Zoology* 56:1768–1772.

Stoinski, T.S., and B.B. Beck. 2001. Spontaneous tool use in captive, free-ranging golden lion tamarins (*Leontopithecus rosalia rosalia*). *Primates* 42 (4):319–326.

Stone, R.C. 1954. "Anting" by wryneck. *British Birds* 47:312.

Stopka, P., and D.W. Macdonald. 2003. Way-marking behaviour: An aid to spatial navigation in the wood mouse (*Apodemus sylvaticus*). *BMC Ecology* 3:3.

Strong, R. 1914. On the habits and behavior of the herring gull, *Larus argentatus* Pont. *Auk* 31:22–49, 178–199.

Struhsaker, T. 1975. *The red colobus monkey*. Chicago: University of Chicago Press.

Struhsaker, T., and P. Hunkeler. 1971. Evidence of tool-using by chimpanzees in the Ivory Coast. *Folia Primatologica* 15:212–219.

Struhsaker, T., and L. Leland. 1977. Palm-nut smashing by *Cebus a. apella* in Columbia. *Biotropica* 9:124–126.

Sugiyama, Y. 1969. Social behavior of chimpanzees in the Budongo Forest, Uganda. *Primates* 10:197–225.

———. 1981. Observations on the population dynamics and behavior of wild chimpanzees at Bossou, Guinea, in 1979–1980. *Primates* 22 (4):435–444.

———. 1985. The brush-stick of chimpanzees found in south-west Cameroon and their cultural characteristics. *Primates* 26 (4):361–374.

———. 1989a. Description of some characteristic behaviors and discussion on their propagation process among chimpanzees of Bossou, Guinea. In *Behavioral studies of wild chimpanzees at Bossou, Guinea*, edited by Y. Sugiyama. Inuyama, Japan: Kyoto University Primate Research Institute, 43–76.

———. 1989b. Local variation of tool and tool behavior among wild chimpanzee populations. In *Behavioral studies of wild chimpanzees at Bossou, Guinea*, edited by Y. Sugiyama. Inuyama, Japan: Kyoto University Primate Research Institute, 1–15.

———. 1993. Local variation of tools and tool use among wild chimpanzee populations. In *The use of tools by human and non-human primates*, edited by A. Berthelet and J. Chavaillon. Oxford: Clarendon Press, 175–187.

———. 1994. Tool use by wild chimpanzees. *Nature* 367 (6461):327.

———. 1995a. Drinking tools of wild chimpanzees at Bossou. *American Journal of Primatology* 37 (3):263–269.

———. 1995b. Tool-use for catching ants by chimpanzees at Bossou and Monts Nimba, West Africa. *Primates* 36 (2):193–205.

———. 1997. Social tradition and the use of tool-composites by wild chimpanzees. *Evolutionary Anthropology* 6 (1):23–27.

———. 1998. Local variation of tool-using repertoire in wild chimpanzees. In *Comparative study of the behavior of the genus* Pan *by compiling video ethogram.*, edited by T. Nishida. Kyoto: Nissindo, 82–91.

Sugiyama, Y., and M. Kawai. 1990. [Nut-cracking culture of chimpanzees]. In *Jinrui izen no shakaigaku: Afurkia ni reichorui o saguru*, edited by M. Kawai. Higashimurayama, Tokyo: Kyoikusha, 435–453.

Sugiyama, Y., and J. Koman. 1979. Tool-using and -making behavior in wild chimpanzees at Bossou, Guinea. *Primates* 20 (4):513–524.

———. 1987. A preliminary list of chimpanzees' alimentation at Bossou, Guinea. *Primates* 28 (1):133–147.

Sugiyama, Y., J. Koman, and M.B. Sow. 1988. Ant-catching wands of wild chimpanzees at Bossou, Guinea. *Folia Primatologica* 51 (1):56–60.

Sugiyama, Y., T. Fushimi, O. Sakura, and T. Matsuzawa.

1993. Hand preference and tool use in wild chimpanzees. *Primates* 34 (2):151–159.
Sumita, K., J. Kitahara-Frisch, and K. Norikoshi. 1985. The acquisition of stone-tool use in captive chimpanzees. *Primates* 26 (2):168–181.
Sutton, M. 1982. Tool-using by a capuchin monkey. *Ratel* 9 (2):11–12.
Suzuki, A. 1966. On the insect-eating habits among wild chimpanzees living in the savanna woodland of western Tanzania. *Primates* 7:481–487.
———. 1969. An ecological study of chimpanzees in savanna woodland. *Primates* 10:103–148.
Suzuki, S., S. Kuroda, and T. Nishihara. 1995. Tool-set for termite-fishing by chimpanzees in the Ndoki forest, Congo. *Behaviour* 132 (3–4):219–235.
Swartz, K.B., S.A. Himmanen, and R.W. Shumaker. 2007. Response strategies in list learning by orangutans (*Pongo pygmaeus* × *P. abelii*). *Journal of Comparative Psychology* 121 (3):260–269.
Takahata, Y. 1982. *Termite-fishing observed in the M group chimpanzees*. Mahale Mountains Chimpanzee Research Project, Ecological Report 18.
Takasaki, H. 1992. Tool-using scenes of Mahale M group chimpanzees. *Bulletin of the Chicago Academy of Sciences* 15 (1):9–10.
Takemoto, H. 2004. [The use of tool-composites by wild chimpanzees for obtaining termites in Campo, Cameroon]. *Reichorui Kenkyu* 20 (S):28.
Takemoto, H., S. Hirata, and Y. Sugiyama. 2005. The formation of the brush-sticks: Modification of chimpanzees or the by-product of folding? *Primates* 46 (3): 183–189.
Takenoshita, Y. 1996. Chimpanzee research in the Ndoki Forest. *Pan Africa News* 3 (2):7–8.
———. 2002. [Tool use for honey extraction by the chimpanzees of the Gambia complex, Gabon]. *Reichorui Kenkyu* 18 (3):372.
Takenoshita, Y., J. Yamagiwa, and T. Nishida. 1998. Branch-drop display of a female chimpanzee (*Pan troglodytes troglodytes*) of Petit Loango, Gabon. *Pan Africa News* 5 (2):15–16.
Takeshita, H., and J.A.R.A.M. van Hooff. 1996. Tool use by chimpanzees (*Pan troglodytes*) of the Arnhem Zoo community. *Japanese Psychological Research* 38 (3): 163–173.
———. 2001. Tool use by chimpanzees (*Pan troglodytes*) of the Arnhem Zoo community. In *Primate origins of human cognition and behavior*, edited by T. Matsuzawa. Tokyo: Springer, 519–536.
Tanaka, I. 1995. [Tool-using and social transmission of behavior in monkeys]. *Reichorui Kenkyu* 11 (3):225–230.
Tanaka, I., E. Tokida, H. Takefushi, and T. Hagiwara. 2001. Tube test in free-ranging Japanese macaques: Use of sticks and stones to obtain fruit from a transparent pipe. In *Primate origins of human cognition and behavior*, edited by T. Matsuzawa. Tokyo: Springer, 509–518.
Tanaka, T., and Y. Ono. 1975. Tool-using in feeding-behavior of ant-lions. *Annual of Animal Psychology* 25 (2):103–118.
———. 1978. The tool use by foragers of *Aphaenogaster famelica*. *Japanese Journal of Ecology* 28 (1):49–58.
Tauber, C.A., and T. de León. 2001. Systematics of green lacewings (Neuroptera: Chrysopidae): Larvae of *Ceraeochrysa* from Mexico. *Annals of the Entomological Society of America* 94:197–209.
Tauber, C.A., T. de León, N.D. Penny, and M.J. Tauber. 2000. The genus *Ceraeochrysa* (Neuroptera: Chrysopidae) of America north of Mexico: Larvae, adults, and comparative biology. *Annals of the Entomological Society of America* 93:1195–1221.
Tayler, C., and G. Saayman. 1973. Imitative behavior by Indian Ocean bottlenose dolphins (*Tursiops aduncus*) in captivity. *Behaviour* 44:286–298.
Taylor, A.H., G.R. Hunt, J.C. Holzhaider, and R.D. Gray. 2007. Spontaneous metatool use by New Caledonian crows. *Current Biology* 17 (17):1504–1507.
Taylor, A.H., G.R. Hunt, F.S. Medina, and R.D. Gray. 2009. Do New Caledonian crows solve physical problems through causal reasoning? *Proceedings of the Royal Society B-Biological Sciences* 276 (1655):247–254.
Tebbich, S., and R. Bshary. 2004. Cognitive abilities related to tool use in the woodpecker finch, *Cactospiza pallida*. *Animal Behaviour* 67 (4):689–697.
Tebbich, S., M. Taborsky, B. Fessl, and D. Blomqvist. 2001. Do woodpecker finches acquire tool-use by social learning? *Proceedings of the Royal Society B-Biological Sciences* 268 (1482):2189–2193.
Tebbich, S., M. Taborsky, B. Fessl, and M. Dvorak. 2002. The ecology of tool-use in the woodpecker finch (*Cactospiza pallida*). *Ecology Letters* 5 (5):656–664.
Tebbich, S., M. Taborsky, B. Fessl, M. Dvorak, and H. Winkler. 2004. Feeding behavior of four arboreal Darwin's finches: Adaptations to spatial and seasonal variability. *Condor* 106 (1):95–105.
Teleki, G. 1973a. Notes on chimpanzee interactions with small carnivores in Gombe National Park, Tanzania. *Primates* (14):407–411.
———. 1973b. The omnivorous chimpanzee. *Scientific American* 228 (1):33–42.
———. 1973c. *The predatory behavior of wild chimpanzees*. Lewisburg, PA: Bucknell University Press.
———. 1974. Chimpanzee subsistence technology: Materials and skills. *Journal of Human Evolution* 3: 575–594.
Temerlin, M.K. 1975. *Lucy: Growing up human*. Palo Alto, CA: Science and Behavior Books.
Tennie, C., J. Call, and M. Tomasello. 2009. Ratcheting up the ratchet: On the evolution of cumulative culture. *Philosophical Transactions of the Royal Society B, Biological Sciences* 364 (1528):2405–2415.
Terdal, E. 2005. Artificial termite tubes as environmental enrichment for captive chimpanzees. *Chimpanzoo Conference Proceedings* 2005:99.
Terrace, H.S. 1979. *Nim*. New York: Knopf.

Terres, J. 1962. Anting behavior of a wood thrush with a snail. *Wilson Bulletin* 74 (2):187.

Thanh, P.D., K. Wada, M. Sato, and Y. Shirayama. 2003. Decorating behaviour by the majid crab *Tiarinia cornigera* as protection against predators. *Journal of the Marine Biological Association of the United Kingdom* 83 (6):1235–1237.

Thierry, B., J.R. Anderson, C. Demaria, C. Desportes, and O. Petit. 1994. Tonkean macaque behaviour from the perspective of the evolution of Sulawesi macaques. In *Social development, learning, and behaviour*, edited by J.J. Roeder, B. Thierry, J.R. Anderson, and N. Herrenschmidt. Vol. 2 of *Current primatology*. Strasbourg, France: Université Louis Pasteur, 103–117.

Thomas, J.W. 1957. Anting performed by scaled quail. *Wilson Bulletin* 69:280.

Thompson, A.L. 1964. Oil gland. In *A new dictionary of birds*, edited by A.L. Thompson. Edinburgh: Thomas Nelson and Sons, 551–552.

Thomsen, L.R., R.D. Campbell, and F. Rosell. 2007. Tool-use in a display behaviour by Eurasian beavers (*Castor fiber*). *Animal Cognition* 10 (4):477–482.

Thorington, R.W, Jr. 1967. Feeding and activity of *Cebus* and *Saimiri* in a Columbian forest. In *Neue ergebnisse der primatologie*, edited by D. Starck, R. Schneider, and H.J. Kuhn. Stuttgart, Germany: Gustav Fischer, 180–184.

Thorpe, S.K.S., R. Holder, and R.H. Crompton. 2009. Orangutans employ unique strategies to control branch flexibility. *Proceedings of the National Academy of Sciences of the United States of America* 106 (31): 12646–12651.

Thorpe, W. 1963. *Learning and instinct in animals*. Cambridge, MA: Harvard University Press.

Thouless, C.R., J.H. Fanshawe, and B.C.R. Bertram. 1989. Egyptian vultures *Neophron percnopterus* and ostrich *Struthio camelus* eggs: The origins of stone-throwing behavior. *Ibis* 131 (1):9–15.

Tilden, J.W. 1953. The digging and provisioning behavior of *Ammophila saeva* Smith (Hymenoptera: Sphecidae). *Pan-Pacific Entomologist* 29:211–218.

Tinbergen, N. 1960. *The herring gull's world*. New York: Basic Books.

Tobias, P.V. 1965. *Australopithecus, Homo habilis*, tool-using, and tool-making. *South African Archaeological Bulletin* 20:167–192.

Tokida, E., I. Tanaka, H. Takefushi, and T. Hagiwara. 1994. Tool-using in Japanese macaques: Use of stones to obtain fruit from a pipe. *Animal Behaviour* 47 (5): 1023–1030.

Tomasello, M. 1990. Cultural transmission in the tool use and communicatory signaling of chimpanzees? In *"Language" and intelligence in monkeys and apes: Comparative developmental perspectives*, edited by S.T. Parker and K.R. Gibson. New York: Cambridge University Press, 274–311.

———. 1999. *The cultural origins of human cognition*. Cambridge, MA: Harvard University Press.

Tomasello, M., and J. Call. 1997. *Primate cognition*. New York: Oxford University Press.

Tomasello, M., A.C. Kruger, and H.H. Ratner. 1993. Cultural learning. *Behavioral and Brain Sciences* 16 (3): 495–552.

Tomasello, M., M. Davis-Dasilva, L. Camak, and K.A. Bard. 1987. Observational learning of tool-use by young chimpanzees. *Human Evolution* 2 (2):175–183.

Tonooka, R. 1994. Coordination of hands in captive chimpanzees during tool-using tasks. In *XVth Congress of the International Primatological Society, August 3–8, Bali, Indonesia*. Indonesia: Indonesian Wildlife Society, 404.

———. 1995. [Leaf folding behavior for drinking water by chimpanzees at Bossou]. *Reichorui Kenkyu* 11 (3):326.

———. 2001. Leaf-folding behavior for drinking water by wild chimpanzees (*Pan troglodytes verus*) at Bossou, Guinea. *Animal Cognition* 4 (3–4):325–334.

Tonooka, R., and N. Inoue-Nakamura. 1993. [Leaf folding behavior for drinking water of chimpanzees in Bossou, Guinea]. *Reichorui Kenkyu* 9 (3):265.

Tonooka, R., N. Inoue-Nakamura, and T. Matsuzawa. 1994. [Leaf-folding behavior for drinking water by wild chimpanzees at Bossou, Guinea: A field experiment and leaf selectivity]. *Reichorui Kenkyu* 10 (3):307–313.

Tonooka, R., M. Tomonaga, and T. Matsuzawa. 1997. Acquisition and transmission of tool making and use for drinking juice in a group of captive chimpanzees (*Pan troglodytes*). *Japanese Psychological Research* 39 (3):253–265.

Toth, N., K.D. Schick, E.S. Savage-Rumbaugh, R.A. Sevcik, and D.M. Rumbaugh. 1993. *Pan* the tool-maker: Investigations into the stone tool-making and tool-using capabilities of a bonobo (*Pan paniscus*). *Journal of Archaeological Science* 20 (1):81–91.

Troscianko, J., L.A. Bluff, and C. Rutz. 2008. Grass-stem tool use in New Caledonian crows (*Corvus moneduloides*). *Ardea* 96 (2):283–285.

Tsukaguchi, S. 1995. *Chrysopidae of Japan*. Osaka: Yutaka Press.

Tsuneki, K. 1968. *The biology of* Ammophila *in East Asia (Hymenoptera, Sphecidae)*. Etizenia, vol. 33. Fukui, Japan: Biological Laboratory, Fukui University.

Tuker, F. 1953. *The chronicle of Private Henry Metcalfe*. London: Cassell.

Tutin, C.E.G., and M. Fernandez. 1992. Insect-eating by sympatric lowland gorillas (*Gorilla g. gorilla*) and chimpanzees (*Pan t. troglodytes*) in the Lopé Reserve, Gabon. *American Journal of Primatology* 28 (1):29–40.

Tutin, C.E.G., R. Ham, and D. Wrogemann. 1995. Tool-use by chimpanzees (*Pan t. troglodytes*) in the Lopé Reserve, Gabon. *Primates* 36 (2):181–192.

Tutin, C.E.G., M. Fernandez, M.E. Rogers, E.A. Williamson, and W.C. McGrew. 1991. Foraging profiles of sympatric lowland gorillas and chimpanzees in the Lopé Reserve, Gabon. *Philosophical Transactions of the Royal Society B, Biological Sciences* 334 (1270):179–186.

Uehara, S. 1982. Seasonal changes in the techniques em-

ployed by wild chimpanzees in the Mahale Mountains, Tanzania, to feed on termites (*Pseudocanthotermes spiniger*). *Folia Primatologica* 37:44–76.
———. 1999. Why don't the chimpanzees of M Group at Mahale fish for termites? *Pan Africa News* 6 (2):22–24.
———. 2002. Evidence of the leaf-clipping behavior by a chimpanzee of an unhabituated group at Mahale. *Pan Africa News* 9 (1):3–4.
Ueno, Y., and K. Fujita. 1998. Spontaneous tool use by a Tonkean macaque (*Macaca tonkeana*). *Folia Primatologica* 69 (5):318–324.
Urbani, B. 1998. An early report on tool use by neotropical primates. *Neotropical Primates* 6 (4):123–124.
———. 1999. Spontaneous use of tools by wedge-capped capuchin monkeys (*Cebus olivaceus*). *Folia Primatologica* 70 (3):172–174.
Urbani, B., and P.A. Garber. 2002. A stone in their hands: Are monkeys tool users? *Anthropologie* 40 (2):183–191.
Utami Atmoko, S.S., I. Singleton, M.A. van Noordwijk, C.P. van Schaik, and T.M. Setia. 2009. Male-male relationships in orangutans. In *Orangutans: Geographic variation in behavioral ecology and conservation*, edited by S.A. Wich, S.S. Utami Atmoko, T.M. Setia, and C.P. van Schaik. New York: Oxford University Press, 225–234.
Vahed, K. 1998. The function of nuptial feeding in insects: A review of empirical studies. *Biological Reviews* 73: 43–78.
Valderrama, X., J.G. Robinson, A.B. Attygalle, and T. Eisner. 2000. Seasonal anointment with millipedes in a wild primate: A chemical defense against insects? *Journal of Chemical Ecology* 26 (12):2781–2790.
van Elsacker, L., and V. Walraven. 1994. The spontaneous use of a pineapple as a recipient by a captive bonobo (*Pan paniscus*). *Mammalia* 58 (1):159–162.
van Hooff, J.A.R.A.M. 1973. The Arnhem Zoo chimpanzee consortium: An attempt to create an ecologically and socially acceptable habitat. *International Zoo Yearbook* 13:195–203.
van Lawick–Goodall, J. 1965. New discoveries among Africa's chimpanzees. *National Geographic* 128:802–831.
———. 1968. *The behaviour of free-living chimpanzees in the Gombe Stream Reserve*. Animal Behaviour Monographs, vol. 1, part 3. London: Baillière, Tindall, & Cassel.
———. 1970. Tool-using in primates and other vertebrates. *Advances in the Study of Behavior* 3:195–249.
———. 1971. *In the shadow of man*. Boston: Houghton Mifflin.
———. 1973. Cultural elements in a chimpanzee community. In *Precultural primate behavior*, edited by E. Menzel. Vol. 1 of *Symposia of the Fourth International Congress of Primatology*. Basel, Switzerland: S. Karger, 144–184.
van Lawick–Goodall, J., and H. van Lawick. 1966. Use of tools by the Egyptian vulture, *Neophron percnopterus*. *Nature* 212:1468–1469.
———. 1968. Tool-using bird: The Egyptian vulture. *National Geographic* 133:631–641.
van Lawick–Goodall, J., H. van Lawick, and C. Packer. 1973. Tool-use in free-living baboons in the Gombe National Park, Tanzania. *Nature* 241:212–213.
van Noordwijk, M.A., and C.P. van Schaik. 2005. Development of ecological competence in Sumatran orangutans. *American Journal of Physical Anthropology* 127 (1):79–94.
van Schaik, C.P. 2004. *Among orangutans: Red apes and the rise of human culture*. Cambridge, MA: Harvard University Press.
———. 2006. Why are some animals so smart? *Scientific American* 294 (4):64–71.
van Schaik, C.P., R.O. Deaner, and M.Y. Merrill. 1999. The conditions for tool use in primates: Implications for the evolution of material culture. *Journal of Human Evolution* 36 (6):719–741.
van Schaik, C.P., and E.A. Fox. 1994. Tool use in wild Sumatran orangutans (*Pongo pygmaeus*). *XVth Congress of the International Primatological Society, August 3–8, Bali, Indonesia*. Indonesia: Indonesian Wildlife Society, 339.
van Schaik, C.P., E.A. Fox, and A.F. Sitompul. 1996. Manufacture and use of tools in wild Sumatran orangutans: Implications for human evolution. *Naturwissenschaften* 83 (4):186–188.
van Schaik, C.P., and C.D. Knott. 2001. Geographic variation in tool use on *Neesia* fruits in orangutans. *American Journal of Physical Anthropology* 114 (4):331–342.
van Schaik, C.P., K.N. Laland, and B.G. Galef. 2009. Geographic variation in the behavior of wild great apes: Is it really cultural? In *The question of animal culture*, edited by K.N. Laland and B.G. Galef. Cambridge, MA: Harvard University Press, 70–98.
van Schaik, C.P., and G.R. Pradhan. 2003. A model for tool-use traditions in primates: Implications for the coevolution of culture and cognition. *Journal of Human Evolution* 44 (6):645–664.
van Schaik, C.P., M.A. van Noordwijk, and S.A. Wich. 2006. Innovation in wild Bornean orangutans (*Pongo pygmaeus wurmbii*). *Behaviour* 143 (7):839–876.
van Schaik, C.P., M. Ancrenaz, G. Borgen, B. Galdikas, C.D. Knott, I. Singleton, A. Suzuki, S.S. Utami, and M. Merrill. 2003. Orangutan cultures and the evolution of material culture. *Science* 299:102–105.
van Schaik, C.P., M. Ancrenaz, R. Djojoasmoro, C.D. Knott, H.C. Morrogh-Bernard, N. Odom, S.S. Utami Atmoko, and M.A. van Noordwijk. 2009. Orangutan cultures revisited. In *Orangutans: Geographic variation in behavioral ecology and conservation*, edited by S.A. Wich, S.S. Utami Atmoko, T.M. Setia, and C.P. van Schaik. New York: Oxford University Press, 299–309.
Vancatova, M.A. 1984. The influence of imitation on tool-using in capuchin monkeys (*Cebus apella*). *Anthropologie* 22 (1):1–2.
———. 2008. Tool use behaviour of western lowland gorillas (*Gorilla gorilla gorilla*) in captivity. *Folia Primatologica* 79 (5):395.
VanderWerf, E.A. 2005. 'Elepaio "anting" with a garlic snail

and a *Schinus* fruit. *Journal of Field Ornithology* 76 (2):134–137.
Veino, C.M., and M.A. Novak. 2003. Tool use in juvenile rhesus macaques (*Macaca mulatta*). *American Journal of Primatology* 60 (S1):47.
———. 2004. The spontaneous use of tools in a captive rhesus macaque (*Macaca mulatta*). *American Journal of Primatology* 62 (S1):118.
Vencl, F.V., and T.C. Morton. 1998. The shield defense of the sumac flea beetle, *Blepharida rhois* (Chrysomelidae: Alticinae). *Chemoecology* 8:25–32.
Verderane, M.P., T. Falótico, B.D. de Resende, M.B. Labruna, P. Izar, and E.B. Ottoni. 2007. Anting in a semi-free-ranging group of *Cebus apella*. *International Journal of Primatology* 28 (1):47–53.
Veselovsky, V.Z. 1970. Einige Beobachtungen des Verhaltens der Vogel zu überoptimalen Ei-Attrappen. *Zoologische Garten* 39:290–294.
Vetter, R.S. 1980. Defensive behavior of the black-widow spider *Latrodectus hesperus* (Araneae: Theridiidae). *Behavioral Ecology and Sociobiology* 7 (3):187–193.
Vetter, R.S., and J. Cokendolpher. 2000. *Homalonychus theologus* (Araneae: Homalonychidae): Description of egg sacs and a possible defensive posture. *Journal of Arachnology* 28:361–363.
Vevers, G.M., and J.S. Weiner. 1963. Use of a tool by a captive capuchin monkey (*Cebus apella*). In *The primates*, edited by J.R. Napier and N.A. Barnicot. Symposia of the Zoological Society of London no. 10. London: Zoological Society of London, 115–117.
Vierke, J. 1973. Das Wasserspucken der Arten der Gattung *Colisa* (Pisces: Anabantidae). *Bonner Zoologische Beiträge* 24:62–104.
Vierke, J., and K. Lüling. 1972. Aquatic archery and the dwarf gourami. *Aquarium Digest* 1:4.
Vincent, F. 1973. [Spontaneous utilization of tools by a mandrill]. *Mammalia* 37:277–280.
Visalberghi, E. 1986. The acquisition of tool-use behavior in two capuchin monkey groups (*Cebus apella*). *Primate Report* 14:226–227.
———. 1987. Acquisition of nut-cracking behaviour by two capuchin monkeys (*Cebus apella*). *Folia Primatologica* 49 (3):168–181.
———. 1997. Success and understanding in cognitive tasks: A comparison between *Cebus apella* and *Pan troglodytes*. *International Journal of Primatology* 18 (5):811–830.
———. 2004. Tool-use in capuchin monkeys: The solution to a mystery? *Folia Primatologica* 75 (S1):15.
———. 2006. The emergence of tool use in captive and wild bearded capuchin monkeys (*Cebus libidinosus*). *Folia Primatologica* 77 (4):330–331.
Visalberghi, E., and F. Antinucci. 1986. Tool use in the exploitation of food resources in *Cebus apella*. In *Primate ecology and conservation*, edited by J.G. Else and P.C. Lee. New York: Cambridge University Press, 57–62.
Visalberghi, E., and D. Fragaszy. 2006. What is challenging about tool use? The capuchin's perspective. In *Comparative cognition: Experimental explorations of animal intelligence*, edited by E. Wasserman and T.R. Zentall. New York: Oxford University Press, 529–552.
Visalberghi, E., D. Fragaszy, and E.S. Savage-Rumbaugh. 1995. Performance in a tool-using task by common chimpanzees (*Pan troglodytes*), bonobos (*Pan paniscus*), an orangutan (*Pongo pygmaeus*), and capuchin monkeys (*Cebus apella*). *Journal of Comparative Psychology* 109 (1):52–60.
Visalberghi, E., and L. Limongelli. 1991. Comprehension of cause-effect relationships in tool-using capuchin monkeys (*Cebus apella*). *International Journal of Anthropology* 6 (4):255.
———. 1992. Comprehension of cause-effect relationships in tool-using capuchin monkeys (*Cebus apella*). In *XIVth Congress of the International Primatological Society*. Strasbourg, France: International Primatological Society, 379.
———. 1994. Lack of comprehension of cause-effect relations in tool-using capuchin monkeys (*Cebus apella*). *Journal of Comparative Psychology* 108 (1):15–22.
———. 1996. Acting and understanding: Tool use revisited through the minds of capuchin monkeys. In *Reaching into thought: The minds of the great apes*, edited by A.R. Russon, K.A. Bard, and S.T. Parker. New York: Cambridge University Press, 57–79.
Visalberghi, E., and L. Trinca. 1989. Tool use in capuchin monkeys: Distinguishing between performing and understanding. *Primates* 30 (4):511–521.
———. 1990. Tool-use in capuchin monkeys (*Cebus apella*): Performance and understanding. *International Journal of Anthropology* 5 (2):148–149.
Visalberghi, E., and A.F. Vitale. 1990. Coated nuts as an enrichment device to elicit tool use in tufted capuchins (*Cebus apella*). *Zoo Biology* 9 (1):65–71.
Visalberghi, E., D. Fragaszy, P. Izar, and E.B. Ottoni. 2006. Tool use in wild bearded capuchin monkeys (*Cebus libidinosus*): New findings and hypotheses. *Folia Primatologica* 77 (4):276–277.
Visalberghi, E., D. Fragaszy, E. Ottoni, P. Izar, M. Gomes de Oliveira, and F.R.D. Andrade. 2007. Characteristics of hammer stones and anvils used by wild bearded capuchin monkeys (*Cebus libidinosus*) to crack open palm nuts. *American Journal of Physical Anthropology* 132 (3):426–444.
Visalberghi, E., G. Sabbatini, N. Spagnoletti, F.R.D. Andrade, E. Ottoni, P. Izar, and D. Fragaszy. 2008. Physical properties of palm fruits processed with tools by wild bearded capuchins (*Cebus libidinosus*). *American Journal of Primatology* 70 (9):884–891.
Visalberghi, E., N. Spagnoletti, E.D.R. da Silva, F.R.D. Andrade, E. Ottoni, P. Izar, and D. Fragaszy. 2009a. Distribution of potential suitable hammers and transport of hammer tools and nuts by wild capuchin monkeys. *Primates* 50 (2):95–104.
Visalberghi, E., E. Addessi, V. Truppa, N. Spagnoletti, E. Ottoni, P. Izar, and D. Fragaszy. 2009b. Selection of effec-

tive stone tools by wild bearded capuchin monkeys. *Current Biology* 19 (3):213–217.
Vitale, A.F., E. Visalberghi, and C. DeLillo. 1991. Responses to a snake model in captive crab-eating macaques (*Macaca fascicularis*) and captive tufted capuchins (*Cebus apella*). *International Journal of Primatology* 12 (3):277–286.
Voelkl, B., S. Rainer, and L. Huber. 2002. Limited understanding of tools in marmosets. In *Caring for primates: Abstracts of the XIXth Congress of the International Primatological Society*. Beijing: Mammalogical Society of China, 336–337.
von Frisch, K. 1974. *Animal architecture*. New York: Harcourt Brace Jovanovich.
Wade, L. 1975. A sea otter possibly feeding on pismo clams. *Journal of Mammalogy* 56:720–721.
Waga, I.C., A.K. Dacier, P.S. Pinha, and M.C.H. Tavares. 2006. Spontaneous tool use by wild capuchin monkeys (*Cebus libidinosus*) in the cerrado. *Folia Primatologica* 77 (5):337–344.
Wagner, H.O. 1956. Freilandbeobachtungen an Klammeraffen. *Zeitschrift für Tierpsychologie* 13:302–313.
Walkup, K.R. 2009. Comprehension of tools by orangutans: Causality, tool properties, and manufacture. Ph.D. diss., Iowa State University.
Walkup, K.R., R.W. Shumaker, and J.D Pruetz. 2009. Comprehension of tool properties by orangutans (*Pongo* spp.): Rigidity and flexibility. *American Journal of Physical Anthropology* 138 (S48):265.
———. 2010. Orangutans (*Pongo* spp.) may prefer tools with rigid properties to flimsy tools. *Journal of Comparative Psychology* 124 (4):351–355.
Wallace, A.F., and V. Lathbury. 1968. Culture and the beaver. *Natural History* 77:59–65.
Wallace, A.R. 1869. *The Malay Archipelago*. New York: Harper.
Walraven, V., L. van Elsacker, and R. Verheyen. 1993. Spontaneous object manipulation in captive bonobos (*Pan paniscus*). In *Bonobo Tidings*. n.p.: Royal Zoological Society of Antwerp, 25–34.
Walsh, J.F., J. Grunewald, and B. Grunewald. 1985. Green-backed herons (*Butorides striatus*) possibly using a lure and using apparent bait. *Journal für Ornithologie* 126 (4):439–442.
Walther, F. 1974. Some reflections on expressive behaviour in combats and courtship of certain horned ungulates. In *The behaviour of ungulates and its relation to management*, edited by V. Geist and F. Walther. Morges, Switzerland: International Union for the Conservation of Nature and Natural Resources [IUCN] 56–106.
Warden, C.J. 1940. The ability of monkeys to use tools. *Transactions of the New York Academy of Sciences* 2: 109–112.
Warden, C.J., A.M. Koch, and H.A. Fjeld. 1940. Instrumentation in cebus and rhesus monkeys. *Journal of Genetic Psychology* 56:297–310.
Warren, J. 1973. Learning in vertebrates. In *Comparative psychology*, edited by D. Dewsbury and D. Rethlingshafer. New York: McGraw-Hill, 471–509.
Washburn, S.L., and P.C. Jay. 1967. More on tool-use among primates. *Current Anthropology* 8:253–254, 257.
Watanabe, K. 1994. Precultural behavior of Japanese macaques: Longitudinal studies of the Koshima troops. In *The ethological roots of culture*, edited by R.A. Gardner, A.B. Chiarelli, B.T. Gardner, and F.X. Plooij. Dordrecht, Netherlands: Kluwer Academic, 81–94.
Watanabe, K., N. Urasopon, and S. Malaivijitnond. 2007. Long-tailed macaques use human hair as dental floss. *American Journal of Primatology* 69 (8):940–944.
Watson, J. 1908. Imitation in monkeys. *Psychological Bulletin* 5:169–178.
Watts, D.P. 2008. Tool use by chimpanzees at Ngogo, Kibale National Park, Uganda. *International Journal of Primatology* 29 (1):83–94.
Wehnelt, S., S. Bird, and A. Lenihan. 2006. Chimpanzee Forest exhibit at Chester Zoo. *International Zoo Yearbook* 40:313–322.
Weinberg, S., and D.K. Candland. 1981. "Stone-grooming" in *Macaca fuscata*. *American Journal of Primatology* 1:465–468.
Weir, A.A.S., J. Chappell, and A. Kacelnik. 2002. Shaping of hooks in New Caledonian crows. *Science* 297 (5583):981.
Weir, A.A.S., and A. Kacelnik. 2006. A New Caledonian crow (*Corvus moneduloides*) creatively re-designs tools by bending or unbending aluminum strips. *Animal Cognition* 9 (4):317–334.
Weir, A.A.S., B. Kenward, J. Chappell, and A. Kacelnik. 2004. Lateralization of tool use in New Caledonian crows (*Corvus moneduloides*). *Proceedings of the Royal Society B-Biological Sciences* 271 (Suppl. 5):S344–S346.
Weldon, P.J. 2004. Defensive anointing: Extended chemical phenotype and unorthodox ecology. *Chemoecology* 14 (1):1–4.
Weldon, P.J., and D.L. Hoffman. 1975. Unique form of tool-using in two gastropod mollusks (Trochidae). *Nature* 256 (5520):720–721.
Weldon, P.J., J.R. Aldrich, J.A. Klun, J.E. Oliver, and M. Debboun. 2003. Benzoquinones from millipedes deter mosquitoes and elicit self-anointing in capuchin monkeys (*Cebus* spp.). *Naturwissenschaften* 90 (7):301–304.
Wells, M.J. 1962. *Brain and behaviour in cephalopods*. Stanford, CA: Stanford University Press.
———. 1978. *Octopus: Physiology and behaviour of an advanced invertebrate*. London: Chapman and Hall.
Wemmer, C. 1969. Impaling behaviour of the loggerhead shrike, *Lanius ludovicianus*. *Zeitschrift für Tierpsychologie* 26:208–224.
Wemmer, C., and G. Johnson. 1976. Egg-breaking behavior in a yellow-throated marten, *Martes flavigula* (Mustelidae: Carnivora). *Zeitschrift für Säugetierekunde* 41: 58–60.
Wemmer, C., B.B. Beck, L. Collins, and B. Rettberg. 1983. The ethogram. In *Père David's deer: The biology and*

management of an extinct species, edited by B.B. Beck and C. Wemmer. Park Ridge, NJ: Noyes, 91–125.

Wenny, D. 1998. Three-striped warbler (*Basileuterus tristriatus*) "anting" with a caterpillar. *Wilson Bulletin* 110 (1):128–131.

Westergaard, G.C. 1988. Lion-tailed macaques (*Macaca silenus*) manufacture and use tools. *Journal of Comparative Psychology* 102 (2):152–159.

———. 1991. Hand preference in the use and manufacture of tools by tufted capuchin (*Cebus apella*) and lion-tailed macaque (*Macaca silenus*) monkeys. *Journal of Comparative Psychology* 105 (2):172–176.

———. 1992. Object manipulation and the use of tools by infant baboons (*Papio cynocephalus anubis*). *Journal of Comparative Psychology* 106 (4):398–403.

———. 1993a. Development of combinatorial manipulation in infant baboons (*Papio cynocephalus anubis*). *Journal of Comparative Psychology* 107 (1):34–38.

———. 1993b. Experimental approaches to the study of tool-use by captive capuchin monkeys. Paper presented at the AAZPA [American Association of Zoological Parks and Aquariums] Regional Conference, Pittsburgh, Pennsylvania.

———. 1993c. Hand preference in the use of tools by infant baboons (*Papio cynocephalus anubis*). *Perceptual and Motor Skills* 76 (2):447–450.

Westergaard, G.C., T.A. Evans, and S. Howell. 2007. Token mediated tool exchange between tufted capuchin monkeys (*Cebus apella*). *Animal Cognition* 10 (4): 407–414.

Westergaard, G.C., and D.M. Fragaszy. 1985. Effects of manipulatable objects on the activity of captive capuchin monkeys (*Cebus apella*). *Zoo Biology* 4 (4):317–327.

———. 1987a. The manufacture and use of tools by capuchin monkeys (*Cebus apella*). *Journal of Comparative Psychology* 101 (2):159–168.

———. 1987b. Self-treatment of wounds by a capuchin monkey (*Cebus apella*). *Human Evolution* 1 (6):557–562.

Westergaard, G.C., H.E. Kuhn, and S.J. Suomi. 1998a. Effects of upright posture on hand preference for reaching vs. the use of probing tools by tufted capuchins (*Cebus apella*). *American Journal of Primatology* 44 (2):147–153.

———. 1998b. Laterality of hand function in tufted capuchin monkeys (*Cebus apella*): Comparison between tool use actions and spontaneous non-tool actions. *Ethology* 104 (2):119–125.

Westergaard, G.C., and T. Lindquist. 1987. Manipulation of objects in a captive group of lion-tailed macaques (*Macaca silenus*). *American Journal of Primatology* 12 (2):231–234.

Westergaard, G.C., and S.J. Suomi. 1993a. Hand preference in the use of nut-cracking tools by tufted capuchin monkeys (*Cebus apella*). *Folia Primatologica* 61 (1):38–42.

———. 1993b. Use of a tool-set by capuchin monkeys (*Cebus apella*). *Primates* 34 (4):459–462.

———. 1994a. Aimed throwing of stones by tufted capuchin monkeys (*Cebus apella*). *Human Evolution* 9 (4):323–329.

———. 1994b. Asymmetrical manipulation in the use of tools by tufted capuchin monkeys (*Cebus apella*). *Folia Primatologica* 63 (2):96–98.

———. 1994c. Hierarchical complexity of combinatorial manipulation in capuchin monkeys (*Cebus apella*). *American Journal of Primatology* 32 (3):171–176.

———. 1994d. A simple stone-tool technology in monkeys. *Journal for Human Evolution* 27:399–404.

———. 1994e. Stone-tool bone-surface modification by monkeys. *Current Anthropology* 35 (4):468–470.

———. 1994f. The use and modification of bone tools by capuchin monkeys. *Current Anthropology* 35 (1):75–77.

———. 1994g. The use of probing tools by tufted capuchins (*Cebus apella*): Evidence for increased right-hand preference with age. *International Journal of Primatology* 15 (4):521–529.

———. 1995a. The manufacture and use of bamboo tools by monkeys: Possible implications for the development of material culture among East Asian hominids. *Journal of Archaeological Science* 22 (5):677–681.

———. 1995b. The production and use of digging tools by monkeys: A nonhuman primate model of a hominid subsistence activity. *Journal of Anthropological Research* 51 (1):1–8.

———. 1995c. Stone-throwing by capuchins (*Cebus apella*): A model of throwing capabilities in *Homo habilis*. *Folia Primatologica* 65 (4):234–238.

———. 1995d. The stone tools of capuchins (*Cebus apella*). *International Journal of Primatology* 16 (6):1017–1024.

———. 1996. Hand preference for stone artifact production and tool-use by monkeys: Possible implications for the evolution of right-handedness in hominids. *Journal of Human Evolution* 30 (4):291–298.

———. 1997a. Capuchin monkey (*Cebus apella*) grips for the use of stone tools. *American Journal of Physical Anthropology* 103 (1):131–135.

———. 1997b. Modification of clay forms by tufted capuchins (*Cebus apella*). *International Journal of Primatology* 18 (3):455–467.

———. 1997c. Transfer of tools and food between groups of tufted capuchins (*Cebus apella*). *American Journal of Primatology* 43 (1):33–41.

Westergaard, G.C., J.A. Greene, M.A. Babitz, and S.J. Suomi. 1995. Pestle use and modification by tufted capuchins (*Cebus apella*). *International Journal of Primatology* 16 (4):643–651.

Westergaard, G.C., J.A. Greene, C. Menuhin-Hauser, and S.J. Suomi. 1996. The use of naturally occurring copper and iron tools by monkeys: Possible implications for the emergence of metal-tool technology in hominids. *Human Evolution* 11 (1):17–25.

Westergaard, G.C., A.L. Lundquist, H.E. Kuhn, and S.J. Suomi. 1997. Ant-gathering with tools by captive tufted capuchins (*Cebus apella*). *International Journal of Primatology* 18 (1):95–103.

Westergaard, G.C., H.E. Kuhn, M.A. Babitz, and S.J. Suomi. 1998a. Aimed throwing as a means of food transfer between tufted capuchins (*Cebus apella*). *International Journal of Primatology* 19 (1):123–131.

Westergaard, G.C., C. Liv, T.J. Chavanne, and S.J. Suomi. 1998b. Token-mediated tool-use by a tufted capuchin monkey (*Cebus apella*). *Animal Cognition* 1 (2):101–106.

Westergaard, G.C., A.L. Lundquist, M.K. Haynie, H.E. Kuhn, and S.J. Suomi. 1998c. Why some capuchin monkeys (*Cebus apella*) use probing tools (and others do not). *Journal of Comparative Psychology* 112 (2): 207–211.

Westergaard, G.C., C. Liv, M.K. Haynie, and S.J. Suomi. 2000. A comparative study of aimed throwing by monkeys and humans. *Neuropsychologia* 38 (11): 1511–1517.

Westergaard, G.C., A. Cleveland, A.M. Rocca, E.L. Wendt, and M.J. Brown. 2003. Throwing behavior and the mass distribution of rock selection in tufted capuchin monkeys (*Cebus apella*). *American Journal of Physical Anthropology* 120 (S36): 223.

Westergaard, G.C., C. Liv, A.M. Rocca, A. Cleveland, and S.J. Suomi. 2004. Tufted capuchins (*Cebus apella*) attribute value to foods and tools during voluntary exchanges with humans. *Animal Cognition* 7 (1):19–24.

Western, S. 1994. Ape opera. *BBC Wildlife* 12 (1):28–36.

Wheatley, B.P. 1988. Cultural behaviour and extractive foraging in *Macaca fascicularis*. *Current Anthropology* 29 (3):516–519.

Wheeler, G.C., and E.H. Wheeler. 1924. The use of a tool by a sphecid wasp. *Science* 59 (1535):486.

Wheeler, W.M. 1910. *Ants: Their structure, development, and behavior*. New York: Columbia University Press.

———. 1930. *Demons of the dust*. New York: W.W. Norton.

Whitaker, L. 1957. A résumé of anting, with particular reference to a captive orchard oriole. *Wilson Bulletin* 69:195–262.

White, F.J., M.T. Waller, A.K. Cobden, and N.M. Malone. 2008. Lomako bonobo population dynamics, habitat productivity, and the question of tool use. *American Journal of Physical Anthropology* 135 (S46):222.

White, L.A. 1942. On the use of tools by primates. *Journal of Comparative Psychology* 34:369–374.

Whiten, A., V. Horner, and F.B.M. de Waal. 2005. Conformity to cultural norms of tool use in chimpanzees. *Nature* 437 (7059):737–740.

Whiten, A., and W.C. McGrew. 2001a. Is this the first portrayal of tool use by a chimp? *Nature* 409 (6816):12.

———. 2001b. Piecing together the history of our knowledge of chimpanzee tool use: Reply. *Nature* 411:413.

Whiten, A., J. Goodall, W.C. McGrew, T. Nishida, V. Reynolds, Y. Sugiyama, C.E.G. Tutin, R.W. Wrangham, and C. Boesch. 1999. Cultures in chimpanzees. *Nature* 399 (6737):682–685.

———. 2001. Charting cultural variation in chimpanzees. *Behaviour* 138 (11–12):1481–1516.

Whiten, A., A. Spiteri, V. Horner, K.E. Bonnie, S.P. Lambeth, S.J. Schapiro, and F.B.M. de Waal. 2007. Transmission of multiple traditions within and between chimpanzee groups. *Current Biology* 17 (12):1038–1043.

Whitesides, G.H. 1985. Nut cracking by wild chimpanzees in Sierra Leone, West Africa. *Primates* 26 (1):91–94.

Whyte, J. 1981. Anting in blue-eared glossy starlings. *Ostrich* 52:185.

Wickler, W., and U. Seibt. 1997. Aimed object-throwing by a wild African elephant in an interspecific encounter. *Ethology* 103 (5):365–368.

Wicksten, M.K. 1978. Attachment of decorating materials in *Loxorhynchus crispatus* (Brachyura: Majidae). *Transactions of the American Microscopical Society* 97 (2): 217–220.

———. 1979. Decorating behaviour in *Loxorhynchus crispatus* Stimpson and *Loxorhynchus grandis* Stimpson (Brachyura: Majidae). *Crustaceana* 5 (S):37–45.

———. 1980. Decorator crabs. *Scientific American* 242 (2):146–154.

———. 1982. Behavior in *Clythrocerus planus* (Rathbun, 1900) (Brachyura: Dorippidae). *Crustaceana* 43 (3): 306–308.

———. 1985. Carrying behavior in the family Homolidae (Decapoda: Brachyura). *Journal of Crustacean Biology* 5 (3):476–479.

———. 1986. Carrying behavior in brachyuran crabs. *Journal of Crustacean Biology* 6 (3):364–369.

———. 1993. A review and a model of decorating behavior in spider crabs (Decapoda: Brachyura: Majidae). *Crustaceana* 64:314–325.

Wilcox, R.S. 1984. Male copulatory guarding enhances female foraging in a water strider. *Behavioral Ecology and Sociobiology* 15:171–174.

Williams, J.H. 1950. *Elephant bill*. Garden City, NY: Doubleday.

Williston, S.W. 1892. Note on the habits of *Ammophila*. *Entomological News* 3:85–86.

Wilson, W.L., and C.C. Wilson. 1968. *Aggressive interactions of captive chimpanzees living in a semi-free-ranging environment*. 6571st Aeromedical Research Laboratory Technical Report No. ARL-TR-68-9. Holloman Air Force Base, New Mexico.

Wimpenny, J.H., A.A.S. Weir, L. Clayton, C. Rutz, and A. Kacelnik. 2009. Cognitive processes associated with sequential tool use in New Caledonian crows. *PLoS One* 4 (8):e6471.

Windholz, G. 1984. Pavlov vs. Köhler: Pavlov's little-known primate research. *Pavlovian Journal of Biological Science* 19 (1):23–31.

Wittiger, L., and J.L. Sunderland-Groves. 2007. Tool use during display behavior in wild Cross River gorillas. *American Journal of Primatology* 69:1–5.

Wojciechowski, S. 2007. A simple enrichment device to facilitate tool use by capuchin monkeys. *Shape of Enrichment* 16 (1–2):11.

Wolfe, L.D. 1981. Display behavior of three troops of Japanese monkeys (*Macaca fuscata*). *Primates* 22:24–32.

Wood, G.A. 1984. Tool use by the palm cockatoo *Probosciger aterrimus* during display. *Corella* 8 (4):94–95.

Wood, J.G. 1877. *Wood's Bible animals*. Guelph, Ontario: J.W. Lyon.

Wood, P. 1986. Fishing greenbacked heron. *Bokmakierie* 38:105.

Wood, R.J. 1984. Spontaneous use of sticks as tools by gorillas at Howletts Zoo Park, England. *International Zoo News* 31 (3):13–18.

Woodruff, G., and D. Premack. 1979. Intentional communication in the chimpanzee: The development of deception. *Cognition* 7:333–362.

Woods, C.M.C., and C.L. McLay. 1994. Use of camouflage materials as a food store by the spider crab *Notomithrax ursus* (Brachyura: Majidae). *New Zealand Journal of Marine and Freshwater Research* 28 (1):97–104.

Woods, C.M.C., and M.J. Page. 1999. Sponge masking and related preferences in the spider crab *Thacanophrys filholi* (Brachyura: Majidae). *Marine and Freshwater Research* 50 (2):135–143.

Woods, S.E. 1991. Maguba moves on. *Gorilla Gazette* 5 (3):6–8.

———. 1992. Implementation and evaluation of a behavioral enrichment program for captive gorillas, with an emphasis on tool behaviors (*Gorilla gorilla gorilla*). Ph.D. diss., University of Colorado at Boulder.

———. 1995. Facilitation of problem solving, tool use, and affiliative social behaviors in captive gorillas. *Association of Zoos and Aquariums Regional Conference Proceedings* 1995:384–388.

Worch, E.A. 2001. Simple tool use by a red-tailed monkey (*Cercopithecus ascanius*) in Kibale Forest, Uganda. *Folia Primatologica* 72 (5):304–306.

Wrangham, R.W. 1974. Artificial feeding of chimpanzees and baboons in their natural habitat. *Animal Behaviour* 22:83–93.

———. 1975. The behavioural ecology of chimpanzees in Gombe National Park, Tanzania. Ph.D. diss., University of Cambridge.

———. 1977. Feeding behaviour of chimpanzees in Gombe National Park, Tanzania. In *Primate ecology: Studies of feeding and ranging behaviour in lemurs, monkeys, and apes*, edited by T.H. Clutton-Brock. New York: Academic Press, 503–538.

Wrangham, R.W., F.B.M. de Waal, and W.C. McGrew. 1994. The challenge of behavioral diversity. In *Chimpanzee cultures*, edited by R.W. Wrangham, W.C. McGrew, F.B.M. de Waal, and P.G. Heltne. Cambridge, MA: Harvard University Press, 1–18.

Wrangham, R.W., and D. Peterson. 1996. *Demonic males: Apes and the origins of human violence*. Boston: Houghton Mifflin.

Wright, B.W. 1994. The effects of behavioral enrichment and the utilization of tools within and between captive groups of western lowland gorillas and Bornean orangutans. *American Journal of Primatology* 33 (3): 252–253.

———. 1995. The effects of behavioral enrichment and the utilization of tools within and between captive groups of western lowland gorillas and Bornean orang-utans. Master's thesis, University of Colorado at Boulder.

Wright, R.V.S. 1972. Imitative learning of a flaked stone technology: The case of an orangutan. *Mankind* 8:296–306.

Xu, Z., D.M. Stoddart, H. Ding, and J. Zhang. 1995. Self-anointing behavior in the rice-field rat, *Rattus rattoides*. *Journal of Mammalogy* 76 (4):1238–1241.

Yamagiwa, J., T. Yumoto, M. Ndunda, and T. Maruhashi. 1988. Evidence of tool-use by chimpanzees (*Pan troglodytes schweinfurthii*) for digging out a bee-nest in the Kahuzi-Biega National Park, Zaïre. *Primates* 29 (3):405–411.

Yamakoshi, G. 1998. Dietary responses to fruit scarcity of wild chimpanzees at Bossou, Guinea: Possible implications for the ecological importance of tool use. *American Journal of Physical Anthropology* 106 (3):283–295.

———. 2001. Ecology of tool use in wild chimpanzees: Toward reconstruction of early hominid evolution. In *Primate origins of human cognition and behavior*, edited by T. Matsuzawa. Tokyo: Springer, 537–556.

Yamakoshi, G., and M. Myowa-Yamakoshi. 2004. New observations of ant-dipping techniques in wild chimpanzees at Bossou, Guinea. *Primates* 45 (1):25–32.

Yamakoshi, G., and Y. Sugiyama. 1995. Pestle-pounding behavior of wild chimpanzees at Bossou, Guinea: A newly observed tool-using behavior. *Primates* 36 (4):489–500.

Yamamoto, S., G. Yamakoshi, T. Humle, and T. Matsuzawa. 2008. Invention and modification of a new tool use behavior: Ant-fishing in trees by a wild chimpanzee (*Pan troglodytes verus*) at Bossou, Guinea. *American Journal of Primatology* 70 (7):699–702.

Yamazaki, Y., H. Namba, and A. Iriki. 2009. Acquisition of an externalized eye by Japanese monkeys. *Experimental Brain Research* 194 (1):131–142.

Yeargan, K.V. 1988. Ecology of a bolas spider, *Mastophora hutchinsoni*: Phenology, hunting tactics, and evidence for aggressive chemical mimicry. *Oecologia* 74 (4): 524–530.

———. 1994. Biology of bolas spiders. *Annual Review of Entomology* 39:81–99.

Yerkes, R.M. 1916. *The mental life of monkeys and apes: A study of ideational behavior*. Behavior Monographs vol. 3, no. 1, serial no. 12. Cambridge, MA: H. Holt.

———. 1927a. *The mind of a gorilla*. Genetic Psychology Monographs vol. 2, nos. 1–2. Worcester, MA: Clark University.

———. 1927b. *The mind of a gorilla: Part II. Mental development*. Genetic Psychology Monographs vol. 2, no. 6. Worcester, MA: Clark University.

———. 1928–1929. *The mind of a gorilla: Part III. Memory*. Comparative Psychology Monographs, vol. 5, no. 2, serial no. 24. Baltimore: Johns Hopkins Press.

———. 1943. *Chimpanzees: A laboratory colony*. New Haven, CT: Yale University Press.

Yerkes, R.M., and B.W. Learned. 1925. *Chimpanzee intelligence and its vocal expressions*. Baltimore: Williams & Wilkins.

Yerkes, R.M., and A.W. Yerkes. 1929. *The great apes: A study of anthropoid life*. New Haven, CT: Yale University Press.

Yoshihara, K. 1985. Use of a stone implement by chimpanzees. *Animals and Zoos* 8:4–9.

Young, H.G. 1984. Herring gulls preying on rabbits. *British Birds* 80:630.

Zach, R. 1978. Selection and dropping of whelks by Northwestern crows. *Behaviour* 67:134–148.

———. 1979. Shell dropping: Decision-making and optimal foraging in Northwestern crows. *Behaviour* 68: 106–117.

Zamma, K. 2002. Leaf-grooming by a wild chimpanzee in Mahale. *Primates* 43 (1):87–90.

Zeledon, R., C.E. Valerio, and J.E. Valerio. 1973. The camouflage phenomenon in several species of Triatominae (Hemiptera: Reduviidae). *Journal of Medical Entomology* 10:209–211.

Zickefoose, J., and W.E. Davis, Jr. 1998. Great blue heron (*Ardea herodias*) uses bread as bait for fish. *Colonial Waterbirds* 21 (1):87–88.

Zimmerman, K. 1952. Werkzeug-Benutzung durch eine Zwergmaus. *Zeitschrift für Tierpsychologie* 9:12.

Ziswiler, V., and D. Farner. 1972. Digestion and the digestive system. In *Avian biology*, edited by D. Farner and J. King. New York: Academic Press, 343–430.

Zito, M., S. Evans, and P.J. Weldon. 2003. Owl monkeys (*Aotus* spp.) self-anoint with plants and millipedes. *Folia Primatologica* 74:159–161.

Zuberbühler, K., L. Gygax, N. Harley, and H. Kummer. 1996. Stimulus enhancement and spread of a spontaneous tool use in a colony of long-tailed macaques. *Primates* 37 (1):1–12.

Zuckerman, S., and B. Zuckerman. 1932. The social life of monkeys and apes. New York: Harcourt, Brace.

INDEX